344
Topics in Current Chemistry

Editorial Board:

K.N. Houk, Los Angeles, CA, USA
C.A. Hunter, Sheffield, UK
M.J. Krische, Austin, TX, USA
J.-M. Lehn, Strasbourg, France
S.V. Ley, Cambridge, UK
M. Olivucci, Siena, Italy
J. Thiem, Hamburg, Germany
M. Venturi, Bologna, Italy
C.-H. Wong, Taipei, Taiwan
H.N.C. Wong, Shatin, Hong Kong

For further volumes:
http://www.springer.com/series/128

Aims and Scope

The series Topics in Current Chemistry presents critical reviews of the present and future trends in modern chemical research. The scope of coverage includes all areas of chemical science including the interfaces with related disciplines such as biology, medicine and materials science.

The goal of each thematic volume is to give the non-specialist reader, whether at the university or in industry, a comprehensive overview of an area where new insights are emerging that are of interest to larger scientific audience.

Thus each review within the volume critically surveys one aspect of that topic and places it within the context of the volume as a whole. The most significant developments of the last 5 to 10 years should be presented. A description of the laboratory procedures involved is often useful to the reader. The coverage should not be exhaustive in data, but should rather be conceptual, concentrating on the methodological thinking that will allow the non-specialist reader to understand the information presented.

Discussion of possible future research directions in the area is welcome.

Review articles for the individual volumes are invited by the volume editors.

Readership: research chemists at universities or in industry, graduate students.

Sunghoon Kim
Editor

Aminoacyl-tRNA Synthetases in Biology and Medicine

With contributions by

T.J. Bullwinkle · V. Dewan · C. Florentz · K.-M. Forsyth ·
P.L. Fox · I. Gruic-Sovulj · M. Guo · J.M. Han · M. Ibba ·
D. Kim · J.H. Kim · S. Kim · N.H. Kwon · S.A. Martinis ·
H. Nechushtan · M.C. Park · T. Passioura · J.J. Perona ·
K. Poruri · E. Razin · J. Reader · H. Schwenzer · M. Sissler ·
S.H. Son · H. Suga · S. Tshori · X.-L. Yang · P. Yao · J. Zoll

Editor
Sunghoon Kim
Department of Molecular Medicine and Biopharmaceutical Sciences
Medicinal Bioconvergence Research Center
College of Pharmacy
Seoul National University
Korea, Republic of (South Korea)

ISSN 0340-1022 ISSN 1436-5049 (electronic)
ISBN 978-94-017-8700-0 ISBN 978-94-017-8701-7 (eBook)
DOI 10.1007/978-94-017-8701-7
Springer Dordrecht Heidelberg New York London

Library of Congress Control Number: 2014936839

© Springer-Verlag Berlin Heidelberg 2014
This work is subject to copyright. All rights are reserved by the Publisher, whether the whole or part of the material is concerned, specifically the rights of translation, reprinting, reuse of illustrations, recitation, broadcasting, reproduction on microfilms or in any other physical way, and transmission or information storage and retrieval, electronic adaptation, computer software, or by similar or dissimilar methodology now known or hereafter developed. Exempted from this legal reservation are brief excerpts in connection with reviews or scholarly analysis or material supplied specifically for the purpose of being entered and executed on a computer system, for exclusive use by the purchaser of the work. Duplication of this publication or parts thereof is permitted only under the provisions of the Copyright Law of the Publisher's location, in its current version, and permission for use must always be obtained from Springer. Permissions for use may be obtained through RightsLink at the Copyright Clearance Center. Violations are liable to prosecution under the respective Copyright Law.
The use of general descriptive names, registered names, trademarks, service marks, etc. in this publication does not imply, even in the absence of a specific statement, that such names are exempt from the relevant protective laws and regulations and therefore free for general use.
While the advice and information in this book are believed to be true and accurate at the date of publication, neither the authors nor the editors nor the publisher can accept any legal responsibility for any errors or omissions that may be made. The publisher makes no warranty, express or implied, with respect to the material contained herein.

Printed on acid-free paper

Springer is part of Springer Science+Business Media (www.springer.com)

Preface

Ever since the structure of the DNA double helix was unveiled more than 60 years ago, every step in the central dogma of life has undergone major scientific scrutiny. As the key components and mechanistic insights into the central dogma have been elucidated, each discovery has greatly impacted our understanding of the basic operational principles of living organisms. In the flow from gene to protein, decoding genetic information with appropriate rate and fidelity is critical to sustaining the vitality of living organisms. Aminoacyl-tRNA synthetases covalently attach amino acids to the ends of tRNAs, thereby forming the first bridge between nucleic acid and protein worlds. The discovery of tRNA synthetase activity Berg and Ofengand (1958) and Schweet et al. (1958) stimulated major efforts to understand the common catalytic activities of these enzymes, including the chemical mechanism of aminoacylation, the architectures of the enzymes and their active sites, substrate recognition, and molecular evolution and proofreading mechanisms. These subjects have formed the fundamental basis for tRNA synthetase research for the last 50 years. In the first three chapters of this volume, Perona, Ibba, Guo, and Yang et al. discuss the past, present, and future of these issues.

As genomic and structural information on different tRNA synthetases from diverse species has accumulated, some unique structural features of eukaryotic tRNA synthetases have been revealed. Relative to their bacterial counterparts, most eukaryotic enzymes were found to have additional unique domains at their extremities or even inserted into their conserved catalytic domains. The structural differences between bacterial and eukaryotic tRNA synthetases distinguished this family of enzymes from the normal evolutionary differences commonly found in other protein families. The eukaryote-specific structural features have provided unique capabilities to form diverse functional protein complexes and have opened new avenues for research. The functional implications for novel protein–protein interactions and complexes mediated by eukarytotic tRNA synthetases are reviewed in the chapter by Han et al.

Sophisticated cell biology studies also revealed a surprising biology for mammalian tRNA synthetases in extracellular space. Although the presence of several different tRNA synthetases or their antibodies in blood has been detected in autoimmune patients, these early observations were considered to be the result of cell necrosis, and the etiology of tRNA synthetases and their autoantibodies in extracellular locations was not well understood. Recent studies have demonstrated diverse physiological implications of mammalian tRNA synthetases in the extracellular milieu, implying that these proteins are likely to result from controlled secretion processes. The number of examples of secreted tRNA synthetases with distinct activities has increased in recent years. Although information on their detailed functions and corresponding receptors is still limited, the emergence of this new family of extracellular signal mediators that is distinct from the more typical cytokines and hormones is of great interest. In this volume, the current state-of-the-art on the secretion and functional implications of human tRNA synthetases in extracellular space is reviewed by Park et al.

Genetic and biochemical studies have uncovered other unexpected roles of tRNA synthetases. As these enzymes are essential for protein synthesis and are thus considered "housekeepers," the involvement of these enzymes in the expression of their own or other genes was not well anticipated. Surprisingly, several different tRNA synthetases regulate gene expression at the levels of transcription, splicing, and translation via non-catalytic and unique mechanisms. Some recent findings related to these novel functions are addressed by Fox and Martinis in this volume. While tRNA synthetases can achieve the regulation of gene expression primarily through their versatile molecular interactions, some of these enzymes can also regulate gene expression through the generation of intriguing small molecules, diadenosine polyphosphates, which are emerging as novel second messengers. This second catalytic activity and its functional implications are addressed by Razin and Nechushtan.

The functional diversity of human tRNA synthetases is also associated with various diseases. For instance, some tRNA synthetases and their autoantibodies are associated with autoimmune diseases such as polymyositis and dermatomyositis, and there is even a link to some cancers. In addition, mutations in tRNA synthetases are strongly implicated in Charcot-Marie-Tooth neuropathy, an inherited disorder of the peripheral nervous system, although the molecular etiology is not yet clearly understood. Unlike other housekeeping genes, variation in transcription and copy numbers of tRNA synthetase genes has been observed in many types of cancer. In addition, many tRNA synthetases appear to interact with factors that are known to play critical roles in the process of tumorigenesis. The pathological association of tRNA synthetases with cancer is relatively new and this emerging topic is addressed by Kwon et al. in this volume.

Eukaryotes have another set of tRNA synthetases that carry out protein synthesis in mitochondria. Since mitochondria serve as the power plants of the cell, their malfunction can lead to critical cellular and organismal defects. Interestingly, similar to their cytoplasmic counterparts, many mitochondrial tRNA synthetases are associated with human diseases. In their chapter, Florentz and Sizzler address the biogenesis of mitochondrial tRNA synthetases, their connections to the mitochondrial translational machinery and respiratory complexes, and pathological implications for mitochondrial disorders.

Preface

Since the aminoacylation reaction is essential for cell viability and proliferation, the inhibition of the catalytic activities of tRNA synthetases has been explored for therapeutic purposes. Moreover, specific synthetases and their cognate tRNAs are packaged into retroviruses including human immunodeficiency virus, and in this context the tRNAs perform a completely different function related to priming of reverse transcription. Thus, the novel role of synthetases and tRNAs in retroviruses and the structural distinction between human and bacterial or fungal tRNA synthetases have provided a basis for the development of anti-infective agents. The application of naturally occurring or synthetic compounds that can inhibit the interactions and catalytic activities of different tRNA synthetases and their potential use as anti-infective agents are reviewed by Musier-Forsyth et al.

Since aminoacyl-tRNA synthetases define the genetic code, they can be used as tools to expand or modify the linkage between amino acids and codons that has naturally evolved. Although reprogramming of the genetic code can be achieved by engineering the specificity of these enzymes for alternate amino acids or tRNAs, a totally different approach that exploits catalytic RNAs has also proved to be useful. An artificially selected ribozyme, named flexizyme, can mediate the covalent linkage of amino acids to the acceptor ends of tRNAs without substrate discrimination. Although this RNA surrogate of tRNA synthetase does not yet work in vivo, it can reprogram the natural genetic code when combined with non-natural amino acids in an appropriate in vitro translation system. This emerging technology is introduced in this volume by Suga et al.

Translation of the genetic code is the central process of life involving the largest number of cellular components, and research in the area of translation typically generates over 20,000 publications annually. Aminoacyl-tRNA synthetases, central players in translation, were discovered more than half a century ago, and it has been assumed that by now all major discoveries related to these enzymes have been uncovered. However, based on the novel biology, chemistry, and medical relevance that have recently been unveiled, it is clear that a renaissance of tRNA synthetase research is underway. During the course of evolution, tRNA synthetases have adopted signal- or metabolite-sensing and novel molecular interaction capabilities. Equipped with new functional domains, they behave as "molecular transformers," changing their structure and function as needed. In particular, efforts to understand the new functions hidden within higher eukaryotic tRNA synthetases are in high gear. While the annual number of publications on tRNA synthetases has varied only slightly over the last decades, research articles on human tRNA synthetases have increased and make up about 50% of the total synthetase-related publications. This trend is expected to continue and indeed soar in the next decade. In this volume, we have selected topics that reflect the recent excitement in tRNA synthetases and emerging research areas.

The name "aminoacyl-tRNA synthetase" was coined by the Paul Berg group in 1958. In this volume, the authors use different abbreviations to indicate aminoacyl-tRNA synthetases including "aaRS" and "ARS." To indicate specific enzymes, the three-letter or single-letter symbol for an amino acid is followed by RS. For instance, glycyl-tRNA synthetase is abbreviated as GlyRS or GRS. The genes encoding each tRNA synthetase are usually indicated by the single letter amino acid code plus ARS. For example, the gene encoding glycyl-tRNA synthetase is indicated as *GARS*. To indicate mitochondrial genes, the authors usually append the number "2" at the end (e.g., *GARS2*).

The non-canonical activities of eukaryotic tRNA synthetases used to be considered as the coincidental products of divergent evolution and were believed to be functionally less significant than their canonical catalytic activities. However, as more pathophysiological functions and structures of eukaryotic and human aminoacyl-tRNA synthetases are revealed, it seems increasingly clear that the non-catalytic regulatory activities of these proteins were systematically acquired to meet the demand of complex higher eukaryotic systems, rather than by random chance during evolution. Thus, aminoacyl-tRNA synthetases appear to form a unique "functionome" working widely throughout diverse cell signaling pathways to maintain system homeostasis. The functionome of aminoacyl-tRNA synthetases is distinguished from other known groups of specific protein networks in a few key respects. First, it is universal with its presence in all locations of cells, including extracellular space. Second, it is constitutively expressed and present at relatively high amounts. Third, its localization and function seem to be primarily controlled by post-translational modification rather than by transcription, although its expression can also be regulated by the latter. Fourth, it can sense cellular nutrition such as amino acid levels and energy status to coordinate protein synthesis with other regulatory processes. Fifth, while all tRNA synthetases work together for protein synthesis, each one works idiosyncratically and with a distinct mechanism to maintain system homeostasis. With these characteristics, tRNA synthetases can respond rapidly to many different types of stimuli and stresses to prevent system disturbance. In this regard, aminoacyl-tRNA synthetases in higher eukaryotes can be considered "guardians for system homeostasis." This volume provides many perspectives on the new biology, chemistry, and medicine derived from these fascinating polyfunctional enzymes and I sincerely thank all the authors for their contributions.

Seoul, Korea   Sunghoon Kim

Contents

Synthetic and Editing Mechanisms of Aminoacyl-tRNA Synthetases 1
John J. Perona and Ita Gruic-Sovulj

Emergence and Evolution 43
Tammy J. Bullwinkle and Michael Ibba

Architecture and Metamorphosis 89
Min Guo and Xiang-Lei Yang

**Protein–Protein Interactions and Multi-component Complexes
of Aminoacyl-tRNA Synthetases** 119
Jong Hyun Kim, Jung Min Han, and Sunghoon Kim

**Extracellular Activities of Aminoacyl-tRNA Synthetases:
New Mediators for Cell–Cell Communication** 145
Sung Hwa Son, Min Chul Park, and Sunghoon Kim

**Non-catalytic Regulation of Gene Expression by Aminoacyl-tRNA
Synthetases** 167
Peng Yao, Kiran Poruri, Susan A. Martinis, and Paul L. Fox

**Amino-Acyl tRNA Synthetases Generate Dinucleotide Polyphosphates
as Second Messengers: Functional Implications** 189
Sagi Tshori, Ehud Razin, and Hovav Nechushtan

Association of Aminoacyl-tRNA Synthetases with Cancer 207
Doyeun Kim, Nam Hoon Kwon, and Sunghoon Kim

**Pathogenic Implications of Human Mitochondrial Aminoacyl-tRNA
Synthetases** 247
Hagen Schwenzer, Joffrey Zoll, Catherine Florentz, and Marie Sissler

Role of Aminoacyl-tRNA Synthetases in Infectious Diseases and Targets for Therapeutic Development 293
Varun Dewan, John Reader, and Karin-Musier Forsyth

Flexizymes, Their Evolutionary History and Diverse Utilities 331
Toby Passioura and Hiroaki Suga

Index 347

Top Curr Chem (2014) 344: 1–42
DOI: 10.1007/128_2013_456
© Springer-Verlag Berlin Heidelberg 2013
Published online: 14 July 2013

Synthetic and Editing Mechanisms of Aminoacyl-tRNA Synthetases

John J. Perona and Ita Gruic-Sovulj

Abstract Aminoacyl-tRNA synthetases (aaRS) ensure the faithful transmission of genetic information in all living cells. The 24 known aaRS families are divided into 2 structurally distinct classes (class I and class II), each featuring a catalytic domain with a common fold that binds ATP, amino acid, and the $3'$-terminus of tRNA. In a common two-step reaction, each aaRS first uses the energy stored in ATP to synthesize an activated aminoacyl adenylate intermediate. In the second step, either the $2'$- or $3'$-hydroxyl oxygen atom of the $3'$-A76 tRNA nucleotide functions as a nucleophile in synthesis of aminoacyl-tRNA. Ten of the 24 aaRS families are unable to distinguish cognate from noncognate amino acids in the synthetic reactions alone. These enzymes possess additional editing activities for hydrolysis of misactivated amino acids and misacylated tRNAs, with clearance of the latter species accomplished in spatially separate post-transfer editing domains. A distinct class of *trans*-acting proteins that are homologous to class II editing domains also perform hydrolytic editing of some misacylated tRNAs. Here we review essential themes in catalysis with a view toward integrating the kinetic, stereochemical, and structural mechanisms of the enzymes. Although the aaRS have now been the subject of investigation for many decades, it will be seen that a significant number of questions regarding fundamental catalytic functioning still remain unresolved.

Keywords Aminoacyl adenylate · Aminoacylation · Enzyme kinetics · Ribonucleo-protein · Substrate-assisted catalysis · Transfer RNA · Transition state

---

J.J. Perona (✉)
Department of Chemistry, Portland State University, P.O. Box 751, Portland, OR 97207, USA

Department of Biochemistry & Molecular Biology, Oregon Health & Sciences University, 3181 SW Sam Jackson Park Road, Portland, OR 97239, USA
e-mail: perona@pdx.edu

I. Gruic-Sovulj
Faculty of Science, Department of Chemistry, University of Zagreb, Horvatovac 102a, 10000 Zagreb, Croatia
e-mail: gruic@chem.pmf.hr

Contents

1 Introduction 2
2 Aminoacyl-tRNA Synthesis 4
   2.1 Enzyme Structures and Overview of the Aminoacylation Reaction 4
   2.2 Aminoacylation by Class I Aminoacyl-tRNA Synthetases 9
   2.3 Aminoacylation by Class II Aminoacyl-tRNA Synthetases 14
3 Editing Mechanisms in Aminoacyl-tRNA Synthesis 19
   3.1 Biological Rationale 19
   3.2 Pathways for Editing Enzymes 19
   3.3 Mechanisms of Pre-transfer Editing 21
   3.4 Mechanisms of Post-transfer Editing 24
   3.5 *Trans* Editing Factors 29
References 30

List of Abbreviations

AA-AMP	Aminoacyl-AMP
aaRS	Aminoacyl-tRNA synthetase
LUCA	Last universal common ancestor
MARS	Multi-tRNA synthetase complex
UAA	Unnatural amino acid

1 Introduction

The enduring fascination of the aminoacyl-tRNA synthetases (aaRS) arises from their central role in mediating information transfer in all cells [1, 2]. In this capacity the aaRS must be highly selective for both amino acid and tRNA substrates, a substantial challenge given the presence of many structurally similar noncognate species of each class (Figs. 1 and 2). Of the two specificities, selection against noncognate tRNAs is more readily accomplished because of the very large protein-RNA contact interfaces that are always present in the complexes. Specificity arises from direct protein readout of base-specific functional groups in the duplex helical and single-stranded regions, from indirect readout of the sugar-phosphate backbone conformation (in turn determined by the precise nucleotide sequence), and, in cases where discrimination against or among class II tRNAs with large variable arms is relevant, from tertiary shape recognition [3]. From a kinetic point of view, discrimination may occur in forming the initial encounter complex between enzyme and tRNA, and also in the first order mutual induced fit rearrangements that occur after binding and en route to a tertiary conformation in which reactive groups in the active site are properly aligned [4]. Extrapolating from studies of chemical model systems and well-studied enzymes [5], it is likely that quite small mispositionings of the tRNA $3'$-A76 ribose in the active site suffice to bring about large decreases in the rate of the chemical steps on the enzyme [6]. This phenomenon may explain the early observation that tRNA discrimination often occurs more at the level of V_{max} than K_m [7].

The problem of amino acid discrimination is more acute, particularly because of the difficulty associated with selecting against noncognate amino acids that are slightly smaller than cognate and that therefore cannot be excluded on steric grounds. The existence of editing mechanisms in the aaRS was first demonstrated in *Escherichia coli* isoleucyl-tRNA synthetase (IleRS), which forms noncognate valyl adenylate and then catalyzes its hydrolysis to valine and AMP in the presence of tRNA [8, 9]. IleRS also catalyzes hydrolysis of Val-tRNA^{Ile}. Other early examples were the hydrolytic clearance of threonine, α-aminobutyric acid, and homocysteine by valyl-tRNA synthetase (ValRS) [10, 11], of homocysteine by methionyl-tRNA synthetase (MetRS) [12, 13], of cysteine by IleRS [12, 13], and of glycine and serine by alanyl-tRNA synthetase (AlaRS) [14]. There are now ten known aaRS that catalyze various types of hydrolytic editing reactions against noncognate amino acids, including those not functioning in coded protein synthesis [2]. Particularly noteworthy are cases such as AlaRS and phenylalanyl-tRNA synthetase (PheRS), where larger noncognate amino acids (Ser and Tyr, respectively) are not sterically excluded by the synthetic domains, thus demanding hydrolytic editing [14, 15]. Enzyme architectures of aaRS carrying out post-transfer editing are generally larger and more complex because of the presence of the additional editing domain and the need for conformational flexibility to move the tRNA between synthetic and editing sites [16].

The discrimination by aaRS against amino acids and tRNAs are often considered separately, but it is crucial to recognize that the enzyme architectures have differentiated to render substrate binding interdependent [6, 17]. This was clearly demonstrated by studies in the *E. coli* glutaminyl-tRNA synthetase (GlnRS) system, which showed that mutation of tRNA identity nucleotides alters K_m for glutamine [18]. Later studies showed that tRNA^{Gln} binding affinity is substantially decreased when noncognate glutamate occupies the amino acid binding site [19]. Further work involving protein engineering of the GlnRS amino acid pocket provided additional support for the notion that tRNA^{Gln} and GlnRS sequences have coevolved in the context of the amino acid specificity for glutamine, and suggested that tRNA identity elements may function to assist in amino acid as well as in tRNA selection [20, 21]. However, the enzyme and RNA determinants responsible for this substrate interdependence remain unknown [1]. This unresolved question gets to the heart of the coding problem and raises once again the issue of how RNA may actively participate in its own aminoacylation [22]. Crosstalk between amino acid and tRNA binding sites is likely to be common to all aaRS. Crucial elements of aaRS protein structure that mediate the allosteric signaling, often over quite long distances, remain unknown and may hold the key to successful protein engineering of the coding apparatus.

The intention of this review is to provide an integrated description of the synthetic and editing reactions of the aaRS, considering the stereochemical mechanisms, the kinetic pathways, and the structural underpinnings as revealed primarily by X-ray crystallography of enzymes and enzyme-ligand complexes. This information forms a crucial foundation upon which to build our understanding of the many noncanonical functions that are now becoming known, particularly in the more complex eukaryotic systems. Some of these new functions further implicate the enzymes in human diseases such as cancers, immune system malfunctions, and neurodegeneration, thus providing considerable new impetus for ongoing studies of both biology and mechanism [23, 24].

2 Aminoacyl-tRNA Synthesis

2.1 Enzyme Structures and Overview of the Aminoacylation Reaction

Class I and class II aaRS comprise 11 and 13 families, respectively, and these are distinguished by a number of structural and functional characteristics (Table 1) [1]. The catalytic domains of all class I aaRS adopt a dinucleotide or Rossmann fold (RF) [25, 26], located at or near the N-terminus and featuring a five-stranded parallel β-sheet connected by α-helices. There is pseudo twofold symmetry with ATP and amino acid binding on opposite sides of the symmetry axis (Fig. 1). In all class I aaRS the RF is split by an inserted domain (connective peptide I (CPI)), which is greatly enlarged in IleRS, ValRS and leucyl-tRNA synthetase (LeuRS) to incorporate a distinct catalytic site for post-transfer editing hydrolysis of misacylated tRNA (Fig. 1) [2, 27]. The carboxyl-terminal domains of most class I aaRS bind the tRNA anticodon region, but their structures are in general divergent, even within subclasses. Most class I aaRS are monomers, and bind tRNA across a large portion of the enzyme surface from the anticodon stem-loop region to the active site (Table 1, Fig. 1) [28]. In these enzymes the RF binds the acceptor stem of the tRNA from the minor groove side, and the 3'-end of the tRNA adopts a hairpin structure to bind in the active site. In the dimeric TyrRS and TrpRS, tRNA binds across the subunits, and the CP1 domain forms the dimer interface [29].

In contrast, the common catalytic domain found in all 13 class II aaRS families is organized instead as a seven-stranded β-sheet flanked by α-helices (Fig. 1) [30, 31]. As in class I aaRS, this common domain binds and properly juxtaposes amino acid, ATP, and the 3'-terminus of tRNA for the catalytic reactions (Fig. 2). Most class II aaRS are homodimers, although examples of monomeric and α_4 and $(\alpha\beta)_2$ tetrameric quaternary organizations are also known (Table 1) [32–38]. Subclass IIA and subclass IIB aaRS possess subclass-specific anticodon binding domains, which occupy a similar position in the tertiary architecture even though located in distinct parts of the primary structures. Subclass IIC is the most heterogeneous of the three subclasses; these enzymes also possess a variety of additional idiosyncratic domains. The structural diversity of class II aaRS is further evident in the variety of tertiary folds found for the post-transfer editing domains in threonyl-tRNA synthetase (ThrRS), prolyl-tRNA synthetase (ProRS), PheRS, and AlaRS [2]. For ThrRS, ProRS, and PheRS, the synthetic and editing domains are positioned to enable translocation of the tRNA 3'-arm without tRNA release [39–43]. However,

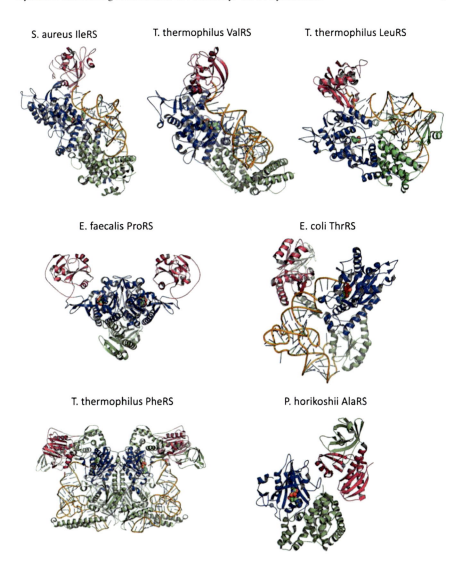

Fig. 1 Crystal structures of aaRS possessing both aminoacylation and post-transfer editing activities. The tRNA in each case is depicted with a *yellow* backbone; a tRNA cocrystal is not yet available for AlaRS while the tRNA bound to ProRS is not depicted because it is bound only through the anticodon. The synthetic domains of all enzymes are shown in *dark blue* and the editing domains are depicted in *magenta*. The structure of the editing domain, and the relative positions of the synthetic and editing domains are conserved in most class I IleRS, ValRS, and LeuRS. However, these characteristics are idiosyncratic in the class II enzymes ProRS, ThrRS, PheRS, and AlaRS. Peripheral domains involved in oligomer formation, tRNA selectivity, and other functions are depicted in *green*. Structures are drawn from the following PDB entries: IleRS; 1FFY, ValRS; 1IVS, LeuRS; 2BYT, ProRS; 2J3L, ThrRS; 1QF6, PheRS; 2IY5, AlaRS; 2ZZG

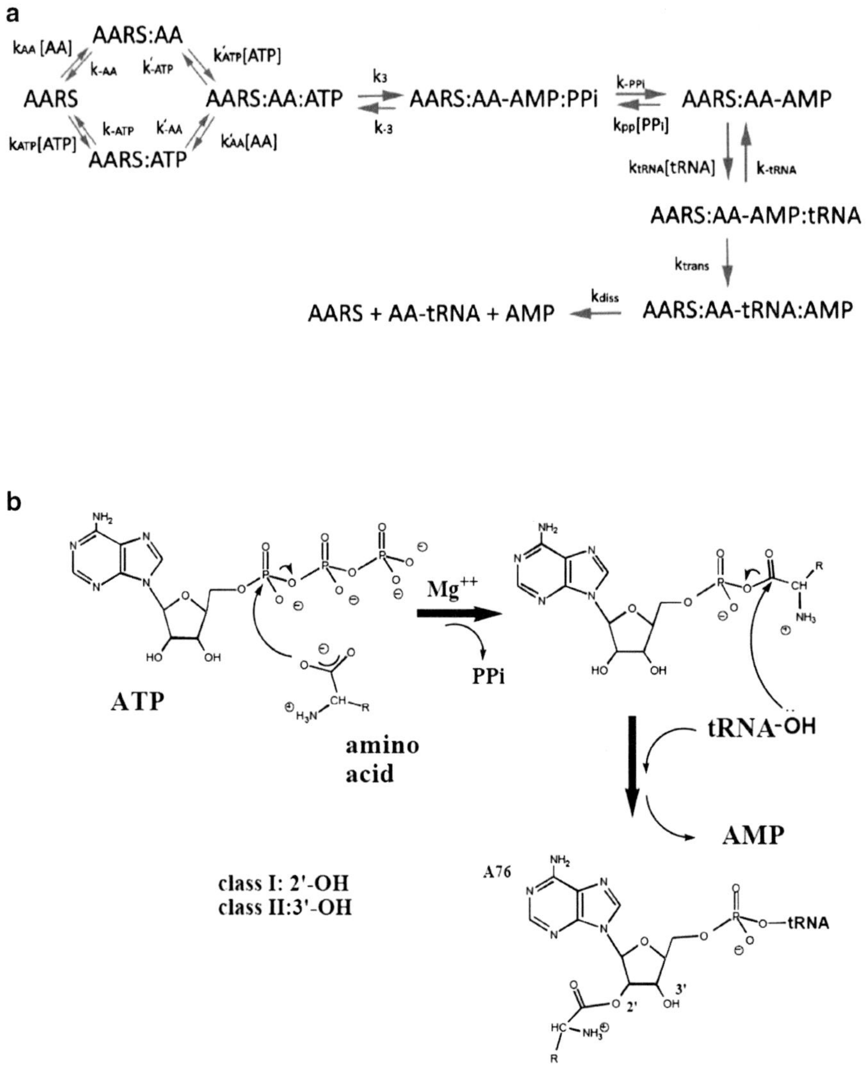

Fig. 2 (a) Kinetic pathway of aminoacyl-tRNA synthesis. Binding of amino acid and ATP is random for both class I and class II enzymes. (b) Stereochemical mechanism of aminoacylation showing formation of the aminoacyl adenylate intermediate (*upper right*) and aminoacyl-tRNA (*bottom right*). The tRNA nucleophile in the second step is the oxygen atom from one of the two ribose hydroxyl groups of the 3'-A76 nucleotide. The identity of the hydroxyl group (2' or 3') is generally correlated with the structural class to which an aaRS belongs

there are indications that AlaRS may bind tRNA differently because the relative position of editing and aminoacylation domains in this enzyme is unusual [44, 45].

With the possible exception of AlaRS, all class II aaRS adopt a common mode of tRNA binding in which the major groove of the acceptor stem duplex is oriented

Synthetic and Editing Mechanisms of Aminoacyl-tRNA Synthetases

Table 1 Classification of aminoacyl-tRNA synthetases[a]

Class I	4° Structure[b]	Editing	Class II	4° Structure[b]	Editing
Subclass IA			*Subclass IIA*		
MetRS	α, α_2	Yes (Hcy)[c]	SerRS	α_2	Yes (Thr, Cys, SerHX)[c]
LeuRS	α	Yes (Nva, Ile, γhL, δhL, γhI, δhI Met)	ProRS	α_2	Yes (Ala, Cys, 4hP)
IleRS	α	Yes (Val, Cys)	ThrRS	α_2	Yes (Ser)
ValRS	α	Yes (Thr, Abu)	GlyRS[d]	α_2	No
			HisRS	α_2	No
Subclass IB					
CysRS	α, α_2	No	*Subclass IIB*		
GlnRS	α	No	AspRS	α_2	No
GluRS	α	No	AsnRS	α_2	No
			LysRS[e]	α_2	Yes (Orn, Hcy, Hse)[c]
Subclass IC					
TyrRS	α	No	*Subclass IIC*		
TrpRS	α_2	No	PheRS	$(\alpha\beta)_2, \alpha$	Yes (Tyr, Ile)
			GlyRS[d]	$(\alpha\beta)_2$	No
Subclass ID			AlaRS	α_2, α	Yes (Ser, Gly)
ArgRS	α_2	No	SepRS[f]	α_4	No
			PylRS[g]	α_2	No
Subclass IE					
LysRS[e]	α	No			

[a]Subclassifications are based on phylogenies constructed from multiple structural alignments [229]. The identities of the most important amino acids edited are indicated in parentheses. Abbreviations for nonstandard amino acids are: *Hcy* homocysteine, *Nva* norvaline, *4hP* 4-hydroxyproline, *SerHX* serine hydroxamate, *Orn* ornithine, *Hse* homoserine, *γhL* γ-hydroxyleucine, *δhL* δ-hydroxyleucine, *γhI* γ-hydroxyisoleucine, *δhI* δ-hydroxyisoleucine, *Abu* α-aminobutyrate

[b]For MetRS, CysRS, PheRS, and AlaRS, distinct quaternary structures are found for enzymes in different organisms or subcellular localizations [1]

[c]For MetRS, SerRS, and class II LysRS, the editing activity is limited to pre-transfer hydrolysis of noncognate aminoacyl adenylates within the synthetic active site [12, 163, 173]

[d]GlyRS is polyphyletic – present as two forms in distinct classes that do not directly share a common ancestor of that specificity [229]

[e]LysRS is the only aaRS that violates the class rule, as enzymes possessing class I or class II architectures are known in distinct organisms [230]. A few methanogens possess both LysRS enzymes

[f]SepRS designates phosphoseryl-tRNA synthetase. This enzyme is found in a small subset of archaea, and aminoacylates tRNA^{Cys} with phosphoserine (Sep). A companion enzyme, SepCysS, converts Sep-tRNA^{Cys} to Cys-tRNA^{Cys} to enable cysteine coding [231]

[g]PylRS designates pyrrolysyl-tRNA synthetase. This enzyme is found only in a subset of methanogens and in some bacteria. It aminoacylates a specific tRNA of unusual tertiary structure, tRNA^{Pyl}, to enable pyrrolysine incorporation into proteins in response to UAG codons [232–235]

toward the catalytic domain. This is the basis for a key class-specific distinction, as the monomeric class I enzymes approach the acceptor stem from the minor groove side (Fig. 1) [30]. Interestingly, the dimeric class I enzymes, tyrosyl-tRNA synthetase (TyrRS) and tryptophanyl-tRNA synthetase (TrpRS), exhibit a class II tRNA binding mode [29]. Beyond the many differences in detail with respect to how rate enhancement is provided, there are several other general mechanistic distinctions of importance between the classes. First, all class I aaRS, including TyrRS and TrpRS, catalyze aminoacylation directly to the 2'-OH of tRNA-A76, while class II enzymes, with the exception of PheRS, catalyze aminoacylation directly to the 3'-OH (Fig. 2). Thus, the initial site of aminoacylation is generally, but not strictly, correlated with an overall orientation of tRNA binding in which the enzyme associates primarily with the minor groove (class I) or the major groove (class II) of the tRNA acceptor stem. Second, class I aaRS are generally rate-limited by release of the aminoacyl-tRNA product (GluRSND and IleRS are exceptions) [46–49], while class II aaRS are rate-limited by an earlier step associated with formation of aminoacyl-tRNA on the enzyme (see discussion below) [50]. Yet a third class-specific distinction involves the precise identity of the attacking oxygen atom in synthesis of the aminoacyl adenylate [51].

All aaRS are thought to follow the two-step mechanism demonstrated explicitly for IleRS and TyrRS [49]:

$$ATP + aa \rightarrow aa\text{-}AMP + PP_i$$
$$aa\text{-}AMP + tRNA^{aa} \rightarrow aa\text{-}tRNA^{aa} + AMP$$

In the first step, one α-carboxylate oxygen of the amino acid attacks the α-phosphorus of Mg-ATP, forming the mixed anhydride linkage in the aminoacyl adenylate (aa-AMP) with release of pyrophosphate (Fig. 2). In the second step, either the 2'- or 3'-hydroxyl group of the *cis*-diol at the 3'-terminal A76 nucleotide of tRNA attacks the carbonyl carbon of the adenylate, forming aminoacyl-tRNA with release of AMP. The order of ATP and amino acid binding has been found to be random in those aaRS (both class I and class II) where the question has been investigated by steady-state kinetics (Fig. 2) [32]. However, GlnRS, glutamyl-tRNA synthetase (GluRS), arginyl-tRNA synthetase (ArgRS), and class I lysyl-tRNA synthetase (LysRS-I) are unable to catalyze aminoacyl adenylate formation in the absence of tRNA [52–55]. While binding of ATP and/or amino acid may be possible without tRNA in these enzymes, productive juxtaposition of the reactive moieties does not occur [56]. Thus, these four aaRS are obligate ribonucleoprotein enzymes with the catalytic activity residing primarily in the protein subunit [57].

Early studies of several class I aaRS by analysis of kinetic isotope effects showed that amino acid activation occurs by in-line attack of an amino acid carboxylate oxygen atom on the electrophilic α-phosphorus of ATP, consistent with structural analyses of both class I and class II enzymes [26, 58, 59]. However, the precise nature of the transition state; that is, whether it more closely resembles a pentacoordinate structure or a metaphosphate, remains unresolved [32]. Recently, computational studies on a large number of class I and class II enzymes suggested that the *syn* oxygen atom of the amino acid carboxylate attacks the α-phosphorus in class I aaRS, while the *anti* oxygen is the equivalent attacking nucleophile in class II aaRS (*syn* vs. *anti* carboxylates adopt different conformations with respect to the α-NH$_3^+$ group of the amino acid) [51]. This distinction is related to differences in the relative binding orientations of amino acid and ATP in class I and class II active sites (Fig. 3). The stereochemistry of the second tRNA aminoacylation step is conserved in all aaRS with the exception of the identity of the attacking nucleophile (2'-OH vs 3'-OH; Fig. 2).

The two-step aminoacylation reaction is thermodynamically driven by hydrolysis of the pyrophosphate product by inorganic pyrophosphatase, so that formation of the aminoacyl-tRNA bond demands cleavage of two high-energy phosphoanhydride bonds. Over evolutionary time, the rate of the two-step reaction has been optimized via selective pressure for equivalent transition-state stabilization for each step, so that the activation energies of the two steps are about equal [60, 61]. For TyrRS, correlating the conservation of active site residues with their known roles in stimulating catalysis of each reaction step suggested that, in the evolutionary development of the enzyme, stabilization of the transition state for amino acid activation preceded that for tRNA transfer [62]. Thus, primordial protein-based aaRS may have been specialized for aminoacyl adenylate formation only, with tRNA aminoacylation performed by an RNA catalyst. This is consistent with proposed substrate-assisted mechanisms for tRNA aminoacylation in contemporary aaRS (see below) [58, 63], and with the finding that minimal enzymes (so-called urzymes) consisting solely of portions of class I and class II aaRS catalytic domains are capable of amino acid activation [64–66].

2.2 Aminoacylation by Class I Aminoacyl-tRNA Synthetases

The RF active site domain of class I aaRS possess two conserved sequence motifs, HIGH and KMSKS. The HIGH sequence is found at the amino-terminal end of the first α-helix in the first half of the Rossmann fold, while KMSKS is located in a loop immediately following the second half of the Rossmann fold (Fig. 3). Only the glycine residue of HIGH is strictly conserved; the backbone at this position adopts a conformation that allows stacking with the adenine ring of ATP, and the strict conservation can be explained by the likely disruption of productive binding that would likely be caused by any β-substituent. The isoleucine residue contributes to internal hydrophobic packing while the histidines form direct and water-mediated hydrogen bonds with the ATP phosphates. Hydrogen-bonding of the enzyme with Watson–Crick moieties of the adenine ring rationalizes the specificity against GTP as high-energy cofactor. The lysine residues of KMSKS, particularly the more highly conserved second lysine of this sequence, make interactions with the ATP phosphates (Fig. 3).

Structures of ATP-bound class I aaRS are available for GlnRS, GluRS, TyrRS, and TrpRS [28, 29, 56, 67–69], but not for members of subclasses IA, ID, or IE (Table 1). By contrast, structures of amino acid substrates bound to class I aaRS are available for all of the enzymes, except that the binary complex structure of lysine bound to class I LysRS likely represents a nonproductive complex given that this enzyme requires tRNA to catalyze lysyl adenylate formation [1, 70]. Except for the

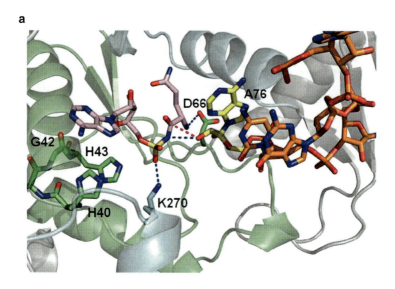

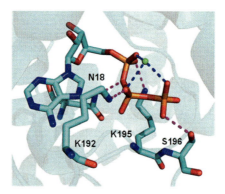

Fig. 3 (a) Structure of *E. coli* GlnRS bound to tRNA[Gln] and the glutaminyl adenylate analog 5′-*O*-[*N*-(L-glutaminyl)sulfamoyl]adenosine (QSI) (PDB 1QTQ) [72]. In QSI the phosphorus is replaced with sulfur and the bridging oxygen substituted by an amine. QSI is depicted in *pink*. The position and interaction of class I motif residues from HIGH (H40; H43) and KMSKS (K270 represents the second Lys) are shown. Hydrogen bonds are shown as *blue dotted lines*. The *red dotted line* depicts the modeled position of 2′-OH attack on the carbonyl carbon of the mixed anhydride of Gln-AMP. The hairpinned tRNA 3′-end, a characteristic of class I monomers, is shown in *orange* at *right* with 3′-A76 in *yellow*. (b) Active site of *B. stearothermophilus* TrpRS bound to Mg-ATP (PDB 1 M83). The *green sphere* represents the position of Mg^{2+} binding to all three ATP phosphates (*dotted blue lines*). Interactions of Asn18 (representing the position of the usually conserved second histidine of the HIGH motif) and of both lysines and second serine of the KMSKS motif with the ATP are also indicated (*dotted magenta lines*)

four enzymes requiring tRNA for amino acid activation, the amino acid pocket generally appears to be preformed, although MetRS is an exception as formation of the methionine binding site is induced by methionine binding [71]. The binding site for the amino acid features a conserved aspartate or glutamate residue that makes a salt bridge with the α-NH_3^+ group, which in most of the enzymes emanates from the second β-strand in the first half of the Rossmann fold (Fig. 3).

For the nonediting class I aaRS, specificity for the amino acid is in most cases easily rationalized based on hydrogen-bonding electrostatic and salt-bridge interactions with the cognate R-group as seen in cocrystal structures [1]. The sole exception is GlnRS, as X-ray structures show that the side-chain amide makes only ambiguous hydrogen bonds with a tyrosine side chain and with water molecules in the pocket [72, 73]. In this case amino acid specificity may rely on a conserved arginine residue functioning as a negative determinant against glutamate binding, and on a hydrogen-bonding interaction of the glutamine substrate side-chain amide with the ATP ribose. This is the only known example among the aaRS of a distal hydrogen-bonding interaction between ATP and the side chain of the amino acid, and it is not formed in the noncognate complex with glutamate [73, 74]. Amino acid specificity determinants in GlnRS are broadly delocalized as demonstrated by the finding that extensive mutagenesis in the second half of the Rossmann fold is necessary to convert the selectivity to glutamate [20, 21].

In four cases, cocrystal structures of class I aaRS bound to tRNA and other substrates or analogs reveal a conformation likely to be productive for aminoacylation. These enzymes are GlnRS, GluRS, ArgRS, and LeuRS [16, 58, 67, 75]. For the other seven class I enzymes, either a tRNA cocrystal is not available or the tRNA 3′-end is not oriented into the synthetic active site [1]. In the catalytically productive cocrystal structures, the HIGH and KMSKS motifs are spatially adjacent and make a number of interactions with each other (Fig. 3), explaining their simultaneous conservation even though many of the enzymes deviate from the canonical sequences. An overview of the tRNA binding interfaces in aaRS and the coupling of tRNA recognition to catalysis has recently appeared [1].

ATP binding stabilizes the structure of the active site in at least the subclass IC aaRS comprising TyrRS and TrpRS (Table 1), which from the standpoint of catalytic mechanism are the best-studied enzymes in either class. The most important conformational change that is induced by ATP binding is the reorientation of the mobile KMSKS loop, which moves approximately 8 Å toward the active site to make interactions that are specific to the transition state for aminoacyl adenylate formation [76, 77]. Extensive studies on TyrRS and TrpRS suggest that the KMSKS loop is not well-ordered in the absence of ligands. At least three distinct conformations of this loop have been visualized in different class I aaRS cocrystal structures: the closed catalytically productive conformation and two distinct conformations in which the loop adopts a more open structure [16, 26, 29, 75, 78, 79]. In TyrRS, a key role of the loop is to destabilize the ternary TyrRS:ATP:Tyr complex in order to promote movement toward the transition state, and this occurs at the expense of ATP binding energy [77]. Further evidence for synergy between ATP and tyrosine binding is provided by recent thermodynamic analyses, which showed that temperature-dependent changes in the enthalpy and entropy of ATP binding occur only when tyrosine is also bound [80]. This work is consistent with extensive studies of TrpRS, where experiments involving X-ray crystallography, kinetic measurements, and molecular dynamics simulations have demonstrated that the binding energy of ATP is converted into an unfavorable high-energy protein conformation, and the stored energy is then used to promote catalysis [68, 81, 82]. The TrpRS structural mechanism also features global induced fit repositioning of domains associated with three distinct conformational states where the protein is visualized bound to distinct ligands [81, 83].

Exhaustive mutational analysis of TyrRS has demonstrated that the synthesis of aminoacyl adenylate is not promoted by either acid–base or covalent catalysis [84]. Instead, rate enhancement derives from use of the enzyme structure to position precisely the reactive groups of ATP and amino acid, and from selective binding of the transition state. Another key element in amino acid activation by most or all class I aaRS is a single magnesium ion that bridges all three ATP phosphates (or just the β- and γ-phosphates) but which apparently does not make direct interactions with protein groups (Fig. 3) [26, 58, 68]. In TrpRS, kinetic measurements made under Mg^{2+}-depleted conditions show that the rate enhancement attributable to Mg^{2+} is approximately 10^5-fold, but a very substantial 10^9-fold rate enhancement remains and is attributable to protein alone [85]. Further work on TrpRS involved measurements of the detailed metal dependencies of Trp-AMP synthesis, analysis of multimutant cycles, correlation of these data with crystal structures in a variety of ligand-bound states, and extensive use of constrained and unconstrained molecular dynamics [86–88]. The insights gained are generally relevant to ATP-utilizing proteins. Mg^{2+} ion binding on the ATP phosphates is energetically coupled to the protein conformational changes, with an important role for a distal cluster of hydrophobic residues (the "D1 switch") that help mediate the global twisting and untwisting motions that are central to stabilizing the transition state conformation [86]. Thus, quite unexpectedly, the basis for rate enhancement by Mg^{2+} derives not only from its local active-site contacts but also from its capacity to couple the global protein conformational change with inducing the transition state structure for synthesis of Trp-AMP.

With respect to the second step of aminoacylation, it so far appears that the reaction mechanism first proposed for GlnRS, in which the phosphate of the aminoacyl adenylate abstracts a proton from the 2′-hydroxyl of tRNA-A76, may be general in class I aaRS (Fig. 4) [16, 58, 89]. This mechanism was deduced on structural grounds from the geometry of the active site when tRNA and ATP are bound, and is consistent with evolution from a ribozyme-based system and with the absence of conserved amino acids among all class I enzymes that are properly positioned to facilitate acid–base chemistry. The mechanism predicts a concerted reaction that features a six-membered ring structure in which the proton on the 2′-OH moiety of tRNA-A76 is transferred to the phosphate group of the aminoacyl adenylate. Corroborating evidence for a concerted reaction was provided by kinetic studies of TyrRS mutants, which showed that catalysis involves relief of strain induced in the scissile bond of the tyrosyl adenylate intermediate [62, 90]. This conclusion was reached based on the finding that enzyme side chains His40,

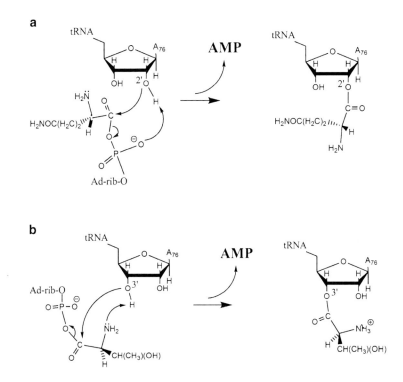

Fig. 4 Proposed mechanisms for transfer of the amino acid portion of the aminoacyl adenylate to tRNA. (**a**) Proposed mechanism for class I GlnRS as deduced from the crystal structure of the *E. coli* enzyme bound to tRNA and ATP, shown in Fig. 3a [58]. (**b**) Proposed mechanism for class II ThrRS based on experimental data and computations [63, 145]

Gln173, Lys82, Arg86 interacting with either A76 of tRNA^{Tyr} or the α-NH$_3$$^+$ group of tyrosyl adenylate are each important for stabilizing the transition state for Tyr-tRNA^{Tyr} formation. Thus, the data are consistent with the notion that scission of the acylphosphate bond of tyrosyl adenylate and formation of the new bond with the tRNA 2'-O moiety are each present in the transition state. Hence, this transition state may be synchronous (concerted) as predicted by the GlnRS structure (Figs. 3 and 4), rather than either tight (associative; reflecting greater bond formation in the transition state) or loose (dissociative; reflecting greater bond cleavage in the transition state) [91]. The plausibility of this reaction mechanism was also supported by energy calculations performed on GluRS [92]. An alternative mechanism for amino acid transfer in *E. coli* GlnRS invoked Glu34 as a general base [72], but this is implausible since mutation of this residue later showed that its carboxylate group is not required in the catalytic steps [89].

The rate-limiting step for most class I aaRS occurs late in the reaction pathway, after formation of aminoacyl-tRNA on the enzyme. This is evident from the observation of burst kinetics in pre-steady state reactions [46, 61, 93–95]. The tight aminoacyl-tRNA product binding by class I aaRS correlates with the ability of

EF-Tu to form a ternary complex with these enzymes, and the further capacity of this protein to enhance release of aminoacylated tRNAs [46]. Exceptions to the observation that a late step is rate-limiting include *Bacillus stearothermophilus* methionyl-tRNA synthetase (MetRS) [96], the nondiscriminating GluRS [48], for which the Glu-tRNA^{Gln} product does not bind EF-Tu, and IleRS [49]. The absence of burst kinetics in IleRS is related to the unusual mechanism of hydrolytic editing exhibited by this enzyme, as compared with the other class I editing enzymes ValRS and LeuRS (see below) [97].

For the majority of class I aaRS that do exhibit burst kinetics, it is likely that the precise step that is rate-limiting is the release of aminoacyl-tRNA. Based on molecular modeling and molecular dynamics simulations, it appears that dissociation of aminoacyl-tRNA from GluRS is controlled by deprotonation of the α-amino group on the amino acid portion of this substrate [92]. Significantly, the proton acceptor is proposed to be a negatively charged residue (Glu41) that is highly conserved as aspartate or glutamate in class I aaRS generally, suggesting that this deprotonation event may be a common feature of many of the enzymes (Fig. 3). Aminoacylation may provide a trigger by which the proton in the Glu41-NH$_3^+$ ion pair shifts onto the carboxylate, thus weakening the binding affinity of Glu-tRNA^{Glu} to GluRS [92].

It is worth noting that, while crystal structures are available for all of the subclasses, usually bound to either or both of the small molecule substrates and tRNA [1], detailed mechanistic work involving characterization of the transition state, particularly for aminoacyl adenylate synthesis, has been primarily carried out only for the class IC enzymes TyrRS and TrpRS. However, the class IC enzymes are unique in several respects: they are dimers binding a single tRNA across subunits, they exhibit complex intersubunit cooperativities, and the orientations of their amino acid binding sites are distinct as compared with the other class I aaRS, which are almost all monomers (Table 1) [29, 98, 99]. Hence, mechanistic conclusions reached based on TyrRS and TrpRS alone may not be fully generalizable across the class I family. In particular, distinct mechanisms for rate enhancement might be of interest to explore in detail for the enzymes possessing a post-transfer editing domain as well as those that require tRNA for aminoacyl adenylate formation.

2.3 *Aminoacylation by Class II Aminoacyl-tRNA Synthetases*

The common catalytic domain of all class II aaRS consists of a seven-stranded β-sheet flanked by α-helices (Fig. 1), and features three conserved motifs not present in the class I aaRS [30, 31, 100]. Motif 1 consists of the sequence + GΦXXΦXΛP$\Phi\Phi$, where + is a positively charged residue, Φ is a hydrophobic residue, Λ is a small amino acid, and X is any amino acid [32, 101]. It forms a long α-helix followed by a short strand that terminates in a conserved proline (underlined), and is involved in dimer interface formation in most of the enzymes [102, 103]. Motifs 2 and 3 consist of the sequences ++$\Phi\Phi$X$\Phi\Lambda$XXFRXEX...

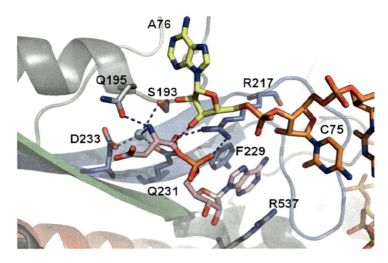

Fig. 5 Crystal structure of class II *T. thermophilus* AspRS bound to tRNA^{Asp} and aspartyl adenylate (PDB 1C0A) [115]. The aspartyl adenylate is in *pink* and the tRNA is in *orange* with the 3′-A76 nucleotide in *yellow*. Hydrogen bonds with the adenylate are shown as *blue dotted lines*. The position and interactions of class II aaRS invariant residues in motif 3 (Arg537) and motif II (Arg217; Phe229) are shown. The *red dotted line* depicts the modeled position of 3′-OH attack on the carbonyl carbon of the mixed anhydride of Asp-AMP

($n = 4$–12)... + ΦXX$\underline{F}$XXΦ and ΛXΦGΦGΦGΦE$\underline{R}$ΦΦΦΦΦ, respectively. Both of these motifs are found directly in the active site in a pair of antiparallel β-strands connected by a loop, and in a strand-helix substructure, respectively (Fig. 5) [31, 100]. The very highly conserved arginine and phenylalanine of motif 2, and the arginine of motif 3 (underlined) play central catalytic roles (see below). In general, motif 3 binds ATP while motif 2 plays a key role in coupling ATP, amino acid, and 3′-tRNA binding [104]. In its ATP binding function motif 3 may be likened to the HIGH and KMSKS motifs of class I enzymes, but there is no analogous motif in class I aaRS that plays the central role attributed to motif 2. Reconstruction of a possibly ancestral urzyme version of histidyl-tRNA synthetase (HisRS), containing only polypeptide segments from the active site domain, suggested that conserved motifs 1 and 2 were presented in a HisRS-type early class II aaRS, and that motif 3 as well as a HisRS insertion domain did not arise until later [64].

Detailed characterization of the transition state for aminoacyl adenylate and aminoacyl-tRNA formation, along the lines of the extensive studies on class IC TyrRS and TrpRS, has yet to be performed for any class II aaRS. Hence, inferences regarding the catalytic mechanism have relied more on X-ray structures and conventional mutational analysis. ATP binds in a common bent conformation in which the α-phosphate is positioned over the adenine base, which stacks on the conserved phenylalanine of motif 2. In the majority of the enzymes the α-phosphate of ATP is anchored by interactions with the invariant motif 2 arginine and a divalent metal ion that bridges the α- and β-phosphates [43, 59, 105–110]. Conserved aspartate and glutamate side chain carboxylates ligate this metal ion. Two additional divalent metal ions that bridge β- and γ-phosphates of ATP are also present in many class II aaRS; these metals help to stabilize the pyrophosphate leaving group. The use of three metal ions in amino acid activation is a characteristic of many of the class II aaRS, providing a clear contrast with the single divalent metal ion used in class I enzymes [34, 111]. The ATP γ-phosphate is also stabilized via contacts with the conserved motif 2 loop and motif 3 arginine residues.

As in class I aaRS, the amino acid binding sites in class II enzymes are generally preformed and the specificity for the amino acid readily rationalized by inspection of cocrystal structures. Amino acid activation proceeds by in-line displacement of the pyrophosphate leaving group by the incoming nucleophile, with inversion of configuration at the ATP α-phosphorus [34, 111]. The invariant motif 2 arginine establishes contacts with both the α-phosphate and the attacking oxygen nucleophile, and also contacts the aminoacyl adenylate intermediate after the first reaction [43, 103, 107–109, 112–116]. It thus appears that this residue may participate in anchoring substrates in the reactive conformation and in stabilizing the transition state in both activation and transfer steps [115]. Based on the positions of the motif 2 arginine and the divalent metal ions, and on experiments in which the motif 2 arginine was mutated, an important contribution for electrophilic catalysis in class II aaRS seems clear [50, 101, 105, 106, 108, 116]. The electron-withdrawing surroundings allow juxtaposition of the negatively charged amino acid carboxylate in close proximity to the negatively charged α-phosphate, further rendering the α-phosphorus more electrophilic and thus more susceptible to nucleophilic attack.

The carbonyl oxygen of the amino acid/aa-AMP is often also contacted by an additional active site residue, which is preferentially glutamine (Fig. 5) [101, 110, 112, 115, 116]. This contact was shown by mutational analysis to be important in amino acid activation and not in the aminoacyl transfer step [50, 63, 117]. The extent to which amino acid activation is compromised by mutation at this residue varies widely in different class II aaRS, however [101, 118].

Given the substantial diversity among the enzymes in the three subclasses, it is not surprising that many important catalytic interactions are specific to particular enzymes. The HisRS-specific Arg259 is important to both amino acid activation and tRNA transfer; the role of this side-chain is also supported by computational studies [50, 116, 119, 120]. Thus amino acid activation by HisRS is mediated by two arginines: Arg259 and the conserved motif 2 arginine. In ThrRS, Lys465 is topologically equivalent to HisRS Arg259, and its presence may compensate for the absence of a divalent metal ion bound to the α-phosphate in that enzyme [63, 121]. Interestingly, crystal structures of PheRS bound to Phe-AMP or a Phe-AMP analog did not reveal any divalent metal ions within the active site [113, 122]. In PheRS the electrophilic role assigned to the metal ions in other class II aaRS may be replaced by a number of polar residues and water molecules [113]. In all class II aaRS, however, the conserved motif 2 arginine establishes contact with the α-phosphate in the ground and presumably transition state of amino acid activation as well.

In addition to the extensive and often idiosyncratic conformational changes associated with tRNA binding, there are also several specific induced-fit rearrangements in the active site that are central to the catalytic mechanisms.

Three distinct loops found in the class II aaRS active sites are involved: the so-called "flipping loop" located between motifs 1 and 2 in the primary sequence, the motif 2 loop, and a loop structure that is exemplified by the "serine ordering loop" in the methanogen-type SerRS enzymes [110, 118]. Loops of this latter type change conformation upon amino acid binding and have been found in a number of the enzymes [108, 123–125]. The conformational change in the motif 2 loop occurs in response to ATP or aminoacyl adenylate binding and has been observed in all class II aaRS so far studied [59, 105–109, 116, 123–126]. In some cases this loop also adopts an alternate conformation when tRNA is bound [127]. The conformational changes in the flipping (helical/ordering) loop have been observed in response to amino acid or tRNA binding, or aminoacyl adenylate formation [106–108, 115, 123–131]. Generally, the flipping loop will close into the active site to form a catalytically productive conformation, upon binding of both amino acid and ATP, but there are several exceptions [108, 124]. It has also been suggested that the flipping loops may be involved in binding the 3′-end of the tRNA [63, 124].

In contrast to most class I aaRS, no burst of aa-tRNA or PP_i is observed under pre-steady-state conditions in the class II enzymes, and it appears that the rate-limiting step in overall aminoacylation is amino acid activation [50, 63, 132, 133]. Interestingly, separate analysis of the activation and transfer steps revealed that both steps proceed with a rate that is faster than overall aminoacylation [50, 118, 126, 132]. The apparent paradox was resolved by the finding that the presence of tRNA slows amino acid activation [50, 63, 132].

The cocrystal structures show that the dimeric AspRS, ThrRS and pyrrolysyl-tRNA synthetase (PylRS) each bind two tRNAs. For the latter two enzymes, the binding of each tRNA involves amino acids from both subunits, while cross-subunit binding is not evident in AspRS [30, 134, 135]. Cocrystal structures of the SerRS–tRNA^{Ser} complex reveal only one tRNA bound across the two subunits [136], but solution studies showed that two tRNAs may bind cooperatively under some conditions [137]. The dimeric nature of most class II aaRS is important to function: even in AspRS, which does not exhibit cross-subunit tRNA interaction, mutations in motif 1 at the dimer interface were detrimental [138]. It was further shown that the reduced catalytic efficiency of bacterial AspRS toward eukaryotic tRNA^{Asp} has its origin in asymmetric binding: this heterologous cocrystal structure, the only one available for any aaRS, shows that only one of the two bound tRNAs is properly oriented in an active site [129].

It is worth noting that asymmetric tRNA binding occurs in cognate tRNA complexes of ProRS and TrpRS [139, 140], and that half-of-sites reactivity is not uncommon in dimeric and tetrameric aaRS [98, 141–143]. Hence, asymmetry in some aaRS–tRNA complexes appears to be a natural feature of their interactions. Using fluorescently labeled HisRS, it was demonstrated that the enzyme exhibits 100-fold different rates of His-AMP synthesis in each subunit, and that a conformational change may be rate-limiting for the reaction. The asymmetry in the HisRS dimer appears to be an inherent feature of the enzyme and may arise when adenylate formation is monitored in the absence of tRNA, because the tRNA binding likely plays a role in coupling the activities in the two subunits. The data are consistent with an alternating site model for the HisRS catalytic cycle, in which an initial "priming" round of catalysis occurs in which His-AMP is synthesized in only one subunit [117]. Subsequently, tRNA binding and aminoacylation to the subunit catalyzing His-AMP formation transmits structural changes, allowing substrate binding and aminoacylation in the opposing subunit. Further efforts will be required to substantiate this model, but the work represents the most detailed effort to define the sequence of events in the catalytic cycle of any multimeric aaRS.

For AspRS, ThrRS, and PheRS, sufficient experimental structures are available to provide confidence that the structural environment in the active site facilitating proper juxtaposition of the substrates for aminoacyl transfer is understood [1]. Detailed mechanisms have been proposed for the second step of aminoacylation in AspRS, ThrRS, and HisRS. First, the reaction mechanism proposed for class I enzymes, in which the phosphate of the aminoacyl adenylate abstracts a proton from the 2'-hydroxyl of tRNA-A76, has been extended to class II as well (Fig. 4) [58, 101]. The original proposal was made for AspRS based on X-ray structures (Fig. 5) [101, 115], and the role of the phosphate group was then elegantly demonstrated by thio substitution experiments on HisRS. A large decrease in the single-turnover rate for His-tRNA^{His} formation from the histidyl adenylate preformed with $\alpha(S_p)$ (α-phosphorothioate-substituted ATP) provided direct evidence for the role of the nonbridging S_p oxygen in aminoacyl transfer [50]. Later, density functional group calculations on HisRS provided corroborating evidence for the plausibility of this mechanism [144]. However, there is no consensus that it necessarily also holds in all other class II aaRS. Indeed, no significant decrease in the rate of Thr-tRNA^{Thr} formation by ThrRS was observed when either S_p or R_p-ATPαS replaced ATP in the reaction [63]. Therefore, use of a nonbridging oxygen atom from the phosphate of the aminoacyl adenylate as a general base is not a universal feature of the aminoacylation reaction.

In ThrRS, a different mechanism was instead proposed in which the vicinal 2'-OH on tRNA-A76 promotes catalysis via proton relay [63]. Based on mutagenesis, a histidine residue in the active site was proposed to accept the proton, and the deprotonated 2'O in turn would accept a proton from the adjacent 3'-OH group to facilitate attack by the 3'-oxygen on Thr-AMP. However, this mechanism was not supported by later computational studies [145]. Instead, it appears more plausible that the proton of 3'-OH is transferred directly to the unprotonated α-NH$_2$ group of Thr-AMP (Fig. 4). Further experimental studies are required to test this new mechanism. Both these proposed mechanisms for ThrRS, as well as the now well-supported mechanism in which the phosphate of the aminoacyl adenylate accepts the proton, involve substrate-assisted catalysis. The existence of further distinct mechanisms also remains possible, since most aaRS of either class have not been closely investigated with a view toward addressing this question.

3 Editing Mechanisms in Aminoacyl-tRNA Synthesis

3.1 Biological Rationale

The rationale for the evolution of editing in the aaRS is presumably the survival advantage associated with a higher proportion of accurately synthesized proteins. Mistranslation arising from an editing-defective AlaRS in mammalian cells gives rise to a severe neurodegenerative phenotype, substantiating this view [146]. In these cells, the basis for disruption of function is triggering of the unfolded protein response due to elevated levels of misfolded proteins incorporating miscoded amino acids. The formation and hydrolysis of unfolded proteins due to mistranslation has also been observed in bacteria [147]. However, some cells are capable of tolerating substantial misincorporation in their proteomes without affecting viability. In these cases, the editing mechanisms apparently become important only when stress conditions such as nutrient limitation are encountered [148]. The loss of editing function in several naturally occurring aaRS, especially in mitochondrial enzymes [149, 150], also emphasizes that the requirement for editing depends on the biological context. Indeed, a capacity for mistranslation may be adaptive under particular conditions, as demonstrated by the triggering of mismethionylation by human MetRS when cells are under oxidative stress [151].

3.2 Pathways for Editing Enzymes

Editing by the aaRS occurs within a common overall kinetic pathway (Fig. 6), with variation among the individual enzymes with respect to which reactions dominate. All ten editing enzymes (Table 1) are capable of synthesizing aminoacyl adenylate in the absence of tRNA, and all misactivate one or more noncognate amino acids at appreciable rates [2]. Editing is usually present when misactivation of the noncognate amino acid occurs with k_{cat}/K_m better than 10^{-3}-fold of the cognate reaction. After misactivation there are several distinct reactions that together constitute pre-transfer editing. The enzyme may directly hydrolyze the noncognate aminoacyl adenylate via an inherent catalytic activity, and this reaction may be either tRNA-independent or tRNA-dependent. Another pre-transfer editing mechanism is selective release into solution where the labile aminoacyl linkage can be hydrolyzed nonenzymatically. (Fig. 6) [2].

Post-transfer editing occurs after the tRNA is misacylated and provides an additional opportunity to clear noncognate amino acids. Seven of the ten editing aaRS possess spatially separate editing domains for hydrolysis of misacylated tRNA (Table 1) (SerRS, MetRS, and class II LysRS lack such a domain and catalyze pre-transfer editing only) [27, 39, 41, 43, 152–157]. In this reaction the 3′-end of the misacylated tRNA first reorients on the enzyme, dissociating from the synthetic site and moving to the editing site in the process of translocation. This step

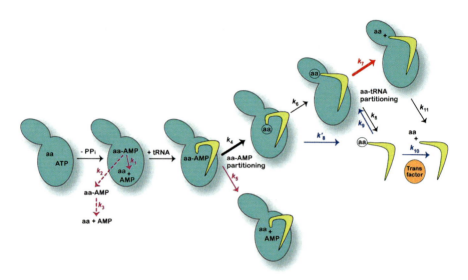

Fig. 6 Overall kinetic scheme for hydrolytic editing by aaRS. The larger lobe of the enzyme depicted in *blue* represents the synthetic domain, and the smaller lobe represents the post-transfer editing domain. The tRNA is depicted in *yellow* with the bending of the 3′-end into the synthetic site representing the situation in class I aaRS. The aminoacyl-AMP (aa-AMP) intermediate may be hydrolyzed via pre-transfer editing in three ways (*purple arrows*): release into solution followed by nonenzymatic hydrolysis (k_2 and k_3), tRNA-independent enzymatic hydrolysis (k_1), and tRNA-dependent enzymatic hydrolysis (k_5). tRNA-dependent pre-transfer editing competes with aminoacylation (k_4) in the first kinetic partitioning. After aminoacylation, the tRNA may undergo a first order translocation on the enzyme (k_6) followed by post-transfer hydrolysis by the editing domain of the enzyme (*red arrow*; k_7) and dissociation of products (k_{11}). Alternatively, the misacylated tRNA may dissociate from the enzyme in the second kinetic partitioning, and may then rebind to either the aaRS or a free-standing protein (*trans* factor) for subsequent hydrolysis (k_7 or k_{10}). In principle, dissociation of misacylated tRNA might occur before or after the translocation step ($k_{8'}$ and k_8, respectively). Although initial tRNA binding is shown to the synthetic domain for simplicity, this is not well-established in any system, and the initial binding site may differ among the enzymes. Initial binding to the post-transfer editing domain would require translocation in both directions to complete a synthetic and editing cycle

constitutes a first order conformational change of the enzyme-tRNA complex, which is conceptually similar to the movement of single-stranded DNA containing a misincorporated nucleotide into the exonuclease editing active site in DNA polymerases [158]. After translocation the noncognate aminoacyl linkage is positioned in the post-transfer editing site, where hydrolysis to generate uncharged tRNA and amino acid occurs. The tRNA and amino acid then dissociate from the enzyme. Depending on the relative rates of each first order process, in some cases dissociation of misacylated tRNA may instead be a preferred pathway. The released aminoacyl-tRNA may then still be hydrolyzed either by rebinding to the aaRS or to a *trans*-acting editing protein with homology to the editing domain [97, 159, 160]. If neither of these events occurs then the misacylated tRNA is available for binding to elongation factor and subsequent possible misincorporation of the noncognate amino acid into protein on the ribosome.

As this description indicates, editing pathways on the aaRS include partitioning steps, so that the choice to follow a particular pathway depends on the relative magnitudes of rate constants in each direction and, in vivo, possibly on other factors as well. In addition to the kinetic partitioning between dissociation and hydrolysis that occurs during post-transfer editing, such partitioning also occurs in the synthetic site, where the noncognate aminoacyl adenylate is either hydrolyzed or its amino acid portion is transferred to tRNA (Fig. 6) [47, 161]. Significant aaRS class-specific differences exist within this overall framework as described below.

3.3 Mechanisms of Pre-transfer Editing

The existence of pre-transfer editing was first inferred from early pre-steady state kinetics experiments on IleRS, in which an observed transient accumulation of noncognate Val-tRNA^{Ile} was much smaller than expected if the sole pathway were the formation and subsequent hydrolysis of this species [8]. This suggested that hydrolysis of misactivated Val-AMP could account for the observed editing. It was also shown that the half-life of Val-AMP in solution is much longer than the rate of its destruction in these experiments, demonstrating that the proposed reaction is enzyme-catalyzed. This excluded the so-called "kinetic proofreading" or selective release mechanism by which a noncognate aminoacyl adenylate is hydrolyzed nonenzymatically in solution after dissociation from the enzyme (Fig. 6) [12]. Later experiments on class I IleRS, ValRS and LeuRS, and on class II ProRS, confirmed that selective release is also not significant in these enzymes [47, 97, 162]. However, SerRS and mitochondrial ThrRS do employ this mechanism in part to clear Thr-AMP and Ser-AMP, respectively [163–165]. Interestingly, both the canonical SerRS and a more ancient methanogen-type SerRS misactivate and subsequently hydrolyze serine hydroxamate (a serine analog in which the α-carboxylate is replaced by α-hydroxamate) by a tRNA-independent pre-transfer mechanism [110, 163, 166, 167]. This common feature occurs within a context of distinct modes of cognate serine recognition in the two enzymes [31, 110]. The finding of tRNA-independent pre-transfer editing in the methanogen SerRS suggests that primordial aaRS may have generally employed tRNA-independent editing as an early proofreading mechanism.

A number of other early studies demonstrated that tRNA-independent pre-transfer editing, whether by selective release or in an enzyme-catalyzed fashion, occurs in both class I and class II aaRS [12, 14, 155, 161, 162, 168, 169]. However, it is unclear whether this reaction contributes in a physiologically significant manner, at least for the enzymes that also possess post-transfer editing domains. In the class I LeuRS, ValRS and IleRS, for example, quantitative steady-state kinetics shows that tRNA-independent hydrolysis proceeds with $k_{\text{cat}}/K_{\text{m}}$ for AMP production that represents only 3% of the total editing [47, 97].

SerRS, MetRS, and class II LysRS each lack distinct domains for post-transfer editing, so all three of these enzymes are limited to pre-transfer reactions localized in the synthetic active site. The reaction by SerRS proceeds by simple hydrolysis of noncognate mixed anhydrides, and so follows the same pre-transfer editing mechanism found in the other editing aaRS. Yeast SerRS activates threonine, cysteine, and serine hydroxamate; interestingly, the latter amino acid analogue stimulates tenfold more AMP production than do the others [163]. However, MetRS and class II LysRS follow distinct pathways. MetRS possesses a tRNA-independent pathway in which it clears misactivated homocysteine by intramolecular cyclization to form homocysteine thiolactone [12, 13, 170]. Crystal structures of MetRS suggest that homocysteine is excluded from aminoacylation because it fails to induce induced-fit rearrangements upon binding to MetRS, as occurs for cognate methionine binding [171]. Class II LysRS misactivates homoserine and ornithine, and edits these in tRNA-independent pathways via cyclization to a lactone and lactam, respectively [172–174]. Interestingly, the class I and class II LysRS enzymes (Table 1) have diverged in their amino acid recognition properties; the class I enzyme does not misactivate noncognate amino acids at a sufficiently high level to require an editing mechanism.

When editing aaRS function with their cognate amino acids, the molar ratio of ATP consumed (or AMP produced) to aminoacyl-tRNA formed is very nearly 1:1. In contrast, the noncognate amino acid triggers editing pathways resulting in significant elevation of this ratio [9, 161, 175]. For example, in *E. coli* LeuRS the rate of norvaline (Nva)-dependent AMP formation is 100-fold greater than that of Nva-tRNA^{Leu} synthesis [97]. In these editing reactions performed in the presence of tRNA, distinguishing tRNA-dependent pre-transfer editing from post-transfer editing is challenging because both processes consume excess ATP and generate excess AMP. The only approach so far developed for separating these reactions exploits the use of point mutations or larger deletions in the post-transfer editing domains, which disable this activity [47, 162, 176, 177]. However, such approaches must be employed cautiously because the extensive conformational changes of the protein, which occur concomitantly with translocation of the tRNA $3'$-acceptor arm between the synthetic and editing domains, may readily generate pleiotropic effects for any mutation in either active site. The recent cocrystal structures of LeuRS bound to tRNA in both synthetic and editing conformations reveal that portions of the CP1 post-transfer editing domain extend directly into the synthetic active site [16]. This observation makes the conformational coupling between synthetic and editing site domains explicitly apparent.

The best approach for elucidating tRNA-dependent pre-transfer editing was developed by combining assays to measure aa-AMP synthesis and hydrolysis with the use of single turnover kinetics to monitor the aminoacyl transfer step, employing both WT enzymes and enzyme mutants that are disabled in post-transfer editing [47, 57, 97, 161, 162]. Direct measurement of pre-transfer editing is possible by using α-^{32}P ATP, which permits independent quantitation of ^{32}P-labeled aa-AMP and AMP by thin-layer chromatography [57, 162]. An independent assay for post-transfer hydrolysis [178], based on direct mixing of enzyme with misacylated tRNA, allows confirmation that excess AMP production in mutated enzymes indeed arises entirely from hydrolysis of aa-AMP. Comparison of AMP formation rates in tRNA-independent and tRNA-dependent reactions by enzymes disabled in post-transfer editing provides estimates of the extent to which tRNA may enhance pre-transfer editing.

In class I editing aaRS, the application of these approaches showed that tRNA-dependent pre-transfer editing is significant in IleRS, but not in LeuRS or ValRS [47, 97]. For the latter enzymes, point mutants in the common structurally homologous CP1 editing domain that fully disable post-transfer editing of norvaline and threonine, respectively, also abolished all tRNA-dependent editing as measured by the lack of ^{32}P-AMP accumulation above levels observed in the tRNA-independent assay and/or tRNA aminoacylation. In contrast, the analogous disabling point mutant in the IleRS post-transfer editing site still permitted robust tRNA-dependent editing. k_{cat} for tRNA-dependent pre-transfer editing in IleRS is approximately tenfold higher than for the tRNA-independent reaction [47].

A kinetic partitioning model for class I aaRS editing was then developed based on the observation that the rates of aminoacyl transfer to tRNA (k_{trans}) in IleRS, ValRS, and LeuRS are inversely related to the amount of tRNA-dependent pre-transfer editing observed in the CP1 editing domain mutants disabled in post-transfer editing. In ValRS and LeuRS, k_{trans} is high and the tRNA-A76 2$'$-OH nucleophile outcompetes water for attack on the noncognate aminoacyl adenylate. However, k_{trans} in IleRS is 100-fold lower and is similar in magnitude to k_{cat} for tRNA-dependent production of ^{32}P-AMP in the same post-transfer editing-defective mutant. Therefore, synthetic site enzyme–catalyzed tRNA-enhanced pre-transfer editing occurs in IleRS because water can compete effectively with tRNA for attack. k_{trans} in IleRS is anomalously low as compared to most other aaRS for which its magnitude has been determined, but the structural basis for this remains unknown.

The essential determination is the comparison between the rates of aa-AMP synthesis and hydrolysis, with that of aminoacyl transfer in the synthetic site. In LeuRS, deletion of the entire CP1 domain produced an active enzyme capable of tRNA-dependent pre-transfer editing, a finding apparently at odds with the conclusion reached based on the use of a point mutant in the post-transfer editing domain [97, 176]. However, these studies are easily reconciled if the rate of aminoacyl transfer in the deletion enzyme is reduced, as seems plausible given the structural interdigitation of the editing and synthetic domains [16].

Studies on class II aaRS are consistent with the kinetic partitioning model and with the notion that pre-transfer editing is a synthetic site activity. In ThrRS, pre-transfer editing was enhanced in a designed mutant with a lowered rate of aminoacyl transfer, providing an independent observation that the rates of these reactions are inversely related [161]. ThrRS and ProRS exhibit robust pre-transfer editing of serine and alanine, respectively, in the absence of tRNA, and enzyme homologs in both systems that naturally lack editing domains are also capable of this activity [161, 162, 179, 180]. As in LeuRS [176], deletion of the post-transfer editing domain in ThrRS also preserves pre-transfer editing [161]. These observations each support a synthetic site-based activity. However, there is little evidence to suggest that pre-transfer editing contributes substantially when tRNA is present for any class II enzyme that possesses a separate post-transfer editing domain [14, 155, 157, 161, 162, 181]. Therefore, among all editing aaRS, IleRS may be unique with respect to this property [8, 47]. Further detailed structure–function studies of IleRS are needed to understand how tRNA enhancement of pre-transfer editing functions.

An earlier model for tRNA-dependent pre-transfer editing, developed for editing of valine by IleRS, invoked a complex mechanism by which noncognate aminoacyl adenylates are shuttled to the CP1 domain editing site, so that this active site would catalyze both pre-transfer and post-transfer editing [182–185]. However, some data supporting this model could not be reproduced using more quantitative approaches [47], and the intramolecular tunnel that would be required for sequestering the labile Val-AMP in its journey to the CP1 domain has not been found in any tRNA cocrystal structures [16, 152, 186]. Further, while the isolated CP1 domain readily catalyzes post-transfer editing, no pre-transfer activity associated with this domain has been reported [187, 188]. Therefore, this channeling model is not well-supported by the experimental evidence.

Interestingly, naturally occurring editing enzymes that have lost the capacity for post-transfer editing (for example, by deletion of the domain housing this activity) did not develop more efficient active site-based pre-transfer editing, but instead appear to have evolved a greater level of specificity in amino acid activation [148, 150, 162, 189, 190]. Optimization of a pre-transfer hydrolytic activity within the synthetic active site may be challenging, since the synthetic reaction will compete. This may account in part for why only IleRS retains tRNA-dependent pre-transfer editing. There are practical ramifications to this question in the engineering of aaRS to accept unnatural amino acids (UAAs), since pre-transfer editing of an activated amino acid may limit the efficiency of incorporation. Such editing may be present in an engineered GluRS enzyme that possesses an enhanced ATPase function [191].

3.4   Mechanisms of Post-transfer Editing

After synthesis of noncognate aminoacyl-tRNA in the synthetic active site, two general models have been proposed for post-transfer editing (Fig. 6). First, the misacylated tRNA may dissociate from the enzyme and then rebind in the dedicated editing domain [97, 159]. Alternatively, a first order rearrangement of the enzyme-tRNA complex may occur to shuttle the $3'$ acceptor end between sites [16, 97]. The location of post-transfer editing on the enzymes is not in doubt since several of the proposed editing domains of both class I and class II editing aaRS can independently catalyze the reaction as isolated proteins [187, 192]. Point mutants that selectively disable post-transfer editing have also been constructed in all of the editing enzymes, providing further evidence of the positions of the editing active sites and offering clues to the chemical mechanism of catalysis [2].

The three class I enzymes that catalyze post-transfer editing each possess a common mixed α/β domain (CP1) inserted into the Rossmann fold (Fig. 1, Table 1) [27]. In mitochondrial and bacterial LeuRS, the precise position of CP1 insertion differs as compared with IleRS, ValRS, and archaeal/eukaryotic LeuRS, but the spatial separation of the synthetic and editing active sites is approximately 30–40 Å in all the enzymes [78, 152, 153, 193]. In contrast, the class II aaRS editing domains are idiosyncratic to each enzyme in both their structures and positions along the primary sequence (Fig. 1). In ThrRS, PheRS, and ProRS, the relative positions of the synthetic and editing domains suggest that first order conformational rearrangement to translocate the tRNA between sites is a plausible mechanism, as is also the case for the class I enzymes [39, 40, 43, 131]. However, AlaRS may be distinct because crystal structures of the enzyme reveal a unique architecture that does not suggest a straightforward structural mechanism by which the flexible end of the acceptor arm can rearrange to move between the two active sites [44, 45]. It is possible then that AlaRS may rely primarily or entirely on the dissociation–reassociation pathway.

The experimental evidence in favor of the translocation model has primarily rested on comparisons of editing enzyme structures when bound in distinct liganded states ([2]. The best example is *E. coli* LeuRS, which represents the first case in which cocrystal structures bound to the full tRNA are available in both synthetic and editing conformations for the same enzyme [16]. Comparison of these structures provides the basis for a detailed translocation model, in which a key role is played by a 60-residue LeuRS-specific domain that follows the second half of the Rossmann fold. This domain binds the distal tRNA anticodon arm only in the editing complex (Fig. 7). In the synthetic mode, the domain reorients toward the synthetic active site by about 20 Å to interact with several newly ordered loops near the substrate binding cleft. Detailed kinetic studies of *E. coli* LeuRS provided functional evidence for a separable translocation step and allowed estimation of a first order rate constant of about 80 s^{-1}, some fourfold slower than the rate of post-transfer editing hydrolysis in the editing active site [97]. The translocation step was not directly monitored in LeuRS (or any other editing aaRS), but mutational analysis has nonetheless allowed insight into the identity of structural determinants that may control the conformational transitions [194]. Experiments employing an editing site-directed inhibitor of LeuRS suggested that tRNA may bind first to the editing domain [195], implying that the full catalytic cycle involves translocation in both directions. This provocative suggestion awaits further experimental verification.

Evidence for the dissociation and rebinding model was provided by experiments on PheRS, which showed that the editing site can compete with elongation factor for binding to misacylated Tyr-tRNA[Phe] [159]. This pathway may be adopted by other class II aaRS since, in general, product release is not rate-limiting in these enzymes [46]. Dissociation and rebinding is a kinetically competent pathway for class I editing aaRS as shown by studies on a mutant of *E. coli* LeuRS, but this mechanism does not contribute significantly in the wild-type enzyme [97]. This is consistent with slow product release as the rate-limiting step in class I enzymes. These data are consistent with the discovery of free-standing post-transfer editing

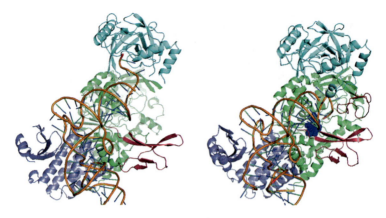

Fig. 7 Structures of the *E. coli* LeuRS–tRNA^{Leu} complex in editing mode bound to a benzoxaborole analog (*left*) and in synthetic mode (*right*) with the tRNA 3′-end bound in the synthetic active site (*green*) next to the Leu-AMS analog (*dark blue*) [142]. The LeuRS-specific domain is in *magenta*. In the editing complex at *left*, the tRNA 3′-end extends to bind in the post-transfer editing domain (CP1 domain depicted in *light blue* at *top*), and the LeuRS-specific domain interacts with the tRNA anticodon arm. In the synthetic complex (*right*), the tRNA 3′-end is bound in the synthetic Rossmann fold domain in a hairpinned conformation (*right*). The LeuRS-specific domain no longer binds the anticodon arm but instead is reoriented to interact with portions of the protein structure that form the active site. Several surface loops shown in *magenta* also become ordered when tRNA binds in the synthetic site

proteins that correspond to a number of class II editing domains but not to the common CP1 editing domain of class I aaRS (see below).

Details of the substrate specificity and catalytic mechanisms in the editing domains are emerging from crystal structures and mutational analysis [2]. Atomic level depictions of the active sites, including the solvent structure, have been provided by high resolution structures of separately expressed editing domains and homologous free-standing proteins (see below) [154, 196–203]. Analogs of the 3′-A76 nucleotide have been cocrystallized with these domains and with the intact enzymes to afford a view of the substrate complexes, from which insights into the stereochemistry of hydrolysis are becoming known [154, 186, 204].

The classic view of post-transfer editing is the "double sieve" mechanism [8], which envisions selectivity based on steric constraints. Amino acids smaller than the cognate substrate but still able to bind adequately in the synthetic site are misacylated, while amino acids larger than cognate are excluded. Then the second sieve or editing site possesses an architecture allowing binding of smaller noncognate amino acids but not the cognate amino acid. This mechanism was proposed based on editing of valine by IleRS [8], and still provides an essentially accurate view. Discrimination against smaller hydrophobic amino acids by LeuRS appears to also be well-described by this scheme. For both of these class I enzymes, crystal structures have identified the amino acids involved in binding the hydrophobic side chains of the noncognate amino acid in the editing site [204, 205]. In LeuRS, mutation of a key interacting threonine residue to smaller amino acids allows for hydrolysis of cognate Leu-tRNA^{Leu}, providing a clear demonstration of the double sieve in operation [206]. Wild-type LeuRS discriminates against norvaline by 10^3-fold at the deacylation step, providing a quantitative measure of the specificity that is intrinsic to the post-transfer editing domain [97].

Hydrolysis of Thr-tRNA^{Val} by ValRS does not follow this classic mechanism, because the cognate and noncognate amino acids are roughly isosteric and of different chemical character. In this case the editing active site plays a more active discriminatory role that includes specific conformational rearrangements [201]. Hydrogen bonds made by the threonine hydroxyl group to several conserved aspartates in the binding site also appear important to specificity [153].

The mechanisms of substrate specificity by class II aaRS are more idiosyncratic. ThrRS is able to exclude valine from the synthetic site based on a key role for a zinc ion in binding the side-chain hydroxyl group [112, 134]. Serine is not excluded, and is cleared by post-transfer editing. In archaeal ThrRS, cognate threonine is excluded by a structural mechanism involving rearrangement of an invariant Tyr-Lys dipeptide in the N2 editing domain, which causes mispositioning of the catalytic water molecule [207]. In ProRS, the INS editing domain discriminates against cognate proline by steric exclusion [41, 208]. Cysteine is also excluded on steric grounds despite the poor discrimination against this amino acid in the synthetic site [189], accounting for the role of the free-standing Ybak enzyme in clearing Cys-tRNA^{Pro} (see below) [160, 209, 210]. Interestingly, ProRS enzymes from archaea and higher eukaryotes do not possess the INS editing domain, and may rely instead on better accuracy in amino acid activation [189, 190].

In addition to Gly-tRNA^{Ala}, AlaRS activates serine and transfers it to tRNA^{Ala} at significant rates [157], and PheRS rapidly catalyzes Tyr-tRNA^{Phe} formation [155, 181], even though it might be thought that these larger and more polar amino acids should be readily excluded from the synthetic pathways. It is thus apparent that the synthetic active sites of these two editing aaRS are not optimized for amino acid discrimination. Crystal structures of PheRS and AlaRS bound to these noncognate amino acids have revealed how the synthetic active sites accommodate the additional substrate hydroxyl groups by hydrogen-bonding, and that the overall orientations of cognate and noncognate substrates are similar [114, 156]. An inherent capacity for some structural plasticity in the PheRS and AlaRS binding pockets, although small, is apparently sufficient to permit misacylation of larger noncognate substrates at rates sufficient to require editing. Since the cognate amino acids of PheRS and AlaRS are smaller than noncognate, the post-transfer editing active sites in these enzymes again must rely on mechanisms other than steric exclusion. The PheRS editing domain makes hydrogen bonding interactions with the hydroxyl group of tyrosine to distinguish this substrate [156], while the editing domain of AlaRS uses a hydrophilic environment and likely a zinc-dependent mechanism to bind serine [45, 197, 211].

The detailed stereochemical mechanism of post-transfer editing is perhaps best described for *E. coli* ThrRS. Crystal structures of the enzyme bound to post-transfer editing substrate analogs suggested that hydrolysis is accomplished by general base catalysis (Fig. 8) [186]. A conserved histidine residue in the active site deprotonates

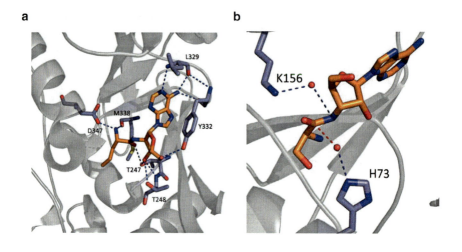

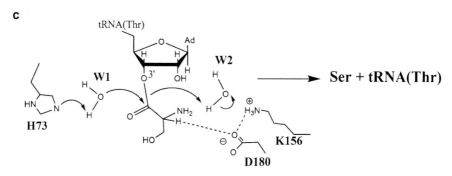

Fig. 8 (a) Structure of the post-transfer editing domain of class I *T. thermophilus* LeuRS bound to the post-transfer editing substrate analog 2′-(L-norvalyl)amino-2-deoxyadenosine (PDB code 1OBC). The conserved Asp347 implicated in the hydrolytic mechanism is at *left*. (b) Structure of a portion of the post-transfer editing domain of class II ThrRS bound to the post-transfer editing substrate analog serine 3′-aminoadenosine (PDB code 1TKY). (c) Schematic description of the proposed stereochemical mechanism of post-transfer editing hydrolysis in ThrRS, based on the crystal structure depicted in panel (**b**)

the nucleophilic water molecule, and a lysine residue participates in water-mediated stabilization of the departing O3′ anion. The mechanism is corroborated by sharp decreases measured in the catalytic rate upon mutation of either of these amino acids [212]. Interestingly, this mechanism is not conserved in the distinct archaebacterial ThrRS editing domain, which possesses structural homology to D-aminoacyl-tRNA deacylases [213–215]. In this case a substrate-assisted mechanism was instead proposed in which the 2′-OH group of A76 plays a role in activating the water nucleophile [199]. Similar substrate-assisted mechanisms based in part on computational studies have been proposed for ProRS, PheRS, and LeuRS [208, 216, 217].

3.5 Trans *Editing Factors*

Free-standing editing proteins have been discovered that catalyze post-transfer hydrolysis of misacylated tRNA after its dissociation from the aaRS (Fig. 6). In all cases these proteins are homologs of the editing domains of class II aaRS, and constitute a third sieve in the editing process [160]. The finding that product release is rate-limiting for class I editing enzymes, and that class II enzymes are instead rate-limited by an early step in the pathway, helps to rationalize why *trans* editing generally does not occur in class I systems [46, 97, 159]. However, TyrRS can misacylate tRNA^{Tyr} with D-tyrosine [218–220], and the D-Tyr-tRNA^{Tyr} deacylase functions as a *trans* editing protein to clear this species. Interestingly, this enzyme is structurally related to the editing domain of archaeal ThrRS, which hydrolyzes Ser-tRNA^{Thr} either as a component of the full-length ThrRS or as a free-standing protein [199, 214]. The free-standing tRNA^{Thr} editing domain is found only in a subset of archaea; in bacteria and eukaryotes, editing of serine by ThrRS is carried out by an unrelated editing domain found at the N-terminus of the enzyme [39].

Free-standing editing domains are more widespread in the proline and alanine systems. A free-standing homolog of the ProRS editing domain (INS), known as Ybak, is dedicated to Cys-tRNA^{Pro} hydrolysis. This reaction cannot be performed by the INS domain found in *cis* as part of the ProRS enzyme, which is instead specific to post-transfer editing of Ala-tRNA^{Pro} [160, 189, 209, 221, 222]. Formation of a ternary complex between YbaK, ProRS, and Cys-tRNA^{Pro} has been inferred from chemical cross-linking, and it appears that YbaK thus acts in collaboration with ProRS prior to release of misacylated tRNA^{Pro} into solution [221]. A second subclass of INS domain paralogs (PrdX proteins) is specific for Ala-tRNA^{Pro} hydrolysis. The presence of a PrdX free-standing protein in a particular organism is correlated with the absence of an editing domain in the corresponding ProRS enzyme [210, 222]. Ybak functions by substrate-assisted deacylation involving a cyclization pathway that relies on the thiol group of the misacylated cysteine [223, 224]. This mechanism is distinct from the proposed substrate-assisted mechanism for alanine editing by the INS domain [224, 225].

The AlaX proteins are free-standing homologs of the AlaRS editing domain and are conserved in all three domains of life [210]. They are classified into several subtypes and are devoted to the hydrolysis of tRNA^{Ala} misacylated with serine or glycine [226]. Some AlaX proteins display specific activity toward Ser-tRNA^{Ala} only [197], suggesting that Ser-tRNA^{Ala} and not Gly-tRNA^{Ala} is the preferred substrate [37, 114, 227]. This notion is supported by the finding that AlaX is able to rescue a bacterial cell from serine toxicity [228]. The need for Ser-tRNA^{Ala} editing arises because the particular design of the AlaRS synthetic site does not allow for serine discrimination on steric grounds [114]. Interestingly, one subgroup of the AlaX proteins possesses homology to the C-terminal C-Ala domain of full-length AlaRS [227].

The finding of the *trans* acting domains allows editing to be conceptualized as a "triple sieve" involving the synthetic active site, the *cis*-acting post-transfer editing domain, and the *trans*-acting domain. As we have seen, each of these protein domains possesses inherent hydrolytic editing activity at a distinct position along the overall pathway (Fig. 6). Taken together, the complexity of editing mechanisms that have evolved surely reflects the adaptive value associated with the capacity of the cell to maintain a level of protein synthesis fidelity that is appropriate in a particular organism, cellular compartment, or condition of environmental stress. The impact of editing on cellular physiology is clearly of great significance, and additional ramifications of the process no doubt await discovery in other biological settings.

Acknowledgments We thank Nevena Cvetesic and Andrew Hadd for assistance with figures. This work was supported by the National Institutes of Health (GM63713 and 1RO3TW008024) and by the Croatian Science Foundation (grant 09.01/293). I.G.S. thanks the Adris foundation for support.

References

1. Perona JJ, Hadd A (2012) Structural diversity and protein engineering of the aminoacyl-tRNA synthetases. Biochemistry 51:8705–8729
2. Yadavalli SS, Ibba M (2012) Quality control in aminoacyl-tRNA synthesis its role in translational fidelity. Adv Protein Chem Struct Biol 86:1–43
3. Perona JJ, Hou YM (2007) Indirect readout of tRNA for aminoacylation. Biochemistry 46:10419–10432
4. Post CB, Ray WJ Jr (1995) Reexamination of induced fit as a determinant of substrate specificity in enzymatic reactions. Biochemistry 34:15881–15885
5. Bruice TC (2002) A view at the millennium: the efficiency of enzymatic catalysis. Acc Chem Res 35:139–148
6. Ibba M, Sever S, Praetorius-Ibba M, Soll D (1999) Transfer RNA identity contributes to transition state stabilization during aminoacyl-tRNA synthesis. Nucleic Acids Res 27:3631–3637
7. Ebel JP, Giege R, Bonnet J, Kern D, Befort N, Bollack C, Fasiolo F, Gangloff J, Dirheimer G (1973) Factors determining the specificity of the tRNA aminoacylation reaction. Non-absolute specificity of tRNA-aminoacyl-tRNA synthetase recognition and particular importance of the maximal velocity. Biochimie 55:547–557
8. Fersht AR (1977) Editing mechanisms in protein synthesis. Rejection of valine by the isoleucyl-tRNA synthetase. Biochemistry 16:1025–1030
9. Baldwin AN, Berg P (1966) Transfer ribonucleic acid-induced hydrolysis of valyladenylate bound to isoleucyl ribonucleic acid synthetase. J Biol Chem 241:839–845
10. Fersht AR, Kaethner MM (1976) Enzyme hyperspecificity. Rejection of threonine by the valyl-tRNA synthetase by misacylation and hydrolytic editing. Biochemistry 15:3342–3346
11. Jakubowski H (1978) Valyl-tRNA synthetase from yellow lupin seeds. Instability of enzyme-bound noncognate adenylates versus cognate adenylate. FEBS Lett 95:235–238
12. Jakubowski H, Fersht AR (1981) Alternative pathways for editing non-cognate amino acids by aminoacyl-tRNA synthetases. Nucleic Acids Res 9:3105–3117
13. Fersht AR, Dingwall C (1979) An editing mechanism for the methionyl-tRNA synthetase in the selection of amino acids in protein synthesis. Biochemistry 18:1250–1256

14. Tsui WC, Fersht AR (1981) Probing the principles of amino acid selection using the alanyl-tRNA synthetase from *Escherichia coli*. Nucleic Acids Res 9:4627–4637
15. Ibba M, Kast P, Hennecke H (1994) Substrate specificity is determined by amino acid binding pocket size in *Escherichia coli* phenylalanyl-tRNA synthetase. Biochemistry 33:7107–7112
16. Palencia A, Crepin T, Vu MT, Lincecum TL Jr, Martinis SA, Cusack S (2012) Structural dynamics of the aminoacylation and proofreading functional cycle of bacterial leucyl-tRNA synthetase. Nat Struct Mol Biol 19:677–684
17. Gruic-Sovulj I, Landeka I, Soll D, Weygand-Durasevic I (2002) tRNA-dependent amino acid discrimination by yeast seryl-tRNA synthetase. Eur J Biochem 269:5271–5279
18. Ibba M, Hong KW, Sherman JM, Sever S, Soll D (1996) Interactions between tRNA identity nucleotides and their recognition sites in glutaminyl-tRNA synthetase determine the cognate amino acid affinity of the enzyme. Proc Natl Acad Sci USA 93:6953–6958
19. Uter NT, Gruic-Sovulj I, Perona JJ (2005) Amino acid-dependent transfer RNA affinity in a class I aminoacyl-tRNA synthetase. J Biol Chem 280:23966–23977
20. Bullock TL, Rodriguez-Hernandez A, Corigliano EM, Perona JJ (2008) A rationally engineered misacylating aminoacyl-tRNA synthetase. Proc Natl Acad Sci USA 105:7428–7433
21. Rodriguez-Hernandez A, Bhaskaran H, Hadd A, Perona JJ (2010) Synthesis of Glu-tRNA (Gln) by engineered and natural aminoacyl-tRNA synthetases. Biochemistry 49:6727–6736
22. de Duve C (1988) Transfer RNAs: the second genetic code. Nature 333:117–118
23. Park SG, Schimmel P, Kim S (2008) Aminoacyl tRNA synthetases and their connections to disease. Proc Natl Acad Sci USA 105:11043–11049
24. Guo M, Yang XL, Schimmel P (2010) New functions of aminoacyl-tRNA synthetases beyond translation. Nat Rev Mol Cell Biol 11:668–674
25. Irwin MJ, Nyborg J, Reid BR, Blow DM (1976) The crystal structure of tyrosyl-transfer RNA synthetase at 2–7 A resolution. J Mol Biol 105:577–586
26. Brick P, Bhat TN, Blow DM (1989) Structure of tyrosyl-tRNA synthetase refined at 2.3 A resolution. Interaction of the enzyme with the tyrosyl adenylate intermediate. J Mol Biol 208:83–98
27. Schmidt E, Schimmel P (1995) Residues in a class I tRNA synthetase which determine selectivity of amino acid recognition in the context of tRNA. Biochemistry 34:11204–11210
28. Rould MA, Perona JJ, Soll D, Steitz TA (1989) Structure of *E. coli* glutaminyl-tRNA synthetase complexed with tRNA(Gln) and ATP at 2.8 A resolution. Science 246:1135–1142
29. Yaremchuk A, Kriklivyi I, Tukalo M, Cusack S (2002) Class I tyrosyl-tRNA synthetase has a class II mode of cognate tRNA recognition. EMBO J 21:3829–3840
30. Ruff M, Krishnaswamy S, Boeglin M, Poterszman A, Mitschler A, Podjarny A, Rees B, Thierry JC, Moras D (1991) Class II aminoacyl transfer RNA synthetases: crystal structure of yeast aspartyl-tRNA synthetase complexed with tRNA(Asp). Science 252:1682–1689
31. Cusack S, Berthet-Colominas C, Hartlein M, Nassar N, Leberman R (1990) A second class of synthetase structure revealed by X-ray analysis of *Escherichia coli* seryl-tRNA synthetase at 2.5 A. Nature 347:249–255
32. First EA (2005) Catalysis of the tRNA aminoacylation reaction. In: Michael Ibba CFaSC (ed) The aminoacyl-tRNA synthetases. Georgetown: Landes Bioscience/Eurekah.com, pp 328–352
33. Ibba M, Soll D (2000) Aminoacyl-tRNA synthesis. Annu Rev Biochem 69:617–650
34. Arnez JG, Moras D (1997) Structural and functional considerations of the aminoacylation reaction. Trends Biochem Sci 22:211–216
35. Cusack S (1995) Eleven down and nine to go. Nat Struct Biol 2:824–831
36. Cusack S (1997) Aminoacyl-tRNA synthetases. Curr Opin Struct Biol 7:881–889
37. Guo M, Schimmel P (2012) Structural analyses clarify the complex control of mistranslation by tRNA synthetases. Curr Opin Struct Biol 22:119–126
38. Tang SN, Huang JF (2005) Evolution of different oligomeric glycyl-tRNA synthetases. FEBS Lett 579:1441–1445

39. Dock-Bregeon A, Sankaranarayanan R, Romby P, Caillet J, Springer M, Rees B, Francklyn CS, Ehresmann C, Moras D (2000) Transfer RNA-mediated editing in threonyl-tRNA synthetase. The class II solution to the double discrimination problem. Cell 103:877–884
40. Cusack S, Yaremchuk A, Krikliviy I, Tukalo M (1998) tRNA(Pro) anticodon recognition by *Thermus thermophilus* prolyl-tRNA synthetase. Structure 6:101–108
41. Wong FC, Beuning PJ, Nagan M, Shiba K, Musier-Forsyth K (2002) Functional role of the prokaryotic proline-tRNA synthetase insertion domain in amino acid editing. Biochemistry 41:7108–7115
42. Goldgur Y, Mosyak L, Reshetnikova L, Ankilova V, Lavrik O, Khodyreva S, Safro M (1997) The crystal structure of phenylalanyl-tRNA synthetase from *Thermus thermophilus* complexed with cognate tRNAPhe. Structure 5:59–68
43. Crepin T, Yaremchuk A, Tukalo M, Cusack S (2006) Structures of two bacterial prolyl-tRNA synthetases with and without a *cis*-editing domain. Structure 14:1511–1525
44. Naganuma M, Sekine S, Fukunaga R, Yokoyama S (2009) Unique protein architecture of alanyl-tRNA synthetase for aminoacylation, editing, and dimerization. Proc Natl Acad Sci USA 106:8489–8494
45. Sokabe M, Ose T, Nakamura A, Tokunaga K, Nureki O, Yao M, Tanaka I (2009) The structure of alanyl-tRNA synthetase with editing domain. Proc Natl Acad Sci USA 106:11028–11033
46. Zhang CM, Perona JJ, Ryu K, Francklyn C, Hou YM (2006) Distinct kinetic mechanisms of the two classes of aminoacyl-tRNA synthetases. J Mol Biol 361:300–311
47. Dulic M, Cvetesic N, Perona JJ, Gruic-Sovulj I (2010) Partitioning of tRNA-dependent editing between pre- and post-transfer pathways in class I aminoacyl-tRNA synthetases. J Biol Chem 285:23799–23809
48. Bhaskaran H, Perona JJ (2011) Two-step aminoacylation of tRNA without channeling in archaea. J Mol Biol 411:854–869
49. Fersht AR, Kaethner MM (1976) Mechanism of aminoacylation of tRNA. Proof of the aminoacyl adenylate pathway for the isoleucyl- and tyrosyl-tRNA synthetases from *Escherichia coli* K12. Biochemistry 15:818–823
50. Guth E, Connolly SH, Bovee M, Francklyn CS (2005) A substrate-assisted concerted mechanism for aminoacylation by a class II aminoacyl-tRNA synthetase. Biochemistry 44:3785–3794
51. Banik SD, Nandi N (2012) Mechanism of the activation step of the aminoacylation reaction: a significant difference between class I and class II synthetases. J Biomol Struct Dyn 30:701–715
52. Ibba M, Losey HC, Kawarabayasi Y, Kikuchi H, Bunjun S, Soll D (1999) Substrate recognition by class I lysyl-tRNA synthetases: a molecular basis for gene displacement. Proc Natl Acad Sci USA 96:418–423
53. Kern D, Lapointe J (1979) Glutamyl transfer ribonucleic acid synthetase of *Escherichia coli*. Study of the interactions with its substrates. Biochemistry 18:5809–5818
54. Mehler AH, Mitra SK (1967) The activation of arginyl transfer ribonucleic acid synthetase by transfer ribonucleic acid. J Biol Chem 242:5495–5499
55. Ravel JM, Wang SF, Heinemeyer C, Shive W (1965) Glutamyl and glutaminyl ribonucleic acid synthetases of *Escherichia coli* W. Separation, properties, and stimulation of adenosine triphosphate-pyrophosphate exchange by acceptor ribonucleic acid. J Biol Chem 240:432–438
56. Sekine S, Nureki O, Dubois DY, Bernier S, Chenevert R, Lapointe J, Vassylyev DG, Yokoyama S (2003) ATP binding by glutamyl-tRNA synthetase is switched to the productive mode by tRNA binding. EMBO J 22:676–688
57. Gruic-Sovulj I, Uter N, Bullock T, Perona JJ (2005) tRNA-dependent aminoacyl-adenylate hydrolysis by a nonediting class I aminoacyl-tRNA synthetase. J Biol Chem 280:23978–23986
58. Perona JJ, Rould MA, Steitz TA (1993) Structural basis for transfer RNA aminoacylation by *Escherichia coli* glutaminyl-tRNA synthetase. Biochemistry 32:8758–8771

59. Desogus G, Todone F, Brick P, Onesti S (2000) Active site of lysyl-tRNA synthetase: structural studies of the adenylation reaction. Biochemistry 39:8418–8425
60. Avis JM, Fersht AR (1993) Use of binding energy in catalysis: optimization of rate in a multistep reaction. Biochemistry 32:5321–5326
61. Avis JM, Day AG, Garcia GA, Fersht AR (1993) Reaction of modified and unmodified tRNA (Tyr) substrates with tyrosyl-tRNA synthetase (*Bacillus stearothermophilus*). Biochemistry 32:5312–5320
62. Xin Y, Li W, Dwyer DS, First EA (2000) Correlating amino acid conservation with function in tyrosyl-tRNA synthetase. J Mol Biol 303:287–298
63. Minajigi A, Francklyn CS (2008) RNA-assisted catalysis in a protein enzyme: the 2′-hydroxyl of tRNA(Thr) A76 promotes aminoacylation by threonyl-tRNA synthetase. Proc Natl Acad Sci USA 105:17748–17753
64. Li L, Weinreb V, Francklyn C, Carter CW Jr (2011) Histidyl-tRNA synthetase urzymes: class I and II aminoacyl tRNA synthetase urzymes have comparable catalytic activities for cognate amino acid activation. J Biol Chem 286:10387–10395
65. Pham Y, Kuhlman B, Butterfoss GL, Hu H, Weinreb V, Carter CW Jr (2010) Tryptophanyl-tRNA synthetase urzyme: a model to recapitulate molecular evolution and investigate intramolecular complementation. J Biol Chem 285:38590–38601
66. Pham Y, Li L, Kim A, Erdogan O, Weinreb V, Butterfoss GL, Kuhlman B, Carter CW Jr (2007) A minimal TrpRS catalytic domain supports sense/antisense ancestry of class I and II aminoacyl-tRNA synthetases. Mol Cell 25:851–862
67. Sekine S, Shichiri M, Bernier S, Chenevert R, Lapointe J, Yokoyama S (2006) Structural bases of transfer RNA-dependent amino acid recognition and activation by glutamyl-tRNA synthetase. Structure 14:1791–1799
68. Retailleau P, Huang X, Yin Y, Hu M, Weinreb V, Vachette P, Vonrhein C, Bricogne G, Roversi P, Ilyin V, Carter CW Jr (2003) Interconversion of ATP binding and conformational free energies by tryptophanyl-tRNA synthetase: structures of ATP bound to open and closed, pre-transition-state conformations. J Mol Biol 325:39–63
69. Shen N, Zhou M, Yang B, Yu Y, Dong X, Ding J (2008) Catalytic mechanism of the tryptophan activation reaction revealed by crystal structures of human tryptophanyl-tRNA synthetase in different enzymatic states. Nucleic Acids Res 36:1288–1299
70. Terada T, Nureki O, Ishitani R, Ambrogelly A, Ibba M, Soll D, Yokoyama S (2002) Functional convergence of two lysyl-tRNA synthetases with unrelated topologies. Nat Struct Biol 9:257–262
71. Schmitt E, Tanrikulu IC, Yoo TH, Panvert M, Tirrell DA, Mechulam Y (2009) Switching from an induced-fit to a lock-and-key mechanism in an aminoacyl-tRNA synthetase with modified specificity. J Mol Biol 394:843–851
72. Rath VL, Silvian LF, Beijer B, Sproat BS, Steitz TA (1998) How glutaminyl-tRNA synthetase selects glutamine. Structure 6:439–449
73. Bullock TL, Uter N, Nissan TA, Perona JJ (2003) Amino acid discrimination by a class I aminoacyl-tRNA synthetase specified by negative determinants. J Mol Biol 328:395–408
74. Corigliano EM, Perona JJ (2009) Architectural underpinnings of the genetic code for glutamine. Biochemistry 48:676–687
75. Konno M, Sumida T, Uchikawa E, Mori Y, Yanagisawa T, Sekine S, Yokoyama S (2009) Modeling of tRNA-assisted mechanism of Arg activation based on a structure of Arg-tRNA synthetase, tRNA, and an ATP analog (ANP). FEBS J 276:4763–4779
76. Fersht AR (1987) Dissection of the structure and activity of the tyrosyl-tRNA synthetase by site-directed mutagenesis. Biochemistry 26:8031–8037
77. First EA, Fersht AR (1993) Mutational and kinetic analysis of a mobile loop in tyrosyl-tRNA synthetase. Biochemistry 32:13658–13663
78. Cusack S, Yaremchuk A, Tukalo M (2000) The 2 A crystal structure of leucyl-tRNA synthetase and its complex with a leucyl-adenylate analogue. EMBO J 19:2351–2361

79. Kobayashi T, Takimura T, Sekine R, Kelly VP, Kamata K, Sakamoto K, Nishimura S, Yokoyama S (2005) Structural snapshots of the KMSKS loop rearrangement for amino acid activation by bacterial tyrosyl-tRNA synthetase. J Mol Biol 346:105–117
80. Sharma G, First EA (2009) Thermodynamic analysis reveals a temperature-dependent change in the catalytic mechanism of *bacillus stearothermophilus* tyrosyl-tRNA synthetase. J Biol Chem 284:4179–4190
81. Kapustina M, Carter CW Jr (2006) Computational studies of tryptophanyl-tRNA synthetase: activation of ATP by induced-fit. J Mol Biol 362:1159–1180
82. Retailleau P, Weinreb V, Hu M, Carter CW Jr (2007) Crystal structure of tryptophanyl-tRNA synthetase complexed with adenosine-5′ tetraphosphate: evidence for distributed use of catalytic binding energy in amino acid activation by class I aminoacyl-tRNA synthetases. J Mol Biol 369:108–128
83. Laowanapiban P, Kapustina M, Vonrhein C, Delarue M, Koehl P, Carter CW Jr (2009) Independent saturation of three TrpRS subsites generates a partially assembled state similar to those observed in molecular simulations. Proc Natl Acad Sci USA 106:1790–1795
84. Leatherbarrow RJ, Fersht AR, Winter G (1985) Transition-state stabilization in the mechanism of tyrosyl-tRNA synthetase revealed by protein engineering. Proc Natl Acad Sci USA 82:7840–7844
85. Weinreb V, Carter CW Jr (2008) Mg^{2+}-free *Bacillus stearothermophilus* tryptophanyl-tRNA synthetase retains a major fraction of the overall rate enhancement for tryptophan activation. J Am Chem Soc 130:1488–1494
86. Weinreb V, Li L, Carter CW Jr (2012) A master switch couples Mg(2)(+)-assisted catalysis to domain motion in *B. stearothermophilus* tryptophanyl-tRNA Synthetase. Structure 20:128–138
87. Weinreb V, Li L, Campbell CL, Kaguni LS, Carter CW Jr (2009) Mg^{2+}-assisted catalysis by *B. stearothermophilus* TrpRS is promoted by allosteric effects. Structure 17:952–964
88. Cammer S, Carter CW Jr (2010) Six Rossmannoid folds, including the class I aminoacyl-tRNA synthetases, share a partial core with the anti-codon-binding domain of a class II aminoacyl-tRNA synthetase. Bioinformatics 26:709–714
89. Uter NT, Perona JJ (2006) Active-site assembly in glutaminyl-tRNA synthetase by tRNA-mediated induced fit. Biochemistry 45:6858–6865
90. Xin Y, Li W, First EA (2000) Stabilization of the transition state for the transfer of tyrosine to tRNA(Tyr) by tyrosyl-tRNA synthetase. J Mol Biol 303:299–310
91. Lassila JK, Zalatan JG, Herschlag D (2011) Biological phosphoryl-transfer reactions: understanding mechanism and catalysis. Annu Rev Biochem 80:669–702
92. Black Pyrkosz A, Eargle J, Sethi A, Luthey-Schulten Z (2010) Exit strategies for charged tRNA from GluRS. J Mol Biol 397:1350–1371
93. Uter NT, Perona JJ (2004) Long-range intramolecular signaling in a tRNA synthetase complex revealed by pre-steady-state kinetics. Proc Natl Acad Sci USA 101:14396–14401
94. Fersht AR, Gangloff J, Dirheimer G (1978) Reaction pathway and rate-determining step in the aminoacylation of tRNAArg catalyzed by the arginyl-tRNA synthetase from yeast. Biochemistry 17:3740–3746
95. Liu C, Sanders JM, Pascal JM, Hou YM (2012) Adaptation to tRNA acceptor stem structure by flexible adjustment in the catalytic domain of class I tRNA synthetases. RNA 18:213–221
96. Mulvey RS, Fersht AR (1978) Mechanism of aminoacylation of transfer RNA. A pre-steady-state analysis of the reaction pathway catalyzed by the methionyl-tRNA synthetase of *Bacillus stearothermophilus*. Biochemistry 17:5591–5597
97. Cvetesic N, Perona JJ, Gruic-Sovulj I (2012) Kinetic partitioning between synthetic and editing pathways in class I aminoacyl-tRNA synthetases occurs at both pre-transfer and post-transfer hydrolytic steps. J Biol Chem 287:25381–25394
98. Ward WH, Fersht AR (1988) Tyrosyl-tRNA synthetase acts as an asymmetric dimer in charging tRNA. A rationale for half-of-the-sites activity. Biochemistry 27:5525–5530

99. Doublie S, Bricogne G, Gilmore C, Carter CW Jr (1995) Tryptophanyl-tRNA synthetase crystal structure reveals an unexpected homology to tyrosyl-tRNA synthetase. Structure 3:17–31

100. Eriani G, Delarue M, Poch O, Gangloff J, Moras D (1990) Partition of tRNA synthetases into two classes based on mutually exclusive sets of sequence motifs. Nature 347:203–206

101. Cavarelli J, Eriani G, Rees B, Ruff M, Boeglin M, Mitschler A, Martin F, Gangloff J, Thierry JC, Moras D (1994) The active site of yeast aspartyl-tRNA synthetase: structural and functional aspects of the aminoacylation reaction. EMBO J 13:327–337

102. Dignam JD, Guo J, Griffith WP, Garbett NC, Holloway A, Mueser T (2011) Allosteric interaction of nucleotides and tRNA(ala) with *E. coli* alanyl-tRNA synthetase. Biochemistry 50:9886–9900

103. Cusack S (1993) Sequence, structure and evolutionary relationships between class 2 aminoacyl-tRNA synthetases: an update. Biochimie 75:1077–1081

104. Cavarelli J, Rees B, Ruff M, Thierry JC, Moras D (1993) Yeast tRNA(Asp) recognition by its cognate class II aminoacyl-tRNA synthetase. Nature 362:181–184

105. Belrhali H, Yaremchuk A, Tukalo M, Berthet-Colominas C, Rasmussen B, Bosecke P, Diat O, Cusack S (1995) The structural basis for seryl-adenylate and Ap4A synthesis by seryl-tRNA synthetase. Structure 3:341–352

106. Berthet-Colominas C, Seignovert L, Hartlein M, Grotli M, Cusack S, Leberman R (1998) The crystal structure of asparaginyl-tRNA synthetase from *Thermus thermophilus* and its complexes with ATP and asparaginyl-adenylate: the mechanism of discrimination between asparagine and aspartic acid. EMBO J 17:2947–2960

107. Schmitt E, Moulinier L, Fujiwara S, Imanaka T, Thierry JC, Moras D (1998) Crystal structure of aspartyl-tRNA synthetase from *Pyrococcus kodakaraensis* KOD: archaeon specificity and catalytic mechanism of adenylate formation. EMBO J 17:5227–5237

108. Arnez JG, Dock-Bregeon AC, Moras D (1999) Glycyl-tRNA synthetase uses a negatively charged pit for specific recognition and activation of glycine. J Mol Biol 286:1449–1459

109. Swairjo MA, Schimmel PR (2005) Breaking sieve for steric exclusion of a noncognate amino acid from active site of a tRNA synthetase. Proc Natl Acad Sci USA 102:988–993

110. Bilokapic S, Maier T, Ahel D, Gruic-Sovulj I, Soll D, Weygand-Durasevic I, Ban N (2006) Structure of the unusual seryl-tRNA synthetase reveals a distinct zinc-dependent mode of substrate recognition. EMBO J 25:2498–2509

111. Arnez JG, Sankaranarayanan R, Dock-Bregeon AC, Francklyn CS, Moras D (2000) Aminoacylation at the atomic level in class IIa aminoacyl-tRNA synthetases. J Biomol Struct Dyn 17:23–27

112. Sankaranarayanan R, Dock-Bregeon AC, Rees B, Bovee M, Caillet J, Romby P, Francklyn CS, Moras D (2000) Zinc ion mediated amino acid discrimination by threonyl-tRNA synthetase. Nat Struct Biol 7:461–465

113. Fishman R, Ankilova V, Moor N, Safro M (2001) Structure at 2.6 A resolution of phenylalanyl-tRNA synthetase complexed with phenylalanyl-adenylate in the presence of manganese. Acta Crystallogr D Biol Crystallogr 57:1534–1544

114. Guo M, Chong YE, Shapiro R, Beebe K, Yang XL, Schimmel P (2009) Paradox of mistranslation of serine for alanine caused by AlaRS recognition dilemma. Nature 462:808–812

115. Eiler S, Dock-Bregeon A, Moulinier L, Thierry JC, Moras D (1999) Synthesis of aspartyl-tRNA(Asp) in *Escherichia coli* – a snapshot of the second step. EMBO J 18:6532–6541

116. Arnez JG, Augustine JG, Moras D, Francklyn CS (1997) The first step of aminoacylation at the atomic level in histidyl-tRNA synthetase. Proc Natl Acad Sci USA 94:7144–7149

117. Guth EC, Francklyn CS (2007) Kinetic discrimination of tRNA identity by the conserved motif 2 loop of a class II aminoacyl-tRNA synthetase. Mol Cell 25:531–542

118. Dulic M, Pozar J, Bilokapic S, Weygand-Durasevic I, Gruic-Sovulj I (2011) An idiosyncratic serine ordering loop in methanogen seryl-tRNA synthetases guides substrates through seryl-tRNASer formation. Biochimie 93:1761–1769

119. Arnez JG, Harris DC, Mitschler A, Rees B, Francklyn CS, Moras D (1995) Crystal structure of histidyl-tRNA synthetase from *Escherichia coli* complexed with histidyl-adenylate. EMBO J 14:4143–4155

120. Banik SD, Nandi N (2010) Aminoacylation reaction in the histidyl-tRNA synthetase: fidelity mechanism of the activation step. J Phys Chem B 114:2301–2311

121. Ng JD, Sauter C, Lorber B, Kirkland N, Arnez J, Giege R (2002) Comparative analysis of space-grown and earth-grown crystals of an aminoacyl-tRNA synthetase: space-grown crystals are more useful for structural determination. Acta Crystallogr D Biol Crystallogr 58:645–652

122. Reshetnikova L, Moor N, Lavrik O, Vassylyev DG (1999) Crystal structures of phenylalanyl-tRNA synthetase complexed with phenylalanine and a phenylalanyl-adenylate analogue. J Mol Biol 287:555–568

123. Torres-Larios A, Sankaranarayanan R, Rees B, Dock-Bregeon AC, Moras D (2003) Conformational movements and cooperativity upon amino acid, ATP and tRNA binding in threonyl-tRNA synthetase. J Mol Biol 331:201–211

124. Yaremchuk A, Tukalo M, Grotli M, Cusack S (2001) A succession of substrate induced conformational changes ensures the amino acid specificity of *Thermus thermophilus* prolyl-tRNA synthetase: comparison with histidyl-tRNA synthetase. J Mol Biol 309:989–1002

125. Yanagisawa T, Ishii R, Fukunaga R, Kobayashi T, Sakamoto K, Yokoyama S (2008) Crystallographic studies on multiple conformational states of active-site loops in pyrrolysyl-tRNA synthetase. J Mol Biol 378:634–652

126. Bovee ML, Pierce MA, Francklyn CS (2003) Induced fit and kinetic mechanism of adenylation catalyzed by *Escherichia coli* threonyl-tRNA synthetase. Biochemistry 42:15102–15113

127. Cusack S, Yaremchuk A, Tukalo M (1996) The crystal structure of the ternary complex of *T. thermophilus* seryl-tRNA synthetase with tRNA(Ser) and a seryl-adenylate analogue reveals a conformational switch in the active site. EMBO J 15:2834–2842

128. Qiu X, Janson CA, Blackburn MN, Chhohan IK, Hibbs M, Abdel-Meguid SS (1999) Cooperative structural dynamics and a novel fidelity mechanism in histidyl-tRNA synthetase. Biochemistry 38:12296–12304

129. Moulinier L, Eiler S, Eriani G, Gangloff J, Thierry JC, Gabriel K, McClain WH, Moras D (2001) The structure of an AspRS-tRNA(Asp) complex reveals a tRNA-dependent control mechanism. EMBO J 20:5290–5301

130. Onesti S, Desogus G, Brevet A, Chen J, Plateau P, Blanquet S, Brick P (2000) Structural studies of lysyl-tRNA synthetase: conformational changes induced by substrate binding. Biochemistry 39:12853–12861

131. Moor N, Kotik-Kogan O, Tworowski D, Sukhanova M, Safro M (2006) The crystal structure of the ternary complex of phenylalanyl-tRNA synthetase with tRNAPhe and a phenylalanyl-adenylate analogue reveals a conformational switch of the CCA end. Biochemistry 45:10572–10583

132. Dibbelt L, Pachmann U, Zachau HG (1980) Serine activation is the rate limiting step of tRNASer aminoacylation by yeast seryl tRNA synthetase. Nucleic Acids Res 8:4021–4039

133. Dibbelt L, Zachau HG (1981) On the rate limiting step of yeast tRNAPhe aminoacylation. FEBS Lett 129:173–176

134. Sankaranarayanan R, Dock-Bregeon AC, Romby P, Caillet J, Springer M, Rees B, Ehresmann C, Ehresmann B, Moras D (1999) The structure of threonyl-tRNA synthetase-tRNA(Thr) complex enlightens its repressor activity and reveals an essential zinc ion in the active site. Cell 97:371–381

135. Nozawa K, O'Donoghue P, Gundllapalli S, Araiso Y, Ishitani R, Umehara T, Soll D, Nureki O (2009) Pyrrolysyl-tRNA synthetase-tRNA(Pyl) structure reveals the molecular basis of orthogonality. Nature 457:1163–1167

136. Biou V, Yaremchuk A, Tukalo M, Cusack S (1994) The 2.9 A crystal structure of *T. thermophilus* seryl-tRNA synthetase complexed with tRNA(Ser). Science 263:1404–1410

137. Borel F, Vincent C, Leberman R, Hartlein M (1994) Seryl-tRNA synthetase from *Escherichia coli*: implication of its N-terminal domain in aminoacylation activity and specificity. Nucleic Acids Res 22:2963–2969
138. Eriani G, Cavarelli J, Martin F, Dirheimer G, Moras D, Gangloff J (1993) Role of dimerization in yeast aspartyl-tRNA synthetase and importance of the class II invariant proline. Proc Natl Acad Sci USA 90:10816–10820
139. Yaremchuk A, Cusack S, Tukalo M (2000) Crystal structure of a eukaryote/archaeon-like protyl-tRNA synthetase and its complex with tRNAPro(CGG). EMBO J 19:4745–4758
140. Yang XL, Otero FJ, Ewalt KL, Liu J, Swairjo MA, Kohrer C, RajBhandary UL, Skene RJ, McRee DE, Schimmel P (2006) Two conformations of a crystalline human tRNA synthetase-tRNA complex: implications for protein synthesis. EMBO J 25:2919–2929
141. Hauenstein SI, Hou YM, Perona JJ (2008) The homotetrameric phosphoseryl-tRNA synthetase from *Methanosarcina mazei* exhibits half-of-the-sites activity. J Biol Chem 283:21997–22006
142. Ambrogelly A, Kamtekar S, Stathopoulos C, Kennedy D, Soll D (2005) Asymmetric behavior of archaeal prolyl-tRNA synthetase. FEBS Lett 579:6017–6022
143. Coleman DE, Carter CW Jr (1984) Crystals of *Bacillus stearothermophilus* tryptophanyl-tRNA synthetase containing enzymatically formed acyl transfer product tryptophanyl-ATP, an active site maker for the $3'$ CCA terminus of tryptophanyl-tRNATrp. Biochemistry 23:381–385
144. Liu H, Gauld JW (2008) Substrate-assisted catalysis in the aminoacyl transfer mechanism of histidyl-tRNA synthetase: a density functional theory study. J Phys Chem B 112:16874–16882
145. Huang W, Bushnell EA, Francklyn CS, Gauld JW (2011) The alpha-amino group of the threonine substrate as the general base during tRNA aminoacylation: a new version of substrate-assisted catalysis predicted by hybrid DFT. J Phys Chem A 115:13050–13060
146. Lee JW, Beebe K, Nangle LA, Jang J, Longo-Guess CM, Cook SA, Davisson MT, Sundberg JP, Schimmel P, Ackerman SL (2006) Editing-defective tRNA synthetase causes protein misfolding and neurodegeneration. Nature 443:50–55
147. Ruan B, Palioura S, Sabina J, Marvin-Guy L, Kochhar S, Larossa RA, Soll D (2008) Quality control despite mistranslation caused by an ambiguous genetic code. Proc Natl Acad Sci USA 105:16502–16507
148. Reynolds NM, Ling J, Roy H, Banerjee R, Repasky SE, Hamel P, Ibba M (2010) Cell-specific differences in the requirements for translation quality control. Proc Natl Acad Sci USA 107:4063–4068
149. Roy H, Ling J, Alfonzo J, Ibba M (2005) Loss of editing activity during the evolution of mitochondrial phenylalanyl-tRNA synthetase. J Biol Chem 280:38186–38192
150. Lue SW, Kelley SO (2005) An aminoacyl-tRNA synthetase with a defunct editing site. Biochemistry 44:3010–3016
151. Netzer N, Goodenbour JM, David A, Dittmar KA, Jones RB, Schneider JR, Boone D, Eves EM, Rosner MR, Gibbs JS, Embry A, Dolan B, Das S, Hickman HD, Berglund P, Bennink JR, Yewdell JW, Pan T (2009) Innate immune and chemically triggered oxidative stress modifies translational fidelity. Nature 462:522–526
152. Silvian LF, Wang J, Steitz TA (1999) Insights into editing from an ile-tRNA synthetase structure with tRNAile and mupirocin. Science 285:1074–1077
153. Fukai S, Nureki O, Sekine S, Shimada A, Tao J, Vassylyev DG, Yokoyama S (2000) Structural basis for double-sieve discrimination of L-valine from L-isoleucine and L-threonine by the complex of tRNA(Val) and valyl-tRNA synthetase. Cell 103:793–803
154. Lincecum TL Jr, Tukalo M, Yaremchuk A, Mursinna RS, Williams AM, Sproat BS, Van Den Eynde W, Link A, Van Calenbergh S, Grotli M, Martinis SA, Cusack S (2003) Structural and mechanistic basis of pre- and posttransfer editing by leucyl-tRNA synthetase. Mol Cell 11:951–963
155. Roy H, Ling J, Irnov M, Ibba M (2004) Post-transfer editing in vitro and in vivo by the beta subunit of phenylalanyl-tRNA synthetase. EMBO J 23:4639–4648

156. Kotik-Kogan O, Moor N, Tworowski D, Safro M (2005) Structural basis for discrimination of L-phenylalanine from L-tyrosine by phenylalanyl-tRNA synthetase. Structure 13:1799–1807
157. Beebe K, Ribas De Pouplana L, Schimmel P (2003) Elucidation of tRNA-dependent editing by a class II tRNA synthetase and significance for cell viability. EMBO J 22:668–675
158. Francklyn CS (2008) DNA polymerases and aminoacyl-tRNA synthetases: shared mechanisms for ensuring the fidelity of gene expression. Biochemistry 47:11695–11703
159. Ling J, So BR, Yadavalli SS, Roy H, Shoji S, Fredrick K, Musier-Forsyth K, Ibba M (2009) Resampling and editing of mischarged tRNA prior to translation elongation. Mol Cell 33:654–660
160. An S, Musier-Forsyth K (2004) *Trans*-editing of Cys-tRNAPro by Haemophilus influenzae YbaK protein. J Biol Chem 279:42359–42362
161. Minajigi A, Francklyn CS (2010) Aminoacyl transfer rate dictates choice of editing pathway in threonyl-tRNA synthetase. J Biol Chem 285:23810–23817
162. Splan KE, Ignatov ME, Musier-Forsyth K (2008) Transfer RNA modulates the editing mechanism used by class II prolyl-tRNA synthetase. J Biol Chem 283:7128–7134
163. Gruic-Sovulj I, Rokov-Plavec J, Weygand-Durasevic I (2007) Hydrolysis of non-cognate aminoacyl-adenylates by a class II aminoacyl-tRNA synthetase lacking an editing domain. FEBS Lett 581:5110–5114
164. Ling J, Peterson KM, Simonovic I, Soll D, Simonovic M (2012) The mechanism of pre-transfer editing in yeast mitochondrial threonyl-tRNA synthetase. J Biol Chem 287:28518–28525
165. Rokov-Plavec J, Lesjak S, Gruic-Sovulj I, Mocibob M, Dulic M, Weygand-Durasevic I (2013) Substrate recognition and fidelity of maize seryl-tRNA synthetases. Arch Biochem Biophys 529:122–130
166. Gruic-Sovulj I, Dulic M, Weygand-Durasevic I (2011) Pre-transfer editing of serine hydroxamate within the active site of methanogenic-type seryl-tRNA synthetase. Croat Chim Acta 84:179–184
167. Gruic-Sovulj I, Dulic M, Cvetesic N, Majsec K, Weygand-Durasevic I (2010) Efficiently activated serine analog is not transferred to yeast tRNA(Ser). Croat Chim Acta 83:163–169
168. Jakubowski H (1980) Valyl-tRNA synthetase form yellow lupin seeds: hydrolysis of the enzyme-bound noncognate aminoacyl adenylate as a possible mechanism of increasing specificity of the aminoacyl-tRNA synthetase. Biochemistry 19:5071–5078
169. Zhu B, Yao P, Tan M, Eriani G, Wang ED (2009) tRNA-independent pretransfer editing by class I leucyl-tRNA synthetase. J Biol Chem 284:3418–3424
170. Jakubowski H (1991) Proofreading in vivo: editing of homocysteine by methionyl-tRNA synthetase in the yeast *Saccharomyces cerevisiae*. EMBO J 10:593–598
171. Serre L, Verdon G, Choinowski T, Hervouet N, Risler JL, Zelwer C (2001) How methionyl-tRNA synthetase creates its amino acid recognition pocket upon L-methionine binding. J Mol Biol 306:863–876
172. Jakubowski H (1999) Misacylation of tRNALys with noncognate amino acids by lysyl-tRNA synthetase. Biochemistry 38:8088–8093
173. Jakubowski H (1997) Aminoacyl thioester chemistry of class II aminoacyl-tRNA synthetases. Biochemistry 36:11077–11085
174. Levengood J, Ataide SF, Roy H, Ibba M (2004) Divergence in noncognate amino acid recognition between class I and class II lysyl-tRNA synthetases. J Biol Chem 279:17707–17714
175. Hopfield JJ, Yamane T, Yue V, Coutts SM (1976) Direct experimental evidence for kinetic proofreading in amino acylation of tRNAIle. Proc Natl Acad Sci USA 73:1164–1168
176. Boniecki MT, Vu MT, Betha AK, Martinis SA (2008) CP1-dependent partitioning of pretransfer and posttransfer editing in leucyl-tRNA synthetase. Proc Natl Acad Sci USA 105:19223–19228
177. Williams AM, Martinis SA (2006) Mutational unmasking of a tRNA-dependent pathway for preventing genetic code ambiguity. Proc Natl Acad Sci USA 103:3586–3591

178. Eldred EW, Schimmel PR (1972) Rapid deacylation by isoleucyl transfer ribonucleic acid synthetase of isoleucine-specific transfer ribonucleic acid aminoacylated with valine. J Biol Chem 247:2961–2964
179. Hati S, Ziervogel B, Sternjohn J, Wong FC, Nagan MC, Rosen AE, Siliciano PG, Chihade JW, Musier-Forsyth K (2006) Pre-transfer editing by class II prolyl-tRNA synthetase: role of aminoacylation active site in "selective release" of noncognate amino acids. J Biol Chem 281:27862–27872
180. Ling J, Peterson KM, Simonovic I, Cho C, Soll D, Simonovic M (2012) Yeast mitochondrial threonyl-tRNA synthetase recognizes tRNA isoacceptors by distinct mechanisms and promotes CUN codon reassignment. Proc Natl Acad Sci USA 109:3281–3286
181. Lin SX, Baltzinger M, Remy P (1984) Fast kinetic study of yeast phenylalanyl-tRNA synthetase: role of tRNAPhe in the discrimination between tyrosine and phenylalanine. Biochemistry 23:4109–4116
182. Nomanbhoy TK, Hendrickson TL, Schimmel P (1999) Transfer RNA-dependent translocation of misactivated amino acids to prevent errors in protein synthesis. Mol Cell 4:519–528
183. Farrow MA, Schimmel P (2001) Editing by a tRNA synthetase: DNA aptamer-induced translocation and hydrolysis of a misactivated amino acid. Biochemistry 40:4478–4483
184. Hendrickson TL, Nomanbhoy TK, de Crecy-Lagard V, Fukai S, Nureki O, Yokoyama S, Schimmel P (2002) Mutational separation of two pathways for editing by a class I tRNA synthetase. Mol Cell 9:353–362
185. Bishop AC, Nomanbhoy TK, Schimmel P (2002) Blocking site-to-site translocation of a misactivated amino acid by mutation of a class I tRNA synthetase. Proc Natl Acad Sci USA 99:585–590
186. Dock-Bregeon AC, Rees B, Torres-Larios A, Bey G, Caillet J, Moras D (2004) Achieving error-free translation; the mechanism of proofreading of threonyl-tRNA synthetase at atomic resolution. Mol Cell 16:375–386
187. Lin L, Hale SP, Schimmel P (1996) Aminoacylation error correction. Nature 384:33–34
188. Betha AK, Williams AM, Martinis SA (2007) Isolated CP1 domain of *Escherichia coli* leucyl-tRNA synthetase is dependent on flanking hinge motifs for amino acid editing activity. Biochemistry 46:6258–6267
189. Beuning PJ, Musier-Forsyth K (2001) Species-specific differences in amino acid editing by class II prolyl-tRNA synthetase. J Biol Chem 276:30779–30785
190. SternJohn J, Hati S, Siliciano PG, Musier-Forsyth K (2007) Restoring species-specific posttransfer editing activity to a synthetase with a defunct editing domain. Proc Natl Acad Sci USA 104:2127–2132
191. Guo LT, Helgadottir S, Soll D, Ling J (2012) Rational design and directed evolution of a bacterial-type glutaminyl-tRNA synthetase precursor. Nucleic Acids Res 40:7967–7974
192. Wong FC, Beuning PJ, Silvers C, Musier-Forsyth K (2003) An isolated class II aminoacyl-tRNA synthetase insertion domain is functional in amino acid editing. J Biol Chem 278:52857–52864
193. Nureki O, Vassylyev DG, Tateno M, Shimada A, Nakama T, Fukai S, Konno M, Hendrickson TL, Schimmel P, Yokoyama S (1998) Enzyme structure with two catalytic sites for double-sieve selection of substrate. Science 280:578–582
194. Mascarenhas AP, Martinis SA (2009) A glycine hinge for tRNA-dependent translocation of editing substrates to prevent errors by leucyl-tRNA synthetase. FEBS Lett 583:3443–3447
195. Rock FL, Mao W, Yaremchuk A, Tukalo M, Crepin T, Zhou H, Zhang YK, Hernandez V, Akama T, Baker SJ, Plattner JJ, Shapiro L, Martinis SA, Benkovic SJ, Cusack S, Alley MR (2007) An antifungal agent inhibits an aminoacyl-tRNA synthetase by trapping tRNA in the editing site. Science 316:1759–1761
196. Fukunaga R, Yokoyama S (2007) Structure of the AlaX-M *trans*-editing enzyme from *Pyrococcus horikoshii*. Acta Crystallogr D Biol Crystallogr 63:390–400
197. Sokabe M, Okada A, Yao M, Nakashima T, Tanaka I (2005) Molecular basis of alanine discrimination in editing site. Proc Natl Acad Sci USA 102:11669–11674

198. Sasaki HM, Sekine S, Sengoku T, Fukunaga R, Hattori M, Utsunomiya Y, Kuroishi C, Kuramitsu S, Shirouzu M, Yokoyama S (2006) Structural and mutational studies of the amino acid-editing domain from archaeal/eukaryal phenylalanyl-tRNA synthetase. Proc Natl Acad Sci USA 103:14744–14749

199. Hussain T, Kruparani SP, Pal B, Dock-Bregeon AC, Dwivedi S, Shekar MR, Sureshbabu K, Sankaranarayanan R (2006) Post-transfer editing mechanism of a D-aminoacyl-tRNA deacylase-like domain in threonyl-tRNA synthetase from archaea. EMBO J 25:4152–4162

200. Murayama K, Kato-Murayama M, Katsura K, Uchikubo-Kamo T, Yamaguchi-Hirafuji M, Kawazoe M, Akasaka R, Hanawa-Suetsugu K, Hori-Takemoto C, Terada T, Shirouzu M, Yokoyama S (2005) Structure of a putative *trans*-editing enzyme for prolyl-tRNA synthetase from *Aeropyrum pernix* K1 at 1.7 A resolution. Acta Crystallogr Sect F Struct Biol Cryst Commun 61:26–29

201. Fukunaga R, Yokoyama S (2005) Structural basis for non-cognate amino acid discrimination by the valyl-tRNA synthetase editing domain. J Biol Chem 280:29937–29945

202. Fukunaga R, Fukai S, Ishitani R, Nureki O, Yokoyama S (2004) Crystal structures of the CP1 domain from *Thermus thermophilus* isoleucyl-tRNA synthetase and its complex with L-valine. J Biol Chem 279:8396–8402

203. Seiradake E, Mao W, Hernandez V, Baker SJ, Plattner JJ, Alley MR, Cusack S (2009) Crystal structures of the human and fungal cytosolic leucyl-tRNA synthetase editing domains: a structural basis for the rational design of antifungal benzoxaboroles. J Mol Biol 390:196–207

204. Fukunaga R, Yokoyama S (2006) Structural basis for substrate recognition by the editing domain of isoleucyl-tRNA synthetase. J Mol Biol 359:901–912

205. Liu Y, Liao J, Zhu B, Wang ED, Ding J (2006) Crystal structures of the editing domain of *Escherichia coli* leucyl-tRNA synthetase and its complexes with Met and Ile reveal a lock-and-key mechanism for amino acid discrimination. Biochem J 394:399–407

206. Mursinna RS, Lincecum TL Jr, Martinis SA (2001) A conserved threonine within *Escherichia coli* leucyl-tRNA synthetase prevents hydrolytic editing of leucyl-tRNALeu. Biochemistry 40:5376–5381

207. Hussain T, Kamarthapu V, Kruparani SP, Deshmukh MV, Sankaranarayanan R (2010) Mechanistic insights into cognate substrate discrimination during proofreading in translation. Proc Natl Acad Sci USA 107:22117–22121

208. Kumar S, Das M, Hadad CM, Musier-Forsyth K (2012) Substrate specificity of bacterial prolyl-tRNA synthetase editing domain is controlled by a tunable hydrophobic pocket. J Biol Chem 287:3175–3184

209. Ahel I, Stathopoulos C, Ambrogelly A, Sauerwald A, Toogood H, Hartsch T, Soll D (2002) Cysteine activation is an inherent in vitro property of prolyl-tRNA synthetases. J Biol Chem 277:34743–34748

210. Ahel I, Korencic D, Ibba M, Soll D (2003) *Trans*-editing of mischarged tRNAs. Proc Natl Acad Sci USA 100:15422–15427

211. Pasman Z, Robey-Bond S, Mirando AC, Smith GJ, Lague A, Francklyn CS (2011) Substrate specificity and catalysis by the editing active site of alanyl-tRNA synthetase from *Escherichia coli*. Biochemistry 50:1474–1482

212. Waas WF, Schimmel P (2007) Evidence that tRNA synthetase-directed proton transfer stops mistranslation. Biochemistry 46:12062–12070

213. Beebe K, Merriman E, Ribas De Pouplana L, Schimmel P (2004) A domain for editing by an archaebacterial tRNA synthetase. Proc Natl Acad Sci USA 101:5958–5963

214. Korencic D, Ahel I, Schelert J, Sacher M, Ruan B, Stathopoulos C, Blum P, Ibba M, Soll D (2004) A freestanding proofreading domain is required for protein synthesis quality control in archaea. Proc Natl Acad Sci USA 101:10260–10265

215. Dwivedi S, Kruparani SP, Sankaranarayanan R (2005) A D-amino acid editing module coupled to the translational apparatus in archaea. Nat Struct Mol Biol 12:556–557

216. Hagiwara Y, Field MJ, Nureki O, Tateno M (2010) Editing mechanism of aminoacyl-tRNA synthetases operates by a hybrid ribozyme/protein catalyst. J Am Chem Soc 132:2751–2758

217. Ling J, Roy H, Ibba M (2007) Mechanism of tRNA-dependent editing in translational quality control. Proc Natl Acad Sci USA 104:72–77
218. Sheoran A, Sharma G, First EA (2008) Activation of D-tyrosine by *Bacillus stearothermophilus* tyrosyl-tRNA synthetase: 1. Pre-steady-state kinetic analysis reveals the mechanistic basis for the recognition of D-tyrosine. J Biol Chem 283:12960–12970
219. Sheoran A, First EA (2008) Activation of D-tyrosine by *Bacillus stearothermophilus* tyrosyl-tRNA synthetase: 2. Cooperative binding of ATP is limited to the initial turnover of the enzyme. J Biol Chem 283:12971–12980
220. Calendar R, Berg P (1967) D-Tyrosyl RNA: formation, hydrolysis and utilization for protein synthesis. J Mol Biol 26:39–54
221. An S, Musier-Forsyth K (2005) Cys-tRNA(Pro) editing by *Haemophilus influenzae* YbaK via a novel synthetase.YbaK.tRNA ternary complex. J Biol Chem 280:34465–34472
222. Ruan B, Soll D (2005) The bacterial YbaK protein is a Cys-tRNAPro and Cys-tRNA Cys deacylase. J Biol Chem 280:25887–25891
223. So BR, An S, Kumar S, Das M, Turner DA, Hadad CM, Musier-Forsyth K (2011) Substrate-mediated fidelity mechanism ensures accurate decoding of proline codons. J Biol Chem 286:31810–31820
224. Kumar S, Das M, Hadad CM, Musier-Forsyth K (2013) Aminoacyl-tRNA substrate and enzyme backbone atoms contribute to translational quality control by YbaK. J Phys Chem B 117:4521–4527
225. Sanford B, Cao B, Johnson JM, Zimmerman K, Strom AM, Mueller RM, Bhattacharyya S, Musier-Forsyth K, Hati S (2012) Role of coupled dynamics in the catalytic activity of prokaryotic-like prolyl-tRNA synthetases. Biochemistry 51:2146–2156
226. Beebe K, Mock M, Merriman E, Schimmel P (2008) Distinct domains of tRNA synthetase recognize the same base pair. Nature 451:90–93
227. Guo M, Chong YE, Beebe K, Shapiro R, Yang XL, Schimmel P (2009) The C-Ala domain brings together editing and aminoacylation functions on one tRNA. Science 325:744–747
228. Chong YE, Yang XL, Schimmel P (2008) Natural homolog of tRNA synthetase editing domain rescues conditional lethality caused by mistranslation. J Biol Chem 283:30073–30078
229. O'Donoghue P, Luthey-Schulten Z (2003) On the evolution of structure in aminoacyl-tRNA synthetases. Microbiol Mol Biol Rev 67:550–573
230. Ibba M, Morgan S, Curnow AW, Pridmore DR, Vothknecht UC, Gardner W, Lin W, Woese CR, Soll D (1997) A euryarchaeal lysyl-tRNA synthetase: resemblance to class I synthetases. Science 278:1119–1122
231. Sauerwald A, Zhu W, Major TA, Roy H, Palioura S, Jahn D, Whitman WB, Yates JR 3rd, Ibba M, Soll D (2005) RNA-dependent cysteine biosynthesis in archaea. Science 307:1969–1972
232. Hao B, Gong W, Ferguson TK, James CM, Krzycki JA, Chan MK (2002) A new UAG-encoded residue in the structure of a methanogen methyltransferase. Science 296:1462–1466
233. Srinivasan G, James CM, Krzycki JA (2002) Pyrrolysine encoded by UAG in archaea: charging of a UAG-decoding specialized tRNA. Science 296:1459–1462
234. Polycarpo C, Ambrogelly A, Berube A, Winbush SM, McCloskey JA, Crain PF, Wood JL, Soll D (2004) An aminoacyl-tRNA synthetase that specifically activates pyrrolysine. Proc Natl Acad Sci USA 101:12450–12454
235. Blight SK, Larue RC, Mahapatra A, Longstaff DG, Chang E, Zhao G, Kang PT, Green-Church KB, Chan MK, Krzycki JA (2004) Direct charging of tRNA(CUA) with pyrrolysine in vitro and in vivo. Nature 431:333–335

Top Curr Chem (2014) 344: 43–88
DOI: 10.1007/128_2013_423
© Springer-Verlag Berlin Heidelberg 2013
Published online: 12 March 2013

Emergence and Evolution

Tammy J. Bullwinkle and Michael Ibba

Abstract The aminoacyl-tRNA synthetases (aaRSs) are essential components of the protein synthesis machinery responsible for defining the genetic code by pairing the correct amino acids to their cognate tRNAs. The aaRSs are an ancient enzyme family believed to have origins that may predate the last common ancestor and as such they provide insights into the evolution and development of the extant genetic code. Although the aaRSs have long been viewed as a highly conserved group of enzymes, findings within the last couple of decades have started to demonstrate how diverse and versatile these enzymes really are. Beyond their central role in translation, aaRSs and their numerous homologs have evolved a wide array of alternative functions both inside and outside translation. Current understanding of the emergence of the aaRSs, and their subsequent evolution into a functionally diverse enzyme family, are discussed in this chapter.

Keywords aaRS · Amino acid · Aminoacyl-tRNA synthetase · Enzyme specificity · Proofreading · Transamidation · Translation · tRNA

Contents

1	Introduction	44
2	The Aminoacyl-tRNA Synthetase Class System	45
	2.1 Class I aaRS	49
	2.2 Class II aaRS	50
	2.3 Examples of Where aaRSs Are Missing from Particular Genomes	51
3	Non-canonical Aminoacyl-tRNA Synthetases	55
	3.1 LysRS I	55
	3.2 Pyrolysyl-tRNA Synthetase (PylRS)	56
	3.3 Phosphoseryl-tRNA Synthetase (SepRS)	58

T.J. Bullwinkle and M. Ibba (✉)
Department of Microbiology, The Ohio State University, 484 West 12th Avenue,
Columbus, OH 43210, USA
e-mail: ibba.1@osu.edu

4	Functional Evolution of Synthetases	59
	4.1 Specificity of Synthetases	60
	4.2 Adapted and Changing Domains	63
5	Emergence of Non-canonical Functions in Aminoacyl-tRNA Synthetases	67
	5.1 Fused Domains Having Non-canonical Functions	67
	5.2 Paralogs of Synthetases	68
6	Conclusion	71
	References	71

1 Introduction

Aminoacyl-tRNA synthetases bridge the transition from RNA to protein during translation by correctly pairing a particular tRNA with its cognate amino acid to be incorporated into a growing peptide chain by the ribosome. How aaRSs contributed to the transition from an RNA-only world to one involving protein synthesis has been the subject of considerable conjecture. Polypeptide formation programmed by a genetic code likely existed early in evolution and the earliest process of translation arose at a time when functions that predate present-day proteins were being performed by other, protein-free systems. Similar to the largely RNA directed functions of the ribosome, there is a general consensus that aminoacyl-tRNA synthesis may have begun as a process catalyzed by RNA [1–3]. Many parts of the ribosome and its factors resemble mini RNA helices, which could have served as substrates for many of the aaRSs [4–8]. Aminoacyl-minihelices can be used by contemporary ribosomes, providing a link between a strictly RNA world and one involving protein synthesis [9]. It is also thought that the early genetic code was comparatively simple, perhaps consisting of only a few amino acids, and the first aaRS functionalities possibly emerged as an early part of the primordial protein synthesis machinery. As the protein code became more complex the synthetases, along with tRNAs, separated from the early ribosome leading to a precursor more similar to the contemporary protein synthesis machinery [10].

Present day aminoacyl-tRNA synthetase enzymes are universally required for protein synthesis in all organisms and show a high degree of evolutionary conservation. AaRSs have undergone extensive horizontal gene transfer, particularly during early evolution, and such events have been identified across all taxonomic levels and are well supported by phylogenic evidence [11]. Synthetases apparently emerged before the tree of life evolved into three domains, during the time of the last universal common ancestor (LUCA), as can be seen by the universal distribution of aaRSs across all branches of life. Additionally, aaRS phylogenies provide evidence that each family of synthetase was present in LUCA [12]. Ancient emergence of aaRSs as well as early and abundant gene transfer events led to deep lineages and low sequence similarity between modern synthetase homologs [13]. Additionally, the evolution of the synthetase enzyme family has included gene duplications and the subsequent emergence of numerous paralogs with new functions (Sect. 5).

Several studies suggest a correlation between how the aaRSs are evolutionarily related to each other and the overall amino acid order or structure of the genetic code. It is not generally believed that the evolution of the aaRS family is responsible for shaping the genetic code, but rather is somewhat converged with the evolution of the code as both are driven by similar properties of the corresponding amino acids [12, 13]. The evolution of the universal genetic code for the 20 canonical amino acids likely occurred early in the history of life and appears to predate the emergence or distinction between class I and class II aaRSs, as the aaRSs had already evolved amino acid specificities by the time of the LUCA. This order of events further suggests that the extant aaRS aminoacylation machinery may have somehow displaced an ancient, now extinct, aminoacylation process [11, 14].

The contemporary aminoacyl-tRNA synthetase enzymes are modular in structure and contain a core catalytic domain responsible for ATP dependent aminoacyl-adenylate formation and subsequent ester bond ligation of the activated amino acid onto the 3′ ribose of tRNA. In addition to the core catalytic domain, each enzyme contains a variety of other modules that function primarily to maintain translational accuracy via substrate and tRNA binding and recognition as well as proofreading of misacylated products. In some instances other domains appended to synthetases are responsible for activities outside of tRNA charging including transcriptional and translational regulation, DNA replication, and cell signaling (Sects. 4 and 5). This chapter will focus on the conventional classification of this large enzyme family, as well the emergence of non-canonical aaRSs, alternative pathways for tRNA aminoacylation and aaRS homologs and paralogs that function both in and apart from tRNA aminoacylation. How further adaptive evolution has allowed for widespread adjustments to the core functions of aaRSs providing selective advantages specific to different organisms and their environments will also be discussed.

2 The Aminoacyl-tRNA Synthetase Class System

AaRSs are grouped into two unrelated structural classes based on the conserved architecture of the catalytic core domains of the enzymes [15, 16]. This classification is independent of other appended domains, with roughly half the aaRSs in each of the two classes (Table 1). The categorization of synthetases into one of two classes is almost completely universal across all three domains. In other words, the synthetase class assignment for a particular cognate amino acid is the same regardless of the origin of the enzyme. There is one recently discovered exception, LysRS,[1] which is found in both classes (Sect. 3.1). Although the two aaRSs classes are evolutionarily and structurally very different from each other, the overall chemistry of the tRNA

---

[1] Specific aminoacyl-tRNA synthetases are denoted by their three-letter amino acid designation, e.g., LysRS for lysyl-tRNA synthetase. Lysine tRNA or tRNA^{Lys} denote uncharged tRNA specific for lysine; lysyl-tRNA or Lys-tRNA denote tRNA aminoacylated with lysine.

Table 1 Classes of aminoacyl-tRNA synthetases and structures

Class[a]		Source	Ligands	Reference
Class I				
Subclass Ia				
CysRS	α	*Escherichia coli*	Cys	[48]
IleRS	α	*Staphylococcus aureus, Thermus thermophilus*	Ile-AMP and Val-AMP analogs, Val-tRNA, tRNA, mupirocin	[29, 49, 50]
LeuRS	α	*E. coli, T. thermophilus*	Leu, norvaline, analogs, tRNA, inhibitors	[51–54]
MetRS	α, α$_2$	*E. coli, Mycobacterium smegmatis, Pyrococcus abyssi, Trypanosoma brucei, Aquifex aeolicus, Leishmania major*	Met, Met-AMP, PP$_i$, tRNA, analogs, inhibitors	[55–61]
ValRS	α	*T. thermophilus*	Val-AMP, tRNA, analogs	[31, 62]
Subclass Ib				
ArgRS	α	*Saccharomyces cerevisiae, T. thermophilus, Pyrococcus horikoshii*	Arg, ATP, tRNA, analog	[63–66]
GlnRS	α	*E. coli, Deinococcus radiodurans*	Gln-AMP analog, tRNA, ATP	[67–72]
GluRS	α	*T. thermophilus, Methanothermobacter thermautotrophicus, Thermosynechococcus elongatus, Thermotoga maritimus*	Glu-AMP, ATP, tRNA, Glu-AMS, inhibitor	[73–78]
LysRS[b]	α	*P. horikoshii*	–	[20]
Subclass Ic				
TrpRS	α$_2$	*Geobacillus stearothermophilus, Homo sapiens, D. radiodurans, S. cerevisiae, Giardia lamblia, T. maritima, T. brucei, Cryptosporidium parvum*	Trp, Trp-AMP, ATP, tRNA, analogs	[79–88]
TyrRS	α$_2$	*T. thermophilus, E. coli, P. horikoshii, A. pernix, H. sapiens (mitochondrial)*	Tyr, Tyr-AMP, Tyrp-AMP analog, ATP, tRNA	[89–92]
Class II				
Subclass IIa				
GlyRS[c]	α$_2$	*T. thermophilus H. sapiens*	Gly-AMP, ATP	[76, 93–95]
HisRS	α$_2$	*E. coli, T. thermophilus, S. aureus, T. brucei, Trypanosoma cruzi, H. sapiens*	His-AMP, ATP, analog	[96–101]
ProRS	α$_2$	*T. thermophilus, M. thermautotrophicus, Methanocaldococcus janaschii, G. lamblia, Enterococcus faecalis, Rhodopseudomonas palustris, H. sapiens*	Pro, Pro-AMP, Cys-AMP and Ala-AMP analogs, ATP, tRNA, inhibitor	[102–106]

SerRS	α_2	*T. thermophilus, P. horikoshii, Candida albicans, Bos taurus (mitochondrial)*	Ser-AMP, Ser-AMP analog, ATP, tRNA	[107–111]
ThrRS	α_2	*E. coli, S. aureus, S. cerevisiae (mitochondrial)*	Thr, Thr-AMP, Thr-AMP and Ser-AMP analogs, ATP, AMP, tRNA, *thrS* mRNA	[112–116]
Subclass IIb				
AsnRS	α_2	*T. thermophilus, P. horikoshii, Brugia malayi, P. abyssi*	Asn, Asp, Asn-AMP, Asn-AMP analog, ATP, PP$_i$	[43, 117–119]
AspRS	α_2	*Pyrococcus kodakarensis, T. thermophilus, S. cerevisae, E. coli, Entamoeba histolytica, H. sapiens (mitochondrial)*	Asp, Asp-AMP, Asp-AMP analog, ATP, tRNA	[120–127]
LysRS	α_2	*E. coli, B. stearothermophilus, T. thermophilus, H. sapiens*	Lys, Lys-AMP analog, ATP, tRNA	[44, 128–131]
Subclass IIc				
AlaRS	α, α_4	*E. coli, P. horikoshii, Archaeoglobus fulgidus, A. aeolicus*	Ala, Ser, Gly, Ala-AMP analog, ATP	[132–135]
GlyRS[c]	$(\alpha\beta)_2$	*Campylobacter jejuni, T. meritima*	Gly, ATP	[136, 137]
PheRS	$(\alpha\beta)_2, \alpha$	*T. thermophilus, E. coli, H. sapiens (mitochondrial)*	Phe, Tyr, *m*-Tyr, DOPA, Phe-AMP analog, *tRNA*	[138–143]
PylRS	α_2	*Methanosarcina mazei, Desulfitobacterium hafniense*	Pyl, analog, Pyl-AMP, AMPPNP, ATP, PP$_i$, tRNA	[144–146]
SepRS	α_4	*M. jannacshii, Methanococcus maripaludis, A. fugidus*	Sep, tRNA	[147, 148]

[a]Classes and subclasses are as previously defined [21]. Slightly different subclass groupings have been made based on variations in phylogenetic analysis [16, 18, 27]

[b]LysRS is found in both class I and II aaRS

[c]GlyRS exists in two unrelated forms

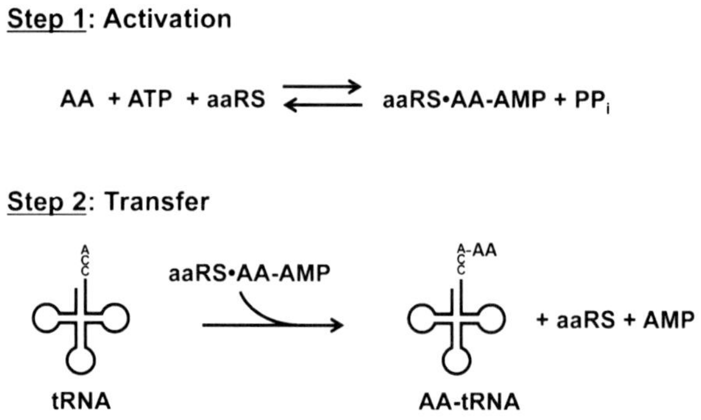

Step 1: Activation

$$AA + ATP + aaRS \rightleftharpoons aaRS \cdot AA\text{-}AMP + PP_i$$

Step 2: Transfer

aaRS·AA-AMP

tRNA → AA-tRNA + aaRS + AMP

Fig. 1 Two-step aminoacylation reaction by aaRSs. In the first step of the reaction, nucleophilic attack by the α-carboxylate carbon of the amino acid on the α-phosphate of ATP leads to the formation of an enzyme-bound aminoacyl-adenylate and pyrophosphate release. The second step of catalysis involves transfer of the aminoacyl-adenylate to tRNA. Nucleophilic attack by either the $2'$ or $3'$ OH of A76 of the tRNA (depending on class) on the α-carbonyl carbon of the aminoacyl-adenylate and subsequent release of AMP drives the tRNA transfer step of the reaction

aminoacylation reaction is similar in both (Fig. 1). The existence of two entirely unrelated classes of synthetases provides a strong example of convergence, with two independent structural solutions evolving to achieve the same enzymatic goal; efficient aminoacylation of tRNA [17]. It has also been suggested that the existence of two independent aaRS classes is indicative of multiple origins of protein synthesis where each used its own set of amino acids and then the two systems fused, or possibly competed, to form eventually one translation process [11].

A docking study of aaRSs on the tRNA acceptor stem found that subclasses of synthetases within each of the two major classes correlate to each other with respect to amino acid specificity [18]. Because two synthetases from the two different classes normally bind the tRNA acceptor stem from opposite sides, it is possible to model the simultaneous docking of two aaRSs onto a single tRNA. Upon doing this, the authors found specific class I/class II aaRS pairs that could be docked without creating major steric clashes and this pairing occurs because the position of the active site in relation to the tRNA acceptor stem varies for aaRSs within each subclass. For each pair of class I and II aaRSs that were co-docked, the corresponding amino acids had similar structural and steric characteristics. From these results, the authors propose the possibility that ancestral aaRSs with the same amino acid specificities evolved into two separate enzymes with different class architectures, but the same amino acid substrates. LysRS provides a modern example of this, where the amino acid substrates are exactly the same for two enzymes of different aaRS classes [19, 20]. In most cases, however, the aaRS pairs have evolved divergent amino acid specificities and new codons designated to distinguish similar amino acids from each other [18].

Fig. 2 Active site domains of (**a**) class I aminoacyl-tRNA synthetase, e.g., GlnRS, and (**b**) class II aminoacyl-tRNA synthetase, e.g., AspRS. Shown are ATP and the acceptor ends of cognate tRNAs (*red*). The locations of the characteristic motifs are indicated: in (**a**), MSK (*dark blue*), HIGH (*red*): in (**b**), motif 1 (*red*), motif 2 (*light blue*), and motif 3 (*dark blue*) (reprinted from [17], Copyright 1997, with permission from Elsevier Science)

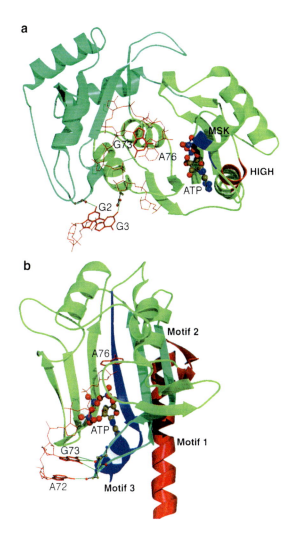

2.1 Class I aaRS

The catalytic domains of class I synthetases are structurally very similar to each other and contain a conserved Rossmann fold domain. This conserved domain is responsible for nucleotide binding and contains two conserved sequences, "HIGH" and "KMSKS" located near the α-phosphate of the bound ATP (Fig. 2) [21]. Also unique to the class I synthetases is the attachment of the activated amino acid at the 2′ OH of tRNA. Class I aaRSs bind the tRNA acceptor helix on the minor groove side and, although many of the synthetases in class I are capable of aminioacylating both the 2′ and 3′ OH of their respective tRNA, the 2′ OH is a much more efficient target for catalysis [22, 23]. Release of the aminoacyl-tRNA is rate limiting for many members of this class of synthetase [24–26]. It should be noted that the analysis of subclass grouping of both class I and class II synthetases can vary slightly based on the structures and sequences available for phylogenic comparison [18, 21, 27, 28].

There are three class I synthetase subclasses. Subclass Ia contains synthetases whose cognate amino acids have hydrophobic aliphatic groups or are sulfur containing. This group shares a well conserved overall structure, including a common tRNA stem-contact fold (SC-fold), an α-helical anticodon binding domain, and a connective peptide I (CPI) domain which is a globular insertion domain located between two parts of the nucleotide binding fold. The CPI domain is attributed to the post-transfer editing function, or proofreading, of misacylated tRNAs in three of the class Ia synthetases, LeuRS, IleRS, and ValRS [29–31] reviewed in [32]. Phylogenetic reconstruction supports previous theories that these three closely related synthetases, which are all cognate for aliphatic amino acids and charge tRNAs that decode NUN codons, arose from a common ancestor which was unable to discriminate between Leu, Ile, and Val [13, 33].

Subclass Ib includes ArgRS, GluRS, GlnRS, and an atypical class I LysRS, all of which recognize cognate charged amino acids. Interestingly these four synthetases are the only exceptions to the distinct two-step reaction, as they all require cognate tRNA binding before catalysis of the pyrophosphate exchange reaction occurs [34–36]. GlnRS likely evolved from GluRS and both are involved in the formation of Gln-tRNA either directly or indirectly depending on the particular organism or organelle (discussed below). GluRS in eukaryotes is more closely related to GlnRS than it is to GluRS in bacteria, suggesting GlnRS emerged from a gene duplication of an ancestral eukaryl GluRS that was then transferred to bacteria [37] reviewed in [38].

The third subclass, Ic, contains the structurally similar TrpRS and TyrRS, both of which have cognate aromatic amino acid substrates. The subclass 1c enzymes are both dimers, in contrast to all other class I synthetases that function as monomers. Based on early sequence comparisons that showed TrpRS and TyrRS from eukaryotes and Archaea were more similar to each other than compared to their counterparts in eubacteria, it was proposed that Trp and Tyr were added more recently to the genetic code with their cognate synthetases diverging after the three domains of life had split [39]. More recent analyses, however, have shown that TrpRS and TyrRS form two monophyletic groups, more in line with other synthetase evolutionary groupings and supporting a much earlier gene duplication event [40, 41].

2.2 Class II aaRS

Class II synthetases tend to work as multimers, as most are homodimers while some forms of PheRS, AlaRS, and the bacterial form of GlyRS function act as tetramers. Class II aaRSs are much less conserved than class I and have a structurally distinct catalytic core that is made up of a characteristic seven-stranded antiparallel β-sheet surrounded by a number of α-helices (Fig. 2) [21]. There are three loosely conserved sequence motifs (1,2,3) found in class II synthetases; motif 1 is found

Emergence and Evolution

at the dimer interface while motifs 2 and 3 participate in substrate binding in the catalytic site. In contrast to class I, class II aaRSs bind the acceptor helix of the tRNA from the major groove side and generally attach the activated amino acid to the 3′ OH of tRNA. The only exception to this last point is PheRS which, like a class I synthetase, aminoacylates the 2′ OH of tRNA.

Subclasses of class II aaRS are defined by differences in primary sequence, subunit organization (dimer, heterodimer, etc.), and location and composition of the anticodon-binding domain. HisRS, ProRS, SerRS, and ThrRS are usually grouped as subclass IIa synthetases. These enzymes have the canonical class II catalytic site and are grouped together due to the similarity in the sequences of their C-termini. With the exception of SerRS, the synthetases within this group have similar C-terminal tRNA anticodon binding domains, which contain an α/β fold responsible for recognizing determinants in the tRNA anticodon loop [16]. Interestingly, this domain is also found in the archaeal/eukaryotic type GlyRS (see below).

Subclass IIb is composed of three synthetases, AspRS, AsnRS, and LysRS, that share several regions of sequence and structural homology indicating these enzymes all originated from a common ancestor. The structural organization of the subclass IIb is highly conserved, in particular the presence of an oligonucleotide binding (OB) fold containing an N-terminal extension that acts as an anticodon-binding module, which contacts tRNA on the minor groove side of its anticodon stem [42–44]. The anticodon stem loops of the cognate tRNAs of the class IIb synthetases all have a conserved central uracil base which makes two contacts with the aaRS [16].

The class IIc synthetases AlaRS, GlyRS, PheRS, PylRS, and SepRS only contain class II motifs 2 and 3 and have less well conserved amino acid and tRNA binding elements than other class II aaRSs. The members of this aaRS subclass mostly exist as tetrameric structures, as opposed to dimers like the other two class II subgroups [45]. There are some exceptions to this such as mitochondrial PheRS, which is a monomer lacking the β-subunit and editing domain [46]. Two forms of GlyRS exist, a homodimer found in archaea, eukaryotes, and some bacteria and a heterotetramer found only in bacteria (see Table 1). The two forms of GlyRS are unrelated in both sequence and structure and the heterotetramer form is not closely related to any of the other class II aaRS [16]. The distribution of GlyRS types in different bacteria does not correlate with the evolutionary emergence of these bacteria [47]. GlyRS is a clear example of a synthetase that does not follow the rule of one conserved aaRS across all domains for each amino acid [11].

2.3 Examples of Where aaRSs Are Missing from Particular Genomes

Not all organisms have a full set of 20 canonical aaRSs to synthesize aa-tRNA from all 20 canonical amino acids. Initial analyses of complete archaeal genomes revealed missing open reading frames encoding several synthetases, complicating

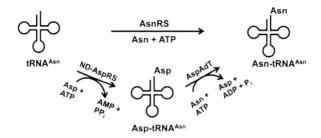

Fig. 3 Direct aminoacylation vs the transamidation pathway for Asn-tRNA^{Asn} formation. In species that encode AsnRS, Asn-tRNA^{Asn} formation occurs directly (*top*). Transamidation (*bottom*) involves Asp-tRNA^{Asn} formation using a non-discriminating AspRS (ND-AspRS). Asp-tRNA^{Asn} is then converted to Asn-tRNA^{Asn} with an amidotransferase (Adt). Indirect aminoacylation of Gln-tRNA^{Gln} occurs similarly, using ND-GluRS and Glu-AdT

the understanding of tRNA aminoacylation at that time [149, 150]. Subsequent studies showed that previously unknown aaRSs and indirect aminoacylation pathways are prevalent in archaea and bacteria. The indirect aminoacylation pathways involve a non-discriminating (ND) synthetase with expanded specificity to form a mischarged canonical amino acid-tRNA pair, which is then further modified by RNA dependent enzymes, changing the tRNA-bound amino acid.

2.3.1 AsnRS and GlnRS

The aaRSs most often missing from certain organisms are those for the direct aminoacylation of Asn-tRNA^{Asn} and Gln-tRNA^{Gln}. There is no known GlnRS encoded in any sequenced archaeal genome, and most bacterial genomes and eukaryotic organelles also lack GlnRS. Additionally, many archaea and prokaryotes do not contain an AsnRS [28]. For organisms lacking these aaRSs, Glu-tRNA^{Gln} or Asp-tRNA^{Asn} is first formed by ND-GluRS or ND-AspRS, respectively. The mischarged tRNA species is then amidated by the appropriate amidotransferase (AdT), requiring ATP as well as an amide source (Fig. 3) [151–153]. Structural and biochemical data suggest aminoacylation and amidation enzymes are able to form a complex known as the transamidosome, which provides channeling of substrates [154–157]. A more recent study has now shown that formation of a transamidosome is not essential in all cases as rapid kinetic channeling of intermediates can still occur without direct protein association [158].

Two different, but related, tRNA-dependent AdTs exist, GatCAB and GatDE. The presence of a particular form and its activity in vivo varies depending on the domain of life as well as whether one or both GlnRS and AsnRS are missing [151], reviewed in [159]. For example, the GatCAB AdT functions both as a Glu-AdT and Asp-AdT, while GatDE functions strictly as a Glu-AdT. GatCAB is found in both bacteria and archaea, but only archaea lacking AsnRS. GatDE is found only in Archaea. Recently it was shown that a unique situation exists in yeast where the cytoplasmic GluRS is imported into the mitochondria and functions there as the ND-GluRS that generates mitochondrial Glu-tRNA^{Gln} [160]. This charged tRNA substrate is then converted to Gln-tRNA^{Gln} using a novel trimeric Adt, GatFAB.

The transamidation pathways likely evolved by the adoption of Asn and Gln biosynthesis pathways by the aminoacyl-tRNA formation machinery [161]. For example, the GatD domain of the Archaeal Glu-AdT originates from an asparaginase, and the GatA domain of the multi-domain AdT is related to amidases responsible for amide bond cleavage [153, 159]. GlnRS and AsnRS were not present in LUCA, and therefore it is likely that Gln-tRNA^{Gln} and Asn-tRNA^{Asn} were first formed by indirect pathways. Where these synthetases do appear, the phylogenies lack any typical patterns, further supporting their recent origin [11, 28]. The fact that so many organisms have not acquired the appropriate aaRS for amide amino acids and have lost the corresponding Adts may reflect the essential role of amidotransferase enzymes in metabolism. For example, Gln is a major source of amides for many biosynthetic pathways. Also, most bacteria that have acquired AsnRS still have an indirect Asn-tRNA formation pathway, which is used as the only source of Asn biosynthesis in these organisms [162, 163].

2.3.2 Formylmethionyl-tRNA

Another example of an aaRS "missing" from genomes involves a unique aa-tRNA that is needed to initiate protein synthesis in bacteria, mitochondria, and chloroplasts. This tRNA, formylmethionyl-tRNA^{fMet}, is aminoacylated indirectly as there are no genes encoding an fMetRS known to date. First, aminoacylation of initiator tRNA^{fMet} with methionine by methionyl-tRNA synthetase (MetRS) occurs followed by formylation of the methionine moiety by methionyl-tRNA transformylase [164, 165]. The initiator tRNA contains sequence elements and modifications that distinguish it from elongator tRNA^{Met} and helps it evade binding to elongation factors and instead bind directly to the ribosomal P site with the help of initiation factors (reviewed in [166]). In *Trypanosoma brucei* mitochondria Met-tRNA^{Met} is imported from the cytoplasm and a fraction is then formylated and used for translation initiation [167]. The formyl modification of methionine is important for the initiator tRNA to function in translation, as it is specifically recognized by bacterial initiation factor 2 (IF2), ensuring the appropriate tRNA is in place for initiation [168].

2.3.3 Selenocysteine-tRNA

The amino acid selenocysteine (Sec) is found in all three domains of life, but not in all organisms, and was the first discovered outside of the original 20 amino acids encoded by the universal genetic code. However, no SecRS or enzyme able to aminoacylate directly tRNA^{Sec} with Sec has been identified. Selenocysteine is similar to cysteine, the difference being the thiol group is replaced by a selenium-containing

Fig. 4 Indirect Sec-tRNA^{Sec} formation. SerRS first aminoacylates tRNA^{Sec} with serine (Ser) and then Ser-tRNA^{Sec} is converted to Sec-tRNA^{Sec} by the enzymes selenocysteine synthase (SelA) in bacteria and O-phosphoseryl-tRNA kinase (PSTK) followed by Sep-tRNA:Sec-tRNA synthase (SepSecS) in eukaryotes and Archaea. Both of these enzymes are dependent on a selenium donor and pyridoxal phosphate (PLP). PSTK also requires ATP

selenol moiety. Selenol has a lower redox potential and a lower pK_a than a thiol group and is ionized and more reactive at physiological pH. Proteins that contain Sec are often enzymes involved in redox reactions and these Sec residues are most often found within the active site [169]. Sec is formed from serine after tRNA charging and before polypeptide insertion. SerRS first aminoacylates tRNA^{Sec} with serine (Ser) and Ser-tRNA^{Sec} is then converted to Sec-tRNA^{Sec} by the enzymes selenocysteine synthase (SelA) in bacteria and O-phosphoseryl-tRNA kinase (PSTK) followed by Sep-tRNA:Sec-tRNA synthase (SepSecS) in eukaryotes and Archaea (Fig. 4) [170, 171]. Similar to tRNA^{Ser}, tRNA^{Sec} species contain particularly long variable arms, a conserved structure of these tRNAs needed for SerRS recognition (reviewed in [172]). However, the structure of tRNA^{Sec} is sufficiently different that aminoacylation of this tRNA is less efficient than that of the cognate tRNA^{Ser} [173, 174]. Incorporation of Sec into the growing peptide occurs at particular stop codons (UGA), which are identified by a nearby *cis* element – a stem-loop structure in the mRNA (bacteria) or a structure in the 3′ untranslated region (archaea and eukaryotes) [175–177]. An additional RNA binding protein is needed to recognize the *cis* element in the RNA to signal the translational machinery for proper encoding of Sec. Unique elongation factors (SelB in bacteria and eEFSec in eukaryotes) replace the function of EF-Tu and deliver Sec-tRNA^{Sec} to the ribosome. The details of these unique mechanisms of ribosomal decoding in archaea and eukaryotes are still under investigation [178, 179]. The fact that the incorporation of selenocysteine into proteins required the development of an alternative route, rather than addition of a new synthetase and a simple change to the existing code, supports the notion that the contemporary genetic code and existing amino acid set are difficult to change as they are to some extent constrained by amino acid metabolism [180].

2.3.4 CysRS

In several methanogenic archaea, including *M. jannaschii* and *M. thermoautotrophicum*, CysRS is not present and the mechanism of Cys-tRNA^{Cys} formation in these organisms was initially unclear [181]. Initial biochemical studies suggested that Cys-tRNA^{Cys} was formed in these organisms by a prolyl-tRNA synthetase

(ProRS) with a dual specificity for both Pro and Cys [182]. However, it was subsequently shown that the absent CysRS is actually replaced by the activity of O-phosphoseryl-tRNA synthetase (SepRS), a newly discovered synthetase which will be discussed below. This synthetase forms Sep-tRNA^{Cys}, which is then converted to Cys-tRNA^{Cys} by Sep-tRNA:Cys-tRNA synthase (SepCysS) [183]. The mechanism for Cys-tRNA^{Cys} synthesis is similar to Sec-tRNA^{Sec} synthesis in archaea. Sep-tRNA is the intermediate in both pathways and serves as a substrate for either SepSecS or SepCysS, and the two enzymes share many similarities [159]. Additionally, it has been proposed from phylogenetic studies that the indirect pathways for Cys-tRNA^{Cys} formation and Sec incorporation in bacteria, Archaea, and eukaryotes were all present at the time of LUCA [171, 184].

3 Non-canonical Aminoacyl-tRNA Synthetases

In addition to the 20 well-characterized canonical aaRSs, there exist several recently discovered enzymes that either fall outside the normal class rules or charge tRNA with amino acids that are not among the 20 encoded by the universal genetic code. Phylogenetic analyses show that these enzymes likely arose early in the evolution of aaRSs and were not retained in most organisms. In most cases they only still appear in small groups of archaeons and dispersed bacteria [144].

3.1 LysRS I

The only known synthetase to date that breaks the class rule and contains enzymes in both class I and class II is LysRS. Class I LysRS (LysRS1) was discovered relatively recently [19] and is found mostly in Archaea, a few dispersed bacteria, and no eukaryotes. Class II LysRS (LysRS2), however, is found in eukaryotes and most bacteria. Most organisms contain one class of LysRS or the other, with the exception of the archaeal group, *Methanosarcinaceae*, and a few isolated species in other genera, such as *Nitrosococcus oceani* [185] and *Bacillus cereus* [186, 187] where both LysRS genes are present [188]. The existence of the same aaRS in two distinct structural classes provides an example of convergent evolution in which divergent mechanisms achieve the same functional goal. In this case, the end result of each enzyme's emergence is the formation of lysylated tRNA^{Lys}. LysRS2 likely existed prior to LysRS1 as it demonstrates deep evolutionary connections to AspRS and AsnRS based on sequence and phylogenetic analyses [11]. Similar phylogenetic associations are lacking between LysRS1 and any other extant synthetase. LysRS1 emerged relatively early in the archaeal lineage and horizontal gene transfer to a few bacteria appears to have come from a pyrococcal progenitor [189]. LysRS1 enzymes from different domains are not deeply rooted and do not group together, whereas other enzymes that are present in both archaea and bacteria do. This LysRS1 distribution pattern is consistent with recent horizontal gene transfer events that possibly occurred more than once [11, 36]. There is a robust correlation between the phylogeny of class I LysRS sequences and the distribution of AsnRS, which may reflect competition for overlapping anticodon sequences during tRNA recognition [190].

Sequence and structural comparisons indicate some distant relationship between LysRS1 and the class I synthetases CysRS, ArgRS, GluRS, and GlnRS (see above) and, similar to three of these synthetases, LysRS1 requires binding of tRNA for formation of the aminoacyl-adenylate [20, 34–36]. Structural and functional data suggest tRNA^{Lys} anticodon recognition by LysRS1 requires fewer interactions than by LysRS2, supporting a less significant role of the anticodon in tRNA recognition by the class I enzyme [191]. LysRS1 has an alpha-helix cage anticodon binding domain, which is similar only to GluRS, suggesting tRNA^{Lys} anticodon specificity may have evolved from the analogous domain of an ancestral GluRS enzyme [191]. In addition to differences in tRNA recognition, LysRS1 and LysRS2 also show divergent resistance to near-cognate amino acids, which may have also impacted the retention of a particular form of the enzyme in different lineages. Lysine recognition differs between the two enzymes and specificity is greater in the LysRS1 active site compared to that of LysRS2, which is more catalytically efficient [192–194]. The need for either strong active site discrimination or efficient catalysis likely depends on the organism and the environment in which it lives, leading to variations in the pressure to retain a particular form of LysRS encoded in a genome.

3.2 Pyrolysyl-tRNA Synthetase (PylRS)

Although natural proteins contain more than 140 different amino acids, the majority of these are the result of posttranslational modifications that occur after protein synthesis [195]. There are only two known additions to the standard 20 amino acid set that are decoded during protein synthesis. These two non-canonical amino acids are selenocysteine (Sect. 2.3.3) and pyrrolysine which, unlike selenocysteine, exists as a free metabolite that requires a unique aaRS to charge it directly onto tRNA. Pyl is encoded in proteins often needed for methylamine utilization and was first identified in a group of archaeal methanogens [196]. More than 20 Pyl-decoding organisms have been identified with roughly half of these being archaeal methanogens of the Methanosarcina family and the rest diverse species of bacteria including *Acetohalobium arabaticum*, *Desulfitobacterium hafniense*, *Desulfitobacterium dehalogenans*, and a symbiontic δ-proteobacterium bacteria of the worm *Olavius algarvensis*, [197, 198].

The mechanism of pyrrolysine insertion into proteins was initially not clear, and thought to require a modification of Lys-tRNA^{Pyl}, which can be formed by the class II LysRS [199]. Additional in vitro studies showed that tRNA^{Pyl} is efficiently aminoacylated with Lys in the presence of both class I and class II LysRSs of

Methanosarcina barkeri [200]. More recently, however, the use of in vitro synthesized pyrrolysine demonstrated that the dedicated tRNA synthetase, pyrrolysyl-tRNA synthetase (PylS), is responsible for charging tRNA^{Pyl} and is unable to use lysine as a substrate [201, 202]. The formation of Pyl-tRNA^{Pyl} and the production of Pyl-containing proteins have been investigated for a handful of pyrrolysine-encoding organisms and this amino acid is found to be inserted into certain proteins at specific UAG stop codons. Although Pyl-tRNA^{Pyl} is recognized by EF-Tu without the help of *trans*-acting factors, a downstream pyrrolysine insertion sequence (PYLIS) promotes incorporation of pyrrolysine over translation termination [201, 203]. Therefore pyrrolysine and selenocysteine insertion are similar in that both require *cis* elements for ribosomal decoding.

The carboxy-terminus of PylRS resembles a typical class II catalytic domain; the amino-terminal domain, however, looks somewhat different compared to other canonical synthetases and is responsible for tRNA^{Pyl} recognition [204]. The genes encoding the carboxy- and amino-termini of PylRS are separated by two genes in the bacterium *D. hafniense*, which differs from the archaeal PylRS-encoding gene arrangement [205]. The amino-terminus of archeal PylRS is dispensable in vitro but required in vivo [206]. The *D. hafniense* PylRS structure demonstrates how the tRNA binding surface is well conserved between all PylRSs and results in an aaRS–tRNA interaction surface that is distinct from those observed in other known aaRS–tRNA complexes [146]. This is thought to be due to the early emergence of PylRS which led to the evolution of unique structural features in both the protein itself and tRNA^{Pyl}. Based on the Archaeal *M. mazei* structure and phylogenetic analysis, PylRS is considered to be a class IIc aaRS along with GlyRS, PheRS, and AlaRS. With the exception of GlyRS, all the synthetases in this subclass share a homologous quaternary architecture; thus it is possible that Pyl exists as a tetramer as well. Although the structural results show a dimeric PylRS bound to two tRNA^{Pyl} molecules, modeling of a potential PylRS tetramer shows conserved residues along the interface of the tetramer, suggesting this is the correct oligomerization state [144]. These structural studies were also successful in deducing the amino acid binding pocket of PylRS, which contains a deep hydrophobic pocket for Pyl binding. The specificity elements of PylRS for its substrates are residue side chains that extend into the amino acid binding pocket. This mode of recognition enables the development of aaRSs that can aminoacylate novel amino acids and arise either by evolution, as with Pyl, or by enzyme design experiments [144, 146].

Although PylRS is an uncommon synthetase with a distribution limited to a small subset of organisms, phylogenetic analyses link its emergence with other class II aaRSs prior to the LUCA [13]. Because the insertion of Pyl into proteins is seen for only a small number of disperse species, it is predicted that the Pyl encoding operon was likely acquired by ancient horizontal gene transfer events between now extinct groups that had a greater use for this amino acid [13, 207, 208]. These gene transfer events were then followed by limited retention of the Pyl encoding operon in extant organisms. Interestingly Pyl is synthesized solely from Lys, connecting amino acid metabolism and synthetase evolution [209]. Pyl insertion at UAG codons is regulated differently in archaea versus some of the bacterial

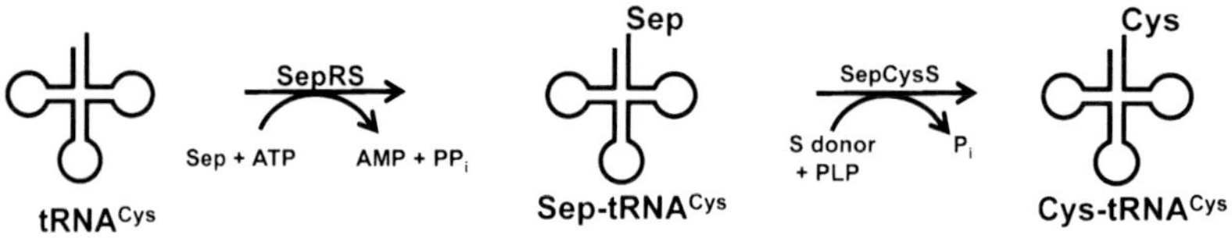

Fig. 5 Indirect Cys-tRNA^{Cys} formation. SepRS first aminoacylates tRNA^{Cys} with O-phosphoserine (Sep) and then Sep-tRNA^{Cys} is converted to Cys-tRNA^{Cys} with Sep-tRNA:Cys-tRNA synthase (SepCysS) in the presence of a sulfur donor and PLP

examples looked at thus far. Pyl-decoding archaea constitutively encode Pyl and have adapted to this by having fewer TAG codons in their genes, whereas bacteria that use Pyl, such as *Acetohalobium arabaticum*, regulate Pyl encoding at the level of transcription of the Pyl operon under particular growth conditions [198].

3.3 Phosphoseryl-tRNA Synthetase (SepRS)

In organisms that lack CysRS, another, non-canonical synthetase has been found responsible for indirect aminoacylation of Cys-tRNA^{Cys}. Initially it was unclear how Cys-tRNA^{Cys} was formed in CysRS lacking organisms and a dual specific ProRS was thought to be responsible (Sect. 2.3.4). Since then, a non-canonical class II aaRS, O-phosphoseryl-tRNA synthetase (SepRS), was found in most methanogenic archaea and is responsible for charging tRNA^{Cys} with o-phosphoserine (Fig. 5). The o-phosphoseryl-tRNA^{Cys} intermediate is then further modified by Cys-tRNA synthase (SepCysS) [183].

The SepRS/SepCysS genes are only found in archaeal genomes that contain the methanogenesis genes for generating or oxidizing methane [184]. This linkage suggests a strong evolutionary connection between the indirect Cys-tRNA^{Cys} pathway and methanogenesis. This tRNA dependent indirect pathway is also the sole method of free cysteine biosynthesis in some organisms, and, in the case of *Methanosarcina mazei*, *cysE*, one of the bacterial genes for cysteine biosynthesis, was apparently lost while the more ancient SepRS/SepCysS system was retained [184, 210]. In a few organisms, namely several *Methanosarcina* species, genes for both the traditional class I CysRS and the indirect SepRS/SepCysS tRNA-charging pathway exist [184]. It appears that both pathways have physiological significance and depend on differing selectivity of various tRNA^{Cys} isoacceptors with the help of particular tRNA modifications. However, the exact role of the redundancy in Cys-tRNA^{Cys} formation in some organisms remains unclear, but is likely closely linked to sulfur and energy metabolism in methanogens [210]. As more is revealed about these unique aminoacylation systems, evolutionary links between protein synthesis, amino acid synthesis and cellular metabolism will become more apparent.

SepRS has a very ancient lineage, stemming at least as far back as the origin of the archaeal branch. Phylogenetic analysis indicates that, although both PylRS and

SepRS evolved much before LUCA, these enzymes were only retained in a handful of organisms, demonstrating the unique metabolic requirements for these amino acids. Alternatively, the emergence of GlnRS and AsnRS in some organisms occurred post-LUCA [11], replacing the more primitive indirect charging pathway for Asn and Gln, which are required in the proteomes of all organisms. Interestingly, both PylRS and SepRS are classified as class IIc synthetases and appear to be distant homologs of PheRS. Phylogenetic evidence shows SepRS evolved from α-PheRS, while PylRS evolved much earlier, before the differentiation of PheRS into a heterotetramer, and likely evolved from an ancestral PheRS as a result of gene duplication [144, 184].

Phylogenetic evidence suggests synthetases evolved after the genetic code was established [11, 14] and therefore, not surprisingly, Sep and Pyl, which are not encoded directly by the code, emerged from an earlier evolved synthetase and required some flexibility of the existing code rather than expansions of the genetic code itself. The discovery of these additional aaRSs and tRNA charging pathways suggests that with further knowledge of uncharted organisms, in terms of sequence and proteome composition, other unidentified aaRSs might exist. Such discoveries could expand the genetic code beyond the current 22 amino acids or uncover new pathways for tRNA aminoacylation. The roles of pyrrolysine in proteins required for methanogen growth on methylamines and selenocysteine in enzymes requiring strong redox capacities indicate the evolutionary selective pressures that underlie the retention of these non-canonical amino acids. Although Crick's adaptor hypothesis [211] is satisfied partially by the discovery of 20 distinctive aaRSs, his later proposed "frozen accident" theory [212] is not. This theory states that the genetic code is "frozen" and that any changes to it would be strongly selected against, if not lethal. The non-canonical examples discussed here demonstrate the code is not "frozen" because these amino acids emerged after establishment of the genetic code and in certain organisms the capacity for coding and incorporation of these amino acids has been lost.

4 Functional Evolution of Synthetases

The early process of deciphering the genetic code for protein synthesis was almost certainly more ambiguous than in extant organisms, likely involving incorporation of a particular "type" of amino acid at codons [213, 214]. Therefore some of the earliest proteins may have been defined more by the general properties of their chemical makeup rather than by the presence of specific chemical groups at specific locations. Aminoacylation accuracy by modern synthetases is challenged by the similarity between many of the substrates used by each of the 23 known aaRS enzymes. The structural and chemical similarities between tRNAs present challenges for accurate recognition of the correct isoacceptors, but even more problematic is the high similarity of some amino acids, some of which can vary by a single methyl group as is the case for isoleucine and valine. Despite these similarities, most aaRSs have a misacylation rate of less than 1 in 5,000 [215]. Aminoacyl-tRNA synthetases have adapted to maximize substrate specificity in order to maintain or control fidelity during protein synthesis. Two different mechanisms are responsible for ensuring highly accurate substrate recognition in different aaRSs. The first depends on the high specificity of the particular enzyme for its amino acid and tRNA substrates. The second way aaRSs can achieve higher substrate specificity, and therefore greater accuracy for protein synthesis, is through editing non-cognate amino acids.

4.1 Specificity of Synthetases

Faithful translation at the level of aaRSs starts with proper identification and pairing of particular tRNAs with their cognate amino acid. Synthetase specificity, or how each enzyme selects the correct tRNA isoacceptors for its cognate amino acid, is often referred to as the second genetic code [216]. Specificity of the aminoacylation reaction is largely dependent on proper recognition of the cognate amino acid and proper tRNAs from the large cellular pool of metabolites and isoacceptors, respectively. Aminoacyl-tRNA specificities can vary between synthetase variants and in some cases appear to have evolved in order to adapt to the particular environments of organisms or the properties of individual cellular compartments [217–219]. Progress in genetic, structural, and biochemical studies has helped shape the underlying principles behind tRNA recognition and amino acid selection of aaRSs, and has provided insight into how these enzymes have adapted to specific evolutionary forces [214, 220, 221].

4.1.1 tRNA Recognition

The primary force for tRNA–aaRS binding is displacement of bound water molecules by the phosphate backbone of the tRNA, and therefore the initial binding event is somewhat non-specific. Structural modeling studies of PheRS, ThrRS, and IleRS showed how electrostatic interactions contribute to the first stages of tRNA binding [222]. It was determined that positive patches on these aaRSs, formed by non-conserved interaction residues, and supplementary domains are most important for determining the long-range potential of the enzyme. These regions are unrelated to the conserved catalytic motifs of aaRSs and determine the ability to attract the tRNA molecule from a distance and direct it to its binding site.

After long distance interactions are made between an aaRS and tRNA, more specific recognition at short distances occur and rely strongly on the conserved catalytic modules. Short distance binding and recognition are also established by similar structural determinants of tRNAs. Differential binding affinity is not sufficient to ensure the correct recognition of the cognate tRNA and therefore kinetic discrimination is used to overcome these limitations and help the aaRS distinguish between cognate and non-cognate tRNAs. Aminoacylation of the correct tRNA is influenced more by k_{cat} effects than by K_m effects [223]. Through various structural and pre-steady state kinetic studies of several tRNA–aaRS pairs, a general model of tRNA binding and recognition has been elucidated [224–226]. The first stage of tRNA binding is fast and the thermodynamic stability of this initial complex depends on interactions with the anticodon or variable arm. These close interactions are followed by a slow conformational change and accommodation step that occurs only when the cognate tRNA is bound. Interactions with the acceptor stem of the cognate tRNA are important for this accommodation and facilitating an efficient rate of aminoacylation and transfer. The precise details of tRNA binding likely vary somewhat even within each class of aaRS as tRNA binding determinants and structural motifs vary between different tRNA–aaRS pairs.

Transfer RNA identity elements are necessary for proper recognition of a particular group of isoacceptors by an aaRS. Some of these elements are positive determinants that promote binding of the cognate tRNA and some are negative (anti-determinants) that prevent acceptance of a non-cognate tRNA [221]. For both class I and class II isoacceptors recognition elements are located on the periphery of the tRNA, in the acceptor arm, and in most cases the anticodon stem loop. Major discrimination occurs at N73, distal base pairs of the acceptor arm, and base 35 of the anticodon stem loop. The anticodon region is not essential for aaRS–tRNA recognition in the three *Escherichia coli* tRNAs specific for Leu, Ser, and Ala. In the case of tRNA^{Ser}, the anticodon nucleotides are different in the six isoacceptors and the acceptor stem, D-loop, and long variable arm unique to these isoacceptors are needed for recognition [227]. Anti-determinants are often modified bases; however, in the cases of Glu, Ile, and Lys modified bases in the anticodon loop are used as positive elements [221]. Anti-determinants of tRNAs from one class of aaRSs tend to be against binding by members of the other aaRS class [227]. Additionally, organisms that lack a particular aaRS will have tRNAs with positive and negative identity elements driven by this absence.

Other minor elements, located throughout the tRNA and in its core region, are more specific to each synthetase system and domain of life. Identity elements found in the core region of most tRNAs tend to be specific and contribute to architectural differences in the tRNA. Specificity elements found in the variable loop, TψC arm, and the D stem often contribute indirectly to binding by providing the necessary tertiary interactions for proper tRNA structure and folding. In the case of the initiator tRNA^{Met} in yeast, changes in the elbow of the tRNA (A20 and A60) result in loss of methylation, while aaRS binding is retained, demonstrating how elements in this region can differentially affect tRNA structure and function [228].

Once folded into their L-shaped structure, tRNAs are basically comprised of two distinct domains, one being the acceptor helix stacked with the TψC arm and the other made up of the anticodon stem aligning with the D stem. These two domains interact with separate regions of the aaRS and are thought to have emerged independently from each other [229]. There are a few unique cases of mitochondrial tRNAs which lack the TψC and D arms and these tRNAs can only be charged by the corresponding mitochondrial aaRSs [230]. The aaRS active site domain, which defines the classification of a particular synthetase by its sequence and structure, interacts with the acceptor helix-TψC arm domain of the tRNA. Interactions made with the tRNA in this region vary depending on the synthetase class and unique features of different enzyme subgroups.

Synthetase interactions made with the anticodon-D-loop domain of the tRNA are carried out through additional enzyme regions that are separate from the "class-defining" catalytic core. These anticodon binding domains of aaRSs are much less conserved and can vary significantly within each class. As shown in the cases of GluRS, GlnRS, and AspRS [231, 232], binding to the anticodon results in large conformational changes in the tRNA, which then transmit changes to the active site. The two separate domains of tRNAs likely evolved separately as did the synthetase domains that recognize them. Modern aaRSs and tRNAs likely arose from ancestors with a simpler mode of tRNA–aaRS recognition solely involving the tRNA acceptor stem and aaRS class-defining catalytic domain [233]. The demonstration that minimalist tRNAs, or minihelices, are aaRS substrates supports this theory and such experiments provide insights into elements that were important for recognition prior to the emergence of larger contemporary tRNAs. Class II synthetases are thought to have appeared first in evolution as these enzymes are best able to aminoacylate minimalist tRNAs and, as mentioned above, some aaRSs of this class completely lack tRNA anticodon recognition elements [221].

4.1.2 Amino Acid Specificity

Amino acid recognition by synthetases takes place in the catalytic site prior to activation and formation of the aminoacyl-adenylate. The mode of amino acid binding varies between different classes of synthetases. Analysis of the CysRS crystal structure and those of other class I synthetases indicate that amino acid binding occurs when the conserved KMSKS motif is in an "open" confirmation. This binding occurs prior to ATP binding and adenylate formation, at which point the loop closes [29, 33, 48, 61, 63, 75, 231, 234]. Class II tRNA synthetases have evolved to discriminate among their amino acid substrates primarily by altering the amino acid side chains in the binding pocket as opposed to changing the position of protein backbone or secondary structure elements [144]. In addition, the size of the amino acid binding pocket may be important, as in the PheRS synthetic site where a conserved Ala residue helps determine specificity of phenylalanine over tyrosine [235]. Interestingly some PheRS variants, such as cytoplasmic PheRS in yeast and humans, contain a glycine at this position, resulting in significantly lower cognate amino acid specificity [217]. ThrRS contains a zinc ion in its active site that contributes to amino acid specificity by recognizing the hydroxyl at the β position of threonine and discriminating against alanine and serine [236]. Although modern synthetases have evolved differentiated structures for proper substrate recognition, the active site architectures of some of these enzymes are unable to distinguish between very similar amino acids with high enough stringency to ensure accurate translation. In these cases editing or proofreading mechanisms are found in many synthetases to aid in the elimination or hydrolysis of misactivated amino acids (Sect. 4.2.2).

4.2 Adapted and Changing Domains

AaRSs are thought to have evolved additional modules to help maintain accurate protein synthesis as the genetic code increased to include more amino acids and the number of isoacceptors increased. Such adapted domains include sites of post-transfer editing, and RNA recognition domains needed for tRNA anti-codon binding and structural stabilization. In addition to the core catalytic and various adapted domains, several aaRS modules evolved into free standing proteins either with synthetase-like functions, such as trans-editing domains, or with other roles in the cell (Sect. 5).

4.2.1 RNA Recognition

As mentioned above, domains outside the synthetase catalytic core can be involved in tRNA recognition. Such domains evolved much later and can vary significantly between synthetase enzymes. Many aaRSs contain additional tRNA recognition domains outside of the region of the catalytic site. For example class IIb aaRSs contain a conserved, lysine rich N-terminal anticodon binding domain (ABD). Structural and biochemical data for yeast AspRS illustrate how the N-terminus of this synthetase participates in tRNA binding, as the presence of this extension considerably increases the stability of the complex between AspRS and its homologous tRNA [42]. Aside from providing stability to the aaRS–tRNA complex for aminoacylation directly, these additional RNA binding domains can be used to provide tRNA stability and even sometimes to facilitate transport. For example, cytoplasmic LysRS in humans is selectively packaged along with the $tRNA^{Lys}$ isoacceptors to help transport the $tRNA^{Lys}$ replication primer into the HIV-1 viron [237]. The viral Gag polyprotein is required for this packaging event. Human LysRS also binds a portion of the HIV genome that contains a $tRNA^{Lys}$ anticodon-like element possibly to release LysRS from $tRNA^{Lys}$, enabling this RNA to anneal to viral RNA for priming [238]. A second example of a trafficking role involves human TyrRS, where the nuclear localization signal is located in the same region of the protein needed for tRNA binding, thereby regulating TyrRS localization to the nucleus [239]. More generally, nuclear pools of synthetases in eukaryotes are predicted to serve as "proofreaders" for properly processed, functional tRNAs before these tRNAs are exported into the cytoplasm for their use in translation [240].

4.2.2 aaRS Editing

In order to maintain faithful translation, particularly in the case of similar amino acids where only so much specificity can be achieved by substrate discrimination, aaRSs have adapted methods to proofread or "edit" misacylated or incorrectly paired amino acid/tRNA pairs selectively. Editing activities can be found in approximately half of the aaRSs and both structural and biochemical studies have helped advance our understanding of how editing processes work in different aaRSs. The catalysis of aminoacylation by synthetases is a highly conserved mechanism; however, the editing mechanisms performed by these enzymes is much more variable. The high degree of diversity in proofreading further exemplifies the long evolutionary pathway of these enzymes as well as the role convergent evolution has played in their emergence. Both pre- and post-transfer editing mechanisms by aaRSs exist and are defined by the substrate. Pre-transfer editing targets the misactivated aminoacyladenylate and occurs within the active site of the aaRS itself. Post-transfer editing involves clearance of misacylated tRNAs and occurs in appended enzymatic domains that emerged later in evolution [241].

The presence of separate catalytic and editing sites in one enzyme, as predicted based on biochemical evidence [242], was first supported by structural studies of IleRS [29, 49]. Since then, editing by dedicated post-transfer editing CP1 domains in class I IleRS, ValRS, and LeuRS have been well characterized, in addition to many other editing systems. IleRS, ValRS, and LeuRS have a high degree of conservation in their CP1 domains that suggests early emergences and selective pressure to maintain editing in these enzymes. It has recently been shown that the rebinding and trans editing of a released misacylated tRNA is a possible post-transfer editing mechanism for these class I aaRSs; however, the relative importance of this pathway is not known [243]. A trans editing model has also been shown for the class II PheRS where the post-transfer hydrolysis of a misacylated tRNA occurs after rebinding and is thought to be a significant editing pathway [244].

Kinetic studies show LeuRS and ValRS mainly rely on post-transfer editing to prevent misincorporation of non-cognate amino acids [245, 246]. IleRS, however, also uses a distinct tRNA-dependent pre-transfer editing activity in its synthetic site [49, 247]. In the case of some LeuRS enzymes, there is less robust post-transfer editing activity against particular amino acids. Yeast cytoplasmic LeuRS is able to clear misacylated Ile-tRNA^{Leu} efficiently; however, the enzyme's post-transfer hydrolytic activity against Met-tRNA^{Leu} is much weaker [248]. It was hypothesized that yeast cytoplasmic LeuRS can shift between pre- and post-editing pathways depending on the identity of the non-cognate amino acid. Human cytoplasmic LeuRS also shows modular pathways for editing different non-cognate amino acids. Norvaline, for example, is predominantly cleared by post-transfer editing while α-amino butyrate is the target of the pre-transfer mechanism [249]. Interestingly, when the yeast mitochondrial CP1 domain from LeuRS was isolated from the full-length enzyme it was unable to hydrolyze misacylated Ile-tRNA^{Leu}, which is in contrast to the isolated *E. coli* LeuRS CP1 domain [250]. This isolated yeast CP1

Emergence and Evolution

domain still retained its intron splicing activity (Sect. 5.1), suggesting that this LeuRS has functionally diverged to have a robust splicing activity, which has come at some expense to aaRS functionality in aminoacylation and editing [250]. The only other class I synthetase with editing activity is the related MetRS. Homocysteine, an intermediate of methionine biosynthesis, is activated by MetRS and subsequently edited prior to transfer to the tRNA. Unusually, this proofreading occurs within the active site of the enzyme and involves cyclization of the adenylate to form homocysteine thiolactone and AMP [251–253].

Separate, adapted post-transfer editing domains are found in class II PheRS, ThrRS, AlaRS, and ProRS and pre-transfer editing activity has been demonstrated in the active sites of ProRS, SerRS, and LysRS II [241]. Class II aaRS post-transfer editing domains are much less conserved than those in the class I enzymes. This variability coincides with the trend of class II synthetases, which tend to share less conservation between different aaRSs. Some of the class II synthetases, where product release is rapid and not rate limiting, also have homologous free standing *trans* acting editing domains, a phenomenon that to date has not been described for class I enzymes. PheRS is among the least well conserved class II aaRSs, and exists in various forms in different domains and cellular compartments [217]. PheRS post-transfer editing takes place in the β subunit of the enzyme 40 Å away from the site of aminoacylation and is responsible for clearing misacylated Tyr-tRNA^{Phe} [235, 254, 255]. Structure based alignments of the PheRS editing domain show considerable divergence as many archaeal/eukaryal PheRSs lack conservation of the critical residues found in bacterial PheRS [256]. Mitochondrial PheRSs exist as a monomer from which the β subunit and post-transfer editing are completely absent [46].

Post-transfer editing in ThrRS is necessary to hydrolyze mischarged Ser-tRNA^{Thr} and takes place in an adapted N2 domain of the N terminus, which shares homology to the same region of AlaRS. Mitochondrial ThrRS lacks the N2 domain and archaeal ThrRSs often contain an unrelated N-terminal domain, and in some cases the editing domain acts in *trans* [257, 258]. In-depth structural analyses of bacterial ThrRS have elucidated the post-transfer editing mechanism and show two water molecules to be involved in the hydrolysis reaction, one of which is excluded when Thr is in the editing site vs Ser [259, 260]. This editing mechanism is based on more than how well the misacylated substrate "fits" into the editing active site and is thought possibly to be similar in other post-transfer editing sites such as those of PheRS and LeuRS [260]. Also, the freestanding protein ThrXp in Archaea is homologous to the editing domain of ThrRS and is able to clear Ser-tRNA^{Thr} in vitro [258]. The editing domains of some ThrRS enzymes from archaea also share sequence and structural homologies with D-Tyr-tRNA^{Tyr} deacylases (DTD) [261–263]. DTDs contain trans editing activity against mischarged D-Tyr-tRNA^{Tyr}, which can be synthesized by TyrRS, and are found across the three domains of life [264, 265]. This activity is essential to cell viability, as D-amino acids could dramatically alter protein folding and function. Interestingly, changing one particular residue in *E. coli* DTD to that found in ThrRS changed the specificity from D-amino acids to L-Ser, supporting the evolutionary linkage between DTDs and ThrRS [260].

Class II AlaRS has a flexible Ala binding pocket and as a result the enzyme has to be able to clear misactivated Gly and Ser [135]. An appended post-transfer editing domain is used to clear both Ser-tRNA^{Ala} and Gly-tRNA^{Ala}, while a trans editing domain, AlaXp is also used to clear the large non-cognate residue, Ser. The appended post-transfer editing domain of AlaRS is thought to have evolved from a primordial AlaXp that later fused to the aminoacylation domain of AlaRS [135]. Interestingly, all three domains of life contain the additional free standing editing domain, AlaXp, which is mainly responsible for clearing mischarged Ser-tRNA^{Ala}. There is strong evolutionary pressure to retain AlaXp in addition to AlaRS editing, as demonstrated in mice where reduced Ser-tRNA^{Ala} editing was linked to protein misfolding in neuronal cells [266].

For ProRS there exist several different mechanisms of synthetase proofreading, many of which include trans editing domains collectively known as ProX enzymes. The insertion domain (INS) is one which exists as an appendage to the core synthetase for most bacteria and is responsible for clearing mischarged Ala-tRNA^{Pro}. Some species, including *Clostridium sticklandii*, lack an INS domain and encode a freestanding domain PrdX that is used to hydrolyze Ala-tRNA^{Pro} [267]. The synthetic site of ProRS is also capable of mischarging Cys-tRNA^{Pro}, which can be cleared by a freestanding editing domain YbaK that is itself homologous to the INS domain [268–270]. Human encoded ProX has recently been shown to deacylate mischarged Ala-tRNA^{Pro}, but not Cys-tRNA^{Pro}, by specifically recognizing the Ala moiety of Ala-tRNA^{Pro} [271]. Additional free-standing ProRS trans editing domain homologs, such as YeaK and PA2301, have also been identified based on sequence similarity to INS and YbaK; however, their function is still not clear [272].

Post-transfer editing domains are not universally conserved, and in several aaRSs these domains have actually been lost. Examples of editing domains that appear to have been lost during evolution include a number of mitochondrial synthetases, such as human mitochondrial ProRS, human mitochondrial LeuRs, human and yeast mitochondrial PheRS, as well as the cytosolic ProRS in higher eukaryotes, and most archaeal and mitochondrial versions of ThrRS. [46, 257–259, 273–276]. Many archaeal ThrRS enzymes have an N-terminal domain that is unrelated to the conserved N2 editing domain and in some cases a trans editing domain is encoded separately from the aminoacylation site [257]. Mycoplasma PheRS and LeuRS are also unable to post-transfer edit effectively as PheRS lacks conserved residues in the β subunit required for editing and LeuRS is missing the CP1 domain altogether [218, 219]. As a result of the error-prone activities of these two synthetases, the *Mycoplasma mobile* proteome naturally contains elevated levels of mistranslation. The evolutionary advantage of these error-prone proteomes is unclear, but it has been proposed that, because many of these organisms are obligate intracellular pathogens, misincorporation of similar amino acids increases antigen diversity without completely losing structural and function integrity of proteins [218]. Understanding the role of translational fidelity is important as editing by synthetases can vary greatly between different domains of life and even within a particular organism [217]. Whether or not reduced quality

Emergence and Evolution

control within a particular organism or cellular compartment is beneficial and what environmental conditions or stresses dictate such benefits or disadvantages, is important to understanding what drives the evolution of synthetase fidelity.

The numerous examples of stable robust freestanding editing domains and their homology to fused synthetase domains strongly suggests that the free-standing variants may have first existed as independent proteins that were later fused with their respective aaRS. This theory is supported by the observation that the editing domains of several synthetases contain a conserved CXXC motif, which is often found in mobile elements that have been incorporated into larger proteins [267]. Interestingly, the trans editing domain YbaK functions most effectively as a stable complex with ProRS, outcompeting EF-Tu, suggesting one possible evolutionary path for the transition from freestanding to fused editing domains [269]. Domain acquisition and movement of editing domains are thought to have occurred more than once during the evolution of some aaRSs [267, 277]. Editing modules in ProRS, for example, have diverse contexts and can include insertions, N-terminal additions, and independent protein forms [267, 278, 279]. The editing domains of AlaRS and ThrRS are similar in sequence yet are located in different regions of the synthetase and could possibly have been acquired at different points during evolution. The lack of structural conservation within the ThrRS editing domains suggests that they were possibly acquired from more than one ancestor where they had emerged early in evolution, or the archaeal ThrRS editing site may have evolved rapidly after the divide of the eukaryote and archaeal lineages [257, 280].

5 Emergence of Non-canonical Functions in Aminoacyl-tRNA Synthetases

5.1 Fused Domains Having Non-canonical Functions

In addition to adapted aaRS domains involved in tRNA aminoacylation, several examples of fused aaRS domains used for non-canonical functions have emerged throughout evolution. There are many N- or C-terminal domains found only in eukaryotic synthetases that are not needed for aminoacylation activity. For example, in archaea and higher eukaryotes fused terminal domains are thought to play a role in forming the multisynthetase complex (MSC). The MSC is a complex of several synthetases and auxiliary proteins including eukaryotic initiation factor 1-α (eIF1-α) that is hypothesized to have a role in promoting synthetase activity and channeling translation components to the ribosome [281, 282]. Other fused aaRS domains, particularly prevalent in eukaryotes, are often involved in cell signaling pathways [283]. One example of a fused aaRS domain used for signaling is the WHEP domain found in chordate TrpRS, which regulates the angiostatic signaling activity of this synthetase [284, 285]. TyrRS in higher eukaryotes contains fused ELR and EMAPII domains that are both used for angiogenesis related signaling. The EMAPII domain blocks the signaling function of the ELR domain, and upon cleavage of the ELR domain post secretion the EMAPII is accessible for signaling. The remaining N-terminal fragment containing the aminoacylation domain (mini-TyrRS) is also able to promote leukocyte migration [286]. Another recent example of how synthetases play a role in signaling involves LeuRS and its key role in linking cellular amino acid levels to the TORC1 response pathway. In both human and yeast cells, leucine-bound LeuRS was found to interact with system specific GTPases through the conserved CP1 editing domain. This interaction in turn promotes lysosomal recruitment and activation of TORC1, which is responsible for regulating protein synthesis, ribosome biogenesis, nutrient uptake, and autophagy [287, 288]. Outside of cell signaling, such domains fused to aaRSs can be involved in the regulation of gene expression as in the cases of AlaRS DNA binding and transcriptional regulation of its own gene, PheRS regulation through transcriptional attenuation, and ThrRS for regulation of translation [289–291]. In *E. coli*, ThrRS is able to bind the leader region of *thrS* mRNA and prevent binding of the 30S ribosomal subunit [292]. Hairpin recognition of the mRNA is similar to anticodon stem loop recognition by ThrRS. Gene regulation by fused domains of aaRSs also occurs in eukaryotes. For example, the synthesis of ribosomal RNA in humans appears to be regulated by the C-terminus of MetRS, which is needed for nucleolar localization and possible nucleic acid interactions [286].

AaRSs have also been found to have functions in tRNA and mRNA transport and processing. In yeast, the CP1 domain of mitochondrial LeuRS is involved in Group I intron splicing and the C-termini of mitochondrial TyrRSs from *S. cerevisiae* and *Neurospora crassa* have been implicated in rRNA splicing [250, 286, 293, 294]. Additionally, TyrRS in yeast appears to be important for tRNA export from the nucleus [240] and LysRS and AlaRS in certain eukaryotes are needed for mitochondrial import of tRNA [286]. Lastly, in most actinomycetes LysRS is fused directly to a multiple peptide resistance virulence factor, which uses specific aminoacylated tRNAs as substrates to aminoacylate and alter the properties and composition of membrane lipids [295–297].

5.2 Paralogs of Synthetases

As more sequence data is becoming available for organisms from all three domains of life it is becoming evident that paralogs to aaRSs are encoded in most genomes, and genetic and biochemical studies have just begun to unravel the function of some of the corresponding gene products. Interestingly there are numerous examples of these paralogs that do not aminoacylate tRNA but are rather used for cellular activities outside of protein synthesis. These examples of free standing enzymes with new functionalities take advantage of many structural and sequence characteristics of aaRSs and further demonstrate the evolutionary importance of the enzymes.

Emergence and Evolution   69

5.2.1 Class I Paralogs

There are a number of class I aaRS paralogs that have been found to function as isoenzymes, or duplications of an enzyme with a different amino acid sequence that still performs the same chemical reaction. For example, two different forms of a synthetase used in the cytoplasm and mitochondria are considered isoenzymes. Synthetase isoenzymes that have evolved new functions have the same amino acid and tRNA specificity, but commonly have a slightly different active site relative to the canonical aaRS. These differences in specificity can be exploited by host organisms to provide resistance to natural inhibitors, as perhaps best exemplified by the IleRS isoenzyme that confers mupirocin resistance to a number of drug-resistant isolates of bacterial pathogens such as *Staphylococcus aureus* [298].

A number of aaRS paralogs have been found to function as peptide synthetases, transferring activated amino acids to carrier proteins in non-ribosomal peptide synthesis. AlbC is an example of a class I paralog with high similarity to an aaRS catalytic domain, in this case TrpRS, that functions as a cyclodipeptide synthase, transferring Phe from Phe-tRNA^{Phe} to an activated serine residue [299]. MshC is a CysRS paralog that catalyzes Cys attachment to the amino group of a mycothiol precursor (not the hydroxyl like aaRSs) [300]. CPDSs are often derived from the class I catalytic domain, demonstrating the high amount of divergence that has occurred in these enzymes. There has also been a series of truncated class II SerRS homologs identified that activate and transfer amino acids to carrier proteins (Sect. 5.2.2) [301].

GluX (yadB in *E. coli*) is a truncated form of class I GluRS that lacks the entire C-terminal anticodon-binding domain. Early studies showed yadB was not essential in *E. coli* and it was thought to be a pseudogene without a known function. [37, 302]. Subsequently, yadB was shown to have a conserved prokaryotic function in posttranscriptionally modifying tRNA^{Asp} on the modified nucleoside queuosine, which is inserted at the wobble position of the anticodon-loop and has been renamed glutamyl-Q-tRNA^{Asp} synthetase [303]. This GluRS catalytic paralog has unusual tRNA binding that goes against the classic idea of the aaRS catalytic site–tRNA acceptor stem interaction. Structural mimicry between the anticodon-stem and loop of tRNA^{Asp} and the amino acid acceptor-stem of tRNA^{Glu} partly explains this unusual and unexpected mode of RNA binding.

5.2.2 Class II Paralogs

Paralogs of class II synthetases perform a wide variety of cellular roles outside of tRNA aminoacylation. For example, PoxA is a class II LysRS paralog found in many bacteria that posttranslationally adds β-lysine to a conserved lysine residue on translation elongation factor P [304]. Other notable examples include two paralogs of HisRS, HisZ, and GCN2, that have very different functionalities. HisZ, which was first identified in *Lactococcus lactis*, is homologous to class II HisRS proteins and is required for the first step in histidine biosynthesis [305]. As an essential subunit of HisG, an ATP phosphoribosyltransferase, HisZ catalyzes the transfer of ATP to 5-phosphoribosyl 1-pyrophosphate (PRPP) producing the substrate for nine additional steps in the histidine biosynthesis pathway. This enzyme provides another evolutionary link between amino acid biosynthesis and the aminoacylation reaction. HisZ also has some non-specific RNA binding activity, but whether this is functionally significant is not known [305]. GCN2 is found in eukaryotes and is used to sense amino acid levels and subsequently regulate translation. GCN2 enzymes contain a Ser-Thr kinase domain and a HisRS-like domain that binds uncharged tRNA and prevents kinase activity in the absence of tRNA binding [306]. HisZ and GNC2 are functionally distinct and products of separate evolutionary events. HisZ is the result of an early gene duplication event in bacteria while GCN2 is the result of a later gene duplication event in eukaryotes [305]. Asparagine synthetase A (AsnA) shows homology to the catalytic domain of Asp/AsnRS and is responsible for catalyzing asparagine synthesis using aspartate, ATP, and ammonia as substrates [119, 307]. Structural and phylogenetic data suggest AsnA evolved from a duplication of the ancestral AspRS gene, leading to the archaeal/eukaryal AspRS and a precursor to AsnRS and AsnA. Upon duplication of this AsnRS precursor gene, one copy evolved Asn activation and tRNA^{Asn} binding activity, while the other copy lost its anticodon-binding domain and evolved a new catalytic site to become AsnA [119]. The biotin protein ligase, BirA in $E.$ $coli$, is a paralog of SerRS [308] at its catalytic site and functions to activate biotin to form biotinyl-5'-adenylate and then catalyze the covalent attachment of this biotin to a subunit of acetyl-CoA carboxylase at a lysine residue [309]. As more structural data became available BirA was also shown to resemble the class II PheRS β subunit [310]. Similarities between the structures of BirA and PheRS were found in a region separate from the catalytic domains of these proteins that resemble Src-homology 3 (SH3)-like DNA binding domains. This region of BirA is responsible for regulating transcription of the biotin operator [311]. Both AsnA and BirA catalyze reactions that involve the formation of an adenylated intermediate, which is not the case for the HisZ enzyme. It was suggested that this absence of adenylation by HisZ and its role in binding and regulating histidine indicate early aaRSs may have been simple amino acid binding proteins [305].

Other SerRS paralogs include SLIMP and homologs that acylate carrier proteins for non-ribosomal protein synthesis. SLIMP is found to be localized in the mitochondria of insects and is needed for proper development in flies [312]. The function of this protein is unknown; however, it shows a general affinity to RNA and may be involved in mRNA processing and/or gene expression. There has also been a series of truncated SerRS homologs identified in a number of bacteria, which are similar in structure to the catalytic region of an atypical SerRS (aSerRS) that is found in methanogenic archaea [313]. These homologs lack tRNA binding and canonical aminoacylation activity, but rather activate and transfer amino acids to a phosphopantetheine prosthetic group on carrier proteins. The functions of these carrier proteins have yet to be identified, but it is thought that they possibly play a role in non-ribosomal protein synthesis [301].

One last example of a class II aaRS paralog with a cellular function outside of tRNA aminoacylation is found in mitochondrial DNA polymerase. The Polγβ subunit, which is responsible for the enzyme's processivity, has a domain that is similar to the catalytic domain of class IIa synthetases. The regions of similarity include the aaRS active site that binds the amino acid, ATP, and the acceptor stem of the tRNA [314]. Polγβ also has a C-terminal domain that is similar to the tRNA anticodon binding domain of the dimeric GlyRS [315]. Despite these similarities, Polγβ has important differences and lacks critical residues necessary for tRNA anticodon binding and therefore does not retain the function of an aaRS. This example does show strong evolutionary links between this polymerase subunit and aaRSs, particularly in their nucleic acid binding properties.

6 Conclusion

Almost 60 years after evidence of aminoacyl synthetase activity first emerged [316, 317], the field is continuing to grow and provide insights into evolution, the fidelity of protein synthesis, and the workings of other biological systems. The discoveries of PylRS and SepRS demonstrate how the genetic code is less rigid than once thought and has been adaptive to changes in environmental demands [183, 201]. The vast increase in aaRS structural information within the last several years has increased our resources for phylogenetic analysis as well as helped explain biochemical mechanisms. Also, the structure of PylRS provides a great example of substrate orthogonality and is paving the way for advances in protein engineering [146]. Lastly, the immense amounts of recent genomic sequencing data have uncovered numerous aaRS accessory domains and paralogs whose functions have connected aaRSs to cellular development and disease and are targets of new therapeutic development [318, 319].

References

1. Illangasekare M, Yarus M (1999) A tiny RNA that catalyzes both aminoacyl-RNA and peptidyl-RNA synthesis. RNA 5(11):1482–1489
2. Lee N, Bessho Y, Wei K, Szostak JW, Suga H (2000) Ribozyme-catalyzed tRNA aminoacylation. Nat Struct Biol 7(1):28–33
3. Kumar RK, Yarus M (2001) RNA-catalyzed amino acid activation. Biochemistry 40(24):6998–7004
4. Francklyn C, Schimmel P (1989) Aminoacylation of RNA minihelices with alanine. Nature 337(6206):478–481
5. Musier-Forsyth K, Schimmel P (1994) Acceptor helix interactions in a class II tRNA synthetase: photoaffinity cross-linking of an RNA miniduplex substrate. Biochemistry 33(3):773–779
6. Schimmel P, Frugier M, Glasfeld E (1997) Peptides for RNA discrimination and for assembly of enzymes that act on RNA. Nucleic Acids Symp Ser (36):1

7. Ibba M, Curnow AW, Soll D (1997) Aminoacyl-tRNA synthesis: divergent routes to a common goal. Trends Biochem Sci 22(2):39–42
8. Ramaswamy K, Wei K, Suga H (2002) Minihelix-loop RNAs: minimal structures for aminoacylation catalysts. Nucleic Acids Res 30(10):2162–2171
9. Tamura K (2011) Ribosome evolution: emergence of peptide synthesis machinery. J Biosci 36(5):921–928
10. Xiao JF, Yu J (2007) A scenario on the stepwise evolution of the genetic code. Genomics Proteomics Bioinformatics 5(3–4):143–151
11. Woese CR, Olsen GJ, Ibba M, Soll D (2000) Aminoacyl-tRNA synthetases, the genetic code, and the evolutionary process. Microbiol Mol Biol Rev 64(1):202–236
12. Nagel GM, Doolittle RF (1995) Phylogenetic analysis of the aminoacyl-tRNA synthetases. J Mol Evol 40(5):487–498
13. Fournier GP, Andam CP, Alm EJ, Gogarten JP (2011) Molecular evolution of aminoacyl tRNA synthetase proteins in the early history of life. Orig Life Evol Biosph 41(6):621–632
14. Hohn MJ, Park HS, O'Donoghue P, Schnitzbauer M, Soll D (2006) Emergence of the universal genetic code imprinted in an RNA record. Proc Natl Acad Sci U S A 103(48):18095–18100
15. Eriani G, Delarue M, Poch O, Gangloff J, Moras D (1990) Partition of tRNA synthetases into two classes based on mutually exclusive sets of sequence motifs. Nature 347(6289):203–206
16. Cusack S (1995) Eleven down and nine to go. Nat Struct Biol 2(10):824–831
17. Arnez JG, Moras D (1997) Structural and functional considerations of the aminoacylation reaction. Trends Biochem Sci 22(6):211–216
18. Ribas de Pouplana L, Schimmel P (2001) Two classes of tRNA synthetases suggested by sterically compatible dockings on tRNA acceptor stem. Cell 104(2):191–193
19. Ibba M, Morgan S, Curnow AW, Pridmore DR, Vothknecht UC, Gardner W, Lin W, Woese CR, Soll D (1997) A euryarchaeal lysyl-tRNA synthetase: resemblance to class I synthetases. Science 278(5340):1119–1122
20. Terada T, Nureki O, Ishitani R, Ambrogelly A, Ibba M, Soll D, Yokoyama S (2002) Functional convergence of two lysyl-tRNA synthetases with unrelated topologies. Nat Struct Biol 9(4):257–262
21. First EA (2005) Catalysis of the tRNA aminoacylation reaction. In: Ibba M, Francklyn C, Cusack S (eds) The aminoacyl-tRNA synthetases. Landes Bioscience, Georgetown, pp 328–352
22. Bedouelle H (2005) Tyrosyl-tRNA synthetases. In: Ibba M, Francklyn C, Cusack S (eds) The aminoacyl-tRNA synthetases. Landes Bioscience, Georgetown
23. Sprinzl M, Cramer F (1975) Site of aminoacylation of tRNAs from Escherichia coli with respect to the 2′- or 3′-hydroxyl group of the terminal adenosine. Proc Natl Acad Sci U S A 72(8):3049–3053
24. Fersht AR, Gangloff J, Dirheimer G (1978) Reaction pathway and rate-determining step in the aminoacylation of tRNAArg catalyzed by the arginyl-tRNA synthetase from yeast. Biochemistry 17(18):3740–3746
25. Zhang CM, Perona JJ, Ryu K, Francklyn C, Hou YM (2006) Distinct kinetic mechanisms of the two classes of aminoacyl-tRNA synthetases. J Mol Biol 361(2):300–311
26. Kaminska M, Shalak V, Mirande M (2001) The appended C-domain of human methionyl-tRNA synthetase has a tRNA-sequestering function. Biochemistry 40(47):14309–14316
27. O'Donoghue P, Luthey-Schulten Z (2003) On the evolution of structure in aminoacyl-tRNA synthetases. Microbiol Mol Biol Rev 67(4):550–573
28. Ibba M, Soll D (2001) The renaissance of aminoacyl-tRNA synthesis. EMBO Rep 2(5):382–387
29. Nureki O, Vassylyev DG, Tateno M, Shimada A, Nakama T, Fukai S, Konno M, Hendrickson TL, Schimmel P, Yokoyama S (1998) Enzyme structure with two catalytic sites for double-sieve selection of substrate. Science 280(5363):578–582

30. Chen JF, Guo NN, Li T, Wang ED, Wang YL (2000) CP1 domain in Escherichia coli leucyl-tRNA synthetase is crucial for its editing function. Biochemistry 39(22):6726–6731
31. Fukai S, Nureki O, Sekine S, Shimada A, Tao J, Vassylyev DG, Yokoyama S (2000) Structural basis for double-sieve discrimination of L-valine from L-isoleucine and L-threonine by the complex of tRNA(Val) and valyl-tRNA synthetase. Cell 103(5):793–803
32. Lincecum TL (2005) Leucyl-tRNA synthetases. In: Ibba M, Francklyn C, Cusack S (eds) The aminoacyl-tRNA synthetases. Landes Bioscience, Georgetown
33. Cusack S, Yaremchuk A, Tukalo M (2000) The 2 A crystal structure of leucyl-tRNA synthetase and its complex with a leucyl-adenylate analogue. EMBO J 19(10):2351–2361
34. Ravel JM, White MN, Shive W (1965) Activation of tyrosine analogs in relation to enzyme repression. Biochem Biophys Res Commun 20(3):352–359
35. Mitra SK, Mehler AH (1967) The arginyl transfer ribonucleic acid synthetase of Escherichia coli. J Biol Chem 242(23):5490–5494
36. Ibba M, Losey HC, Kawarabayasi Y, Kikuchi H, Bunjun S, Soll D (1999) Substrate recognition by class I lysyl-tRNA synthetases: a molecular basis for gene displacement. Proc Natl Acad Sci U S A 96(2):418–423
37. Lamour V, Quevillon S, Diriong S, N'Guyen VC, Lipinski M, Mirande M (1994) Evolution of the Glx-tRNA synthetase family: the glutaminyl enzyme as a case of horizontal gene transfer. Proc Natl Acad Sci U S A 91(18):8670–8674
38. Dubois DY (2005) Glutamyl-tRNA synthetases. In: Ibba M, Francklyn C, Cusack S (eds) The aminoacyl-tRNA synthetases. Landes Bioscience, Georgetown, pp 89–98
39. Ribas de Pouplana L, Frugier M, Quinn CL, Schimmel P (1996) Evidence that two present-day components needed for the genetic code appeared after nucleated cells separated from eubacteria. Proc Natl Acad Sci U S A 93(1):166–170
40. Andam CP, Williams D, Gogarten JP (2010) Biased gene transfer mimics patterns created through shared ancestry. Proc Natl Acad Sci U S A 107(23):10679–10684
41. Brown JR, Robb FT, Weiss R, Doolittle WF (1997) Evidence for the early divergence of tryptophanyl- and tyrosyl-tRNA synthetases. J Mol Evol 45(1):9–16
42. Frugier M, Moulinier L, Giege R (2000) A domain in the N-terminal extension of class IIb eukaryotic aminoacyl-tRNA synthetases is important for tRNA binding. EMBO J 19(10):2371–2380
43. Berthet-Colominas C, Seignovert L, Hartlein M, Grotli M, Cusack S, Leberman R (1998) The crystal structure of asparaginyl-tRNA synthetase from Thermus thermophilus and its complexes with ATP and asparaginyl-adenylate: the mechanism of discrimination between asparagine and aspartic acid. EMBO J 17(10):2947–2960
44. Commans S, Plateau P, Blanquet S, Dardel F (1995) Solution structure of the anticodon-binding domain of Escherichia coli lysyl-tRNA synthetase and studies of its interaction with tRNA(Lys). J Mol Biol 253(1):100–113
45. Delarue M, Moras D (1993) The aminoacyl-tRNA synthetase family: modules at work. Bioessays 15(10):675–687
46. Roy H, Ling J, Alfonzo J, Ibba M (2005) Loss of editing activity during the evolution of mitochondrial phenylalanyl-tRNA synthetase. J Biol Chem 280(46):38186–38192
47. Shiba K (2005) Glycyl-tRNA synthetases. In: Ibba C, Francklyn C, Cusack S (eds) The aminoacyl-tRNA synthetases. Landes Bioscience, Georgetown, pp 125–134
48. Newberry KJ, Hou YM, Perona JJ (2002) Structural origins of amino acid selection without editing by cysteinyl-tRNA synthetase. EMBO J 21(11):2778–2787
49. Silvian LF, Wang J, Steitz TA (1999) Insights into editing from an ile-tRNA synthetase structure with tRNAile and mupirocin. Science 285(5430):1074–1077
50. Nakama T, Nureki O, Yokoyama S (2001) Structural basis for the recognition of isoleucyl-adenylate and an antibiotic, mupirocin, by isoleucyl-tRNA synthetase. J Biol Chem 276 (50):47387–47393

51. Tukalo M, Yaremchuk A, Fukunaga R, Yokoyama S, Cusack S (2005) The crystal structure of leucyl-tRNA synthetase complexed with tRNALeu in the post-transfer-editing conformation. Nat Struct Mol Biol 12(10):923–930
52. Rock FL, Mao W, Yaremchuk A, Tukalo M, Crepin T, Zhou H, Zhang YK, Hernandez V, Akama T, Baker SJ, Plattner JJ, Shapiro L, Martinis SA, Benkovic SJ, Cusack S, Alley MR (2007) An antifungal agent inhibits an aminoacyl-tRNA synthetase by trapping tRNA in the editing site. Science 316(5832):1759–1761
53. Lincecum TL Jr, Tukalo M, Yaremchuk A, Mursinna RS, Williams AM, Sproat BS, Van Den Eynde W, Link A, Van Calenbergh S, Grotli M, Martinis SA, Cusack S (2003) Structural and mechanistic basis of pre- and posttransfer editing by leucyl-tRNA synthetase. Mol Cell 11(4):951–963
54. Palencia A, Crepin T, Vu MT, Lincecum TL Jr, Martinis SA, Cusack S (2012) Structural dynamics of the aminoacylation and proofreading functional cycle of bacterial leucyl-tRNA synthetase. Nat Struct Mol Biol 19(7):677–684
55. Larson ET, Kim JE, Zucker FH, Kelley A, Mueller N, Napuli AJ, Verlinde CL, Fan E, Buckner FS, Van Voorhis WC, Merritt EA, Hol WG (2011) Structure of Leishmania major methionyl-tRNA synthetase in complex with intermediate products methionyladenylate and pyrophosphate. Biochimie 93(3):570–582
56. Nakanishi K, Ogiso Y, Nakama T, Fukai S, Nureki O (2005) Structural basis for anticodon recognition by methionyl-tRNA synthetase. Nat Struct Mol Biol 12(10):931–932
57. Shibata S, Gillespie JR, Ranade RM, Koh CY, Kim JE, Laydbak JU, Zucker FH, Hol WG, Verlinde CL, Buckner FS, Fan E (2012) Urea-based inhibitors of Trypanosoma brucei methionyl-tRNA synthetase: selectivity and in vivo characterization. J Med Chem 55(14):6342–6351
58. Crepin T, Schmitt E, Blanquet S, Mechulam Y (2004) Three-dimensional structure of methionyl-tRNA synthetase from Pyrococcus abyssi. Biochemistry 43(9):2635–2644
59. Ingvarsson H, Unge T (2010) Flexibility and communication within the structure of the Mycobacterium smegmatis methionyl-tRNA synthetase. FEBS J 277(19):3947–3962
60. Crepin T, Schmitt E, Mechulam Y, Sampson PB, Vaughan MD, Honek JF, Blanquet S (2003) Use of analogues of methionine and methionyl adenylate to sample conformational changes during catalysis in Escherichia coli methionyl-tRNA synthetase. J Mol Biol 332(1):59–72
61. Mechulam Y, Schmitt E, Maveyraud L, Zelwer C, Nureki O, Yokoyama S, Konno M, Blanquet S (1999) Crystal structure of Escherichia coli methionyl-tRNA synthetase highlights species-specific features. J Mol Biol 294(5):1287–1297
62. Fukai S, Nureki O, Sekine S, Shimada A, Vassylyev DG, Yokoyama S (2003) Mechanism of molecular interactions for tRNA(Val) recognition by valyl-tRNA synthetase. RNA 9(1):100–111
63. Delagoutte B, Moras D, Cavarelli J (2000) tRNA aminoacylation by arginyl-tRNA synthetase: induced conformations during substrates binding. EMBO J 19(21):5599–5610
64. Konno M, Sumida T, Uchikawa E, Mori Y, Yanagisawa T, Sekine S, Yokoyama S (2009) Modeling of tRNA-assisted mechanism of Arg activation based on a structure of Arg-tRNA synthetase, tRNA, and an ATP analog (ANP). FEBS J 276(17):4763–4779
65. Shimada A, Nureki O, Goto M, Takahashi S, Yokoyama S (2001) Structural and mutational studies of the recognition of the arginine tRNA-specific major identity element, A20, by arginyl-tRNA synthetase. Proc Natl Acad Sci U S A 98(24):13537–13542
66. Cavarelli J, Delagoutte B, Eriani G, Gangloff J, Moras D (1998) L-Arginine recognition by yeast arginyl-tRNA synthetase. EMBO J 17(18):5438–5448
67. Perona JJ, Swanson RN, Rould MA, Steitz TA, Soll D (1989) Structural basis for misaminoacylation by mutant E. coli glutaminyl-tRNA synthetase enzymes. Science 246(4934):1152–1154
68. Rould MA, Perona JJ, Soll D, Steitz TA (1989) Structure of E. coli glutaminyl-tRNA synthetase complexed with tRNA(Gln) and ATP at 2.8 A resolution. Science 246(4934):1135–1142

Emergence and Evolution

69. Deniziak M, Sauter C, Becker HD, Paulus CA, Giege R, Kern D (2007) Deinococcus glutaminyl-tRNA synthetase is a chimer between proteins from an ancient and the modern pathways of aminoacyl-tRNA formation. Nucleic Acids Res 35(5):1421–1431
70. Rath VL, Silvian LF, Beijer B, Sproat BS, Steitz TA (1998) How glutaminyl-tRNA synthetase selects glutamine. Structure 6(4):439–449
71. Bullock TL, Uter N, Nissan TA, Perona JJ (2003) Amino acid discrimination by a class I aminoacyl-tRNA synthetase specified by negative determinants. J Mol Biol 328(2):395–408
72. Rould MA, Perona JJ, Steitz TA (1991) Structural basis of anticodon loop recognition by glutaminyl-tRNA synthetase. Nature 352(6332):213–218
73. Sekine S, Nureki O, Dubois DY, Bernier S, Chenevert R, Lapointe J, Vassylyev DG, Yokoyama S (2003) ATP binding by glutamyl-tRNA synthetase is switched to the productive mode by tRNA binding. EMBO J 22(3):676–688
74. Sekine S, Shichiri M, Bernier S, Chenevert R, Lapointe J, Yokoyama S (2006) Structural bases of transfer RNA-dependent amino acid recognition and activation by glutamyl-tRNA synthetase. Structure 14(12):1791–1799
75. Nureki O, Vassylyev DG, Katayanagi K, Shimizu T, Sekine S, Kigawa T, Miyazawa T, Yokoyama S, Morikawa K (1995) Architectures of class-defining and specific domains of glutamyl-tRNA synthetase. Science 267(5206):1958–1965
76. Nureki O, O'Donoghue P, Watanabe N, Ohmori A, Oshikane H, Araiso Y, Sheppard K, Soll D, Ishitani R (2010) Structure of an archaeal non-discriminating glutamyl-tRNA synthetase: a missing link in the evolution of Gln-tRNAGln formation. Nucleic Acids Res 38(20):7286–7297
77. Schulze JO, Masoumi A, Nickel D, Jahn M, Jahn D, Schubert WD, Heinz DW (2006) Crystal structure of a non-discriminating glutamyl-tRNA synthetase. J Mol Biol 361(5):888–897
78. Ito T, Kiyasu N, Matsunaga R, Takahashi S, Yokoyama S (2010) Structure of nondiscriminating glutamyl-tRNA synthetase from Thermotoga maritima. Acta Crystallogr D Biol Crystallogr 66(Pt 7):813–820
79. Doublie S, Bricogne G, Gilmore C, Carter CW Jr (1995) Tryptophanyl-tRNA synthetase crystal structure reveals an unexpected homology to tyrosyl-tRNA synthetase. Structure 3(1):17–31
80. Retailleau P, Yin Y, Hu M, Roach J, Bricogne G, Vonrhein C, Roversi P, Blanc E, Sweet RM, Carter CW Jr (2001) High-resolution experimental phases for tryptophanyl-tRNA synthetase (TrpRS) complexed with tryptophanyl-5′AMP. Acta Crystallogr D Biol Crystallogr 57(Pt 11):1595–1608
81. Retailleau P, Huang X, Yin Y, Hu M, Weinreb V, Vachette P, Vonrhein C, Bricogne G, Roversi P, Ilyin V, Carter CW Jr (2003) Interconversion of ATP binding and conformational free energies by tryptophanyl-tRNA synthetase: structures of ATP bound to open and closed, pre-transition-state conformations. J Mol Biol 325(1):39–63
82. Retailleau P, Weinreb V, Hu M, Carter CW Jr (2007) Crystal structure of tryptophanyl-tRNA synthetase complexed with adenosine-5′ tetraphosphate: evidence for distributed use of catalytic binding energy in amino acid activation by class I aminoacyl-tRNA synthetases. J Mol Biol 369(1):108–128
83. Buddha MR, Crane BR (2005) Structure and activity of an aminoacyl-tRNA synthetase that charges tRNA with nitro-tryptophan. Nat Struct Mol Biol 12(3):274–275
84. Yang XL, Otero FJ, Ewalt KL, Liu J, Swairjo MA, Kohrer C, RajBhandary UL, Skene RJ, McRee DE, Schimmel P (2006) Two conformations of a crystalline human tRNA synthetase-tRNA complex: implications for protein synthesis. EMBO J 25(12):2919–2929
85. Arakaki TL, Carter M, Napuli AJ, Verlinde CL, Fan E, Zucker F, Buckner FS, Van Voorhis WC, Hol WG, Merritt EA (2010) The structure of tryptophanyl-tRNA synthetase from Giardia lamblia reveals divergence from eukaryotic homologs. J Struct Biol 171(2):238–243

86. Han GW, Yang XL, McMullan D, Chong YE, Krishna SS, Rife CL, Weekes D, Brittain SM, Abdubek P, Ambing E, Astakhova T, Axelrod HL, Carlton D, Caruthers J, Chiu HJ, Clayton T, Duan L, Feuerhelm J, Grant JC, Grzechnik SK, Jaroszewski L, Jin KK, Klock HE, Knuth MW, Kumar A, Marciano D, Miller MD, Morse AT, Nigoghossian E, Okach L, Paulsen J, Reyes R, van den Bedem H, White A, Wolf G, Xu Q, Hodgson KO, Wooley J, Deacon AM, Godzik A, Lesley SA, Elsliger MA, Schimmel P, Wilson IA (2010) Structure of a tryptophanyl-tRNA synthetase containing an iron-sulfur cluster. Acta Crystallogr Sect F Struct Biol Cryst Commun 66(Pt 10):1326–1334
87. Zhou M, Dong X, Shen N, Zhong C, Ding J (2010) Crystal structures of Saccharomyces cerevisiae tryptophanyl-tRNA synthetase: new insights into the mechanism of tryptophan activation and implications for anti-fungal drug design. Nucleic Acids Res 38(10):3399–3413
88. Merritt EA, Arakaki TL, Gillespie R, Napuli AJ, Kim JE, Buckner FS, Van Voorhis WC, Verlinde CL, Fan E, Zucker F, Hol WG (2011) Crystal structures of three protozoan homologs of tryptophanyl-tRNA synthetase. Mol Biochem Parasitol 177(1):20–28
89. Yaremchuk A, Kriklivyi I, Tukalo M, Cusack S (2002) Class I tyrosyl-tRNA synthetase has a class II mode of cognate tRNA recognition. EMBO J 21(14):3829–3840
90. Kobayashi T, Takimura T, Sekine R, Kelly VP, Kamata K, Sakamoto K, Nishimura S, Yokoyama S (2005) Structural snapshots of the KMSKS loop rearrangement for amino acid activation by bacterial tyrosyl-tRNA synthetase. J Mol Biol 346(1):105–117
91. Kuratani M, Sakai H, Takahashi M, Yanagisawa T, Kobayashi T, Murayama K, Chen L, Liu ZJ, Wang BC, Kuroishi C, Kuramitsu S, Terada T, Bessho Y, Shirouzu M, Sekine S, Yokoyama S (2006) Crystal structures of tyrosyl-tRNA synthetases from Archaea. J Mol Biol 355(3):395–408
92. Bonnefond L, Frugier M, Touze E, Lorber B, Florentz C, Giege R, Sauter C, Rudinger-Thirion J (2007) Crystal structure of human mitochondrial tyrosyl-tRNA synthetase reveals common and idiosyncratic features. Structure 15(11):1505–1516
93. Logan DT, Mazauric MH, Kern D, Moras D (1995) Crystal structure of glycyl-tRNA synthetase from Thermus thermophilus. EMBO J 14(17):4156–4167
94. Arnez JG, Dock-Bregeon AC, Moras D (1999) Glycyl-tRNA synthetase uses a negatively charged pit for specific recognition and activation of glycine. J Mol Biol 286(5):1449–1459
95. Cader MZ, Ren J, James PA, Bird LE, Talbot K, Stammers DK (2007) Crystal structure of human wildtype and S581L-mutant glycyl-tRNA synthetase, an enzyme underlying distal spinal muscular atrophy. FEBS Lett 581(16):2959–2964
96. Arnez JG, Harris DC, Mitschler A, Rees B, Francklyn CS, Moras D (1995) Crystal structure of histidyl-tRNA synthetase from Escherichia coli complexed with histidyl-adenylate. EMBO J 14(17):4143–4155
97. Arnez JG, Augustine JG, Moras D, Francklyn CS (1997) The first step of aminoacylation at the atomic level in histidyl-tRNA synthetase. Proc Natl Acad Sci U S A 94(14):7144–7149
98. Qiu X, Janson CA, Blackburn MN, Chhohan IK, Hibbs M, Abdel-Meguid SS (1999) Cooperative structural dynamics and a novel fidelity mechanism in histidyl-tRNA synthetases. Biochemistry 38(38):12296–12304
99. Aberg A, Yaremchuk A, Tukalo M, Rasmussen B, Cusack S (1997) Crystal structure analysis of the activation of histidine by Thermus thermophilus histidyl-tRNA synthetase. Biochemistry 36(11):3084–3094
100. Merritt EA, Arakaki TL, Gillespie JR, Larson ET, Kelley A, Mueller N, Napuli AJ, Kim J, Zhang L, Verlinde CL, Fan E, Zucker F, Buckner FS, van Voorhis WC, Hol WG (2010) Crystal structures of trypanosomal histidyl-tRNA synthetase illuminate differences between eukaryotic and prokaryotic homologs. J Mol Biol 397(2):481–494
101. Xu Z, Wei Z, Zhou JJ, Ye F, Lo WS, Wang F, Lau CF, Wu J, Nangle LA, Chiang KP, Yang XL, Zhang M, Schimmel P (2012) Internally deleted human tRNA synthetase suggests evolutionary pressure for repurposing. Structure 20(9):1470–1477

Emergence and Evolution

102. Yaremchuk A, Tukalo M, Grotli M, Cusack S (2001) A succession of substrate induced conformational changes ensures the amino acid specificity of Thermus thermophilus prolyl-tRNA synthetase: comparison with histidyl-tRNA synthetase. J Mol Biol 309(4):989–1002

103. Kamtekar S, Kennedy WD, Wang J, Stathopoulos C, Soll D, Steitz TA (2003) The structural basis of cysteine aminoacylation of tRNAPro by prolyl-tRNA synthetases. Proc Natl Acad Sci U S A 100(4):1673–1678

104. Larson ET, Kim JE, Napuli AJ, Verlinde CL, Fan E, Zucker FH, Van Voorhis WC, Buckner FS, Hol WG, Merritt EA (2012) Structure of the prolyl-tRNA synthetase from the eukaryotic pathogen Giardia lamblia. Acta Crystallogr D Biol Crystallogr 68(Pt 9):1194–1200

105. Crepin T, Yaremchuk A, Tukalo M, Cusack S (2006) Structures of two bacterial prolyl-tRNA synthetases with and without a cis-editing domain. Structure 14(10):1511–1525

106. Zhou H, Sun L, Yang XL, Schimmel P (2013) ATP-directed capture of bioactive herbal-based medicine on human tRNA synthetase. Nature 494:121–124

107. Fujinaga M, Berthet-Colominas C, Yaremchuk AD, Tukalo MA, Cusack S (1993) Refined crystal structure of the seryl-tRNA synthetase from Thermus thermophilus at 2.5 A resolution. J Mol Biol 234(1):222–233

108. Belrhali H, Yaremchuk A, Tukalo M, Larsen K, Berthet-Colominas C, Leberman R, Beijer B, Sproat B, Als-Nielsen J, Grubel G et al (1994) Crystal structures at 2.5 Angstrom resolution of seryl-tRNA synthetase complexed with two analogs of seryl adenylate. Science 263(5152):1432–1436

109. Chimnaronk S, Gravers Jeppesen M, Suzuki T, Nyborg J, Watanabe K (2005) Dual-mode recognition of noncanonical tRNAs(Ser) by seryl-tRNA synthetase in mammalian mitochondria. EMBO J 24(19):3369–3379

110. Itoh Y, Sekine S, Kuroishi C, Terada T, Shirouzu M, Kuramitsu S, Yokoyama S (2008) Crystallographic and mutational studies of seryl-tRNA synthetase from the archaeon Pyrococcus horikoshii. RNA Biol 5(3):169–177

111. Rocha R, Pereira PJ, Santos MA, Macedo-Ribeiro S (2011) Unveiling the structural basis for translational ambiguity tolerance in a human fungal pathogen. Proc Natl Acad Sci U S A 108 (34):14091–14096

112. Sankaranarayanan R, Dock-Bregeon AC, Romby P, Caillet J, Springer M, Rees B, Ehresmann C, Ehresmann B, Moras D (1999) The structure of threonyl-tRNA synthetase-tRNA(Thr) complex enlightens its repressor activity and reveals an essential zinc ion in the active site. Cell 97(3):371–381

113. Torres-Larios A, Dock-Bregeon AC, Romby P, Rees B, Sankaranarayanan R, Caillet J, Springer M, Ehresmann C, Ehresmann B, Moras D (2002) Structural basis of translational control by Escherichia coli threonyl tRNA synthetase. Nat Struct Biol 9(5):343–347

114. Torres-Larios A, Sankaranarayanan R, Rees B, Dock-Bregeon AC, Moras D (2003) Conformational movements and cooperativity upon amino acid, ATP and tRNA binding in threonyl-tRNA synthetase. J Mol Biol 331(1):201–211

115. Ling J, Peterson KM, Simonovic I, Cho C, Soll D, Simonovic M (2012) Yeast mitochondrial threonyl-tRNA synthetase recognizes tRNA isoacceptors by distinct mechanisms and promotes CUN codon reassignment. Proc Natl Acad Sci U S A 109(9):3281–3286

116. Ling J, Peterson KM, Simonovic I, Soll D, Simonovic M (2012) The mechanism of pre-transfer editing in yeast mitochondrial threonyl-tRNA synthetase. J Biol Chem 287(34):28518–28525

117. Iwasaki W, Sekine S, Kuroishi C, Kuramitsu S, Shirouzu M, Yokoyama S (2006) Structural basis of the water-assisted asparagine recognition by asparaginyl-tRNA synthetase. J Mol Biol 360(2):329–342

118. Crepin T, Peterson F, Haertlein M, Jensen D, Wang C, Cusack S, Kron M (2011) A hybrid structural model of the complete Brugia malayi cytoplasmic asparaginyl-tRNA synthetase. J Mol Biol 405(4):1056–1069

119. Blaise M, Frechin M, Olieric V, Charron C, Sauter C, Lorber B, Roy H, Kern D (2011) Crystal structure of the archaeal asparagine synthetase: interrelation with aspartyl-tRNA and asparaginyl-tRNA synthetases. J Mol Biol 412(3):437–452
120. Ruff M, Krishnaswamy S, Boeglin M, Poterszman A, Mitschler A, Podjarny A, Rees B, Thierry JC, Moras D (1991) Class II aminoacyl transfer RNA synthetases: crystal structure of yeast aspartyl-tRNA synthetase complexed with tRNA(Asp). Science 252(5013):1682–1689
121. Schmitt E, Moulinier L, Fujiwara S, Imanaka T, Thierry JC, Moras D (1998) Crystal structure of aspartyl-tRNA synthetase from Pyrococcus kodakaraensis KOD: archaeon specificity and catalytic mechanism of adenylate formation. EMBO J 17(17):5227–5237
122. Poterszman A, Delarue M, Thierry JC, Moras D (1994) Synthesis and recognition of aspartyl-adenylate by Thermus thermophilus aspartyl-tRNA synthetase. J Mol Biol 244(2):158–167
123. Sauter C, Lorber B, Cavarelli J, Moras D, Giege R (2000) The free yeast aspartyl-tRNA synthetase differs from the tRNA(Asp)-complexed enzyme by structural changes in the catalytic site, hinge region, and anticodon-binding domain. J Mol Biol 299(5):1313–1324
124. Moulinier L, Eiler S, Eriani G, Gangloff J, Thierry JC, Gabriel K, McClain WH, Moras D (2001) The structure of an AspRS-tRNA(Asp) complex reveals a tRNA-dependent control mechanism. EMBO J 20(18):5290–5301
125. Rees B, Webster G, Delarue M, Boeglin M, Moras D (2000) Aspartyl tRNA-synthetase from Escherichia coli: flexibility and adaptability to the substrates. J Mol Biol 299(5):1157–1164
126. Merritt EA, Arakaki TL, Larson ET, Kelley A, Mueller N, Napuli AJ, Zhang L, Deditta G, Luft J, Verlinde CL, Fan E, Zucker F, Buckner FS, Van Voorhis WC, Hol WG (2010) Crystal structure of the aspartyl-tRNA synthetase from Entamoeba histolytica. Mol Biochem Parasitol 169(2):95–100
127. Neuenfeldt A, Lorber B, Ennifar E, Gaudry A, Sauter C, Sissler M, Florentz C (2013) Thermodynamic properties distinguish human mitochondrial aspartyl-tRNA synthetase from bacterial homolog with same 3D architecture. Nucleic Acids Res 41(4):2698–2708
128. Cusack S, Yaremchuk A, Tukalo M (1996) The crystal structures of T. thermophilus lysyl-tRNA synthetase complexed with E. coli tRNA(Lys) and a T. thermophilus tRNA(Lys) transcript: anticodon recognition and conformational changes upon binding of a lysyl-adenylate analogue. EMBO J 15(22):6321–6334
129. Onesti S, Desogus G, Brevet A, Chen J, Plateau P, Blanquet S, Brick P (2000) Structural studies of lysyl-tRNA synthetase: conformational changes induced by substrate binding. Biochemistry 39(42):12853–12861
130. Guo M, Ignatov M, Musier-Forsyth K, Schimmel P, Yang XL (2008) Crystal structure of tetrameric form of human lysyl-tRNA synthetase: implications for multisynthetase complex formation. Proc Natl Acad Sci U S A 105(7):2331–2336
131. Sakurama H, Takita T, Mikami B, Itoh T, Yasukawa K, Inouye K (2009) Two crystal structures of lysyl-tRNA synthetase from Bacillus stearothermophilus in complex with lysyladenylate-like compounds: insights into the irreversible formation of the enzyme-bound adenylate of L-lysine hydroxamate. J Biochem 145(5):555–563
132. Sokabe M, Ose T, Nakamura A, Tokunaga K, Nureki O, Yao M, Tanaka I (2009) The structure of alanyl-tRNA synthetase with editing domain. Proc Natl Acad Sci U S A 106 (27):11028–11033
133. Naganuma M, Sekine S, Fukunaga R, Yokoyama S (2009) Unique protein architecture of alanyl-tRNA synthetase for aminoacylation, editing, and dimerization. Proc Natl Acad Sci U S A 106(21):8489–8494
134. Swairjo MA, Schimmel PR (2005) Breaking sieve for steric exclusion of a noncognate amino acid from active site of a tRNA synthetase. Proc Natl Acad Sci U S A 102(4):988–993
135. Guo M, Chong YE, Shapiro R, Beebe K, Yang XL, Schimmel P (2009) Paradox of mistranslation of serine for alanine caused by AlaRS recognition dilemma. Nature 462(7274):808–812
136. http://www.rcsb.org/pdb/explore/explore.do?structureId=3RF1 DOI:10.2210/pdb3rf1/pdb
137. http://www.rcsb.org/pdb/explore/explore.do?structureId=1J5W DOI:10.2210/pdb1j5w/pdb

Emergence and Evolution

138. Reshetnikova L, Moor N, Lavrik O, Vassylyev DG (1999) Crystal structures of phenylalanyl-tRNA synthetase complexed with phenylalanine and a phenylalanyl-adenylate analogue. J Mol Biol 287(3):555–568

139. Goldgur Y, Mosyak L, Reshetnikova L, Ankilova V, Lavrik O, Khodyreva S, Safro M (1997) The crystal structure of phenylalanyl-tRNA synthetase from thermus thermophilus complexed with cognate tRNAPhe. Structure 5(1):59–68

140. Moor N, Kotik-Kogan O, Tworowski D, Sukhanova M, Safro M (2006) The crystal structure of the ternary complex of phenylalanyl-tRNA synthetase with tRNAPhe and a phenylalanyl-adenylate analogue reveals a conformational switch of the CCA end. Biochemistry 45(35):10572–10583

141. Mermershtain I, Finarov I, Klipcan L, Kessler N, Rozenberg H, Safro MG (2011) Idiosyncrasy and identity in the prokaryotic Phe-system: crystal structure of E. coli phenylalanyl-tRNA synthetase complexed with phenylalanine and AMP. Protein Sci 20(1):160–167

142. Klipcan L, Moor N, Kessler N, Safro MG (2009) Eukaryotic cytosolic and mitochondrial phenylalanyl-tRNA synthetases catalyze the charging of tRNA with the meta-tyrosine. Proc Natl Acad Sci U S A 106(27):11045–11048

143. Klipcan L, Moor N, Finarov I, Kessler N, Sukhanova M, Safro MG (2012) Crystal structure of human mitochondrial PheRS complexed with tRNA(Phe) in the active "open" state. J Mol Biol 415(3):527–537

144. Kavran JM, Gundllapalli S, O'Donoghue P, Englert M, Soll D, Steitz TA (2007) Structure of pyrrolysyl-tRNA synthetase, an archaeal enzyme for genetic code innovation. Proc Natl Acad Sci U S A 104(27):11268–11273

145. Yanagisawa T, Ishii R, Fukunaga R, Kobayashi T, Sakamoto K, Yokoyama S (2008) Crystallographic studies on multiple conformational states of active-site loops in pyrrolysyl-tRNA synthetase. J Mol Biol 378(3):634–652

146. Nozawa K, O'Donoghue P, Gundllapalli S, Araiso Y, Ishitani R, Umehara T, Soll D, Nureki O (2009) Pyrrolysyl-tRNA synthetase-tRNA(Pyl) structure reveals the molecular basis of orthogonality. Nature 457(7233):1163–1167

147. Fukunaga R, Yokoyama S (2007) Structural insights into the first step of RNA-dependent cysteine biosynthesis in archaea. Nat Struct Mol Biol 14(4):272–279

148. Kamtekar S, Hohn MJ, Park HS, Schnitzbauer M, Sauerwald A, Soll D, Steitz TA (2007) Toward understanding phosphoseryl-tRNACys formation: the crystal structure of Methanococcus maripaludis phosphoseryl-tRNA synthetase. Proc Natl Acad Sci U S A 104(8):2620–2625

149. Bult CJ, White O, Olsen GJ, Zhou L, Fleischmann RD, Sutton GG, Blake JA, FitzGerald LM, Clayton RA, Gocayne JD, Kerlavage AR, Dougherty BA, Tomb JF, Adams MD, Reich CI, Overbeek R, Kirkness EF, Weinstock KG, Merrick JM, Glodek A, Scott JL, Geoghagen NS, Venter JC (1996) Complete genome sequence of the methanogenic archaeon, Methanococcus jannaschii. Science 273(5278):1058–1073

150. Smith DR, Doucette-Stamm LA, Deloughery C, Lee H, Dubois J, Aldredge T, Bashirzadeh R, Blakely D, Cook R, Gilbert K, Harrison D, Hoang L, Keagle P, Lumm W, Pothier B, Qiu D, Spadafora R, Vicaire R, Wang Y, Wierzbowski J, Gibson R, Jiwani N, Caruso A, Bush D, Reeve JN et al (1997) Complete genome sequence of Methanobacterium thermoautotrophicum deltaH: functional analysis and comparative genomics. J Bacteriol 179(22):7135–7155

151. Tumbula DL, Becker HD, Chang WZ, Soll D (2000) Domain-specific recruitment of amide amino acids for protein synthesis. Nature 407(6800):106–110

152. Gagnon Y, Lacoste L, Champagne N, Lapointe J (1996) Widespread use of the glu-tRNAGln transamidation pathway among bacteria. A member of the alpha purple bacteria lacks glutaminyl-trna synthetase. J Biol Chem 271(25):14856–14863

153. Curnow AW, Hong K, Yuan R, Kim S, Martins O, Winkler W, Henkin TM, Soll D (1997) Glu-tRNAGln amidotransferase: a novel heterotrimeric enzyme required for correct decoding of glutamine codons during translation. Proc Natl Acad Sci U S A 94(22):11819–11826

154. Bailly M, Blaise M, Lorber B, Becker HD, Kern D (2007) The transamidosome: a dynamic ribonucleoprotein particle dedicated to prokaryotic tRNA-dependent asparagine biosynthesis. Mol Cell 28(2):228–239
155. Rampias T, Sheppard K, Soll D (2010) The archaeal transamidosome for RNA-dependent glutamine biosynthesis. Nucleic Acids Res 38(17):5774–5783
156. Ito T, Yokoyama S (2010) Two enzymes bound to one transfer RNA assume alternative conformations for consecutive reactions. Nature 467(7315):612–616
157. Blaise M, Bailly M, Frechin M, Behrens MA, Fischer F, Oliveira CL, Becker HD, Pedersen JS, Thirup S, Kern D (2010) Crystal structure of a transfer-ribonucleoprotein particle that promotes asparagine formation. EMBO J 29(18):3118–3129
158. Bhaskaran H, Perona JJ (2011) Two-step aminoacylation of tRNA without channeling in Archaea. J Mol Biol 411(4):854–869
159. Sheppard K, Yuan J, Hohn MJ, Jester B, Devine KM, Soll D (2008) From one amino acid to another: tRNA-dependent amino acid biosynthesis. Nucleic Acids Res 36(6):1813–1825
160. Frechin M, Senger B, Braye M, Kern D, Martin RP, Becker HD (2009) Yeast mitochondrial Gln-tRNA(Gln) is generated by a GatFAB-mediated transamidation pathway involving Arc1p-controlled subcellular sorting of cytosolic GluRS. Genes Dev 23(9):1119–1130
161. Di Giulio M (1993) Origin of glutaminyl-tRNA synthetase: an example of palimpsest? J Mol Evol 37(1):5–10
162. Min B, Pelaschier JT, Graham DE, Tumbula-Hansen D, Soll D (2002) Transfer RNA-dependent amino acid biosynthesis: an essential route to asparagine formation. Proc Natl Acad Sci U S A 99(5):2678–2683
163. Becker HD, Kern D (1998) Thermus thermophilus: a link in evolution of the tRNA-dependent amino acid amidation pathways. Proc Natl Acad Sci U S A 95(22):12832–12837
164. Marcker K, Sanger F (1964) N-Formyl-methionyl-S-Rna. J Mol Biol 8:835–840
165. Lee CP, Seong BL, RajBhandary UL (1991) Structural and sequence elements important for recognition of Escherichia coli formylmethionine tRNA by methionyl-tRNA transformylase are clustered in the acceptor stem. J Biol Chem 266(27):18012–18017
166. Kolitz SE, Lorsch JR (2010) Eukaryotic initiator tRNA: finely tuned and ready for action. FEBS Lett 584(2):396–404
167. Tan TH, Bochud-Allemann N, Horn EK, Schneider A (2002) Eukaryotic-type elongator tRNAMet of Trypanosoma brucei becomes formylated after import into mitochondria. Proc Natl Acad Sci U S A 99(3):1152–1157
168. Tomsic J, Vitali LA, Daviter T, Savelsbergh A, Spurio R, Striebeck P, Wintermeyer W, Rodnina MV, Gualerzi CO (2000) Late events of translation initiation in bacteria: a kinetic analysis. EMBO J 19(9):2127–2136
169. Johansson L, Chen C, Thorell JO, Fredriksson A, Stone-Elander S, Gafvelin G, Arner ES (2004) Exploiting the 21st amino acid-purifying and labeling proteins by selenolate targeting. Nat Methods 1(1):61–66
170. Bock A (2005) Selenocysteine. In: Ibba M, Francklyn C, Cusack S (eds) The aminoacyl-tRNA synthetases. Landes Bioscience, Georgetown, pp 320–327
171. Yuan J, Palioura S, Salazar JC, Su D, O'Donoghue P, Hohn MJ, Cardoso AM, Whitman WB, Soll D (2006) RNA-dependent conversion of phosphoserine forms selenocysteine in eukaryotes and archaea. Proc Natl Acad Sci U S A 103(50):18923–18927
172. Commans S, Bock A (1999) Selenocysteine inserting tRNAs: an overview. FEMS Microbiol Rev 23(3):335–351
173. Baron C, Heider J, Bock A (1990) Mutagenesis of selC, the gene for the selenocysteine-inserting tRNA-species in E. coli: effects on in vivo function. Nucleic Acids Res 18(23):6761–6766
174. Palioura S, Sherrer RL, Steitz TA, Soll D, Simonovic M (2009) The human SepSecS-tRNASec complex reveals the mechanism of selenocysteine formation. Science 325(5938):321–325

175. Forchhammer K, Leinfelder W, Bock A (1989) Identification of a novel translation factor necessary for the incorporation of selenocysteine into protein. Nature 342(6248):453–456
176. Zinoni F, Heider J, Bock A (1990) Features of the formate dehydrogenase mRNA necessary for decoding of the UGA codon as selenocysteine. Proc Natl Acad Sci U S A 87(12):4660–4664
177. Berry MJ, Banu L, Chen YY, Mandel SJ, Kieffer JD, Harney JW, Larsen PR (1991) Recognition of UGA as a selenocysteine codon in type I deiodinase requires sequences in the 3′ untranslated region. Nature 353(6341):273–276
178. Allmang C, Wurth L, Krol A (2009) The selenium to selenoprotein pathway in eukaryotes: more molecular partners than anticipated. Biochim Biophys Acta 1790(11):1415–1423
179. Caban K, Copeland PR (2012) Selenocysteine insertion sequence (SECIS)-binding protein 2 alters conformational dynamics of residues involved in tRNA accommodation in 80 S ribosomes. J Biol Chem 287(13):10664–10673
180. Ambrogelly A, Palioura S, Soll D (2007) Natural expansion of the genetic code. Nat Chem Biol 3(1):29–35
181. Doolittle RF (1998) Microbial genomes opened up. Nature 392(6674):339–342
182. Stathopoulos C, Li T, Longman R, Vothknecht UC, Becker HD, Ibba M, Soll D (2000) One polypeptide with two aminoacyl-tRNA synthetase activities. Science 287(5452):479–482
183. Sauerwald A, Zhu W, Major TA, Roy H, Palioura S, Jahn D, Whitman WB, Yates JR 3rd, Ibba M, Soll D (2005) RNA-dependent cysteine biosynthesis in archaea. Science 307(5717):1969–1972
184. O'Donoghue P, Sethi A, Woese CR, Luthey-Schulten ZA (2005) The evolutionary history of Cys-tRNACys formation. Proc Natl Acad Sci U S A 102(52):19003–19008
185. Klotz MG, Arp DJ, Chain PS, El-Sheikh AF, Hauser LJ, Hommes NG, Larimer FW, Malfatti SA, Norton JM, Poret-Peterson AT, Vergez LM, Ward BB (2006) Complete genome sequence of the marine, chemolithoautotrophic, ammonia-oxidizing bacterium Nitrosococcus oceani ATCC 19707. Appl Environ Microbiol 72(9):6299–6315
186. Ivanova N, Sorokin A, Anderson I, Galleron N, Candelon B, Kapatral V, Bhattacharyya A, Reznik G, Mikhailova N, Lapidus A, Chu L, Mazur M, Goltsman E, Larsen N, D'Souza M, Walunas T, Grechkin Y, Pusch G, Haselkorn R, Fonstein M, Ehrlich SD, Overbeek R, Kyrpides N (2003) Genome sequence of Bacillus cereus and comparative analysis with Bacillus anthracis. Nature 423(6935):87–91
187. Ataide SF, Jester BC, Devine KM, Ibba M (2005) Stationary-phase expression and aminoacylation of a transfer-RNA-like small RNA. EMBO Rep 6(8):742–747
188. Deppenmeier U, Johann A, Hartsch T, Merkl R, Schmitz RA, Martinez-Arias R, Henne A, Wiezer A, Baumer S, Jacobi C, Bruggemann H, Lienard T, Christmann A, Bomeke M, Steckel S, Bhattacharyya A, Lykidis A, Overbeek R, Klenk HP, Gunsalus RP, Fritz HJ, Gottschalk G (2002) The genome of Methanosarcina mazei: evidence for lateral gene transfer between bacteria and archaea. J Mol Microbiol Biotechnol 4(4):453–461
189. Ambrogelly A, Korencic D, Ibba M (2002) Functional annotation of class I lysyl-tRNA synthetase phylogeny indicates a limited role for gene transfer. J Bacteriol 184(16):4594–4600
190. Soll D, Becker HD, Plateau P, Blanquet S, Ibba M (2000) Context-dependent anticodon recognition by class I lysyl-tRNA synthetases. Proc Natl Acad Sci U S A 97(26):14224–14228
191. Levengood JD, Roy H, Ishitani R, Soll D, Nureki O, Ibba M (2007) Anticodon recognition and discrimination by the alpha-helix cage domain of class I lysyl-tRNA synthetase. Biochemistry 46(39):11033–11038
192. Jester BC, Levengood JD, Roy H, Ibba M, Devine KM (2003) Nonorthologous replacement of lysyl-tRNA synthetase prevents addition of lysine analogues to the genetic code. Proc Natl Acad Sci U S A 100(24):14351–14356
193. Wang S, Praetorius-Ibba M, Ataide SF, Roy H, Ibba M (2006) Discrimination of cognate and noncognate substrates at the active site of class I lysyl-tRNA synthetase. Biochemistry 45(11):3646–3652
194. Ataide SF, Ibba M (2004) Discrimination of cognate and noncognate substrates at the active site of class II lysyl-tRNA synthetase. Biochemistry 43(37):11836–11841

195. Uy R, Wold F (1977) Posttranslational covalent modification of proteins. Science 198 (4320):890–896
196. Hao B, Gong W, Ferguson TK, James CM, Krzycki JA, Chan MK (2002) A new UAG-encoded residue in the structure of a methanogen methyltransferase. Science 296(5572):1462–1466
197. Zhang Y, Gladyshev VN (2007) High content of proteins containing 21st and 22nd amino acids, selenocysteine and pyrrolysine, in a symbiotic deltaproteobacterium of gutless worm Olavius algarvensis. Nucleic Acids Res 35(15):4952–4963
198. Prat L, Heinemann IU, Aerni HR, Rinehart J, O'Donoghue P, Soll D (2012) Carbon source-dependent expansion of the genetic code in bacteria. Proc Natl Acad Sci U S A 109(51):21070–21075
199. Theobald-Dietrich A, Frugier M, Giege R, Rudinger-Thirion J (2004) Atypical archaeal tRNA pyrrolysine transcript behaves towards EF-Tu as a typical elongator tRNA. Nucleic Acids Res 32(3):1091–1096
200. Polycarpo C, Ambrogelly A, Ruan B, Tumbula-Hansen D, Ataide SF, Ishitani R, Yokoyama S, Nureki O, Ibba M, Soll D (2003) Activation of the pyrrolysine suppressor tRNA requires formation of a ternary complex with class I and class II lysyl-tRNA synthetases. Mol Cell 12(2):287–294
201. Blight SK, Larue RC, Mahapatra A, Longstaff DG, Chang E, Zhao G, Kang PT, Green-Church KB, Chan MK, Krzycki JA (2004) Direct charging of tRNA(CUA) with pyrrolysine in vitro and in vivo. Nature 431(7006):333–335
202. Polycarpo C, Ambrogelly A, Berube A, Winbush SM, McCloskey JA, Crain PF, Wood JL, Soll D (2004) An aminoacyl-tRNA synthetase that specifically activates pyrrolysine. Proc Natl Acad Sci U S A 101(34):12450–12454
203. Longstaff DG, Blight SK, Zhang L, Green-Church KB, Krzycki JA (2007) In vivo contextual requirements for UAG translation as pyrrolysine. Mol Microbiol 63(1):229–241
204. Jiang R, Krzycki JA (2012) PylSn and the homologous N-terminal domain of pyrrolysyl-tRNA synthetase bind the tRNA that is essential for the genetic encoding of pyrrolysine. J Biol Chem 287(39):32738–32746
205. Srinivasan G, James CM, Krzycki JA (2002) Pyrrolysine encoded by UAG in Archaea: charging of a UAG-decoding specialized tRNA. Science 296(5572):1459–1462
206. Herring S, Ambrogelly A, Gundllapalli S, O'Donoghue P, Polycarpo CR, Soll D (2007) The amino-terminal domain of pyrrolysyl-tRNA synthetase is dispensable in vitro but required for in vivo activity. FEBS Lett 581(17):3197–3203
207. Krzycki JA (2004) Function of genetically encoded pyrrolysine in corrinoid-dependent methylamine methyltransferases. Curr Opin Chem Biol 8(5):484–491
208. Fournier GP, Huang J, Gogarten JP (2009) Horizontal gene transfer from extinct and extant lineages: biological innovation and the coral of life. Philos Trans R Soc Lond B Biol Sci 364(1527):2229–2239
209. Gaston MA, Zhang L, Green-Church KB, Krzycki JA (2011) The complete biosynthesis of the genetically encoded amino acid pyrrolysine from lysine. Nature 471(7340):647–650
210. Hauenstein SI, Perona JJ (2008) Redundant synthesis of cysteinyl-tRNACys in Methanosarcina mazei. J Biol Chem 283(32):22007–22017
211. Crick FH (1958) On protein synthesis. Symp Soc Exp Biol 12:138–163
212. Crick FHC (1968) The origin of the genetic code. J Mol Biol 38(3):367–379
213. Jimenez-Sanchez A (1995) On the origin and evolution of the genetic code. J Mol Evol 41(6):712–716
214. Guo M, Schimmel P (2012) Structural analyses clarify the complex control of mistranslation by tRNA synthetases. Curr Opin Struct Biol 22(1):119–126
215. Loftfield RB, Vanderjagt D (1972) The frequency of errors in protein biosynthesis. Biochem J 128(5):1353–1356
216. Schimmel P, Giege R, Moras D, Yokoyama S (1993) An operational RNA code for amino acids and possible relationship to genetic code. Proc Natl Acad Sci U S A 90(19):8763–8768

Emergence and Evolution

217. Reynolds NM, Ling J, Roy H, Banerjee R, Repasky SE, Hamel P, Ibba M (2010) Cell-specific differences in the requirements for translation quality control. Proc Natl Acad Sci U S A 107(9):4063–4068
218. Li L, Boniecki MT, Jaffe JD, Imai BS, Yau PM, Luthey-Schulten ZA, Martinis SA (2011) Naturally occurring aminoacyl-tRNA synthetases editing-domain mutations that cause mistranslation in Mycoplasma parasites. Proc Natl Acad Sci U S A 108(23):9378–9383
219. Yadavalli SS, Ibba M (2013) Selection of tRNA charging quality control mechanisms that increase mistranslation of the genetic code. Nucleic Acids Res 41:1104–1112
220. Jakubowski H, Goldman E (1992) Editing of errors in selection of amino acids for protein synthesis. Microbiol Rev 56(3):412–429
221. Giege R, Sissler M, Florentz C (1998) Universal rules and idiosyncratic features in tRNA identity. Nucleic Acids Res 26(22):5017–5035
222. Tworowski D, Feldman AV, Safro MG (2005) Electrostatic potential of aminoacyl-tRNA synthetase navigates tRNA on its pathway to the binding site. J Mol Biol 350(5):866–882
223. Ebel JP, Giege R, Bonnet J, Kern D, Befort N, Bollack C, Fasiolo F, Gangloff J, Dirheimer G (1973) Factors determining the specificity of the tRNA aminoacylation reaction. Non-absolute specificity of tRNA-aminoacyl-tRNA synthetase recognition and particular importance of the maximal velocity. Biochimie 55(5):547–557
224. Guth EC, Francklyn CS (2007) Kinetic discrimination of tRNA identity by the conserved motif 2 loop of a class II aminoacyl-tRNA synthetase. Mol Cell 25(4):531–542
225. Ibba M, Sever S, Praetorius-Ibba M, Soll D (1999) Transfer RNA identity contributes to transition state stabilization during aminoacyl-tRNA synthesis. Nucleic Acids Res 27(18):3631–3637
226. Uter NT, Perona JJ (2004) Long-range intramolecular signaling in a tRNA synthetase complex revealed by pre-steady-state kinetics. Proc Natl Acad Sci U S A 101(40):14396–14401
227. Vasil'eva IA, Moor NA (2007) Interaction of aminoacyl-tRNA synthetases with tRNA: general principles and distinguishing characteristics of the high-molecular-weight substrate recognition. Biochemistry (Mosc) 72(3):247–263
228. Aphasizhev R, Senger B, Fasiolo F (1997) Importance of structural features for tRNA(Met) identity. RNA 3(5):489–497
229. Schimmel P, Ribas de Pouplana L (1995) Transfer RNA: from minihelix to genetic code. Cell 81(7):983–986
230. Wolstenholme DR, Macfarlane JL, Okimoto R, Clary DO, Wahleithner JA (1987) Bizarre tRNAs inferred from DNA sequences of mitochondrial genomes of nematode worms. Proc Natl Acad Sci U S A 84(5):1324–1328
231. Cavarelli J, Rees B, Ruff M, Thierry JC, Moras D (1993) Yeast tRNA(Asp) recognition by its cognate class II aminoacyl-tRNA synthetase. Nature 362(6416):181–184
232. Ibba M, Hong KW, Sherman JM, Sever S, Soll D (1996) Interactions between tRNA identity nucleotides and their recognition sites in glutaminyl-tRNA synthetase determine the cognate amino acid affinity of the enzyme. Proc Natl Acad Sci U S A 93(14):6953–6958
233. Hou YM, Schimmel P (1988) A simple structural feature is a major determinant of the identity of a transfer RNA. Nature 333(6169):140–145
234. Sugiura I, Nureki O, Ugaji-Yoshikawa Y, Kuwabara S, Shimada A, Tateno M, Lorber B, Giege R, Moras D, Yokoyama S, Konno M (2000) The 2.0 A crystal structure of Thermus thermophilus methionyl-tRNA synthetase reveals two RNA-binding modules. Structure 8(2):197–208
235. Ibba M, Kast P, Hennecke H (1994) Substrate specificity is determined by amino acid binding pocket size in Escherichia coli phenylalanyl-tRNA synthetase. Biochemistry 33(23):7107–7112
236. Sankaranarayanan R, Dock-Bregeon AC, Rees B, Bovee M, Caillet J, Romby P, Francklyn CS, Moras D (2000) Zinc ion mediated amino acid discrimination by threonyl-tRNA synthetase. Nat Struct Biol 7(6):461–465
237. Kleiman L, Jones CP, Musier-Forsyth K (2010) Formation of the tRNALys packaging complex in HIV-1. FEBS Lett 584(2):359–365

238. Jones CP, Saadatmand J, Kleiman L, Musier-Forsyth K (2013) Molecular mimicry of human tRNALys anti-codon domain by HIV-1 RNA genome facilitates tRNA primer annealing. RNA 19:219–229
239. Fu G, Xu T, Shi Y, Wei N, Yang XL (2012) tRNA-controlled nuclear import of a human tRNA synthetase. J Biol Chem 287(12):9330–9334
240. Azad AK, Stanford DR, Sarkar S, Hopper AK (2001) Role of nuclear pools of aminoacyl-tRNA synthetases in tRNA nuclear export. Mol Biol Cell 12(5):1381–1392
241. Yadavalli SS, Ibba M (2012) Quality control in aminoacyl-tRNA synthesis its role in translational fidelity. Adv Protein Chem Struct Biol 86:1–43
242. Schmidt E, Schimmel P (1994) Mutational isolation of a sieve for editing in a transfer RNA synthetase. Science 264(5156):265–267
243. Cvetesic N, Perona JJ, Gruic-Sovulj I (2012) Kinetic partitioning between synthetic and editing pathways in class I aminoacyl-tRNA synthetases occurs at both pre-transfer and post-transfer hydrolytic steps. J Biol Chem 287(30):25381–25394
244. Ling J, So BR, Yadavalli SS, Roy H, Shoji S, Fredrick K, Musier-Forsyth K, Ibba M (2009) Resampling and editing of mischarged tRNA prior to translation elongation. Mol Cell 33(5):654–660
245. Englisch S, Englisch U, von der Haar F, Cramer F (1986) The proofreading of hydroxy analogues of leucine and isoleucine by leucyl-tRNA synthetases from E. coli and yeast. Nucleic Acids Res 14(19):7529–7539
246. Dulic M, Cvetesic N, Perona JJ, Gruic-Sovulj I (2010) Partitioning of tRNA-dependent editing between pre- and post-transfer pathways in class I aminoacyl-tRNA synthetases. J Biol Chem 285(31):23799–23809
247. Nordin BE, Schimmel P (2003) Transiently misacylated tRNA is a primer for editing of misactivated adenylates by class I aminoacyl-tRNA synthetases. Biochemistry 42(44):12989–12997
248. Sarkar J, Martinis SA (2011) Amino-acid-dependent shift in tRNA synthetase editing mechanisms. J Am Chem Soc 133(46):18510–18513
249. Chen X, Ma JJ, Tan M, Yao P, Hu QH, Eriani G, Wang ED (2011) Modular pathways for editing non-cognate amino acids by human cytoplasmic leucyl-tRNA synthetase. Nucleic Acids Res 39(1):235–247
250. Sarkar J, Poruri K, Boniecki MT, McTavish KK, Martinis SA (2012) Yeast mitochondrial leucyl-tRNA synthetase CP1 domain has functionally diverged to accommodate RNA splicing at expense of hydrolytic editing. J Biol Chem 287(18):14772–14781
251. Fersht AR, Dingwall C (1979) An editing mechanism for the methionyl-tRNA synthetase in the selection of amino acids in protein synthesis. Biochemistry 18(7):1250–1256
252. Jakubowski H, Fersht AR (1981) Alternative pathways for editing non-cognate amino acids by aminoacyl-tRNA synthetases. Nucleic Acids Res 9(13):3105–3117
253. Kim HY, Ghosh G, Schulman LH, Brunie S, Jakubowski H (1993) The relationship between synthetic and editing functions of the active site of an aminoacyl-tRNA synthetase. Proc Natl Acad Sci U S A 90(24):11553–11557
254. Igloi GL, von der Haar F, Cramer F (1980) A novel enzymatic activity of phenylalanyl transfer ribonucleic acid synthetase from baker's yeast: zinc ion induced transfer ribonucleic acid independent hydrolysis of adenosine triphosphate. Biochemistry 19(8):1676–1680
255. Roy H, Ling J, Irnov M, Ibba M (2004) Post-transfer editing in vitro and in vivo by the beta subunit of phenylalanyl-tRNA synthetase. EMBO J 23(23):4639–4648
256. Sasaki HM, Sekine S, Sengoku T, Fukunaga R, Hattori M, Utsunomiya Y, Kuroishi C, Kuramitsu S, Shirouzu M, Yokoyama S (2006) Structural and mutational studies of the amino acid-editing domain from archaeal/eukaryal phenylalanyl-tRNA synthetase. Proc Natl Acad Sci U S A 103(40):14744–14749
257. Beebe K, Merriman E, Ribas De Pouplana L, Schimmel P (2004) A domain for editing by an archaebacterial tRNA synthetase. Proc Natl Acad Sci U S A 101(16):5958–5963

258. Korencic D, Ahel I, Schelert J, Sacher M, Ruan B, Stathopoulos C, Blum P, Ibba M, Soll D (2004) A freestanding proofreading domain is required for protein synthesis quality control in Archaea. Proc Natl Acad Sci U S A 101(28):10260–10265

259. Dock-Bregeon A, Sankaranarayanan R, Romby P, Caillet J, Springer M, Rees B, Francklyn CS, Ehresmann C, Moras D (2000) Transfer RNA-mediated editing in threonyl-tRNA synthetase. The class II solution to the double discrimination problem. Cell 103(6):877–884

260. Hussain T, Kruparani SP, Pal B, Dock-Bregeon AC, Dwivedi S, Shekar MR, Sureshbabu K, Sankaranarayanan R (2006) Post-transfer editing mechanism of a D-aminoacyl-tRNA deacylase-like domain in threonyl-tRNA synthetase from archaea. EMBO J 25(17):4152–4162

261. Wydau S, van der Rest G, Aubard C, Plateau P, Blanquet S (2009) Widespread distribution of cell defense against D-aminoacyl-tRNAs. J Biol Chem 284(21):14096–14104

262. Rigden DJ (2004) Archaea recruited D-Tyr-tRNATyr deacylase for editing in Thr-tRNA synthetase. RNA 10(12):1845–1851

263. Dwivedi S, Kruparani SP, Sankaranarayanan R (2005) A D-amino acid editing module coupled to the translational apparatus in archaea. Nat Struct Mol Biol 12(6):556–557

264. Calendar R, Berg P (1967) D-Tyrosyl RNA: formation, hydrolysis and utilization for protein synthesis. J Mol Biol 26(1):39–54

265. Ferri-Fioni ML, Fromant M, Bouin AP, Aubard C, Lazennec C, Plateau P, Blanquet S (2006) Identification in archaea of a novel D-Tyr-tRNATyr deacylase. J Biol Chem 281(37):27575–27585

266. Lee JW, Beebe K, Nangle LA, Jang J, Longo-Guess CM, Cook SA, Davisson MT, Sundberg JP, Schimmel P, Ackerman SL (2006) Editing-defective tRNA synthetase causes protein misfolding and neurodegeneration. Nature 443(7107):50–55

267. Ahel I, Korencic D, Ibba M, Soll D (2003) Trans-editing of mischarged tRNAs. Proc Natl Acad Sci U S A 100(26):15422–15427

268. Zhang H, Huang K, Li Z, Banerjei L, Fisher KE, Grishin NV, Eisenstein E, Herzberg O (2000) Crystal structure of YbaK protein from Haemophilus influenzae (HI1434) at 1.8 A resolution: functional implications. Proteins 40(1):86–97

269. An S, Musier-Forsyth K (2004) Trans-editing of Cys-tRNAPro by Haemophilus influenzae YbaK protein. J Biol Chem 279(41):42359–42362

270. Ruan B, Soll D (2005) The bacterial YbaK protein is a Cys-tRNAPro and Cys-tRNA Cys deacylase. J Biol Chem 280(27):25887–25891

271. Ruan LL, Zhou XL, Tan M, Wang ED (2013) Human cytoplasmic ProX edits mischarged tRNAPro with amino acid but not tRNA specificity. Biochem J 450:243–252

272. So BR, An S, Kumar S, Das M, Turner DA, Hadad CM, Musier-Forsyth K (2011) Substrate-mediated fidelity mechanism ensures accurate decoding of proline codons. J Biol Chem 286 (36):31810–31820

273. Musier-Forsyth K, Stehlin C, Burke B, Liu H (1997) Understanding species-specific differences in substrate recognition by Escherichia coli and human prolyl-tRNA synthetases. Nucleic Acids Symp Ser 36:5–7

274. Bullard JM, Cai YC, Demeler B, Spremulli LL (1999) Expression and characterization of a human mitochondrial phenylalanyl-tRNA synthetase. J Mol Biol 288(4):567–577

275. Beuning PJ, Musier-Forsyth K (2001) Species-specific differences in amino acid editing by class II prolyl-tRNA synthetase. J Biol Chem 276(33):30779–30785

276. Lue SW, Kelley SO (2005) An aminoacyl-tRNA synthetase with a defunct editing site. Biochemistry 44(8):3010–3016

277. SternJohn J, Hati S, Siliciano PG, Musier-Forsyth K (2007) Restoring species-specific posttransfer editing activity to a synthetase with a defunct editing domain. Proc Natl Acad Sci U S A 104(7):2127–2132

278. Wong FC, Beuning PJ, Nagan M, Shiba K, Musier-Forsyth K (2002) Functional role of the prokaryotic proline-tRNA synthetase insertion domain in amino acid editing. Biochemistry 41(22):7108–7115

279. Wong FC, Beuning PJ, Silvers C, Musier-Forsyth K (2003) An isolated class II aminoacyl-tRNA synthetase insertion domain is functional in amino acid editing. J Biol Chem 278(52):52857–52864

280. Ling J, Reynolds N, Ibba M (2009) Aminoacyl-tRNA synthesis and translational quality control. Annu Rev Microbiol 63:61–78
281. Raina M, Elgamal S, Santangelo TJ, Ibba M (2012) Association of a multi-synthetase complex with translating ribosomes in the archaeon Thermococcus kodakarensis. FEBS Lett 586(16):2232–2238
282. Mirande M (2010) Processivity of translation in the eukaryote cell: role of aminoacyl-tRNA synthetases. FEBS Lett 584(2):443–447
283. Guo M, Yang XL, Schimmel P (2010) New functions of aminoacyl-tRNA synthetases beyond translation. Nat Rev Mol Cell Biol 11(9):668–674
284. Sajish M, Zhou Q, Kishi S, Valdez DM Jr, Kapoor M, Guo M, Lee S, Kim S, Yang XL, Schimmel P (2012) Trp-tRNA synthetase bridges DNA-PKcs to PARP-1 to link IFN-gamma and p53 signaling. Nat Chem Biol 8(6):547–554
285. Jia J, Arif A, Ray PS, Fox PL (2008) WHEP domains direct noncanonical function of glutamyl-prolyl tRNA synthetase in translational control of gene expression. Mol Cell 29 (6):679–690
286. Smirnova EV, Lakunina VA, Tarassov I, Krasheninnikov IA, Kamenski PA (2012) Noncanonical functions of aminoacyl-tRNA synthetases. Biochemistry (Mosc) 77(1):15–25
287. Han JM, Jeong SJ, Park MC, Kim G, Kwon NH, Kim HK, Ha SH, Ryu SH, Kim S (2012) Leucyl-tRNA synthetase is an intracellular leucine sensor for the mTORC1-signaling pathway. Cell 149(2):410–424
288. Bonfils G, Jaquenoud M, Bontron S, Ostrowicz C, Ungermann C, De Virgilio C (2012) Leucyl-tRNA synthetase controls TORC1 via the EGO complex. Mol Cell 46(1):105–110
289. Putney SD, Schimmel P (1981) An aminoacyl tRNA synthetase binds to a specific DNA sequence and regulates its gene transcription. Nature 291(5817):632–635
290. Mayaux JF, Fayat G, Panvert M, Springer M, Grunberg-Manago M, Blanquet S (1985) Control of phenylalanyl-tRNA synthetase genetic expression. Site-directed mutagenesis of the pheS, T operon regulatory region in vitro. J Mol Biol 184(1):31–44
291. Putzer H, Laalami S, Brakhage AA, Condon C, Grunberg-Manago M (1995) Aminoacyl-tRNA synthetase gene regulation in Bacillus subtilis: induction, repression and growth-rate regulation. Mol Microbiol 16(4):709–718
292. Moine H, Romby P, Springer M, Grunberg-Manago M, Ebel JP, Ehresmann B, Ehresmann C (1990) Escherichia coli threonyl-tRNA synthetase and tRNA(Thr) modulate the binding of the ribosome to the translational initiation site of the thrS mRNA. J Mol Biol 216(2):299–310
293. Rho SB, Lincecum TL Jr, Martinis SA (2002) An inserted region of leucyl-tRNA synthetase plays a critical role in group I intron splicing. EMBO J 21(24):6874–6881
294. Myers CA, Kuhla B, Cusack S, Lambowitz AM (2002) tRNA-like recognition of group I introns by a tyrosyl-tRNA synthetase. Proc Natl Acad Sci U S A 99(5):2630–2635
295. Roy H, Ibba M (2008) RNA-dependent lipid remodeling by bacterial multiple peptide resistance factors. Proc Natl Acad Sci U S A 105(12):4667–4672
296. Maloney E, Stankowska D, Zhang J, Fol M, Cheng QJ, Lun S, Bishai WR, Rajagopalan M, Chatterjee D, Madiraju MV (2009) The two-domain LysX protein of Mycobacterium tuberculosis is required for production of lysinylated phosphatidylglycerol and resistance to cationic antimicrobial peptides. PLoS Pathog 5(7):e1000534
297. Ernst CM, Staubitz P, Mishra NN, Yang SJ, Hornig G, Kalbacher H, Bayer AS, Kraus D, Peschel A (2009) The bacterial defensin resistance protein MprF consists of separable domains for lipid lysinylation and antimicrobial peptide repulsion. PLoS Pathog 5(11): e1000660
298. Thomas CM, Hothersall J, Willis CL, Simpson TJ (2010) Resistance to and synthesis of the antibiotic mupirocin. Nat Rev Microbiol 8(4):281–289
299. Sauguet L, Moutiez M, Li Y, Belin P, Seguin J, Le Du MH, Thai R, Masson C, Fonvielle M, Pernodet JL, Charbonnier JB, Gondry M (2011) Cyclodipeptide synthases, a family of class-I aminoacyl-tRNA synthetase-like enzymes involved in non-ribosomal peptide synthesis. Nucleic Acids Res 39(10):4475–4489

Emergence and Evolution

300. Sareen D, Steffek M, Newton GL, Fahey RC (2002) ATP-dependent L-cysteine:1D-myo-inosityl 2-amino-2-deoxy-alpha-D-glucopyranoside ligase, mycothiol biosynthesis enzyme MshC, is related to class I cysteinyl-tRNA synthetases. Biochemistry 41(22):6885–6890
301. Mocibob M, Ivic N, Bilokapic S, Maier T, Luic M, Ban N, Weygand-Durasevic I (2010) Homologs of aminoacyl-tRNA synthetases acylate carrier proteins and provide a link between ribosomal and nonribosomal peptide synthesis. Proc Natl Acad Sci U S A 107(33):14585–14590
302. Salazar JC, Ahel I, Orellana O, Tumbula-Hansen D, Krieger R, Daniels L, Soll D (2003) Coevolution of an aminoacyl-tRNA synthetase with its tRNA substrates. Proc Natl Acad Sci U S A 100(24):13863–13868
303. Blaise M, Becker HD, Lapointe J, Cambillau C, Giege R, Kern D (2005) Glu-Q-tRNA(Asp) synthetase coded by the yadB gene, a new paralog of aminoacyl-tRNA synthetase that glutamylates tRNA(Asp) anticodon. Biochimie 87(9–10):847–861
304. Roy H, Zou SB, Bullwinkle TJ, Wolfe BS, Gilreath MS, Forsyth CJ, Navarre WW, Ibba M (2011) The tRNA synthetase paralog PoxA modifies elongation factor-P with (R)-beta-lysine. Nat Chem Biol 7(10):667–669
305. Sissler M, Delorme C, Bond J, Ehrlich SD, Renault P, Francklyn C (1999) An aminoacyl-tRNA synthetase paralog with a catalytic role in histidine biosynthesis. Proc Natl Acad Sci U S A 96(16):8985–8990
306. Dong J, Qiu H, Garcia-Barrio M, Anderson J, Hinnebusch AG (2000) Uncharged tRNA activates GCN2 by displacing the protein kinase moiety from a bipartite tRNA-binding domain. Mol Cell 6(2):269–279
307. Cedar H, Schwartz JH (1969) The asparagine synthetase of Escherichia coli. I. Biosynthetic role of the enzyme, purification, and characterization of the reaction products. J Biol Chem 244(15):4112–4121
308. Artymiuk PJ, Rice DW, Poirrette AR, Willet P (1994) A tale of two synthetases. Nat Struct Biol 1(11):758–760
309. Chapman-Smith A, Mulhern TD, Whelan F, Cronan JE Jr, Wallace JC (2001) The C-terminal domain of biotin protein ligase from E. coli is required for catalytic activity. Protein Sci 10(12):2608–2617
310. Safro M, Mosyak L (1995) Structural similarities in the noncatalytic domains of phenylalanyl-tRNA and biotin synthetases. Protein Sci 4(11):2429–2432
311. Buoncristiani MR, Howard PK, Otsuka AJ (1986) DNA-binding and enzymatic domains of the bifunctional biotin operon repressor (BirA) of Escherichia coli. Gene 44(2–3):255–261
312. Guitart T, Leon Bernardo T, Sagales J, Stratmann T, Bernues J, Ribas de Pouplana L (2010) New aminoacyl-tRNA synthetase-like protein in insecta with an essential mitochondrial function. J Biol Chem 285(49):38157–38166
313. Jaric J, Bilokapic S, Lesjak S, Crnkovic A, Ban N, Weygand-Durasevic I (2009) Identification of amino acids in the N-terminal domain of atypical methanogenic-type Seryl-tRNA synthetase critical for tRNA recognition. J Biol Chem 284(44):30643–30651
314. Holm L, Sander C (1995) DNA polymerase beta belongs to an ancient nucleotidyltransferase superfamily. Trends Biochem Sci 20(9):345–347
315. Carrodeguas JA, Theis K, Bogenhagen DF, Kisker C (2001) Crystal structure and deletion analysis show that the accessory subunit of mammalian DNA polymerase gamma, Pol gamma B, functions as a homodimer. Mol Cell 7(1):43–54
316. Hoagland MB (1955) An enzymic mechanism for amino acid activation in animal tissues. Biochim Biophys Acta 16(2):288–289
317. Davie EW, Koningsberger VV, Lipmann F (1956) The isolation of a tryptophan-activating enzyme from pancreas. Arch Biochem Biophys 65(1):21–38
318. Park SG, Schimmel P, Kim S (2008) Aminoacyl tRNA synthetases and their connections to disease. Proc Natl Acad Sci U S A 105(32):11043–11049
319. Kim S, You S, Hwang D (2011) Aminoacyl-tRNA synthetases and tumorigenesis: more than housekeeping. Nat Rev Cancer 11(10):708–718

Top Curr Chem (2014) 344: 89–118
DOI: 10.1007/128_2013_424
© Springer-Verlag Berlin Heidelberg 2013
Published online: 28 March 2013

Architecture and Metamorphosis

Min Guo and Xiang-Lei Yang

Abstract When compared to other conserved housekeeping protein families, such as ribosomal proteins, during the evolution of higher eukaryotes, aminoacyl-tRNA synthetases (aaRSs) show an apparent high propensity to add new sequences, and especially new domains. The stepwise emergence of those new domains is consistent with their involvement in a broad range of biological functions beyond protein synthesis, and correlates with the increasing biological complexity of higher organisms. These new domains have been extensively characterized based on their evolutionary origins and their sequence, structural, and functional features. While some of the domains are uniquely found in aaRSs and may have originated from nucleic acid binding motifs, others are common domain modules mediating protein–protein interactions that play a critical role in the assembly of the multi-synthetase complex (MSC). Interestingly, the MSC has emerged from a miniature complex in yeast to a large stable complex in humans. The human MSC consists of nine aaRSs (LysRS, ArgRS, GlnRS, AspRS, MetRS, IleRS, LeuRS, GluProRS, and bifunctional aaRs) and three scaffold proteins (AIMP1/p43, AIMP2/p38, and AIMP3/p18), and has a molecular weight of 1.5 million Dalton. The MSC has been proposed to have a functional dualism: facilitating protein synthesis and serving as a reservoir of non-canonical functions associated with its synthetase and non-synthetase components. Importantly, domain additions and functional expansions are not limited to the components of the MSC and are found in almost all aaRS proteins. From a structural perspective, multi-functionalities are represented by multiple conformational states. In fact, alternative conformations

---

M. Guo (✉)
Department of Cancer Biology, The Scripps Research Institute, 130 Scripps Way, Jupiter, FL 33410, USA
e-mail: GuoMin@scripps.edu

X.-L. Yang (✉)
Department of Cancer Biology, The Scripps Research Institute, 10550 N. Torrey Pines Road, La Jolla, CA 92037, USA
e-mail: xlyang@scripps.edu

of aaRSs have been generated by various mechanisms from proteolysis to alternative splicing and posttranslational modifications, as well as by disease-causing mutations. Therefore, the metamorphosis between different conformational states is connected to the activation and regulation of the novel functions of aaRSs in higher eukaryotes.

Keywords Domain expansion · Evolution · Novel function · Structural metamorphosis · tRNA synthetase

Contents

1	Introduction	90
2	Stepwise Appearance of New Domains and Sequences in Higher Eukaryotes	91
	2.1 N-Terminal Helix	92
	2.2 EMAPII Domain	93
	2.3 GST Domain	94
	2.4 WHEP Domain	96
	2.5 Leucine Zipper	97
	2.6 Unique Domains	98
3	Emergence of Multi-Synthetase Complex in Higher Eukaryotes	101
	3.1 Miniature Complex in Yeast	101
	3.2 Divergent Complex in *C. elegans*	103
	3.3 Multi-tRNA Synthetase Complex from Fruit Fly to Human	104
4	Metamorphosis of tRNA Synthetases	106
	4.1 Proteolysis-Based Structural Resection	106
	4.2 Alternative Splicing-Based Structural Resection	107
	4.3 Posttranslational Modification-Based Structural Change	109
	4.4 Mutation-Induced Structural Change	110
	4.5 Quaternary Structural Change	111
5	Concluding Remarks	112
References		113

1 Introduction

The first surprise of the human genome project was the discovery of an unexpectedly small number of protein-coding genes [1]. The explanation that has slowly emerged since is that human genes possess more diverse functions and their regulation is more complicated than their counterpart genes in lower organisms. In this regard, the increasing complexity of aminoacyl-tRNA synthetases (aaRSs) – at the sequence, structural, and functional levels – stands out as a prominent example.

Known as an essential component of the translational apparatus, the aaRS family catalyzes the first step reaction in protein synthesis, that is to attach specifically each amino acid to its cognate tRNAs. While preserving this essential role, higher eukaryotic tRNA synthetases have developed other roles during evolution. Human cytoplasmic tRNA synthetases, in particular, mediate diverse functions in different

Architecture and Metamorphosis

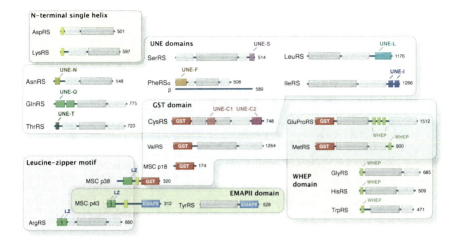

Fig. 1 New domains and sequence additions as found in human aaRSs

pathways including angiogenesis, inflammation, development, and tumorigenesis. The functional expansion of aaRSs is thought to be intimately associated with their continuous addition of new domains and motifs during the evolution of higher eukaryotes. Here we review the current knowledge on how human cytoplasmic tRNA synthetases developed their complexity, in sequence and in structure, to expand their "functionome." Importantly, not all conserved housekeeping protein families possess the increasing complexity of the aaRS family.

2 Stepwise Appearance of New Domains and Sequences in Higher Eukaryotes

Comparing the protein sequences of eukaryotic cytoplasmic aaRSs with their prokaryotic counterparts, it is immediately obvious that eukaryotic cytoplasmic aaRSs are generally larger [2], which is mainly due to extensions or insertions of conserved sequences (Fig. 1). Interestingly, most of these additions are found, at least by structural predictions, to form well-folded structures. Some of them are homologous to structural modules such as leucine-zipper and glutathione *S*-transferase (GST) domains that are widely contained in human proteins, whereas others are only found in aaRSs or aaRS-associated protein factors. Those aaRS-specific domains include two common domains (WHEP and EMAPII) that are found in more than one member of the aaRS protein family, and several unique domains with each found in a single aaRS. In addition to those well-folded domains, some shorter sequences that may not form defined structures are also found to append eukaryotic aaRSs (Fig. 1).

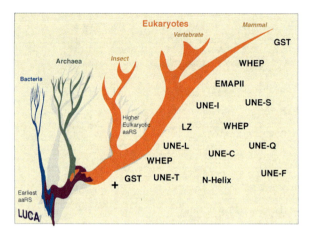

Fig. 2 Elaboration of new domains and sequence additions on aaRSs during the evolution of eukaryotes

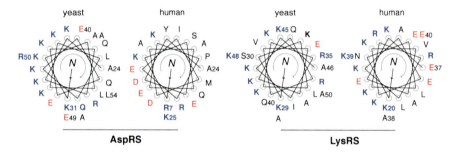

Fig. 3 Evolution of N-helix in eukaryotic AspRS and LysRS

Remarkably, the elaboration of such new domains in aaRS proteins reflects the increasing biological complexity of the higher organisms wherein the proteins reside. The more evolutionarily advanced an organism, the more of these domains and sequences appear in its aaRS proteins. Therefore, the evolutionary process to add the domains and sequences on aaRSs appears to be continuous and almost irreversible (Fig. 2).

2.1 N-Terminal Helix

The simplest form of a eukaryotic domain addition is the N-terminal helix. Single helix sequences are found in the N-terminal region of eukaryotic LysRS and AspRS. These two aaRSs belong to class IIb of the aaRS family and are more closely related to each other than to other aaRS members. Prokaryotic AspRS and LysRS consist of only an OB (oligonucleotide binding)-fold tRNA anticodon-binding domain and an aminoacylation domain. However, during evolution of single-celled eukaryote yeasts to humans, each of the two synthetases acquired a ~30–50 aa long N-terminal extension in front of the tRNA binding domain. The extensions, as suggested by structural prediction, contain a long helix of 20–40 aa (Fig. 3). The helical conformation was confirmed by NMR structure determination of the N-terminal extensions of human AspRS and human LysRS [3, 4].

These N-terminal helices have evolved to mediate biological activities. They are mostly amphiphilic, with charged residues on one side and hydrophobic residues on the other. For LysRS, positively charged residues dominate the hydrophilic side of the N-terminal helix. Early work indicated that this helical extension in human LysRS binds to the elbow region of $tRNA^{Lys}$ and enhances the binding affinity of the synthetase for the tRNA [5, 6]. Human LysRS also plays an important role in HIV infection by delivering $tRNA^{Lys3}$ (which acts as a primer for viral reverse transcription) into the virion. Importantly, the function of LysRS in HIV packaging depends on the N-terminal extension, presumably because of its tRNA binding property [7].

Interestingly, the positively charged N-terminal helix of human LysRS not only interacts with nucleic acids but also with phospholipids and proteins. Recent work has shown that human LysRS, upon phosphorylation, is translocated to the plasma membrane where its N-terminal region, including the N-terminal extension, interacts with the transmembrane region of 67LR laminin receptor. The interaction inhibits the ubiquitin-dependent degradation of 67LR, thereby enhancing laminin-induced cancer cell migration [8].

For yeast AspRS, the N-terminal helix is also positively charged on one side, and has been demonstrated to have tRNA binding properties [9]. Interestingly, compared to the yeast enzyme, human AspRS possesses more negatively charged residues located on the N-terminal helix (Fig. 3), suggesting that the amphiphilic helices may have evolved from facilitating tRNA binding in lower eukaryotes to mediating other interactions in higher eukaryotes [10–12].

2.2 EMAPII Domain

Another appended domain that facilitates tRNA binding is the EMAPII domain. Although originally discovered as a cytokine (endothelial monocyte activating polypeptide II), it is in fact a proteolytic product of MSC p43/AIMP1 – a scaffold protein in the multi-aminoacyl-tRNA synthetase complex (MSC; see below for more information on MSC) [13].

EMAPII domains are only found in aaRSs (*Caenorhabditis elegans* MetRS and metazoan TyrRS) and in aaRS-associated proteins [p43/AIMP1 and yeast Arc1p (a scaffold protein for the yeast aaRS complex, see below)]. This strict distribution of EMAPII domain suggests that its function evolved specifically for tRNA synthetases. The N-terminal portion of EMAPII (~160 aa) shares sequence homology with Trbp111, a 111 aa, free-standing tRNA binding protein found in bacteria [14]. Structural analysis of EMAPII has revealed that the monomeric EMAPII

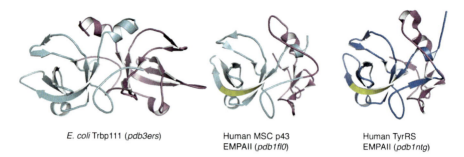

E. coli Trbp111 (pdb3ers)   Human MSC p43 EMAPII (pdb1fl0)   Human TyrRS EMAPII (pdb1ntg)

Fig. 4 The structure of *E. coli* Trbp111 and the structures of the EMAPII domain from human MSC p43 and human TyrRS. The potential cytokine motif in EMAPII domains is labeled in *yellow*

mimics the dimeric structure of Trbp111 by forming a pseudo-dimer interface with its C-terminal sequences (Fig. 4) [13]. Trbp111 specifically recognizes tRNA by binding to the elbow region of all tRNAs [15]. In certain bacteria and plants, a Trbp111-like domain is fused to the C-terminus of MetRS to enhance its aminoacylation activity by facilitating the binding of tRNA [16].

A domain homologous to EMAPII is present in TyrRS from insects to humans. The EMAPII domain in TyrRS, like many other new domains, is dispensable for aminoacylation. However, removal of EMAPII domain (to generate mini-TyrRS) through natural proteolysis outside the cell activates a cytokine-like function embedded in human TyrRS [17, 18]. Removal of EMAPII appears to expose an otherwise masked tri-peptide ELR cytokine motif in the catalytic domain [19]. A separate cytokine activity associated with the EMAPII domain is also activated when it is released from human TyrRS [17]. Interestingly, the ELR motif is not conserved in bacteria and lower eukaryotic TyrRSs, but only starts to appear from insects, concurrent with the addition of the EMAPII domain.

2.3 GST Domain

The origin of aaRS appended domains is not confined to tRNA-binding motifs. For example, *Saccharomyces cerevisiae* MetRS does not have a Trbp111 domain as seen in the C-terminus of a prokaryotic MetRS (such as *Escherichia coli*), but instead has a GST domain at its N-terminus. A GST domain sequence is also present in MetRS from insects to humans and in vertebrate ValRS [20–22]. Removal of the GST domain in yeast MetRS abolished the Arc1p-dependent aminoacylation of tRNA^{Met} [23]. Similarly, cleavage of the GST domain in human MetRS rendered the enzyme to be inactive, indicating the importance of the GST domains in aminoacylation [24].

In addition to MetRS and ValRS, two other class I tRNA synthetases, GluRS and CysRS, and three aaRS-associated proteins (the yeast Arc1p, two of the three auxiliary factors of the MSC: MSC p38/AIMP2 and MSC p18/AIMP3) all contain

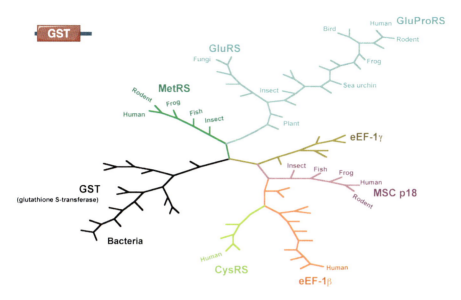

Fig. 5 Phylogenetic analysis of GST domains from aaRSs and other GST-containing proteins

a GST domain [12]. These GST domain-containing proteins are found in complexes with other proteins [25, 26], supporting the general idea that GST domains act as protein–protein binding modules. In fact, GST domains do not necessarily possess enzyme activity and are commonly used for protein assembly and for regulating protein folding (such as the GST domains in the S-crystallins, eukaryotic elongation factors 1-γ (eEF-1γ), 1-β (eEF-1β), and the heat shock protein 26 (HSP26) family of stress-related proteins) [27–29]. Interestingly, the yeast MetRS-Arc1p-GluRS ternary complex is formed through two binary GST–GST domain interactions of MetRS-Acr1p and Arc1p-GluRS (detailed assembly of the complex will be discussed in the next section) [26]. Consistently, human MetRS was shown to interact with two other GST domain-containing scaffold proteins (p38/AIMP2 and p18/AIMP3) through its GST domain [30–32], deletion of which abolished the incorporation of MetRS to MSC in cell [16]. On the other hand, although the yeast GluRS is known to interact with Arc1p through their GST domains, the function of the GST domain in human bifunctional GluProRS remains to be defined. The role of the N-terminal GST domain in mammalian CysRS, one of the latest domain additions in aaRS, is also undefined. Intriguingly, two versions of CysRS – one with the GST domain and one without – are produced by alternative splicing in mouse and human [33, 34].

Sequence analysis indicates that all GST domains in aaRSs and in MSC p18/AIMP3 and MSC p38/AIMP2 share strong homology with the eukaryote-specific elongation factors eEF-1β and eEF-1γ [26, 27]. Phylogenetic analysis of these GST domains showed stronger homology with each other than with other common GST proteins or GST domains found in bacteria (Fig. 5). Possibly the GST domains in the aaRS protein family (including the elongation factors) were generated from the same origin by gene duplication events. It is interesting to note that yeast GluRS, MetRS and Arc1p genes are located on the same chromosome (VII). However, how these gene duplication events were specifically confined to aaRS proteins is currently unknown.

2.4 WHEP Domain

Many appended domains are involved with functions other than aminoacylation, such as the regulations of inflammation, angiogenesis, and even p53 activation. The WHEP domain is one example. Initially discovered in three human aaRSs [TrpRS(W), HisRS(H), GluProRS(EP)], hence its name, the WHEP domain is ~50 aa long and shares apparent sequence homology among different aaRSs [35]. Two other human aaRSs – GlyRS and MetRS – also contain a WHEP domain at their N- and C-termini, respectively. Early studies, based on sequence similarity and intron positions, suggested that the WHEP domain might have first appeared in HisRS in single-cell eukaryotes (e.g., yeast) and then propagated to other aaRSs [35, 36]. Interestingly, the acquisition of the WHEP domain to GluProRS happened concurrently with the fusion of GluRS and ProRS into one gene, an event that took place prior to the divergence of cnidarians and bilaterians [37]. Among these aaRSs, TrpRS and MetRS appear to have the latest WHEP domain acquisition events that did not occur until the first vertebrates [22, 35]. In contrast to other aaRSs that have only one WHEP domain, GluProRS contains a varying number of WHEP domains depending on the species (3–6). In particular, human GluProRS contains three consecutive WHEP domains in between the N-terminal GluRS and the C-terminal ProRS.

The spreading of this short sequence among several aaRSs suggests that the WHEP domain might be a common tRNA-binding motif [38], though experiments testing this hypothesis have not arrived at a clear conclusion. Structure and functional analyses have indicated that the WHEP domains fold as a simple helix-turn-helix structure and act as a unique RNA recognition motif (Fig. 6) [39]. However, the WHEP domain does not significantly affect, at least in in vitro studies, the aminoacylation efficiency or the tRNA binding affinity of their host aaRSs including TrpRS, GlyRS, and GluProRS [40–42]. Recent studies revealed that the WHEP domains in human GluProRS perform non-canonical functions through protein–protein and protein–RNA interactions. For instance, the WHEP domains directly mediate the interaction between the synthetase and NSAP1 (NS1-associated protein), L13a, and GAPDH (glyceraldehyde 3-phosphate dehydrogenase) to form a gamma-IFN-activated inhibitor of translation (GAIT) complex [43–45], which interacts with eIF4G to block 43S recruitment and mRNA translation [46]. Using their RNA binding property, the WHEP domains are also responsible for recognizing the GAIT element located on the $5'$-UTR of target mRNAs [45].

Although not needed for aminoacylation, the WHEP domain appears to be a regulator for the non-canonical functions of human TrpRS. In human TrpRS the

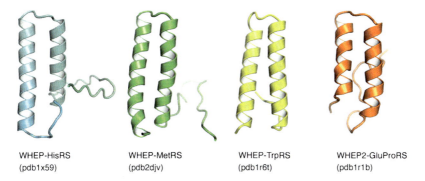

Fig. 6 Structure of WHEP domains in aaRSs

N-terminal fused single WHEP domain undergoes a conformational change when the synthetase is bound to Trp-AMP (aminoacylation reaction intermediate), which places the WHEP domain close to the active site pocket [47]. The WHEP domain can be specifically removed by proteolysis or alternative splicing to generate fragments of TrpRS (T2-TrpRS and mini-TrpRS, respectively) that exhibit angiostatic activity through interaction with the extracellular domain of VE-cadherin on the surface of endothelial cells [48–50]. The WHEP domain of TrpRS also mediates direct interactions with DNA-PK (DNA-dependent protein kinase) and PARP-1 (poly(ADP-ribose) polymerase 1) in the nucleus to activate p53 [51]. Finally, it is interesting to note that, like EMAPII domain, WHEP domain only exists within aaRS genes. This aaRS-specific domain expansion is suggestive of a special selective pressure to develop new functions for aaRS genes during the evolution of higher eukaryotes. Understanding such pressure may be instructive for understanding the various expanded functions of aaRSs, and vice versa.

2.5 Leucine Zipper

Certain expanded domains in aaRS proteins are only involved in protein–protein interactions. As such, a standard module for protein assembly, the leucine zipper, is often found to mediate protein–protein interactions in many biological processes, such as in forming the snare complex in vesicle trafficking and in forming the trimeric structure of gp41 to facilitate the entry of HIV to target cells [52, 53]. The leucine zipper is a helical motif that has leucine residues (or other hydrophobic residues) at every fourth position of the heptad repeats, so that the protruding isobutyl side chains are lined up on one side of the helix. This design creates a hydrophobic spine that interlocks with its partner to form a coiled-coil zipper.

Leucine zippers exist in ArgRS (a component of the MSC) in higher eukaryotes from insects to humans, but not in lower eukaryotes such as yeast. Leucine zippers also exist in MSC scaffold proteins p43/AIMP1 and p38/AIMP2, suggesting that

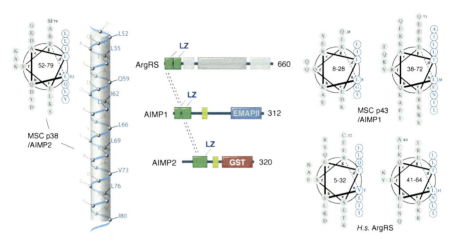

Fig. 7 Leucine zippers in human ArgRS and aaRS-associated MSC p43/AIMP1 and p38/AIMP2

they play critical roles in the assembly of the MSC (Fig. 7). The leucine zipper motifs in ArgRS interact with the leucine zippers in p43/AIMP1, which in turn interact with the leucine zipper motif in p38/AIMP2 [54]. Therefore these three proteins may form an ArgRS-p43-p38 subcomplex of MSC through several coiled-coil structures. Interestingly, a shorter form of human cytoplasmic ArgRS, which is produced from an alternative translation initiation site on the same mRNA [55–57], lacks the N-terminal leucine zipper motifs (residues 1–72) and the ability to associate with the MSC.

Leucine zippers are completely absent from aaRSs that are not associated with the MSC, suggesting that leucine zippers in aaRSs are exclusively used for the assembly of the MSC. A distinct feature of the leucine zipper motif, compared with other protein–protein interaction modules, is its linear and extended geometry, which may be an important characteristic for providing a framework to support the structure of the MCS.

2.6 Unique Domains

Besides the above-mentioned appended domains that are shared among aaRSs or aaRS-associating factors, other domains/sequences that have evolved within eukaryotic aaRSs share no detectable sequence similarity to common structural modules and, for the most part, each domain is present in only one specific aaRS. Because of their uniqueness, these domains are named UNE-X, where X represents the amino acid-specificity (in single letter code) of the aaRS to which each domain is appended [22].

Recently, several studies have suggested the importance of these unique domains in both canonical and non-canonical functions of aaRSs. Eukaryotic

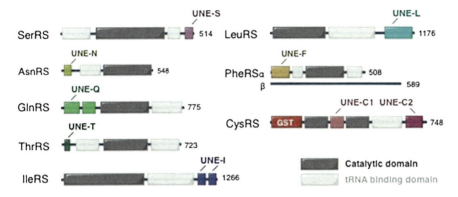

Fig. 8 UNE domains as found in human aaRSs

IleRS contains two large additions at the C-terminus (named as UNE-I$_1$ and UNE-I$_2$). UNE-I$_1$ (residue 968–1064) is found throughout eukaryotes (from yeast to human), while UNE-I$_2$ (residue 1065–1266) exists only in vertebrates (Fig. 8) [58]. Sequence analysis shows that UNE-I$_2$ contains two repetitive sequences each of ~90 aa. This region interacts with the WHEP domains of GluProRS and therefore may play a role in retaining IleRS in the MSC [31, 59].

In higher eukaryotes (from *C. elegans* to human), LeuRS contains a unique domain at the C-terminus with a size of ~110 aa (Fig. 8) [31]. UNE-L is predicted to have rich secondary structures (beta-strands and alpha helices) that presumably fold into a discrete 3D structure. Human LeuRS was recently found to function as a leucine sensor for the mTOR pathway [60–62]. Importantly, human LeuRS associates and activates the RagD GTPase of mTORC1 in a leucine-dependent manner. Removal of the C-terminal ~220 residues (including UNE-L and a LeuRS-specific domain) abolished the interaction with RagD [61]. Interestingly, the yeast LeuRS, which does not contain the UNE-L, also controls the TOR pathway. However, in contrast to the human LeuRS, the N-terminal CP1 (editing) domain of yeast LeuRS was proposed to be the binding site for the GTPase, suggesting that the mechanism of LeuRS in regulating the mTOR pathway might be substantially different in yeast as compared to that in mammals [60, 62]. It remains to be determined whether the presence of UNE-L has a role in shifting the RagD binding site from the editing domain in yeast to the C-terminus of LeuRS in human.

UNE-F is found at the N-terminus of PheRS α-subunit in eukaryotes (Fig. 8). The structure of human PheRS reveals that UNE-F folds into three continuous DNA-binding fold domains (DBD-1, -2, -3) with intervening sequences [63]. Each DBD contains three α helices folded against a three-stranded antiparallel β-sheet. The topology of the DBDs is found in many DNA-binding proteins as well as in double-stranded RNA adenosine deaminase [63]. Modeling of tRNA^{Phe} onto human PheRS suggested that UNE-F interacts with the D, T loops and the anti-codon stem of the tRNA. Deletion of UNE-F abolishes the aminoacylation activity of PheRS, consistent with its predicted role in binding and recognition of tRNA^{Phe}.

Interestingly, *T. thermophilus* PheRS, with an N-terminal structure distinct from the eukaryotic DBD, binds to a specific DNA sequence on its own gene [64]. The presence of three DNA-binding modules in UNE-F suggests that human PheRS might have non-canonical functions involving dsDNA/dsRNA binding such as in transcriptional and translational regulations.

Metazoan and fungal AsnRS differ from their bacterial homologues by the addition of a conserved N-terminal extension of ~110 aa (UNE-N) (Fig. 8). Recent structural characterization showed the UNE-N contains a structured region with a novel fold (residues 1–73) that is connected to the remainder of the enzyme by an unstructured linker (residues 74–110) [65]. Shown by NMR, the folded portion of UNE-N features a lysine-rich helix that interacts with tRNA. Whether UNE-N is also involved with non-canonical functions of AsnRS remains to be determined.

GlnRS is predominately found in eukaryotes, whereas, in most prokaryotes, Gln-tRNA^{Gln} is synthesized by an indirect pathway to form first Glu-tRNA^{Gln} by GluRS, followed by the conversion of Glu-tRNA^{Gln} to Gln-tRNA^{Gln} by a tRNA-dependent amidotransferase. Compared to the few existing bacterial GlnRSs, the eukaryotic enzymes contain an N-terminal extension of ~200 aa (UNE-Q) (Fig. 8). Yeast mutants lacking UNE-Q exhibit growth defects and have reduced complementarity for tRNA^{Gln} and glutamine [66]. Structural analysis shows that UNE-Q consists of two subdomains that resemble the two adjacent tRNA specificity-determining domains in the GatB subunit of GatCAB, the trimeric amidotransferase that can use both Glu-tRNA^{Gln} and Asp-tRNA^{Asn} as substrates to form Gln-tRNA^{Gln} and Asn-tRNA^{Asn}, respectively. The two subdomains of UNE-Q are connected by a conserved hinge region which, when mutated, reduced yeast GlnRS' affinity for tRNA^{Gln}. UNE-Q gives another example that the domain addition or exchange is highly selective for aaRSs or aaRS-related genes.

In addition to the N-terminal GST domain that seems to exist only in mammals, all eukaryotic CysRS contains two other sequence additions (UNE-C$_1$ and UNE-C$_2$) (Fig. 8). UNE-C$_1$ is inserted in human CysRS between residues 108 and 223, and in front of the well-known CP1 insertion (residue 273–419). UNE-C$_2$ is a C-terminal extension of ~150 aa. Both UNE-C$_1$ and UNE-C$_2$ are unique to CysRS, and show no apparent sequence homology to other domains or motifs.

The smallest UNE domain is the UNE-S motif (~30–40 aa) that is located at the C-terminus of vertebrate SerRS (Fig. 8). UNE-S was found to be essential for vascular development [67]. Deletion of the C-terminal sequences of SerRS, including UNE-S, led to abnormal blood vessel formation that resulted in premature death in zebrafish [68, 69]. Removal of UNE-S has little effect on the aminoacylation activity of human SerRS. Further studies discovered a significant portion of SerRS is in the nucleus of human umbilical vein endothelial cells, and the localization is directed by a classical nuclear localization signal (NLS) embedded in the UNE-S. Interestingly, SerRS regulates the expression of VEGFA in the nucleus through an unknown mechanism. This novel function is independent of the canonical aminoacylation activity of SerRS, as an aminoacylation-defective form of full length SerRS could fully rescue the vascular phenotype in zebrafish [68]. Therefore, acquisition of UNE-S appears to be important for the non-canonical function of vertebrate SerRS.

In summary, higher eukaryotic aaRSs continue to evolve with the additions of new domains/sequences. Many of these domains are related to their canonical function, especially their tRNA binding properties, while others are dispensable for aminoacylation. Although the timing of each domain addition may be different, one common feature is that, once a domain is added to an aaRS, the process is generally irreversible and the domain is conserved from then on as an integral part of that aaRS. Therefore, the timing of domain acquisition may be linked to the new biology associated with increasing complexity of the organism and could provide important hints for their potential functions beyond the canonical aminoacylation function of their prokaryotic homologues [22]. We predict that more non-canonical functions will be discovered that link to the acquisitions of the UNE domains and other appended domains of aaRSs.

3 Emergence of Multi-Synthetase Complex in Higher Eukaryotes

Prokaryotic aaRSs are a large family of ~20 proteins that perform a common function, namely, aminoacylation of tRNA. Unlike many other housekeeping machineries, such as the RNA polymerase complex and the ribosome, they do not form a complex. This feature is presumably related to the independence of the aminoacylation activity for each amino acid and their tRNAs. It might even be beneficial not to have aaRSs too near each other in a complex, as they might otherwise sterically hinder each other for tRNA-binding or attract tRNA nonspecifically, which could increase the chance of misbinding and mischarging. Although several small (binary/ternary) aaRS complexes have been found in archaea, they appear to occur serendipitously in only certain species rather than being a common feature of the family [70, 71]. Thus the emergence of a conserved, stable, and large complex of aaRSs in higher eukaryotes is remarkable.

3.1 Miniature Complex in Yeast

In eukaryotes, the complexity of an aaRS complex increases with the complexity of the hosting organism. For instance, in the single celled yeast (*S. cerevisiae*), the aaRS complex is a relatively simple ternary complex comprising MetRS, GluRS, and Arc1p (aminoacyl-tRNA synthetase cofactor 1 protein) (Fig. 9) [72]. With a GST domain at the N-terminus and a Trbp111-like domain at the C-terminus, Arc1p is essentially a fusion protein that links a protein-binding module to a tRNA-binding domain [73]. Concurrently, both MetRS and GluRS in yeast

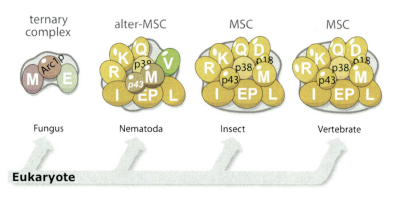

Fig. 9 Evolution of aaRS complexes in eukaryotes

acquired an N-terminal GST domain. It is through the GST domains that Arc1p links MetRS and GluRS together [74].

As mentioned in Sect. 2.2, Trbp111 is a non-specific tRNA-binding domain. The Trbp111-like domain in Acr1p not only facilitates the binding of tRNA^{Met} and tRNA^{Glu} to the complex but also nonspecifically binds to other tRNAs [23]. Although Arc1p is not essential in vivo, it enhances the aminoacylation activity by two orders of magnitude for MetRS and by one order of magnitude for GluRS as shown by in vitro kinetic studies [75]. It is worth noting that the *E. coli* MetRS has a Trbp111 domain at its C-terminus. Therefore, the same strategy for enhancing MetRS activity in *Escherichia coli* is also used in yeast through the formation of a MetRS-Arc1p-GluRS complex [73]. Interestingly, a recent study reported that the MetRS-Arc1p-GluRS complex can bind and mismethionylates many tRNA species in vitro [76]. Moreover, a similar effect on tRNA mismethionylation was achieved by fusing the Trbp111 domain of Arc1p to the yeast MetRS [77].

The MetRS-Arc1p-GluRS complex has a number of other functions. One is to regulate the cytoplasmic localization of MetRS and GluRS. Disruption of the complex by deletion of the GST-domain of Arc1p resulted in strong nuclear localization of all three components [78]; GluRS and Arc1p were found to associate with apurinic/apyrimidinic sites of damaged DNA in the nucleus [79]. In addition, the MetRS-Arc1p-GluRS complex is also important for GluRS trafficking into the mitochondria, which is important for providing an alternative pathway to the GlnRS activity that is lacking in yeast mitochondria [80, 81]. In most bacteria, archaea, and organelles of many eukaryotes, including yeast mitochondria, Gln-tRNA^{Gln} is generated through an indirect pathway, whereby a nonspecific GluRS synthesized Glu-tRNA^{Gln}, and then an aminotransferase converts it to Gln-tRNA^{Gln}. However, the yeast mitochondrial GluRS cannot mischarge tRNA^{Gln} [82]. Instead, the cytoplasmic GluRS is transported to the mitochondria to charge the mitochondrial tRNA^{Gln} with glutamate. The cellular level of Arc1p controls the translocation of GluRS. Interestingly, the expression of Arc1p is downregulated when yeast switches from fermentation to respiratory metabolism, which comes with a high demand for mitochondrial protein synthesis [80].

In addition, Arc1p is potentially involved with tRNA maturation and exportation from the nucleus. Deletion of Arc1p showed synthetic lethality when combined with the deletion of tRNA transporter Los1, which is responsible for transporting mature tRNAs from the nucleus to the cytosol [23, 83]. Overall, formation of this prototypic aaRS complex is to anchor GluRS and MetRS in the cytosol and to facilitate the aminoacylation function of the two synthetases by providing them with an additional tRNA-binding module – similar roles are found in further developed complexes in higher species.

3.2 Divergent Complex in C. elegans

In ascending the tree of life, eukaryotes evolved from single-celled to multicellular organisms, which allows for the differentiation of cells to have specialized roles. Compared to fungi, the sizes of aaRSs in C. elegans are largely increased by the addition of appended domains. Extensive interactions between new domains present in C. elegans, but not in fungi, appear to facilitate the formation of the aaRS complex in C. elegans, which shares seven aaRS components with the most highly evolved mammalian MSC. A "pull down" assay revealed that MetRS associates with a complex that contains an MSC p38/AIMP2 homologue and eight aaRSs including LeuRS, IleRS, GluRS, ValRS, MetRS, GlnRS, ArgRS, and LysRS (Fig. 9) [84]. The only exception is ValRS, which is found in the C. elegans complex but not in the mammalian MSC. No appended domain is found in C. elegans ValRS, and what enables ValRS to be associated with the C. elegans aaRS complex remains unclear. On the other hand, two aaRS components of the modern MSC (AspRS and ProRS) and two accessory factors (p43/AIMP1 and p18/AIMP3) are missing in the C. elegans complex. Interestingly, C. elegans MetRS has a C-terminal extension (~320 aa) that contains a leucine zipper and an EMAPII domain that shares extensive sequence homology with mammalian MSC p43/AIMP1. Functional studies have confirmed that human p43 can substitute the C-terminal extension of C. elegans MetRS for its in vivo activity [85]. Therefore, C. elegans MetRS can be viewed as a combination of MetRS and p43/AIMP1.

Although the C. elegans aaRS complex and the modern MSC share a significant degree of homology, some evidence suggests that not all interactions within the complexes have arisen through convergent evolution. Recent phylogenetic analysis of the GluRS and ProRS genes showed that the fusion of the two genes into one bifunctional GluProRS gene appeared before the Bilateria, and that a fission event happened in C. elegans that separated the GluProRS gene back to the two genes [37]. This analysis supported the idea that the distinct C. elegans aaRS complex is, at least with respect to some of its components, the result of divergent evolution [84].

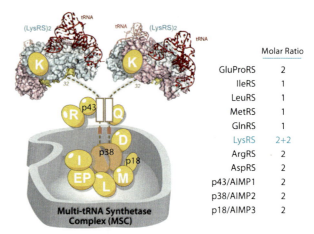

Fig. 10 Multi-aminoacyl-tRNA synthetase complex (MSC) in higher eukaryote

3.3 Multi-tRNA Synthetase Complex from Fruit Fly to Human

It has long been proposed that the MSC is a ubiquitous cellular component of higher eukaryotes [86]. Its presence has been documented in various higher eukaryotes including rabbits, sheep, bovines, mice, rats, and humans, as well as in flies (*Drosophila melanogaster*), and genes encoding the three MSC scaffold proteins – p43, p38, and p18 – are found in all representative species from insects to humans [22]. Also named as aminoacyl-tRNA synthetase interacting multi-functional protein 1, 2, and 3 (AIMP1, 2, and 3), these three scaffold proteins interact with each other and with the nine aaRS components (LysRS, ArgRS, GlnRS, AspRS, MetRS, IleRS, LeuRS, GluProRS and bifunctional aaRs) to promote assembly of the MSC with a molecular weight of 1.5 million Da (Fig. 9). The function(s) of MSC is unclear but has been proposed to facilitate channeling of aminoacylated tRNA to the ribosome during protein synthesis [87], and to provide a cellular reservoir of various non-canonical functions of its synthetase and non-synthetase components to be released in response to certain stimulation or environmental cues [25, 88].

The stoichiometry of the MSC components remains largely consistent, from *Drosophila* to human [86, 89] (Fig. 10). However, the stoichiometry appears to vary. For example, one study reported that the number of MSC p43/AIMP1 increased from 2 to 4 molecules per MSC, when the protein was overexpressed in cell [90]. Also, when the cellular level of methionine decreased, the amount of MetRS in the MSC was observed to double to two molecules per MSC [91].

A number of studies suggest that whereas the MSC serves as a reservoir for almost half of the cellular tRNA synthetases it also controls the flow of synthetases between their canonical and noncanonical functions [88, 92]. As evidence for the latter function, many MSC components have been reported to be released from the complex under certain conditions. For example, LysRS is released upon antigen-IgE induced mast cell activation [93]; GluProRS is released for the resolution of INF-γ related inflammation [43]; GlnRS is released during Fas-ligand triggered apoptosis. Interestingly, even the scaffold proteins can be released [94]. MSC p43/AIMP1 is released and secreted to act as a cytokine under stress conditions [95]; MSC p38/AIMP2 is released upon DNA damage and translocated to the nucleus to activate p53, FBP, and TRAF2 [96–98]. MSC p18/AIMP3 is released under UV radiation and is also translocated to the nucleus [99, 100].

How MSC is assembled through various interactions and how the assembly would allow for specific release of its individual components have been interesting topics for more than two decades. Early studies employing gene knockdowns of each component and crosslinking approaches have mapped out the interactions within the MSC [30, 31, 101–103]. Interestingly, certain interactions found in the early yeast complex are conserved in the MSC. For example, the interactions between the GST domains of MetRS, Arc1p, and GluRS in the yeast complex are likely to be maintained in the MSC between the GST domains of MetRS, p38, and p18 as revealed by crystal structure analysis of human MSC p18/AIMP3 [74]. On the other hand, the MSC p43/AIMP1 and p38/AIMP2 are proteins newly invented in higher eukaryotes, and are responsible for a number of interactions specific to the MSC.

The MSC component p38/AIMP2 is the core of the MSC, and directly interacts with most of the MSC components, including LysRS, GlnRS, AspRS, GluProRS, and IleRS as well as p43 and p18 [30, 90, 101, 104]. Depletion of MSC p38 leads to complete disruption of MSC [30, 105]. Mapping studies indicated that p38/AIMP2 interacts with other MSC components in a linear and sequence-dependent manner. The N-terminal end of p38/AIMP2 interacts with LysRS; the following coiled-coil region interacts with p43/AIMP1 and ArgRS; a less structured linker interacts with GlnRS; and finally, the C-terminal GST domain binds to p18/AIMP3, AspRS, MetRS, and other aaRSs (Fig. 10).

Among those interactions, the LysRS and p38 interaction appears to be most critical for the stability of MSC. Sequence alignment shows that the LysRS interacting residues in p38 are highly conserved, supporting the idea that the MSC assembly, especially the LysRS:p38/AIMP2 interaction, is conserved in all higher eukaryotes [106]. In functional tests, gene knock-downs of LysRS led to MSC disruption and subsequent degradation of MSC components [105]. A further study, using recombinant proteins, reconstituted the subcomplex of LysRS:p38 in vitro. Surprisingly, the LysRS:p38 subcomplex showed a (2:1) × 2 stoichiometry with each p38 subunit binding to one LysRS dimer through its N-terminal sequences [107]. Because p38 forms this dimer through its C-terminal GST domain, this explains the long observed stoichiometry of four subunits of LysRS present in the MSC, the highest among all MSC components [89]. Presumably, this arrangement of mutually independent bindings of two LysRS dimers for p38 allows for specific release of one LysRS dimer under certain responses, while maintaining the stability of the MSC with the other LysRS dimer [93]. This novel stoichiometry was later confirmed by a high resolution co-crystal structure of human LysRS-p38, the first subcomplex structure of MSC. The structure shows that the N-terminal 32 residues of p38 are composed of two short motifs, which sequentially bind to two symmetric grooves of the dimeric LysRS [106].

4 Metamorphosis of tRNA Synthetases

Structural metamorphosis of aaRSs has occurred during evolution by the addition of new domains, and has played a critical role in the activation and regulation of the expanded functions of aaRSs. Different functions of an aaRS may be represented by distinct conformations of the same aaRS, and each aaRS may be viewed as a collection of various conformational states that can be converted from one to the other. The various conformations of aaRS provide the structural basis for obtaining different interaction partners in different cellular or extracellular milieus that dictate its functionality, and provide ways of regulation in response to environmental stimulus or developmental cues.

The conformational change of an aaRS protein can be achieved in various ways. One mechanism is through proteolysis to remove part of the protein sequence, thus exposing new areas that were masked or hindered in the conformation of the full-length protein. Such conformational resection can also be achieved by alternative splicing at the mRNA level, with subsequent translation of the splice variants into truncated protein sequences. In addition, alternative splicing-based changes are, in principle, capable of creating bona fide structures. Other mechanisms to generate alternative conformations without a resection are exemplified as posttranslational modifications or disease-associated missense mutations which have been identified in the human population. A conformational change in protein may further trigger a shift in its structural organization at the quaternary level.

4.1 Proteolysis-Based Structural Resection

Proteolysis has been observed with several human aaRSs at specific regions. Those regions are usually linkers that join the evolutionary conserved aaRS enzyme core with a new domain incorporated at a later stage in evolution. Interestingly, these linkers are usually disordered in crystal structures of aaRSs or, for those without crystal structure information, are predicted to be unstructured. High flexibility of the linkers may render them more accessible for various proteases. Interestingly, proteolysis that removes an appended domain from the core enzyme has been shown to be associated with activation of a non-canonical function of aaRS. For example, the EMAPII-like domain appended to human TyrRS is linked to the core enzyme (named mini-TyrRS) via an unstructured loop of 22 amino acids (D343–I364) that is disordered in the crystal structure [108]. This loop is cleaved by at least two proteases – plasmin and leukocyte elastase – with different sequence specificities (Fig. 11a) [17, 109]. The cleavage at this loop activates the cytokine activities of both mini-TyrRS and EMAPII, which are otherwise mutually inhibited in the context of the full-length protein. Another example is human TrpRS, which is appended with a helix-turn-helix WHEP domain. The WHEP domain is linked to the core enzyme of TrpRS via a 29-residue loop (G55–D83) in which 21 residues

Architecture and Metamorphosis

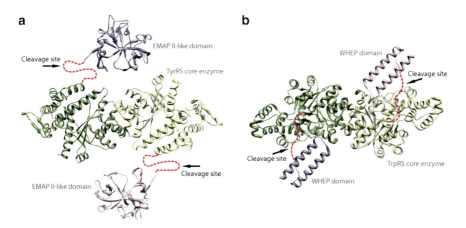

Fig. 11 Appended domains join to TyrRS (**a**) and TrpRS (**b**) enzyme cores via flexible linkers that serve as cleavage sites for the activation of novel functions of TyrRS and TrpRS

(D61–E81) were disordered in the crystal structure. Similar to TyrRS, the loop contains the cleavage site for both plasmin and leukocyte elastase [49, 110], which cleave off the WHEP domain from the core enzyme to activate its angiostatic activity (Fig. 11b).

Addition or removal of an appended domain might not dramatically affect the conformation of the core enzyme. This scenario has been demonstrated by human TrpRS, where the core enzyme adapts essentially the same conformation as a stand-alone protein and as part of the full-length protein [111]. However, as a result of an appended domain being removed, the overall conformation of an aaRS must be affected and certain areas may become exposed. In the case of human TyrRS, removal of the EMAPII domain exposes the ELR motif that is critical for mediating the angiogenic and cytokine-like activity of mini-TyrRS [17]. As for human TrpRS, removal of the WHEP domain opens the active site that has both a Trp and an ATP binding pocket. The active site is used to bind to two Trp residues near the N-terminus of VE-cadherin, and the binding blocks the hemophilic interaction between VE-cadherins that is critical for angiogenesis [48–50]. Importantly, a flexible and long linker region, like that found in TyrRS and TrpRS, would allow for the incorporation of more than one protease recognition site to facilitate cleavage at the linker region.

4.2 Alternative Splicing-Based Structural Resection

Alternative splicing at the mRNA level can also achieve structural resection of aaRSs. For example, an exon-skipping event generates an mRNA variant of human TrpRS that lacks exon 2. The splice variant is translated into a shorter form of TrpRS (mini-TrpRS) that is missing the first 47 residues and the majority of the

Fig. 12 Alternative splicing generates structural resections of aaRSs. (a) Human TrpRS and its splicing isoform. (b) Human HisRS and its splicing isoform. (c, d) Structure of human full-length HisRS (c) and the alternatively spliced isoform (d)

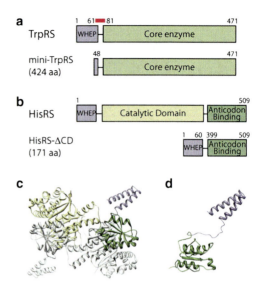

WHEP domain [112–114]. Mini-TrpRS, similar to the other WHEP domain-deleted forms of TrpRS generated by proteolysis, exhibits angiostatic activity that is masked in the full-length protein (Fig. 12a) [115].

Alternative splicing can also achieve internal deletion, which is not possible by proteolysis. A recent example is an exon-skipping event on human HisRS that removes a large segment of mRNA from exon 3 to exon 10. This event results in the precise deletion of the entire catalytic domain (CD) to make a protein product that directly links the N-terminal WHEP domain to the C-terminal anticodon-binding domain [116]. NMR spectroscopy analysis revealed a dumbbell-like conformation of the splice variant with the WHEP and anticodon-binding domains loosely linked together. Although the conformation of each domain is more or less preserved, the overall tertiary and quaternary structures of the splice variant dramatically differ from that of the full-length HisRS (Fig. 12b–d).

In principle, as alternatively spliced mRNAs are translated into new polypeptides they can generate independent new structures. The new conformations could result from new protein sequences that are being created, for example, by intron retention or frame shift events, or from sequence deletions that may affect the conformation of the rest of the proteins. In the case where an internal deletion takes place in the middle of a globular domain (rather than at the domain boundary, as in the HisRS case above), a bona fide new structure may be generated that would not be possible to generate by proteolysis.

4.3 Posttranslational Modification-Based Structural Change

Posttranslational modifications such as phosphorylation have also been found to control the non-canonical functions of aaRS proteins, presumably through induced conformational change. A well-studied example is GluProRS, the dual functional tRNA synthetase comprised of an N-terminal GluRS and C-terminal ProRS that are linked together through three consecutive WHEP domains. GluProRS is a component of the MSC. Upon interferon-γ stimulation, GluProRS is phosphorylated at Ser886 (in between the second and the third WHEP domains) and at Ser999 (after the third WHEP domain and before the C-terminal ProRS), and both phosphorylation events are required to trigger the release GluProRS from the MSC [44]. Although the structural change is undefined, phosphorylated GluProRS, but not its unphosphorylated form, can facilitate the assembly of a heterotetrameric γ-interferon-activated inhibitor of translation (GAIT) complex that binds to mRNAs with GAIT elements to suppress their translation.

Another example is LysRS. Upon immunological challenge, LysRS in mast cells is phosphorylated at Ser207, which triggers the release of LysRS from the MSC and its subsequent nuclear localization to regulate transcription factor MITF [93]. In the absence of Ser207 phosphorylation, LysRS is strongly associated within the MSC in a "closed" form to catalyze the aminoacylation reaction that charges lysine onto tRNA^{Lys} for protein synthesis [106]. However, phosphorylation at Ser207 triggers a distinct conformational change that opens up the structure. As a result, phosphorylated LysRS is released from the MSC and translocated from cytoplasm to the nucleus, where it binds to MITF, and generates diadenosine tetraphosphate (Ap$_4$A) to activate the transcription of MITF target genes. The open conformation can no longer aminoacylate tRNA but has significantly elevated activity in Ap$_4$A production [106]. Therefore, phosphorylation-based conformational change switches the function of LysRS from translation to transcription (Fig. 13). Significantly, phosphorylation at a different site, Thr52, translocates LysRS to cell membrane, where it interacts with the 67LR laminin receptor and enhances laminin-dependent cell migration in breast cancer cells [8]. Therefore the observations that two phosphorylation events on one LysRS protein lead to two completely different signaling cascades further indicates the high potential of aaRS structural metamorphosis in regulating the cellular pathways [8].

In addition to phosphorylation, other types of posttranslational modifications such as acetylation and neddylation have been found on aaRS [117–119]. Although not yet characterized, the modifications could also induce structural and functional changes on their target aaRSs.

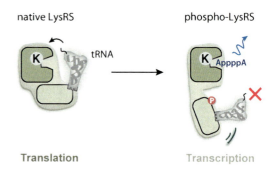

Fig. 13 Post-translational modifications generate structural metamorphosis of LysRS

4.4 Mutation-Induced Structural Change

Mutations in aaRS genes have been associated with various diseases, and most prominently with a genetic disorder named Charcot-Marie-Tooth (CMT) disease. CMT disease is the most common heritable peripheral neuropathy affecting approximately 1 in 2,500 people [120]. More than 40 genes have so far been associated with CMT through mutations that lead to similar clinical presentations that are characterized by loss of muscle tissue and touch sensation in body extremities [121]. Among them, four genes encode aaRSs (i.e., *GARS, YARS, AARS,* and *KARS,* encoding GlyRS, TyrRS, AlaRS, and LysRS, respectively) and thus make aaRS one of the largest gene families associated with CMT.

Among the four aaRSs linked to CMT, *GARS* mutations were most frequently found in CMT patients. Eleven mutations in *GARS* have been linked to an axonal type of CMT (CMT2D), and two separate mutations were found in mice to cause CMT2D-like phenotypes. The mutations are distributed throughout the protein in multiple domains, and do not always affect the enzymatic activity of the synthetase [122]. A gain-of-function mechanism has been clearly demonstrated using the mouse model *Nmf249,* where the CMT2D-like phenotype that is linked to a spontaneous mutation P234KY cannot be rescued by overexpression of WT GlyRS [123].

Crystal structure analysis did not reveal significant conformational change caused by a CMT2D mutation, presumably because the potential conformational change is restrained by the crystal lattice interactions [124]. However, a study using a solution method (i.e., hydrogen-deuterium exchange analysis) found that different CMT2D-causing mutations induce similar and dramatic conformational change that opens up the structure to expose a consensus area that is otherwise masked in the WT protein [125]. Based on this study, this consensus area that is opened up by different mutations is hypothesized to be responsible for mediating a gain-of-function interaction that leads to pathological sequelae found in CMT2D patients (Fig. 14).

Although a gain-of-function mechanism has not yet been demonstrated with other aaRS proteins linked to CMT, it is possible that a similar mutation-induced structural change may be involved. This consideration is based on the fact that

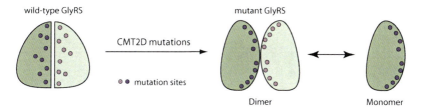

Fig. 14 CMT2D-causing mutations generate structural metamorphosis of GlyRS

almost all aaRS-associated CMT mutations are dominant, consistent with gain-of-function mutations [126–129]. In addition, CMT-causing mutations are predominantly located near subunit or domain interface, a presumably sensitive location to trigger conformational change. For example, all 13 CMT2D mutation-linked residues in GlyRS are located near the dimerization interface and approximately half of them make direct dimer interactions [122]. Similarly, CMT-linked residues in LysRS are distributed at the dimer interface [130, 131].

4.5 Quaternary Structural Change

About half of the aaRS family members form a quaternary structure (dimer, tetramer, or heterodimer) in order to be catalytically active for aminoacylation. The most common form of a aaRS is a dimer. Interestingly, at least for some aaRSs, the dimer interface is considerably smaller in the human proteins as compared to their bacteria counterparts [111]. Presumably the reduced dimer interface corresponds to a higher propensity for subunit dissociation, whereby the resulting monomeric conformation may be associated with functions that are distinct from the dimer form. An example is human TyrRS. Although both the monomer and the dimer form of mini-TyrRS can bind to the cell surface receptor CXCR1/2, only the monomer form can induce the migration of PMN cells [132]. Another example is LysRS. The monomeric form of LysRS is suggested to interact with the capsid Gag protein of HIV, which helps the packing of tRNA^{Lys}, a primer for viral reverse transcription, into the HIV virion [133]. Interestingly, although monomeric LysRS is inactive for aminoacylation [134], disruption of LysRS dimerization seems not to have a major effect on tRNA binding [135]. These examples raised the possibility that monomer-dimer equilibrium is a mechanism to regulate the aminoacylation (dimer) and the novel (monomer) functions of aaRSs.

The catalytically active dimer form can not only dissociate into monomers, but also further associate to form tetramers (Fig. 15). For example, purified human LysRS is found in solution to exist as dimers and as tetramers [107, 130]. Inside the cell, LysRS is a component of the MSC through its interaction with AIMP2/p38, a scaffold protein required for the assembly of the MSC. The binding surface for AIMP2/p38 on LysRS overlaps with the dimer–dimer interface of the LysRS tetramer [107] and, therefore, if such tetramers exist in vivo, they may regulate the assembly of the MSC.

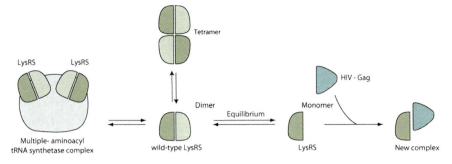

Fig. 15 Structural metamorphosis of aaRSs generated by quaternary structure changes

A change at the quaternary level can also be combined with, or regulated by, conformational changes caused by other mechanisms. For example, the aforementioned splice variant of HisRS can no longer form dimers as the dimer interface is embedded in the catalytic domain which is ablated in the splice variant [116]. Interestingly, the conformational opening of GlyRS by CMT2D-causing mutations alters the dimerization interface and depending on the mutation, either inhibits or promotes dimer formation [125].

5 Concluding Remarks

The number of genes that can be traced back to form the genetic core of a common ancestor is extremely small (~80) [136]. More than half of them are components of the translation machinery, including tRNA synthetases, ribosomal proteins, and their related factors. Therefore, from the point of view of evolutionary conservation, ribosomal proteins and their related factors are the only proteins that match the aaRS family. In comparison to these other conserved protein families, aaRS proteins have shown a high propensity to add new sequences and especially new domains during the evolution of higher eukaryotes. Why is the expansion of new domains/sequences unique among the fundamental protein machinery but universal to the aaRS family? And why are aaRS proteins particularly prominent in their capacities to mediate a broad range of functions beyond translation? These fundamentally intriguing questions remain unanswered. The uniqueness of aaRSs may be connected to some of their special features, such as their long exposure to evolutionary processes, their ubiquitous nature and presence in all life forms, their modular domain architectures, their diverse array of specific amino acid binding pockets, or their unique ability to coordinate translation with other cellular processes. Possibly the answer lies in the combination of all the features above, and more to be discovered in the future.

Acknowledgements This work was supported by US National Institutes of Health (NIH) grants GM 100136 (M.G.) and GM 088278 (X.L.Y.).

References

1. International Human Genome Sequencing Consortium (2004) Finishing the euchromatic sequence of the human genome. Nature 431(7011):931–945
2. Mirande M (1991) Aminoacyl-tRNA synthetase family from prokaryotes and eukaryotes: structural domains and their implications. Prog Nucleic Acid Res Mol Biol 40:95–142
3. Cheong HK et al (2003) Structure of the N-terminal extension of human aspartyl-tRNA synthetase: implications for its biological function. Int J Biochem Cell Biol 35 (11):1548–1557
4. Liu S et al (2013) (1)H, (13)C and (15)N resonance assignment of the N-terminal domain of human lysyl aminoacyl tRNA synthetase. Biomol NMR Assign (in press)
5. Francin M, Mirande M (2003) Functional dissection of the eukaryotic-specific tRNA-interacting factor of lysyl-tRNA synthetase. J Biol Chem 278(3):1472–1479
6. Francin M et al (2002) The N-terminal domain of mammalian Lysyl-tRNA synthetase is a functional tRNA-binding domain. J Biol Chem 277(3):1762–1769
7. Cen S et al (2004) Ability of wild-type and mutant lysyl-tRNA synthetase to facilitate tRNA (Lys) incorporation into human immunodeficiency virus type 1. J Virol 78(3):1595–1601
8. Kim DG et al (2012) Interaction of two translational components, lysyl-tRNA synthetase and p40/37LRP, in plasma membrane promotes laminin-dependent cell migration. FASEB J 26:4142–4159
9. Frugier M, Moulinier L, Giege R (2000) A domain in the N-terminal extension of class IIb eukaryotic aminoacyl-tRNA synthetases is important for tRNA binding. EMBO J 19(10):2371–2380
10. Escalante C, Yang DC (1993) Expression of human aspartyl-tRNA synthetase in Escherichia coli. Functional analysis of the N-terminal putative amphiphilic helix. J Biol Chem 268 (8):6014–6023
11. Jacobo-Molina A, Peterson R, Yang DC (1989) cDNA sequence, predicted primary structure, and evolving amphiphilic helix of human aspartyl-tRNA synthetase. J Biol Chem 264 (28):16608–16612
12. Guo M, Schimmel P, Yang XL (2010) Functional expansion of human tRNA synthetases achieved by structural inventions. FEBS Lett 584(2):434–442
13. Renault L et al (2001) Structure of the EMAPII domain of human aminoacyl-tRNA synthetase complex reveals evolutionary dimer mimicry. EMBO J 20(3):570–578
14. Morales AJ, Swairjo MA, Schimmel P (1999) Structure-specific tRNA-binding protein from the extreme thermophile Aquifex aeolicus. EMBO J 18(12):3475–3483
15. Nomanbhoy T et al (2001) Simultaneous binding of two proteins to opposite sides of a single transfer RNA. Nat Struct Biol 8(4):344–348
16. Kaminska M et al (2000) A recurrent general RNA binding domain appended to plant methionyl-tRNA synthetase acts as a cis-acting cofactor for aminoacylation. EMBO J 19 (24):6908–6917
17. Wakasugi K, Schimmel P (1999) Two distinct cytokines released from a human aminoacyl-tRNA synthetase. Science 284(5411):147–151
18. Wakasugi K, Schimmel P (1999) Highly differentiated motifs responsible for two cytokine activities of a split human tRNA synthetase. J Biol Chem 274(33):23155–23159
19. Lee PS et al (2012) Uncovering of a short internal peptide activates a tRNA synthetase procytokine. J Biol Chem 287(24):20504–20508
20. Bec G, Kerjan P, Waller JP (1994) Reconstitution in vitro of the valyl-tRNA synthetase-elongation factor (EF) 1 beta gamma delta complex. Essential roles of the NH2-terminal extension of valyl-tRNA synthetase and of the EF-1 delta subunit in complex formation. J Biol Chem 269(3):2086–2092
21. Negrutskii BS et al (1999) Functional interaction of mammalian valyl-tRNA synthetase with elongation factor EF-1alpha in the complex with EF-1H. J Biol Chem 274(8):4545–4550

22. Guo M, Yang XL, Schimmel P (2010) New functions of aminoacyl-tRNA synthetases beyond translation. Nat Rev Mol Cell Biol 11(9):668–674
23. Simos G et al (1996) The yeast protein Arc1p binds to tRNA and functions as a cofactor for the methionyl- and glutamyl-tRNA synthetases. EMBO J 15(19):5437–5448
24. He R et al (2009) Two non-redundant fragments in the N-terminal peptide of human cytosolic methionyl-tRNA synthetase were indispensable for the multi-synthetase complex incorporation and enzyme activity. Biochim Biophys Acta 1794(2):347–354
25. Lee SW et al (2004) Aminoacyl-tRNA synthetase complexes: beyond translation. J Cell Sci 117(Pt 17):3725–3734
26. Simader H et al (2006) Structural basis of yeast aminoacyl-tRNA synthetase complex formation revealed by crystal structures of two binary sub-complexes. Nucleic Acids Res 34(14):3968–3979
27. Koonin EV et al (1994) Eukaryotic translation elongation factor 1 gamma contains a glutathione transferase domain – study of a diverse, ancient protein superfamily using motif search and structural modeling. Protein Sci 3(11):2045–2054
28. Tang SS, Chang GG (1996) Kinetic characterization of the endogenous glutathione transferase activity of octopus lens S-crystallin. J Biochem 119(6):1182–1188
29. Kobayashi S, Kidou S, Ejiri S (2001) Detection and characterization of glutathione S-transferase activity in rice EF-1betabeta'gamma and EF-1gamma expressed in Escherichia coli. Biochem Biophys Res Commun 288(3):509–514
30. Kim JY et al (2002) p38 is essential for the assembly and stability of macromolecular tRNA synthetase complex: implications for its physiological significance. Proc Natl Acad Sci U S A 99(12):7912–7916
31. Rho SB et al (1999) Genetic dissection of protein-protein interactions in multi-tRNA synthetase complex. Proc Natl Acad Sci U S A 96(8):4488–4493
32. Quevillon S et al (1999) Macromolecular assemblage of aminoacyl-tRNA synthetases: identification of protein-protein interactions and characterization of a core protein. J Mol Biol 285(1):183–195
33. Kim JE et al (2000) An elongation factor-associating domain is inserted into human cysteinyl-tRNA synthetase by alternative splicing. Nucleic Acids Res 28(15):2866–2872
34. Takeda J et al (2008) Low conservation and species-specific evolution of alternative splicing in humans and mice: comparative genomics analysis using well-annotated full-length cDNAs. Nucleic Acids Res 36(20):6386–6395
35. Shiba K (2002) Intron positions delineate the evolutionary path of a pervasively appended peptide in five human aminoacyl-tRNA synthetases. J Mol Evol 55(6):727–733
36. Brenner S, Corrochano LM (1996) Translocation events in the evolution of aminoacyl-tRNA synthetases. Proc Natl Acad Sci U S A 93(16):8485–8489
37. Ray PS et al (2011) Evolution of function of a fused metazoan tRNA synthetase. Mol Biol Evol 28(1):437–447
38. Rho SB et al (1998) A multifunctional repeated motif is present in human bifunctional tRNA synthetase. J Biol Chem 273(18):11267–11273
39. Maris C, Dominguez C, Allain FH (2005) The RNA recognition motif, a plastic RNA-binding platform to regulate post-transcriptional gene expression. FEBS J 272(9):2118–2131
40. Ting SM, Bogner P, Dignam JD (1992) Isolation of prolyl-tRNA synthetase as a free form and as a form associated with glutamyl-tRNA synthetase. J Biol Chem 267(25):17701–17709
41. Ewalt KL et al (2005) Variant of human enzyme sequesters reactive intermediate. Biochemistry 44(11):4216–4221
42. Cahuzac B et al (2000) A recurrent RNA-binding domain is appended to eukaryotic aminoacyl-tRNA synthetases. EMBO J 19(3):445–452
43. Sampath P et al (2004) Noncanonical function of glutamyl-prolyl-tRNA synthetase: gene-specific silencing of translation. Cell 119(2):195–208
44. Arif A et al (2009) Two-site phosphorylation of EPRS coordinates multimodal regulation of noncanonical translational control activity. Mol Cell 35(2):164–180

45. Jia J et al (2008) WHEP domains direct noncanonical function of glutamyl-Prolyl tRNA synthetase in translational control of gene expression. Mol Cell 29(6):679–690
46. Kapasi P et al (2007) L13a blocks 48S assembly: role of a general initiation factor in mRNA-specific translational control. Mol Cell 25(1):113–126
47. Ilyin VA et al (2000) 2.9 A crystal structure of ligand-free tryptophanyl-tRNA synthetase: domain movements fragment the adenine nucleotide binding site. Protein Sci 9(2):218–231
48. Tzima E et al (2003) Biologically active fragment of a human tRNA synthetase inhibits fluid shear stress-activated responses of endothelial cells. Proc Natl Acad Sci U S A 100 (25):14903–14907
49. Wakasugi K et al (2002) A human aminoacyl-tRNA synthetase as a regulator of angiogenesis. Proc Natl Acad Sci U S A 99(1):173–177
50. Otani A et al (2002) A fragment of human TrpRS as a potent antagonist of ocular angiogenesis. Proc Natl Acad Sci U S A 99(1):178–183
51. Sajish M et al (2012) Trp-tRNA synthetase bridges DNA-PKcs to PARP-1 to link IFN-gamma and p53 signaling. Nat Chem Biol 8:547–554
52. Sutton RB et al (1998) Crystal structure of a SNARE complex involved in synaptic exocytosis at 2.4 A resolution. Nature 395(6700):347–353
53. Chan DC et al (1997) Core structure of gp41 from the HIV envelope glycoprotein. Cell 89(2):263–273
54. Ahn HC, Kim S, Lee BJ (2003) Solution structure and p43 binding of the p38 leucine zipper motif: coiled-coil interactions mediate the association between p38 and p43. FEBS Lett 542(1–3):119–124
55. Deutscher MP, Ni RC (1982) Purification of a low molecular weight form of rat liver arginyl-tRNA synthetase. J Biol Chem 257(11):6003–6006
56. Vellekamp G, Sihag RK, Deutscher MP (1985) Comparison of the complexed and free forms of rat liver arginyl-tRNA synthetase and origin of the free form. J Biol Chem 260 (17):9843–9847
57. Zheng YG et al (2006) Two forms of human cytoplasmic arginyl-tRNA synthetase produced from two translation initiations by a single mRNA. Biochemistry 45(4):1338–1344
58. Shiba K et al (1994) Human cytoplasmic isoleucyl-tRNA synthetase: selective divergence of the anticodon-binding domain and acquisition of a new structural unit. Proc Natl Acad Sci U S A 91(16):7435–7439
59. Rho SB et al (1996) Interaction between human tRNA synthetases involves repeated sequence elements. Proc Natl Acad Sci U S A 93(19):10128–10133
60. Segev N, Hay N (2012) Hijacking leucyl-tRNA synthetase for amino acid-dependent regulation of TORC1. Mol Cell 46(1):4–6
61. Han JM et al (2012) Leucyl-tRNA synthetase is an intracellular leucine sensor for the mTORC1-signaling pathway. Cell 149:410–424
62. Bonfils G et al (2012) Leucyl-tRNA synthetase controls TORC1 via the EGO complex. Mol Cell 46(1):105–110
63. Finarov I et al (2010) Structure of human cytosolic phenylalanyl-tRNA synthetase: evidence for kingdom-specific design of the active sites and tRNA binding patterns. Structure 18 (3):343–353
64. Dou X, Limmer S, Kreutzer R (2001) DNA-binding of phenylalanyl-tRNA synthetase is accompanied by loop formation of the double-stranded DNA. J Mol Biol 305(3):451–458
65. Crepin T et al (2011) A hybrid structural model of the complete Brugia malayi cytoplasmic asparaginyl-tRNA synthetase. J Mol Biol 405(4):1056–1069
66. Grant TD et al (2012) Structural conservation of an ancient tRNA sensor in eukaryotic glutaminyl-tRNA synthetase. Nucleic Acids Res 40(8):3723–3731
67. Xu XL et al (2012) Unique domain appended to vertebrate tRNA synthetase is essential for vascular development. Nat Commun 3
68. Herzog W et al (2009) Genetic evidence for a noncanonical function of seryl-tRNA synthetase in vascular development. Circ Res 104(11):1260–1266

69. Fukui H, Hanaoka R, Kawahara A (2009) Noncanonical activity of seryl-tRNA synthetase is involved in vascular development. Circ Res 104(11):1253–1259
70. Raina M et al (2012) Association of a multi-synthetase complex with translating ribosomes in the archaeon Thermococcus kodakarensis. FEBS Lett 586(16):2232–2238
71. Hausmann CD, Ibba M (2008) Structural and functional mapping of the archaeal multi-aminoacyl-tRNA synthetase complex. FEBS Lett 582(15):2178–2182
72. Frechin M et al (2010) Arc1p: anchoring, routing, coordinating. FEBS Lett 584(2):427–433
73. Karanasios E, Simos G (2010) Building arks for tRNA: structure and function of the Arc1p family of non-catalytic tRNA-binding proteins. FEBS Lett 584(18):3842–3849
74. Kim KJ et al (2008) Determination of three-dimensional structure and residues of the novel tumor suppressor AIMP3/p18 required for the interaction with ATM. J Biol Chem 283 (20):14032–14040
75. Graindorge JS et al (2005) Role of Arc1p in the modulation of yeast glutamyl-tRNA synthetase activity. Biochemistry 44(4):1344–1352
76. Wiltrout E et al (2012) Misacylation of tRNA with methionine in Saccharomyces cerevisiae. Nucleic Acids Res 40:10494–10506
77. Karanasios E, Boleti H, Simos G (2008) Incorporation of the Arc1p tRNA-binding domain to the catalytic core of MetRS can functionally replace the yeast Arc1p-MetRS complex. J Mol Biol 381(3):763–771
78. Galani K et al (2001) The intracellular location of two aminoacyl-tRNA synthetases depends on complex formation with Arc1p. EMBO J 20(23):6889–6898
79. Rieger RA et al (2006) Proteomic approach to identification of proteins reactive for abasic sites in DNA. Mol Cell Proteomics 5(5):858–867
80. Frechin M, Duchene AM, Becker HD (2009) Translating organellar glutamine codons: a case by case scenario? RNA Biol 6(1):31–34
81. Nagao A et al (2009) Biogenesis of glutaminyl-mt tRNAGln in human mitochondria. Proc Natl Acad Sci U S A 106(38):16209–16214
82. Rinehart J et al (2005) Saccharomyces cerevisiae imports the cytosolic pathway for Gln-tRNA synthesis into the mitochondrion. Genes Dev 19(5):583–592
83. Cook AG et al (2009) Structures of the tRNA export factor in the nuclear and cytosolic states. Nature 461(7260):60–65
84. Havrylenko S et al (2011) Caenorhabditis elegans evolves a new architecture for the multi-aminoacyl-tRNA synthetase complex. J Biol Chem 286(32):28476–28487
85. Havrylenko S et al (2010) Methionyl-tRNA synthetase from Caenorhabditis elegans: a specific multidomain organization for convergent functional evolution. Protein Sci 19 (12):2475–2484
86. Kerjan P et al (1994) The multienzyme complex containing nine aminoacyl-tRNA synthetases is ubiquitous from Drosophila to mammals. Biochim Biophys Acta 1199 (3):293–297
87. Kyriacou SV, Deutscher MP (2008) An important role for the multienzyme aminoacyl-tRNA synthetase complex in mammalian translation and cell growth. Mol Cell 29(4):419–427
88. Ray PS, Arif A, Fox PL (2007) Macromolecular complexes as depots for releasable regulatory proteins. Trends Biochem Sci 32(4):158–164
89. Mirande M (2005) Aminoacyl-tRNA synthetases complexes. In: Ibba M, Francklyn C, Cusack S (eds) The aminoacyl-tRNA synthetases. Eurekah, Georgetown, pp 298–308
90. Wolfe CL et al (2005) A three-dimensional working model of the multienzyme complex of aminoacyl-tRNA synthetases based on electron microscopic placements of tRNA and proteins. J Biol Chem 280(46):38870–38878
91. Lazard M, Mirande M, Waller JP (1987) Expression of the aminoacyl-tRNA synthetase complex in cultured Chinese hamster ovary cells. Specific depression of the methionyl-tRNA synthetase component upon methionine restriction. J Biol Chem 262(9):3982–3987

92. Park SG, Ewalt KL, Kim S (2005) Functional expansion of aminoacyl-tRNA synthetases and their interacting factors: new perspectives on housekeepers. Trends Biochem Sci 30(10):569–574
93. Yannay-Cohen N et al (2009) LysRS serves as a key signaling molecule in the immune response by regulating gene expression. Mol Cell 34(5):603–611
94. Ko YG et al (2001) Glutamine-dependent antiapoptotic interaction of human glutaminyl-tRNA synthetase with apoptosis signal-regulating kinase 1. J Biol Chem 276(8):6030–6036
95. Park SG, Choi EC, Kim S (2010) Aminoacyl-tRNA synthetase-interacting multifunctional proteins (AIMPs): a triad for cellular homeostasis. IUBMB Life 62(4):296–302
96. Liu J et al (2011) JTV1 co-activates FBP to induce USP29 transcription and stabilize p53 in response to oxidative stress. EMBO J 30(5):846–858
97. Choi JW et al (2009) AIMP2 promotes TNFalpha-dependent apoptosis via ubiquitin-mediated degradation of TRAF2. J Cell Sci 122(Pt 15):2710–2715
98. Han JM et al (2008) AIMP2/p38, the scaffold for the multi-tRNA synthetase complex, responds to genotoxic stresses via p53. Proc Natl Acad Sci U S A 105(32):11206–11211
99. Oh YS et al (2010) Downregulation of lamin A by tumor suppressor AIMP3/p18 leads to a progeroid phenotype in mice. Aging Cell 9(5):810–822
100. Kwon NH et al (2011) Dual role of methionyl-tRNA synthetase in the regulation of translation and tumor suppressor activity of aminoacyl-tRNA synthetase-interacting multifunctional protein-3. Proc Natl Acad Sci U S A 108(49):19635–19640
101. Kaminska M et al (2009) Dissection of the structural organization of the aminoacyl-tRNA synthetase complex. J Biol Chem 284(10):6053–6060
102. Han JM, Kim JY, Kim S (2003) Molecular network and functional implications of macromolecular tRNA synthetase complex. Biochem Biophys Res Commun 303(4):985–993
103. Norcum MT, Warrington JA (1998) Structural analysis of the multienzyme aminoacyl-tRNA synthetase complex: a three-domain model based on reversible chemical crosslinking. Protein Sci 7(1):79–87
104. Robinson JC, Kerjan P, Mirande M (2000) Macromolecular assemblage of aminoacyl-tRNA synthetases: quantitative analysis of protein-protein interactions and mechanism of complex assembly. J Mol Biol 304(5):983–994
105. Han JM et al (2006) Hierarchical network between the components of the multi-tRNA synthetase complex: implications for complex formation. J Biol Chem 281(50):38663–38667
106. Ofir-Birin Y et al (2013) Structural switch of lysyl-tRNA synthetases between translation and transcription. Mol Cell 49(1):30–42
107. Fang P et al (2011) Structural context for mobilization of a human tRNA synthetase from its cytoplasmic complex. Proc Natl Acad Sci U S A 108(20):8239–8244
108. Yang XL et al (2002) Crystal structure of a human aminoacyl-tRNA synthetase cytokine. Proc Natl Acad Sci U S A 99(24):15369–15374
109. Yang XL et al (2007) Gain-of-function mutational activation of human tRNA synthetase procytokine. Chem Biol 14(12):1323–1333
110. Kapoor M et al (2009) Mutational separation of aminoacylation and cytokine activities of human tyrosyl-tRNA synthetase. Chem Biol 16(5):531–539
111. Yang XL et al (2007) Functional and crystal structure analysis of active site adaptations of a potent anti-angiogenic human tRNA synthetase. Structure 15(7):793–805
112. Tolstrup AB et al (1995) Transcriptional regulation of the interferon-gamma-inducible tryptophanyl-tRNA synthetase includes alternative splicing. J Biol Chem 270(1):397–403
113. Shaw AC et al (1999) Mapping and identification of interferon gamma-regulated HeLa cell proteins separated by immobilized pH gradient two-dimensional gel electrophoresis. Electrophoresis 20(4–5):984–993
114. Liu J et al (2004) A new gamma-interferon-inducible promoter and splice variants of an anti-angiogenic human tRNA synthetase. Nucleic Acids Res 32(2):719–727
115. Zhou Q et al (2010) Orthogonal use of a human tRNA synthetase active site to achieve multifunctionality. Nat Struct Mol Biol 17(1):57–61

116. Xu Z et al (2012) Internally deleted human tRNA synthetase suggests evolutionary pressure for repurposing. Structure 20(9):1470–1477
117. Choudhary C et al (2009) Lysine acetylation targets protein complexes and co-regulates major cellular functions. Science 325(5942):834–840
118. Xirodimas DP et al (2008) Ribosomal proteins are targets for the NEDD8 pathway. EMBO Rep 9(3):280–286
119. Jones J et al (2008) A targeted proteomic analysis of the ubiquitin-like modifier nedd8 and associated proteins. J Proteome Res 7(3):1274–1287
120. Skre H (1974) Genetic and clinical aspects of Charcot-Marie-Tooth's disease. Clin Genet 6 (2):98–118
121. Patzko A, Shy ME (2011) Update on Charcot-Marie-Tooth disease. Curr Neurol Neurosci Rep 11(1):78–88
122. Nangle LA et al (2007) Charcot-Marie-Tooth disease-associated mutant tRNA synthetases linked to altered dimer interface and neurite distribution defect. Proc Natl Acad Sci U S A 104 (27):11239–11244
123. Motley WW et al (2011) Charcot-Marie-Tooth-linked mutant GARS is toxic to peripheral neurons independent of wild-type GARS levels. PLoS Genet 7(12):e1002399
124. Xie W et al (2007) Long-range structural effects of a Charcot-Marie-Tooth disease-causing mutation in human glycyl-tRNA synthetase. Proc Natl Acad Sci U S A 104(24):9976–9981
125. He W et al (2011) Dispersed disease-causing neomorphic mutations on a single protein promote the same localized conformational opening. Proc Natl Acad Sci U S A 108 (30):12307–12312
126. Froelich CA, First EA (2011) Dominant Intermediate Charcot-Marie-Tooth disorder is not due to a catalytic defect in tyrosyl-tRNA synthetase. Biochemistry 50(33):7132–7145
127. Storkebaum E et al (2009) Dominant mutations in the tyrosyl-tRNA synthetase gene recapitulate in Drosophila features of human Charcot-Marie-Tooth neuropathy. Proc Natl Acad Sci U S A 106(28):11782–11787
128. Jordanova A et al (2006) Disrupted function and axonal distribution of mutant tyrosyl-tRNA synthetase in dominant intermediate Charcot-Marie-Tooth neuropathy. Nat Genet 38 (2):197–202
129. Antonellis A et al (2003) Glycyl tRNA synthetase mutations in Charcot-Marie-Tooth disease type 2D and distal spinal muscular atrophy type V. Am J Hum Genet 72(5):1293–1299
130. Guo M et al (2008) Crystal structure of tetrameric form of human lysyl-tRNA synthetase: implications for multisynthetase complex formation. Proc Natl Acad Sci U S A 105 (7):2331–2336
131. McLaughlin HM et al (2010) Compound heterozygosity for loss-of-function lysyl-tRNA synthetase mutations in a patient with peripheral neuropathy. Am J Hum Genet 87 (4):560–566
132. Vo MN, Yang XL, Schimmel P (2011) Dissociating quaternary structure regulates cell-signaling functions of a secreted human tRNA synthetase. J Biol Chem 286 (13):11563–11568
133. Kleiman L, Jones CP, Musier-Forsyth K (2010) Formation of the tRNALys packaging complex in HIV-1. FEBS Lett 584(2):359–365
134. Kovaleski BJ et al (2006) In vitro characterization of the interaction between HIV-1 Gag and human lysyl-tRNA synthetase. J Biol Chem 281(28):19449–19456
135. Guo M et al (2010) Packaging HIV virion components through dynamic equilibria of a human tRNA synthetase. J Phys Chem B 114(49):16273–16279
136. Harris JK et al (2003) The genetic core of the universal ancestor. Genome Res 13(3):407–412

Top Curr Chem (2014) 344: 119–144
DOI: 10.1007/128_2013_479
© Springer-Verlag Berlin Heidelberg 2013
Published online: 27 September 2013

Protein–Protein Interactions and Multi-component Complexes of Aminoacyl-tRNA Synthetases

Jong Hyun Kim, Jung Min Han, and Sunghoon Kim

Abstract Protein–protein interaction occurs transiently or stably when two or more proteins bind together to mediate a wide range of cellular processes such as protein modification, signal transduction, protein trafficking, and structural folding. The macromolecules involved in protein biosynthesis such as aminoacyl-tRNA synthetase (ARS) have a number of protein–protein interactions. The mammalian multi-tRNA synthetase complex (MSC) consists of eight different enzymes: EPRS, IRS, LRS, QRS, MRS, KRS, RRS, and DRS, and three auxiliary proteins: AIMP1/p43, AIMP2/p38, and AIMP/p18. The distinct ARS proteins are also connected to diverse protein networks to carry out biological functions. In this chapter we first show the protein networks of the entire MSC and explain how MSC components interact with or can regulate other proteins. Finally, it is pointed out that the understanding of protein–protein interaction mechanism will provide insight to potential therapeutic application for diseases related to the MSC network.

Keywords Aminoacyl-tRNA synthetase · Multi-tRNA synthetase complex · Protein–protein interaction

Contents

1 Complex Formation of Aminoacyl-tRNA Synthetase 120
   1.1 Multi-tRNA Synthetase Complex 120
   1.2 Functional Implications for the MSC Formation 122

J.H. Kim and S. Kim
Medicinal Bioconvergence Research Center, Graduate School of Convergence Science and Technology, College of Pharmacy, Seoul National University, Seoul 151-742, South Korea
e-mail: kimjohn@snu.ac.kr; sungkim@snu.ac.kr

J.M. Han (✉)
Department of Integrated OMICS for Biomedical Science, College of Pharmacy,
Yonsei University, Seoul 120-749, South Korea
e-mail: jhan74@yonsei.ac.kr

2	Pairwise Interactions of Aminoacyl-tRNA Synthetases and Their Functional Implication	125
	2.1	Interaction Pairs of AIMP1	125
	2.2	Interaction Pairs of AIMP2	127
	2.3	Interaction Pairs of AIMP3	130
	2.4	Interaction Pairs of KRS	132
	2.5	Interaction Pairs of VRS	135
	2.6	Interaction Pairs of QRS	136
	2.7	Interaction Pairs of LRS	136
	2.8	Interaction Pairs of WRS	137
	2.9	Interaction Pairs of EPRS	139
	2.10	Interaction Pairs of GRS	140
	2.11	Roles of HSP90 in the Molecular Interaction of ARSs	140
References			141

1 Complex Formation of Aminoacyl-tRNA Synthetase

1.1 Multi-tRNA Synthetase Complex

ARSs can be classified into two groups based on their structural features. Class I ARSs each possess a Rossman fold in their catalytic domains, whereas class II enzymes contain three homologous motifs with degenerate sequence similarity [1–3]. ARSs can also be grouped on the basis of their ability to form multi-tRNA synthetase complex (MSC) with each other and auxiliary factors, whereby ARSs can form a macromolecular complex. Among the complexes formed by ARSs, the mammalian MSC is the most intriguing [4–6]. This complex is distinctive as compared with other macromolecular protein complexes in that its components are enzymes that carry out similar catalytic reactions simultaneously, and only a subset of ARSs are involved (Fig. 1).

In the yeast *Saccharomyces cerevisiae*, a primitive form of such ARS complexes is present. The yeast MSC is composed of three proteins: ERS, MRS, and the non-enzyme component called Arc1p, which is homologous to mammalian AIMP1 [7]. Arc1p binds to the N-terminal domains of both ERS and MRS, increasing their affinity for the cognate tRNAs, thereby promoting the export of tRNA from nucleus to cytosol [8]. In studies where the yeast Arc1p complexes were analyzed in solution by small-angle X-ray scattering (SAXS), the ternary complex of ERS and MRS with Arc1p displayed a peculiar extended star-like shape. The SAXS analysis revealed that binding of the cognate tRNAs initiated a striking compaction of the pentameric complex as compared to the ternary one. Based on this data, a hybrid low-resolution model of the pentameric complex has been constructed, which rationalizes the compaction effect as occurring through interactions of negatively charge-tRNA backbones with the positively charge-tRNA binding domains of the two enzymes [9, 10].

In the nematode *Caenorhabditis elegans*, a species grouped with arthropods in modern phylogeny, the structural organization of MSC displays significant variation, not containing an EPRS enzyme, as observed in the vertebrate branch

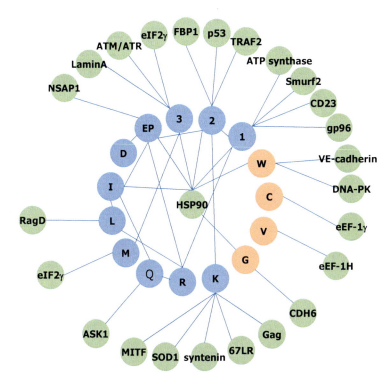

Fig. 1 Protein–protein interaction network of aminoacyl-tRNA synthetases (ARSs). This interaction network shows the possible connections among ARSs, ARS-interacting multi-functional proteins (AIMPs), and their binding partners. Eight different ARSs (EPRS, DRS, IRS, LRS, MRS, QRS, RRS, and KRS), and three auxiliary factors (AIMP1, AIMP2, and AIMP3), form a macromolecular protein complex, which is known as the multisynthetase complex (MSC). MSC-forming and noncomplex components are shown in *black* and *white*, respectively

of deuterostomes and in the arthropod branch of protostomes [11]. The other specific features of this complex are the retention of VRS, and the absence of DRS, AIMP1, and AIMP3 proteins. Thus, metazoan MSC contains RRS, ERS, QRS, IRS, LRS, KRS, MRS, and AIMP2.

The metazoan MSC is well conserved and comprises more proteins compared with that of yeast. The molecular weight of the complex is approximately $1.0–1.2 \times 10^6$ Da and sediments as an 18–30S particle. In negatively stained electron micrographs, the particle appears in several orientations including triangular, rectangular, and square shapes. A working structural model of the complex has been proposed as a cup or elongated "U" shape, based on low resolution electron microscopic images. The earliest electron micrographs of density gradient fractions of the ARS preparations from rat liver showed negatively stained "dumbbell"-shaped and circular particles 11–14 nm in size [12–14].

The mammalian MSC consists of eight different enzymes: EPRS, IRS, LRS, QRS, MRS, KRS, RRS, and DRS, and three auxiliary proteins: AIMP1/p43,

AIMP2/p38, and AIMP3/p18. They are all invariably associated with the MSC [10, 15, 16]. On the basis of previous studies [15, 16], the components of MSC can be grouped into two subdomains based on their association with AIMP2. One subdomain includes QRS, RRS, and AIMP1, which is linked to the N-terminal region of AIMP2, containing the leucine zipper domain. The other subdomain includes KRS, MRS, DRS, AIMP3, and the three high molecular weight proteins EPRS, IRS, and LRS, which is linked to the C-terminal region of AIMP2, containing the GST domain.

Although the spatial arrangement and stoichiometry of the components are undefined, it is obvious that AIMP2 is an essential scaffold molecule for the assembly of the MSC [17]. Systematic depletion studies of MSC components with the specific siRNAs revealed that the protein stability of some components depends on the presence of their neighbors in the complex. Thus the overall stability of the MSC complex is determined by the inclusion of these components. Furthermore, the depletion of AIMPs results in the dissociation of some components such as MRS, QRS, KRS, and RRS from MSC, suggesting that these auxiliary factors are also important for MSC assembly.

When comparing ARSs which participate in MSC formation with those that do not, there are no obvious differences in their size distributions, structural features, post-translational modifications, expression profiles, or chromosomal locations. However, comparison of functional motifs present in ARSs reveals that the presence of the GST domains and leucine zipper domains may be crucial determinants for MSC formation. GST domains are present in the N-terminal extensions of MRS and EPRS of the complex-forming enzymes, and the C-terminal regions of AIMP2 and AIMP3 of the auxiliary factors [18–20]. In addition, leucine zipper domains are present in the N-terminal extensions of RRS, one of the complex-forming enzymes, as well as the N-terminal regions of AIMP1 and AIMP2, two of the auxiliary factors.

1.2 Functional Implications for the MSC Formation

Over the years a number of possible functions have been proposed for the MSC. These include (1) serving as a means to stabilize its own components, (2) acting as a molecular reservoir to regulate the non-canonical activities of ARSs, and (3) channeling and delivery of aminoacyl-tRNAs for protein synthesis. We describe these functions below.

1. With regards to MSC stabilizing its own components, experiments where human MSC components were depleted using siRNAs suggest that the stability characteristics of each component varies depending on the presence of other components in the complex, as well as cell type and growth conditions. Among the components, AIMPs appear to be critical for the stability of the whole MSC. This stability is also dependent on the presence of other enzymes in the complex. However, KRS differs from all other MSC components as it does not require the

other components for its stability, although it is essential for formation of the MSC. In contrast, RRS and MRS appear to be strongly dependent on complex formation for their stability [16]. Some components such as MRS, QRS, KRS, and RRS appear to be easily released from the complex upon appropriate signals. The linkages of these components may play a part in their dissociation from the MSC in response to specific signals.

2. Another hypothesized role for MSC is its function as a molecular reservoir to regulate the non-canonical activities of ARSs. Some mammalian MSC components, such as EPRS, KRS, AIMP2, and AIMP3, are released from the MSC upon specific signals, whereupon they exert non-canonical activities. For instance, EPRS is known to be a component of the hetero-tetrameric gamma-interferon-activated inhibitor of translation (GAIT) complex that binds 3′UTR GAIT elements in multiple interferon-gamma (IFNγ)-inducible mRNAs, and suppresses their translation. IFN-gamma induces sequential phosphorylation of serine 886 and serine 999 in the non-catalytic linker of EPRS, which triggers EPRS release from the MSC. Phosphorylation at serine 886 is required for the interaction of NSAP1, which blocks EPRS binding to target mRNAs. The same phosphorylation event induces subsequent binding of ribosomal protein L13a and GAPDH, and finally restores mRNA binding. Phosphorylation at serine 999 induces the formation of a functional GAIT complex that binds initiation factor eIF4G, and eventually represses translation. This two-site phosphorylation provides structural and functional pliability to EPRS and controls the repertoire of activities that regulates inflammatory gene expression [21–25].

In another example, upon binding of laminin, KRS is phosphorylated at threonine 52 by p38MAPK and dissociated from the MSC. This phosphorylation is required for KRS translocation to the membrane, whereupon it binds to the 67 kDa laminin receptor (67LR). Protein turnover of 67LR in the membrane is controlled by Nedd4-mediated ubiquitination. KRS inhibits ubiquitin-mediated degradation of 67LR, thereby enhancing laminin-induced cell migration [26]. In stimulated mast cells, KRS is also phosphorylated, inducing its release from the MSC, and activating the transcription factor MITF. Phosphorylation at serine 207 induces a striking conformational change, resulting in the inactivation of its translational function, and activation of transcriptional function. Structural analysis shows that KRS is held in the MSC by its binding to the N-terminus of the AIMP2 [27]. Phosphorylation at serine 207 disrupts the binding grooves for AIMP2, resulting in release of KRS and its nuclear translocation. This conformational change also exposes the C-terminal domain of KRS, allowing it to bind to MITF, and triggers the formation of second messenger Ap4A, an endogenous molecule consisting of two adenosine molecules linked by four phosphates. Ap4A is known to be synthesized by KRS in the absence of cognate tRNAs. It binds to Hint and liberates MITF, leading to the activation of MITF-dependent gene expression in immune response [28].

Upon DNA damage response, AIMP2, which is a scaffolding protein required for the assembly of the MSC, is phosphorylated, dissociated from the MSC, and translocated into the nucleus. AIMP2 directly interacts with p53, thereby preventing MDM2-mediated ubiquitination and degradation of p53 in response to DNA damage. One possible explanation for the involvement of AIMP2 in the p53 pathway is that AIMP2 controls tRNAs biogenesis in coordination with p53, since p53 is known to function as a repressor of RNA polymerase III transcription and inhibits the synthesis of essential small RNAs including tRNAs [29]. Furthermore, in mammalian MSC, AIMP3 has been shown to associate with MRS. Upon DNA damage response, the general control nonrepressed-2 (GCN2) kinase phosphorylates MRS at serine 662 and this phosphorylation induces a conformational change of MRS and its dissociation from AIMP3, resulting in translocation to nucleus [30]. Together, these multiple lines of evidence suggest a functional role for MSC formation in a signaling reservoir that regulates the non-canonical activities of ARSs in cellular homeostasis.

3. The structural organization of MSC suggests that aminoacyl-tRNAs may be channeled in cells. In Chinese hamster ovary (CHO) cells, co-electroporation of various combinations of free [^{14}C]amino acids and [^{3}H]aminoacyl-tRNAs, whereby free amino acids serve as precursors for protein synthesis, exogenous aminoacyl-tRNAs are utilized poorly. The lack of incorporation into protein from exogenous aminoacyl-tRNAs is not due to their leakage from the cell, to their instability, or to their damage during electroporation [31, 32]. Furthermore, in contrast to the findings with intact cells, extracts of CHO cells incorporate both free amino acids and aminoacyl-tRNAs into protein with similar efficiencies, suggesting channeling occurs within live cells.

Other studies support the channeling model for mammalian translation. In mammalian cells there are two forms of RRS, both products of the same gene. The high molecular weight form, which is present exclusively as an integral component of the MSC, is used to generate Arg-tRNA^{Arg} for translation, whereas in the low molecular weight, free RRS generates Arg-tRNA^{Arg} for N-terminal arginylation of certain proteins. The high molecular weight form of RRS is essential for normal protein synthesis and growth of CHO cells even when low molecular weight, free RRS is present and Arg-tRNA continues to be synthesized at close to wild-type levels [33]. In the MSC, AIMP3 and MRS are binding neighbors to each other, and form a ternary complex with eukaryotic initiation factor 2γ subunit (eIF2γ) [34]. AIMP3 recruits active eIF2γ to the MRS–AIMP3 complex. AIMP3 also has Met-tRNA^{iMet} -binding ability and mediates the delivery of Met-tRNA$_i^{Met}$ from MRS to the eIF2 complex for accurate and efficient translation initiation. Interactions between EF-1α and various ARSs have been described in eukaryotes. EF-1α is known to form a stable complex with LRS. Complex formation has little effect on EF-1α activity, but increases the k(cat) for Leu-tRNA^{Leu} synthesis approximately eightfold. In addition, EF-1α co-purifies with the archaeal MSC comprised of LRS, KRS, and PRS [35]. These interactions between EF-1α and the archaeal MSC contribute to translational fidelity, both by enhancing the aminoacylation efficiencies of the three ARSs in the complex and by coupling two stages of translation: aminoacylation of cognate tRNAs and their subsequent channeling to the ribosome.

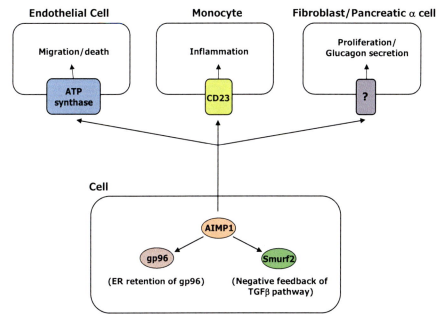

Fig. 2 Protein–protein interaction network of AIMP1. In the cytoplasm, AIMP1 mainly associates with RRS within the MSC. In the ER, AIMP1 binds to gp96 to regulate its ER retention and chaperone activity. AIMP1 translocates to the nucleus and interacts with Smurf2 to control TGFb signaling. AIMP1 is secreted and regulates angiogenesis through ATP synthase on endothelial cells, inflammatory response through CD23 on monocytes, glucose homeostasis through glucagon secretion from pancreatic α cells, and cell proliferation of fibroblasts

2 Pairwise Interactions of Aminoacyl-tRNA Synthetases and Their Functional Implication

2.1 Interaction Pairs of AIMP1

2.1.1 AIMP1 and gp96

AIMP1 consists of 312 amino acids, and binds to RRS within the MSC. Its C-terminal domain (95 amino acids) contains an oligonucleotide-binding motif capable of interacting with tRNA. Although the roles of the C-terminal domain have not yet been clarified, they seem to be involved in specific protein–protein interaction with RRS. AIMP1 performs a variety of functions depending on its cellular distribution and interacting partners (Fig. 2).

One such partner is gp96, an endoplasmic reticulum (ER)-resident chaperone protein and a member of the heat shock protein 90 (HSP90) family [36–38]. AIMP1 interacts with gp96 and mediates gp96-KDEL receptor binding, thereby controlling ER retention of gp96 in the regulation of immune tolerance. Evidence supporting the role of gp96 in immune tolerance includes data from a transgenic mouse model expressing cell surface gp96, which showed significant dendritic cell (DC) activation and maturation, and the development of lupus-like autoimmune disease. As gp96 is expressed on the cell surface of immune cells, it appears to be involved in innate and adaptive immunity. Direct interaction of gp96 with DC induces the maturation of DC, promotes the secretion of proinflammatory cytokine, and regulates the activation of major histocompatibility classes I and II (MHC I and II) [39].

Gp96 contains a C-terminal KDEL sequence, which is involved in retrograde transport from the Golgi compartment to the ER [40, 41]. The KDEL receptor is located at the Golgi, binds with high affinity to KDEL-bearing proteins such as gp96, and is translocated to the ER by chaperone proteins. The association of KDEL ligand with the KDEL receptor in the Golgi or ER is dependent on the pH of compartment lumen, whereby the KDEL ligand is dissociated from its receptor in the ER in high pH conditions. AIMP1 interacts with gp96 and mediates gp96-KDEL receptor binding, thereby controlling ER retention of gp96 in the regulation of immune tolerance. Gp96 has a number of functional domains including a nucleotide binding ATPase domain, acidic domain, and dimerization domain. AIMP1 interacts with the dimerization domain of gp96 and facilitates binding between the gp96 dimer and KDEL receptor dimer. In addition, the central region of AIMP1 interacts with gp96.

2.1.2 AIMP1 and CD23

Although the main portion of AIMP1 is bound to the MSC, a lesser portion is secreted to the extracellular matrix where it works as a cytokine on various target cells such as endothelial cells, macrophage, and pancreatic α cells. CD23 is the low-affinity receptor for immunoglobulin E (IgE) and is also cleaved from the cell surface to produce a soluble CD23 (sCD23) protein with pleiotropic cytokine-like activity [42, 43]. As a membrane protein, CD23 is a type II transmembrane glycoprotein of approximately 45 kDa molecular weight. It is comprised of a large C-terminal globular extracellular domain that is similar to C-type lectins, the stalk region bearing a putative leucine zipper important for CD23 oligomerization, a single hydrophobic membrane-spanning region, and an N-terminal cytoplasmic domain. CD23 interacts with binding partners such as IgE, CD21, major histocompatibility complex (MHC) class II and αv integrin, and mediates a number of cellular responses including IgE synthesis, antigen presentation, proliferation of B cells, and activation of monocytes. It has been suggested that CD 23 can potentially serve as a diagnosis marker in a wide range of disease as well as a target for therapeutic intervention [44]. In immune cells such as THP-1 monocytic cells and human peripheral blood mononuclear cells (PBMCs), CD23 functions as a receptor for AIMP1 and mediates cell migration and TNFα secretion by AIMP1. The C-terminal EMAPII domain of AIMP1 is not involved in CD23-mediated activities; however, the central region of AIMP1 is closely linked to the cytokine effect of AIMP1 by its binding to CD23 [45] (Fig. 2).

2.1.3 AIMP1 and Smurf2

AIMP1 functions as a component of the negative feedback loop of the TGF-β pathway [46, 47]. Smad-mediated ubiquitination regulatory factor 1 (Smurf1) and Smurf2 are HECT (homologous to the E6-accessory protein C-terminus)-type E3 ubiquitin ligases which regulate TGF-β and BMP signaling. Studies suggest that Smurfs exert inhibitory roles on the TGF-β pathway [46]. Smurfs contains an N-terminal C2 domain for membrane binding, a central region containing two or three WW domains for protein–proteins interaction, and a C-terminal HECT domain for protein ubiquitination. Both Smurf1 and Smurf2 interact with Smad7 to induce the ubiquitin-dependent degradation of Smad7 and the associated receptors of the TGF-β family. Although Smurf1 and Smurf2 are localized to the nucleus, their binding to Smad7 induces their export and recruitment to activated receptors, resulting in the degradation of the receptor and Smad7 [46].

Smurf2 was discovered to be an inhibitor of the TGF-β pathway through its degradation of Smad2 [48]. Smurf2 also interacts with Smad7 to target TGFβ receptor I for degradation [49]. Thus, in conjunction with Smad7, Smurf2 serves as a negative feedback mechanism to halt TGF-β signal transduction. Thus, the nature of Smurf2 function is paradoxical, both enhancing TGF-β pathways through the degradation of nuclear corepressors while inhibiting them through the degradation of the receptor. This contradictory mechanism of Smurf2 is explained by its location within cells, whereby nuclear Smurf2 promotes TGF-β signaling and cytoplasmic Smurf2 exhibits the opposite behavior. A recent study suggested that cytoplasmic Smurf2 negatively regulates TGF-β signaling through binding to the C-terminal region of AIMP1, which excludes the EMAPII domain [46] (Fig. 2).

2.2 Interaction Pairs of AIMP2

2.2.1 AIMP2 and FBP1

AIMP2 consists of 320 amino acids with four exons, and in the MSC associates with KRS and AIMP1 through its leucine zipper domain. The protein stability of the far upstream element binding protein 1 (FBP1) is regulated by AIMP2 [50, 51] (Fig. 3). FBP1 was first discovered as a transcriptional regulator of the proto-oncogene transcription factor c-myc, which controls diverse cellular responses including cell proliferation, differentiation and apoptosis [52]. The human *FBP1* gene encodes a 644-amino acid protein with 3 well-defined domains: an amphipathic helix N-terminal domain, a tyrosine-rich C-terminal transactivation domain, and a DNA-binding central domain.

FBP1 interacts with the central region of AIMP2 through its C-terminus, as shown by studies in which AIMP2 does not bind to the C-terminus of FBP2, the C-terminus of FBP3, and the N-terminus of FBP1. AIMP2 promotes FBP1 degradation by post-translational ubiquitination, without altering FBP1 mRNA levels or the overall rate of

FBP1 protein synthesis. An in vivo study showed that knocking out AIMP2 increases the levels of FBP1 protein and c-Myc expression in fetal lungs, leading to a defect in lung differentiation and development of respiratory distress syndrome [50].

2.2.2 AIMP2 and TRAF2

AIMP2 controls the cell fate decision of apoptosis vs proliferation through signaling pathways involving FBP1, p53, and TNF-receptor associated factor 2 (TRAF2) [53] (Fig. 3). As a type II transmembrane protein, TNF ligand forms trimers and can interact with its cognate receptors through its cell-bound or soluble forms. TNF ligand regulates cellular events by binding to its cognate receptor, TNFR. TNFR is a type I transmembrane protein with conserved cysteine-rich domains (CRD), typically consisting of three disulfide bridges [54].

Binding of TNF induces the trimerization of TNFR, which leads to intracellular signaling. Recent studies have shown that TNFR can self-assemble in the absence of a ligand, then undergoing conformational changes upon ligand engagement. Although the extracellular domains of TNFR are similar, the different forms of TNFR can be classified into three subgroups based on motifs in their cytoplasmic tail; death receptor, decoy receptor, and activating receptor of TNFR [54, 55].

TNFR1 serves as the major receptor for TNF-induced signaling pathway, mediating such pleiotropic functions as activation of nuclear factor kappaB (NF-κB) and induction of apoptosis. Ligand-induced reorganization of preassembled receptor complexes enables TNFR1 to recruit the adaptor protein TNF receptor-associated protein with a death domain (TRADD) and TRAF2. TRAF2 consists of an N-terminal RING domain followed by five zinc fingers and a C-terminal TRAF domain, which mediates homo-trimerization and interaction with TRADD. TRAF proteins are defined by their ability to couple TNFR to signaling pathways that govern the cellular effects induced by TNF. TRAF predominantly signals to proteins leading to the activation of the NF-κB and AP-1 transcription factors. AIMP2 also controls apoptosis through ubiquitin-dependent degradation of TRAF2 as mediated by TNFα signaling. In this case, the central region of AIMP2 binds to TRAF2 and increases the association of c-IAP (E3 ubiquitin ligase) with TRAF2 to regulate degradation of TRAF2 [53]. TNFR2 expression is restricted to specific neuronal subtypes, astrocytes, endothelial cells, lymphocyte, cardiac myocytes, and mesenchymal stem cells.

2.2.3 AIMP2 and p53

Although AIMP2 is primarily known to be a scaffold protein of the MSC, it also works as a direct positive regulator of p53. Upon DNA damage, it is phosphorylated and translocated to the nucleus, where it plays a role as a pro-apoptotic mediator through its binding to the p53 tumor suppressor [29] (Fig. 3). p53 is the most frequently inactivated tumor suppressor gene in human cancers, mutated in

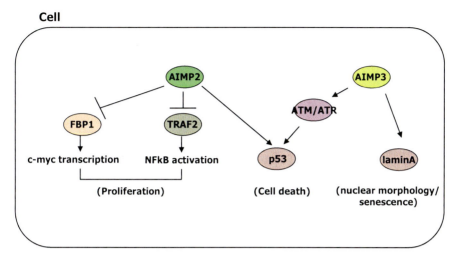

Fig. 3 Protein–protein interaction network of AIMP2 and AIMP3. On genotoxic damage, AIMP2 is translocated to the nucleus where it activates p53 directly. AIMP2 also augments the apoptotic signal of tumor necrosis factor (TNF) through the downregulation of TNF receptor associated factor 2 (TRAF2) and mediates the transforming growth factor-β (TGFβ) anti-proliferative signal through the downregulation of fuse-binding protein (FBP). AIMP3 is also translocated to the nucleus upon damage. AIMP3 activates p53 through the activation of ataxia telangiectasia mutated (ATM) and ATM- and Rad3-related (ATR) kinases. AIMP3 also induces the cellular senescence through the regulation of lamin A

approximately 50% of human cancers based on clinical studies [56, 57]. Although p53 is normally maintained in the inactivated state at low levels in unstressed cells, under stress conditions both the level of p53 and its transcriptional activity are increased to induce diverse cellular processes such as apoptosis, DNA repair, cell cycle, and senescence under stress conditions. The level and transcriptional activity of p53 are controlled by the negative regulator murine double minute 2 (Mdm2), which is discovered on double minute chromosomes in a derivative cell line of NIH 3T3 cells. Mdm2 belongs to the family of ubiquitin E3 ligases that contains a RING (really interesting new gene) finger domain. There are two distinct mechanisms by which Mdm2 suppresses p53 [58, 59]: (1) by binding to the N-terminal domain of p53 and masking its access to transcriptional machinery and (2) by ubiquitinating p53 and targeting it for proteasomal degradation. However, recent research has shown that Mdm2-p53 binding alone is insufficient to suppress p53 activity in the absence of Mdm2 E3 ubiquitin ligase activity. Recently the stability of p53 was reported to be protected from ubiquitination-mediated degradation by Mdm2. In this process, AIMP2 may be involved in the regulation of p53 stability by binding to it. Specifically, the central region of AIMP2 binds to the N-terminus of p53, which is a binding site for Mdm2. Thus, AIMP2 has as a pro-apoptotic activity as well as an anti-proliferative property through the regulation of ubiquitination. Another paper reported that AIMP2 has an alternative splicing variant and its splice variant lacking exon 2 (AIMP2-DX2) is highly expressed in lung cancer cells [60]. In addition, it was reported that AIMP2-DX2 negatively regulates the tumor suppressor activity of AIMP2 by disrupting interaction with p53.

2.3 Interaction Pairs of AIMP3

2.3.1 AIMP3 and ATM/ATR

Consisting of 174 amino acids, AIMP3, known as a haploinsufficient tumor suppressor, is the smallest component in the MSC. It is homologous to translation elongation factor 1 and interacts with MRS. Following UV irradiation, AIMP3 dissociates from MRS through phosphorylation by general control nonrepressed-2 (GCN2) [30]. In response to DNA damage or oncogenic stresses, AIMP3 is translocated to the nucleus, interacts with ataxia telangiectasia mutated (ATM) and ATM- and Rad3-related (ATR), and ultimately regulates p53 [61, 62] (Fig. 3).

ATM/ATR kinases play key roles in the response to a broad spectrum of DNA damage and DNA replication stress. The ATM/ATR kinases are members of the phosphoinositide 3-kinase (PI3K)-like protein kinase (PIKK) family. Both ATM and ATR are large kinases with significant sequence homology and functional similarity [63]. These kinases have a strong preference for phosphorylating serine and threonine residues that are followed by glutamine. However, although ATM is activated in response to rare occurrences of double strand breaks (DSBs), ATR is activated during every S phase to regulate the firing of replication origins and the repair of damaged replication forks. It is no wonder that ATM/ATR regulates p53 activity at multiple levels. Both ATM and ATR directly interact with p53, and enhance its transactivating activity by phosphorylating it at serine 15. ATM is also required for dephosphorylation of serine 376, which can create a binding site for 14-3-3 adaptor protein. The association of p53 with 14-3-3 increases the affinity of p53 for its specific DNA sequence, therefore enhancing its transcriptional activity.

Other ATM sites where p53 can be phosphorylated are serine 6, 9, 46, and threonine 18. Phosphorylation at serine 46 is a critical process for apoptotic activity, whereas other phosphorylation sites are required for acetylation of p53 [64]. In addition, ATM/ATR can regulate p53 through other kinases such as c-abl and chk2. ATM-induced c-abl phosphorylates p53 at serine 20, which is important for the stabilization of p53, since this phosphorylation disrupts the interaction between p53 and Mdm2 [65]. ATM-activated chk2 also phosphorylates the same site, leading to activation of p53. Furthermore, ATM binds and phosphorylates Mdm2, inhibiting p53 degradation and promoting its accumulation within cells. Recently it was reported that AIMP3 directly interacts with ATM/ATR and activates ATM/ATR in response to DNA damage, and is a haploinsufficient tumor suppressor and key factor for ATM/ATR-mediated p53 activation [63].

2.3.2 AIMP3 and Lamin A

AIMP3 is involved in senescence as well as tumorigenesis. Whereas mice that are homozygous null for AIMP3 show early embryonic death, heterozygous mice are born live but show highly increased susceptibility to various cancers. AIMP3 heterozygous cells transformed with oncogenes such as Ras or Myc also display severe nuclear fragmentation and chromosome instability. AIMP3 transgenic mice exhibit a progeroid phenotype and significantly shorter lifespan than wild-type mice, and AIMP3-overexpressing cells show defects in nuclear morphology and accelerated senescence. The mechanism for AIMP3's regulation of senescence may be through its binding to lamin A, which is one of the filament proteins of nuclear lamina (Fig. 3). A recent paper suggested that AIMP3 regulates cellular senescence through lamin A ubiquitination and degradation mediated by the E3 ubiquitin ligase Siah1 [66]. Lamins play important roles in the maintenance of nuclear shape and integrity, organization of chromatin, and distribution of nuclear pore complexes. Mammalian somatic cells contain both A-type (lamins A and C) and B-type (lamins B1 and B2) lamins. A-type lamins are generated by alternative splicing of the RNA transcribed from the LMNA gene, while B-type lamins (lamins B1 and B2) are encoded by distinct transcripts originating from the LMNB1 and LMNB2 genes [67, 68]. Although the expression of A-type lamins is developmentally regulated and is understood to accompany tissue and cellular differentiation, the expression of B-type lamins is essential for cell viability. After transcription and translation of LMNA, the precursor protein prelamin A is formed which is further processed to lamin A in four distinct steps; farnesylation, removal of the last tree amino acids, carboxymethylation, and finally removal of the last 15 amino acids. Most lamniopathies arise from defects in this posttranslational modification process, which leads to an accumulation of intermediate prelamin A isoforms. One good example is systemic disorder, Hutchison-Gilford Progeria Syndrome (HGPS), which is associated with a mutation in the lamins A/C gene (LMNA) causing a deletion of 50 amino acids in exon 11. This inhibits the final cleavage step and leads to the accumulation of a permanently farnesylated mutant lamin A protein, called progerin [69].

2.3.3 AIMP3, MRS and eIF2γ

Among the three AIMPs, AIMP3 strongly interacts with MRS in the MSC, where they interact with eukaryotic initiation factor 2γ (eIF2γ) subunit in a non-competitive manner [70]. A domain mapping study indicates that the GST domains of MRS and AIMP3 are important for their binding to eIF2γ, and, conversely, the GTPase domain of eIF2γ is crucial for its binding to the MRS and AIMP3 complex. This complex formation is responsible for binding with Met-tRNA$_i^{Met}$ for its delivery to the ribosome for initiation of translation. Like all members of the family of large G-proteins, eIF2 activity is regulated by alteration of bound guanine nucleotide.

All are active in a GTP-bound state and inactive when GTP is hydrolyzed to GDP. Although their functions are diverse, G-proteins share common structural features in their nucleotide binding G-domain. The main difference between translation factor G-protein and small G-protein is that they contain multiple additional domains. In the case of eIF2, extra subunits are essential to their functions. EF-Tu has two additional domains (I and II). This protein interacts with aminoacyl-tRNA, its GEF (EF-Ts), and ribosome. The eIF2 has three subunits (α, β, and γ) and the eIF2γ is a nucleotide-binding subunit, a close sequence homologue of EF-Tu. The EF-Tu bound to GTP binds all aminoacylated elongator tRNAs. By analogy, and from analysis of mutations, eIF2γ bound to GTP is major contributor form of binding with Met-tRNA$_i^{Met}$. Macromolecular modeling suggests that a pocket in eIF2γ formed by the position of Sw1 in the GTP form provides a site for the aminoacylated 3'-end of Met-tRNA$_i$. In its GDP-bound form, eIF2 interacts with the GEF eIF2B, which binds all three eIF2 subunits. The eIF2α is a target for the kinase-regulated inhibition of eIF2B. However, eIF2β and eIF2γ interactions are implicated in nucleotide exchange [71].

2.4   Interaction Pairs of KRS

2.4.1   KRS and 67LR

The 67-kDa laminin receptor (67LR) is a non-integrin cell surface receptor of the extracellular matrix, whose expression is increased in neoplastic cells with enhanced invasive and metastatic potential. It is derived from homo- or hetero-dimerization of a smaller precursor, the 37 kDa laminin receptor precursor (37 LRP), although the precise mechanism by which cytoplasmic 37 LRP becomes 67LR, which is embedded in the cell membrane, is unclear [72, 73]. 67LR is acylated by fatty acids, suggesting that 37LRP can dimerize with itself or with another peptide by strong hydrophobic bonds mediated by fatty acids. However, the amino acid compositions of 67LR and 37LRP are identical, suggesting homodimer formation. 67LR also interacts with integrins, dimeric cell membrane proteins that mediate cell adhesion to the ECM and signal transduction to the nucleus. Although 37LRP has a transmembrane domain (a.a. 86–101), it is abundant in the cytoplasm. Interestingly, 37LRP appears to be a multifunctional protein as well as a ribosomal component involved in the translational machinery and has also been observed in the nucleus, where it is tightly associated with nuclear structures. KRS interacts with 67LR and regulates its stability via ubiquitination in membrane, and enhances laminin-induced cancer cell migration and invasion (Fig. 4). The N-terminal extension and anticodon-binding domains of KRS are involved in its binding to 67LR [26].

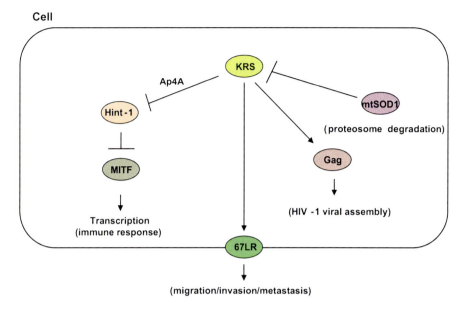

Fig. 4 Protein–protein interaction network of lysyl-tRNA synthetase. In activated mast cells, KRS is phosphorylated at the serine 207, dissociated from the MSC, translocated to the nucleus, and bound to transcription factor such as microphthalmia-associated transcription factor (MITF). When KRS is phosphorylated, it generates Ap4A, which can bind to HINT1, leading to the release of MITF from MITF-HINT1 complex. In migratory cells, KRS is phosphorylated at the threonine 52, dissociated from the MSC, translocated to the membrane, where it binds to 67LR. Finally, KRS controls the laminin-induced cell migration through the regulation of membrane stability of 67LR. In HIV-1 viron, reverse transcription of the HIV genome is primed by a tRNA^{Lys} that is packaged into the virion by KRS and HIV Gag protein. A subset of familial and sporadic amyotrophic lateral sclerosis cases are due to mutations in the gene encoding Cu, Zn superoxide dismutase (SOD1). This mutant SOD1, but not wild-type SOD1, interacts with mitochondrial KRS. In the presence of mutant SOD1, mitochondrial KRS is misfolded and aggregated, becoming a target for proteasome degradation. Impaired mitochondrial KRS is correlated with decreased mitochondrial protein synthesis

2.4.2 KRS and MITF

In immunologically activated mast cells, KRS released from the MSC complex generates the signaling molecule, diadenosine tetraphosphate (Ap4A), which induces the interaction of KRS with MITF [28, 74] (Fig. 4). KRS-generated Ap4A can disrupt the interaction of Hint1 to MITF by direct binding to Hint1 in the nucleus. ERK phosphorylation of KRS at serine 207 also contributes to its association with MITF after triggering by the antigen-IgE. The critical motif of association of KRS to MITF is the basic helix-loop-helix leucine zipper (bHLH-LZ) region suggesting that a truncated mutant of KRS having only the C-terminal aminoacylation domain is required for binding to MITF.

The phosphorylation of KRS induces opening of the N-terminal anticodon-binding domain of KRS, which provides the binding pockets for MITF [27]. This structural opening may increase the association of KRS with MITF, and may induce release of phosphorylated KRS from the MSC, whereupon it can translocate from the cytosol to the nucleus, bind to MITF, and generate Ap4A to activate the transcription of MITF target genes. In conclusion, the open conformation of KRS by phosphorylation is necessary and sufficient to switch its functions from the translation to interaction with transcription factor MITF.

2.4.3 KRS and Gag

Upon cellular infection by human immunodeficiency virus 1 (HIV-1), the viral RNA genome is copied into a double-stranded cDNA by the viral enzyme reverse transcriptase (RT). The major human $tRNA^{Lys}$ isoacceptors, $tRNA^{Lys1,2}$ and $tRNA^{Lys3}$, are selectively packaged into HIV-1 during assembly where $tRNA^{Lys3}$ can act as the primer to initiate reverse transcription [75]. The resultant viral DNA is translocated into the nucleus of the infected cell where it integrates into the host cell's DNA, and codes for viral mRNA and proteins. Proteins comprising the viral structure include both the glycosylated envelope proteins (gp120 and gp41) and mature proteins resulting from the processing of the large precursor protein, Gag. KRS is also packaged into HIV-1 and interacts with Gag [76]. Gel chromatography studies support the formation of a Gag-KRS heterodimer in vitro. Sequences in the catalytic domain of KRS, from amino acids 249 to 309, are required for binding of Gag. However, the stoichiometry of KRS-Gag binding is unclear. In human cells, cytoplasmic KRS and mitochondrial KRS are encoded from the same gene by means of alternative splicing, and it is unclear which form is responsible for Gag binding. Nonetheless, KRS appears to be critical for tRNA primer packaging into HIV-1 and for viral replication (Fig. 4) [77].

2.4.4 KRS and SOD1

SOD1 is a superoxide dismutase as well as a free radical scavenging enzyme. Although SOD1 is abundantly expressed in the cytoplasm, a proportion of mutant SOD1 is also associated with mitochondria, where its aggregation could have pathological consequences. Amyotrophic lateral sclerosis (ALS) is a progressive neurodegenerative disorder of motor neurons that result in paralysis and death within 5 years of diagnosis. Approximately 10% of ALS cases are inherited, of which 20% are associated with mutations in the Cu, Zn-superoxide dismutase, SOD1 [78]. In contrast to the precursor proteins implicated in other neurodegenerative diseases, SOD1, with its immunoglobulin-like structure, is soluble and relatively easy to study.

ALS has been linked to more than 120 mutations in SOD1, providing a unique tool for molecular analysis of the disease mechanisms. Transgenic mice expressing mutant human SOD1 develop dysfunction of mitochondrial respiration and ATP synthesis. In a yeast two-hybrid screen, KRS was shown to interact with mutant SOD1 but not with WT SOD1, and moreover, KRS is expressed in both the cytoplasm and mitochondria (Fig. 4). It is possible that aberrant interactions between mutant SOD1 and mitochondrial KRS can destabilize mitochondria [79, 80]. However, the nature of the molecular interactions between mitochondrial KRS and mutant SOD1 remains to be fully elucidated. One hypothesis is that accumulation of aberrant high molecular weight protein structures on the surface of mitochondria can lead to damage by sequestration of proteins necessary for maintaining mitochondrial integrity and dynamics. Despite the evidence that mutant SOD1 causes mitochondrial dysfunction, the mechanisms underlying the cause of mitochondrial damage remain to be identified.

2.5 Interaction Pairs of VRS

2.5.1 VRS and EF-1 Complex

Valyl-tRNA synthetase (VRS) forms a stable complex with the translational elongation factor EF-1βγδ complex [81, 82]. This complex has also been shown to interact functionally with EF-1α [83, 84]. Since tRNA aminoacylation precedes tRNA binding to eEF-1A, the presence of VRS in complex with eEF-1B supports the channeling hypothesis, in which eukaryotic translation is streamlined by coupling the charging of tRNA to its delivery to the ribosome. In higher eukaryotes, the eEF-1B complex includes a third factor, eEF-1Bβ, which possesses its own GEF activity. The C-terminal domain of eEF-1Bβ and eEF-1Bδ is highly similar to the corresponding region of eEF-1Bα and accounts for its nucleotide exchange activity.

Early characterization of the eEF-1 complex in mammalian cells depended upon in vitro reconstitution experiments with purified rabbit proteins. Using chromatography to test different combinations of eEF-1Bαγ, eEF-1Bδ, and native or truncated VRS, it was shown that the N-terminal extension of VRS is required for binding to eEF-1Bδ. An eEF-1Bαγ prevents eEF-1Bδ from forming a high molecular weight aggregate, thereby giving the complex a defined quaternary structure. The entire complex is formed through dimerization via a leucine heptad repeat within the N-terminal region eEF-1Bδ. The ability of eEF-1Bδ to dimerize along with its distinct-binding interface with eEF1Bγ indicates that eEF-1Bα and eEF-1Bδ may have different regulatory roles. This model is also supported by interactions detected using the yeast two-hybrid systems, which demonstrated eEF-1Bγ can indeed interact with eEF-1Bδ. Yeast two-hybrid screen and Co-IP analysis recently demonstrated that eEF-1Bδ also interacts with SIAH-1, an E3 ubiquitin ligase [85]. Overexpression of eEF-1Bδ inhibited E3 ligase activity, as evidenced by the decrease in ubiquitination of its substrate. Overall, the data suggest that the endogenous interaction between two proteins is regulated by specific cellular cues.

2.6 Interaction Pairs of QRS

2.6.1 QRS and ASK1

Glutaminyl-tRNA synthetase (QRS) belongs to the class I ARS family, and comprises 775 amino acids, consisting of 4 domains with the Rossman fold. Apoptosis signal regulating kinase 1 (ASK1) plays key roles in cancer, cardiovascular diseases and neurodegenerative diseases [86]. Human ASK 1 (hASK1) and mouse ASK1 (mASK1) consist of 1,374, and 1,379 amino acids, respectively. Although the entire tertiary structure of full-length ASK1 is yet to be elucidated, the crystal structure of hASK1 catalytic domain in complex with staurosporine has been solved [87]. ASK1 has a serine/threonine kinase domain in the middle part of the molecule and two coiled-coil domains at the N-termini and C-termini. The C-terminal coiled-coil domain is required for homo-oligomerization and full activation of ASK1. This kinase is also a member of the mitogen-activated protein kinase (MAPK) family that activates downstream MAPKs, such as c-Jun N-terminal kinase (JNK) and p38 MAPK [88]. Phosphorylation of a threonine residue within the kinase domain (Thr 838) is essential for activation of ASK1. In ASK mutants (Thr 838) substituted by alanine, its activity is abolished since this site is a trans-autophosphorylation residue under oxidative stress. ASK1 is activated by a variety of cellular stresses such as oxidative and endoplasmic reticulum stresses, calcium influx, infection, and Fas ligands. In response to an agonistic antibody to Fas, Fas death receptor is activated and induces ASK1 activation, resulting in cell death. The molecular interaction between QRS and ASK1 is dependent on glutamine concentration. QRS inhibits ASK1 activity and has an anti-apoptotic function [89].

2.7 Interaction Pairs of LRS

2.7.1 LRS and RagD GTPase

Leucyl-tRNA synthetase (LRS), a member of the class I ARS family, comprises 1,176 amino acids with 5 structural domains: a catalytic, editing, leucine-specific, anticodon-binding, and C-terminal domain. The mammalian target of rapamycin (mTOR) is an atypical serine/threonine protein kinase that belongs to the phosphoinositide 3-kinase (PI3K)-related kinase family and interacts with several proteins to form two distinct complexes named mTOR complex 1 (mTORC1) and 2 (mTORC2) [90]. Leucine, one of the branched amino acids, is identified as a key nutrient of anabolic signaling via the mTOR complex [91]. mTOR regulates the signaling networks related to translation and cell growth by coordinating upstream inputs such as growth factors, intracellular energy status, and amino acid availability. mTOR directly phosphorylates the translational regulators eukaryotic translation initiation factor 4E (eIF4E)-binding proteins (4E-BP1) and S6 kinase 1 (S6K1), which in turn promote protein synthesis. The activation of S6K1 is mediated by a variety of

Fig. 5 Protein–protein interaction network of leucyl-tRNA synthetase. LRS, a sensor of amino acids, transmits a signal to mTORC1. In cells, leucine-loaded LRS, which is required for protein synthesis, binds to RagD GTPase and functions as a GTPase-activating protein (GAP) for RagD GTPase. mTORC1 activation facilitates translation initiation, but suppresses autophagy

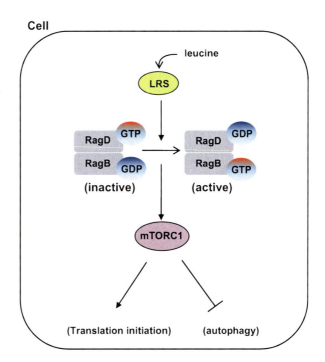

effectors, and results in an increase in mRNA biogenesis, as well as translational initiation and elongation. The mechanism by which LRS activates mTORC1 is by its role in intracellular leucine sensing and as a GTPase-activating protein (GAP) for the RagD GTPase [92]. In mammals, Rag GTPase is necessary for the activation of mTORC1 by leucine [93, 94]. Rag GTPases belong to the Ras GTPase subfamily and are highly conserved from yeast to human. There are four Rag GTPases in humans: RagA, RagB, RagC, and RagD. RagA and RagB are closely related to each other with 98% sequence identity with the exception of 33 additional residues at the N terminus of Rag B. RagC and RagD are closely related to each other with 81% sequence identity. The homology between RagA/B and RagC/D is limited. RagA or RagB interact with RagC or RagD to constitute a heterodimer [95]. According to the current model of mTORC1 activation by leucine through the Rag GTPase, a Rag complex is constitutively anchored on the surface of the lysosome (Fig. 5).

2.8 Interaction Pairs of WRS

2.8.1 WRS, DNA-PKs, and Poly (ADP-ribose) Polymerase-1

The DNA-dependent protein kinase (DNA-PK) is a DNA-activated serine/threonine protein kinase consisting of a catalytic subunit (DNA-PKc) and a heterodimer of Ku (Ku70/Ku80) proteins [96]. DNA-PK is a critical component in the DNA-damage repair pathway and abundantly expressed in almost all mammalian cells. Homozygous knock-out mice of the DNA-PK catalytic subunit (DNA-PKcs$^{-/-}$) are hypersensitive to radiation and chemical treatment. Poly (ADP-ribose) polymerases (PARP) catalyze a reaction in which the ADP-ribose moiety of NAD$^+$ is transferred to a receptor amino acid, building poly (ADP-ribose) (PAR) polymers. PAR polymers are an evolutionarily conserved posttranslational modification affecting a large array of proteins. PARP-1 and PARP-2 are highly conserved proteins, ubiquitously expressed in mammalian tissues and with predominantly nuclear localization. PARP-1 and PARP-2 have historically been described as key DNA damage repair enzymes.

WRS comprises 653 amino acids and when the N-terminal 47 amino acids are removed by alternative splicing, it is designated "mini-WRS" [97, 98]. Mini-WRS is also truncated by proteolytic cleavage with leukocyte elastase to generate T1-WRS and T2-WRS, which lack N-terminal 70 and 93 amino acids, respectively. WRS is connected to p53 signaling by bridging DNA-PKc to PARP-1 pathway through IFNγ. In response to IFNγ, a ternary complex comprising DNA-PKc/WRS/PARP-1 is formed. IFNγ induces DNA-PKc to phosphorylate the PARP-1, then PARP-1 PARylates DNA-PKc in the presence of nuclear WRS. Finally DNA-PKc phosphorylates p53 at the serine 15 to integrate IFNγ signaling. Nuclear WRS provides the link for p53 tumor suppressor to integrate IFNγ signaling and p53 activation [99].

2.8.2 WRS and VE-Cadherin

Mini-WRS is further truncated by proteolytic cleavage by leukocyte elastase to generate T1-WRS and T2-WRS, which lack the N-terminal 70 and 93 amino acids, respectively. Several WRS isoforms work as active cytokines to regulate angiogenesis. Mini-WRS mediates angiostatic signaling in human umbilical vein endothelial cells (HUVEC). Deletion of the tRNA anticodon-binding (TAB) domain insertion, consisting of eight residues in the human WRS, abolishes the enzyme's apoptotic activity for endothelial cells, whereas its translational catalysis and cell-binding activities remained unchanged. Vascular endothelial (VE)-cadherin, a calcium-dependent adhesion molecule, has been identified as a receptor for mini-WRS and an endothelial-specific cell–cell adhesion protein of the adherens junction complex [100]. VE-cadherin plays important roles in endothelial barrier function, monolayer permeability, and angiogenesis. Thus VE-adherin availability and function are tightly regulated [101] through mechanisms that include regulation of VE-cadherin activity and control of the amount of VE-cadherin available for engagement at both the protein and mRNA level. It is generally accepted that phosphorylation of VE-cadherin induces the destabilization of the adherens junction complex and increases monolayer permeability, although the residues that are phosphorylated and which modify VE-cadherin activity remain to be elucidated.

In addition to its adhesive properties, VE-cadherin also participates directly and indirectly in intracellular signaling pathways to control cell dynamics and influences endothelial cell behavior by modulating activity of growth factor receptors, intracellular messengers, and proteins to regulate gene transcription. VE-cadherin associates with two growth factor receptors [102]: VEGFR2 and TGF-βR. In confluent cells, VE-cadherin binds VEGFR2 indirectly through β-catenin. This prevents the tyrosine phosphorylation of VEGFR2 and internalization into clathrin-coated vesicles and reduces MAPK signals and proliferative signals. In contrast, upon VE-cadherin binding to the TGFβR complex, anti-proliferative and anti-migratory signals are enhanced. VE-cadherin promotes the assembly of the TGFβR complex into an active receptor complex which is able to regulate phosphorylation and activation of Smad-dependent transcription. Thus, VE-cadherin exerts opposing effects on these two receptors, but both interactions mediate stabilization of the vasculature. Although it is unknown whether the phosphorylation status of VE-cadherin affects its association with either VEGFR2 or TGFβR receptor complexes, data suggest that regulation of these growth factors may mediate the transition between quiescent and activated cell states [103, 104].

2.9   Interaction Pairs of EPRS

2.9.1   EPRS and GAIT Complex

Human EPRS is a bifunctional ARS of the MSC, in which the two catalytic domains are linked by three tandem repeats. In response to IFNγ, EPRS forms a multi-component complex with other regulatory proteins at a $3'$ UTR region that is involved in the translational silencing of target transcripts, many of which function during an inflammatory response. The translation of vascular endothelial growth factor A (*VEGFA*), which is a crucial factor for angiogenesis, is also controlled by EPRS through a similar mechanism. Thus, EPRS seems to serve as a key gatekeeper of inflammatory gene translation. Gene expression is regulated at multiple steps, including transcription, RNA stability, translation, protein stability, and post-translational modification. In many cases, translational control has evolved as an "off-switch" mechanism to modulate protein expression because it exhibits key regulatory advantages.

Ceruloplasmin (Cp) is a copper-containing plasma protein synthesized by activated macrophage [105]. Synthesis of Cp by macrophage is induced by IFNγ but is subsequently suppressed by a mechanism involving translational silencing. Translational silencing requires binding of the GAIT complex to the Cp mRNA $3'$UTR. EPRS is a component of the GAIT complex that binds the $3'$ UTR GAIT element in multiple proinflammatory transcripts (e.g., VEGF-A) and inhibits their translation in macrophage. The GAIT complex comprises EPRS, NS1-associated protein (NSAP1), ribosomal protein L13a, and GAPDH. The ribosomal protein L13a interacts with the Cp $3'$UTR GAIT element such that L13a is required for translational silencing activity in IFNγ-treated cells [21, 23].

Several kinases are involved in the modification of the GAIT complex. ZIPK-mediated L13a phosphorylation causes its release from the large ribosomal subunit. In addition, the activation of CDK5 by IFNγ results in its phosphorylation of EPRS at serine 886, thereby triggering its release from the MSC and inducing VEGR-A expression. The unidentified AGC kinase also phosphorylates EPRS at serine 999 upon IFNγ stimulation. Thus, GAIT-mediated translational control of inflammatory transcripts (e.g., VEGF-A) may have a crucial role in protecting cells from inflammation and injury associated with tumorigenesis [25]. Genetic defects in components of the GAIT pathway or defects caused by environmental stress may contribute to progression of chronic inflammatory disorders. For instance, prolonged inflammatory gene expression contributes to malignant tumor progression.

2.10 Interaction Pairs of GRS

2.10.1 GRS and Cadherin 6

Glycyl-tRNA synthetase (GRS), an intrinsic component of protein synthesis as a secreted molecule, is also involved in extracellular cancer-immune network and immune surveillance. GRS is secreted from macrophage by Fas ligand which is derived from tumor cells. When the secreted GRS interacts with cadherin 6 (CDH6) on tumor cells, the protein phosphatase 2A (PP2A) is dissociated from CDH6 and then activated. Thus, PP2A suppresses ERK activation through dephosphorylation and finally results in apoptosis of tumor cells [106].

Classical cadherin adhesion receptors exert many of their biological effects through close cooperation with the cytoskeleton [107]. Cadherins comprise a large superfamily with over 350 members. The most salient feature of this superfamily is the presence of a variable number of successive extracellular cadherin (EC) repeat domains, each comprising ~110 amino acids, which are made rigid by binding of 3 Ca^{2+} ions at linker regions between these domains [108]. There are five highly conserved EC domains (EC1–EC5) in the ectodomain of cadherins. The type I and type II classical cadherins were originally named on the basis of the tissues within which they were first identified (e.g., type I, epithelial (E)-cadherin and neural (N)-cadherin, type II vascular endothelial (VE)-cadherin and kidney (K)-cadherin, CDH1, CDH2, CDH5, and CDH6, respectively). CDH6 is classical type II cadherin that mediates calcium-dependent cell–cell adhesion.

2.11 Roles of HSP90 in the Molecular Interaction of ARSs

Heat shock protein 90 (HSP90) is a molecular chaperone protein essential for cellular survival. HSP90α and HSP90β consisted of 732 amino acids and 724 amino acids, are expressed in cytosol and nucleus, and contain an N-terminal

ATP-binding domain that is essential for most of their cellular functions [109, 110]. It appears to mediate protein–protein interactions of mammalian ARSs, and inactivation of HSP90 interferes with the in vivo incorporation of the nascent ARSs into the MSC [111]. HSP90 shows distinguishing characteristics between cancer cells and normal cells. HSP90 plays an essential role in maintaining cellular protein homeostasis by acting as a molecular chaperone to aid in folding as well as in intracellular trafficking of its partners. HSP90 exists as a homo-dimer with an N-terminal domain that is critical for hydrolysis of ATP to ADP. The HSP90 complex cycles from the ADP to the ATP-bound state. The conformational change that occurs with replacement of ADP by ATP stabilizes and activates binding proteins. This process is highly regulated by interactions with a variety of co-chaperones. Co-chaperones assist HSP90 throughout its conformational cycling that is required for its normal functions, act as substrate recognition proteins, and even provide additional enzymatic activity. The predominant class of co-chaperones is the tetratricopeptide repeat (TPR) domain containing proteins, which bind the MEEVD motif found in the C-terminus of HSP90. Among the co-chaperones with a TPR domain are C-terminus of HSP90-interacting protein, HSP70-HSP90 organizing protein (Hop), cyclophilin 40, and FK506-binding protein (FK506-BP). Other co-chaperones that interact with HSP90 via alternative domains are activator of HSP90 ATPase homolog 1 (Aha1), which enhances the function of HSP90 by stimulating its ATPase activity [112–114]. The co-chaperone that is most strongly implicated in facilitating tumorigenesis is cell division cycle 37 (Cdc 37) because it associates with mutant kinase that drive cancer progression. HSP90 also interacts with the complex-forming ARSs including AIMP1, 2, and 3, and weakly interacts with GRS, HRS, and WRS, which are non-complex-forming ARSs. HSP90 also interacts with ERPS and IRS in HSP90 activity-dependent manner, implying that their association with HSP90 is also dependent on its activity [111]. Taken together, HSP90 plays an important role in mediating protein–protein interaction to regulate diverse cellular response.

References

1. Eriani G, Delarue M, Poch O, Gangloff J, Moras D (1990) Nature 347:203
2. Burbaum JJ, Schimmel P (1991) J Biol Chem 266:16965
3. Cusack S, Hartlein M, Leberman R (1991) Nucleic Acids Res 19:3489
4. Robinson JC, Kerjan P, Mirande M (2000) J Mol Biol 304:983
5. Kim JY, Kang YS, Lee JW, Kim HJ, Ahn YH, Park H, Ko YG, Kim S (2002) Proc Natl Acad Sci USA 99:7912
6. Ko YG, Park H, Kim S (2002) Proteomics 2:1304
7. Simos G, Segref A, Fasiolo F, Hellmuth K, Shevchenko A, Mann M, Hurt EC (1996) EMBO J 5:5437
8. Galani K, Grosshans H, Deinert K, Hurt EC, Simos G (2001) EMBO J 20:6889
9. Christine K, Adam R, Hannes S, Dietrich S, Dmitri S (2013) Nucleic Acids Res 41:667
10. Lee SW, Cho BH, Park SG, Kim S (2004) J Cell Sci 117:3725
11. Havrylenko S, Legouis R, Negrutskii B, Mirande M (2011) J Biol Chem 286:28476

12. Norcum MT (1989) J Biol Chem 264:15043
13. Norcum MT, Boisset N (2002) FEBS Lett 512:298
14. Wolfe CL, Warrington JA, Davis S, Green S, Norcum MT (2003) Protein Sci 12:2282
15. Park SG, Schimmel P, Kim S (2008) Proc Natl Acad Sci USA 105:11043
16. Han JM, Lee MJ, Park SG, Lee SH, Razin E, Choi EC, Kim S (2006) J Biol Chem 281:38663
17. Kim JY, Kang YS, Lee JW, Kim HJ, Ahn YH, Park H, Ko YG, Kim S (2002) Proc Natl Acad Sci USA 99:7912
18. Quevillon S, Mirande M (1996) FEBS Lett 395:63
19. Quevillon S, Robinson JC, Berthonneau E, Siatecka M, Mirande M (1999) J Mol Biol 285:183
20. Galani K, Grosshans H, Deinert K, Hurt EC, Simos G (2001) EMBO J 20:6889
21. Sampath P, Mazumder B, Seshadri V, Gerber CA, Chavatte L, Kinter M, Ting SM, Dignam JD, Kim S, Driscoll DM, Fox PL (2004) Cell 119:195
22. Jia J, Arif A, Ray PS, Fox PL (2008) Mol Cell 29:679
23. Ray PS, Jia J, Yao P, Majumder M, Hatzoglou M, Fox PL (2009) Nature 457:915
24. Arif A, Jia J, Moodt RA, DiCorleto PE, Fox PL (2011) Proc Natl Acad Sci USA 108:1415
25. Yao P, Potdar AA, Arif A, Ray PS, Mukhopadhyay R, Willard B, Xu Y, Yan J, Saidel GM, Fox PL (2012) Cell 149:88
26. Kim DG, Choi JW, Lee JY, Kim H, Oh YS, Lee JW, Tak YK, Song JM, Razin E, Yun SH, Kim S (2012) FASEB J 26:4142
27. Ofir-Birin Y, Fang P, Bennett SP, Zhang HM, Wang J, Rachmin I, Shapiro R, Song J, Dagan A, Pozo J, Kim S, Marshall AG, Schimmel P, Yang XL, Nechushtan H, Razin E, Guo M (2013) Mol Cell 49:30
28. Yannay-Cohen N, Carmi-Levy I, Kay G, Yang CM, Han JM, Kemeny DM, Kim S, Nechushtan H, Razin E (2009) Mol Cell 34:603
29. Han JM, Park BJ, Park SG, Oh YS, Choi SJ, Lee SW, Hwang SK, Chang SH, Cho MH, Kim S (2008) Proc Natl Acad Sci USA 105:11206
30. Kwon NH, Kang T, Lee JY, Kim HH, Kim HR, Hong J, Oh YS, Han JM, Ku MJ, Lee SY, Kim S (2011) Proc Natl Acad Sci USA 108:19635
31. Negrutskii BS, Deutscher MP (1991) Proc Natl Acad Sci USA 88:4991
32. Negrutskii BS, Deutscher MP (1992) Proc Natl Acad Sci USA 89:3601
33. Kyriacou SV, Deutscher MP (2008) Mol Cell 29:419
34. Shin BS, Kim JR, Walker SE, Dong J, Lorsch JR, Dever TE (2011) Nat Struct Mol Biol 18:1227
35. Lee JS, Park SG, Park H, Seol W, Lee S, Kim S (2002) Biochem Biophys Res Commun 291:158
36. Facciponte JG, Wang XY, MacDonald IJ, Park JE, Arnouk H, Grimm MJ, Li Y, Kim H, Manjili MH, Easton DP, Subjeck JR (2006) Cancer Immunol Immunother 55:339
37. Hilf N, Singh-Jasuja H, Schild H (2002) Int J Hyperthermia 18:521
38. Manjili MH, Wang XY, Park J, Facciponte JG, Repasky EA, Subjeck JR (2002) Front Biosci 7:d43
39. Han JM, Park SG, Liu B, Park BJ, Kim JY, Jin CH, Song YW, Li Z, Kim S (2007) Am J Pathol 170:2042
40. Schild H, Rammensee HG (2000) Nat Immunol 1:100
41. Nicchitta CV (1998) Curr Opin Immunol 10:103
42. Kijimoto-Ochiai S (2002) Cell Mol Life Sci 59:648
43. Conrad DH, Kilmon MA, Studer EJ, Cho S (1997) Biochem Soc Trans 25:393
44. Mossalayi MD, Arock M, Debré P (1997) Int Rev Immunol 16:129
45. Kwon HS, Park MC, Kim DG, Cho K, Park YW, Han JM, Kim S (2012) J Cell Sci 125:4620
46. Lee YS, Han JM, Son SH, Choi JW, Jeon EJ, Bae SC, Park YI, Kim S (2008) Biochem Biophys Res Commun 371:395
47. Lee SW, Kim G, Kim S (2008) Exp Opin Drug Discov 3:945

48. Bonni S, Wang HR, Causing CG, Kavsak P, Stroschein SL, Luo K, Wrana JL (2001) Nat Cell Biol 3:587
49. Kavsak P, Rasmussen RK, Causing CG, Bonni S, Zhu H, Thomsen GH, Wrana JL (2000) Mol Cell 6:1365
50. Kim MJ, Park BJ, Kang YS, Kim HJ, Park JH, Kang JW, Lee SW, Han JM, Lee HW, Kim S (2003) Nat Genet 34:330
51. Liu J, Chung HJ, Vogt M, Jin Y, Malide D, He L, Dundr M, Levens D (2011) EMBO J 30:846
52. Zhang J, Chen QM (2012) Oncogene 2012 32:2907
53. Choi JW, Kim DG, Park MC, Um JY, Han JM, Park SG, Choi EC, Kim S (2009) J Cell Sci 122:2710
54. Magis C, van der Sloot AM, Serrano L, Notredame C (2012) Trends Biochem Sci 37:353
55. Shen HM, Pervaiz S (2006) FASEB J 20:1589
56. Bosari S, Viale G (1995) Virchows Arch 427:229
57. Louis DN (1994) J Neuropathol Exp Neurol 53:11
58. Mayo LD, Donner DB (2002) Trends Biochem Sci 27:462
59. Daujat S, Neel H, Piette J (2001) Trends Genet 17:459
60. Choi JW, Lee JW, Kim JK, Jeon HK, Choi JJ, Kim DG, Kim BG, Nam DH, Kim HJ, Yun SH, Kim S (2012) J Mol Cell Biol 4:164
61. Park BJ, Kang JW, Lee SW, Choi SJ, Shin YK, Ahn YH, Choi YH, Choi D, Lee KS, Kim S (2005) Cell 120:209
62. Park BJ, Oh YS, Park SY, Choi SJ, Rudolph C, Schlegelberger B, Kim S (2006) Cancer Res 66:6913
63. Zhou J, Lim CU, Li JJ, Cai L, Zhang Y (2006) Cancer Lett 243:9
64. Meulmeester E, Pereg Y, Shiloh Y, Jochemsen AG (2005) Cell Cycle 4:1166
65. Caspari T (2000) Curr Biol 10:R315
66. Oh YS, Kim DG, Kim G, Choi EC, Kennedy BK, Suh Y, Park BJ, Kim S (2010) Aging Cell 9:810
67. Moir RD, Spann TP (2001) Cell Mol Life Sci 58:1748
68. Ho CY, Lammerding J (2012) J Cell Sci 125:2087
69. Pereira S, Bourgeois P, Navarro C, Esteves-Vieira V, Cau P, De Sandre-Giovannoli A, Lévy N (2008) Mech Ageing Dev 129:449
70. Kang T, Kwon NH, Lee JY, Park MC, Kang E, Kim HH, Kang TJ, Kim S (2012) J Mol Biol 423:475
71. Mohammad-Qureshi SS, Jennings MD, Pavitt GD (2008) Biochem Soc Trans 36:658
72. Mbazima V, Da Costa DB, Omar A, Jovanovic K, Weiss SF (2010) Front Biosci 15:1150
73. Rea VE, Rossi FW, De Paulis A, Ragno P, Selleri C, Montuori N (2012) Infez Med 20:8
74. Carmi-Levy I, Yannay-Cohen N, Kay G, Razin E, Nechushtan H (2008) Mol Cell Biol 28:5777
75. Saadatmand J, Kleiman L (2012) Virus Res 169:340
76. Javanbakht H, Halwani R, Cen S, Saadatmand J, Musier-Forsyth K, Gottlinger H, Kleiman L (2003) J Biol Chem 278:27644
77. Stark LA, Hay RT (1998) J Virol 72:3037
78. Sheng Y, Chattopadhyay M, Whitelegge J, Valentine JS (2012) Curr Top Med Chem 12:2560
79. Kunst CB, Mezey E, Brownstein MJ, Patterson D (1997) Nat Genet 15:91
80. Kawamata H, Magrané J, Kunst C, King MP, Manfredi G (2008) J Biol Chem 283:28321
81. Bec G, Kerjan P, Zha XD, Waller JP (1989) J Biol Chem 264:21131
82. Bec G, Kerjan P, Waller JP (1994) J Biol Chem 269:2086
83. Negrutskii BS, Shalak VF, Kerjan P, El'skaya AV, Mirande M (1999) J Biol Chem 274:4545
84. Galani K, Grosshans H, Deinert K, Hurt EC, Simos G (2001) EMBO J 20:6889
85. Wu H, Shi Y, Lin Y, Qian W, Yu Y, Huo K (2011) Mol Cell Biochem 357:209
86. Fujisawa T, Takeda K, Ichijo H (2007) Mol Biotechnol 37:13
87. Takeda K, Matsuzawa A, Nishitoh H, Ichijo H (2003) Cell Struct Funct 28:23
88. Nagai H, Noguchi T, Takeda K, Ichijo H (2007) J Biochem Mol Biol 40:1

89. Ko YG, Kim EY, Kim T, Park H, Park HS, Choi EJ, Kim S (2001) J Biol Chem 276:6030
90. Martin DE, Hall MN (2005) Curr Opin Cell Biol 17:158
91. Anthony JC, Anthony TG, Kimball SR, Jefferson LS (2001) J Nutr 131:856
92. Bonfils G, Jaquenoud M, Bontron S, Ostrowicz C, Ungermann C, De Virgilio C (2012) Mol Cell 46:105
93. Han JM, Jeong SJ, Park MC, Kim G, Kwon NH, Kim HK, Ha SH, Ryu SH, Kim S (2012) Cell 149:410
94. Jewell JL, Russell RC, Guan KL (2013) Nat Rev Mol Cell Biol 14:133
95. Sancak Y, Sabatini DM (2009) Biochem Soc Trans 37:289
96. Collis SJ, DeWeese TL, Jeggo PA, Parker AR (2005) Oncogene 24:949
97. Wakasugi K, Slike BM, Hood J, Otani A, Ewalt KL, Friedlander M, Cheresh DA, Schimmel P (2002) Proc Natl Acad Sci USA 99:173
98. Yang XL, Skene RJ, McRee DE, Schimmel P (2002) Proc Natl Acad Sci USA 99:15369
99. Sajish M, Zhou Q, Kishi S, Valdez DM Jr, Kapoor M, Guo M, Lee S, Kim S, Yang XL, Schimmel P (2012) Nat Chem Biol 8:547
100. Tzima E, Reader JS, Irani-Tehrani M, Ewalt KL, Schwartz MA, Schimmel P (2005) J Biol Chem 280:2405
101. Dejana E, Orsenigo F, Lampugnani MG (2008) J Cell Sci 121:2115
102. Guelte LA, Gavard DJ (2011) J Biol Cell 103:593
103. London NR, Whitehead KJ, Li DY (2009) Angiogenesis 12:149
104. Vestweber D, Winderlich M, Cagna G, Nottebaum AF (2009) Trends Cell Biol 19:8
105. Fox PL, Mazumder B, Ehrenwald E, Mukhopadhyay CK (2000) Free Radic Biol Med 28:1735
106. Park MC, Kang T, Jin D, Han JM, Kim SB, Park YJ, Cho K, Park YW, Guo M, He W, Yang XL, Schimmel P, Kim S (2012) Proc Natl Acad Sci USA 109:640
107. Brieher WM, Yap AS (2013) Curr Opin Cell Biol 25:39
108. Akins MR, Benson DL, Greer CA (2007) J Comp Neurol 501:483
109. Picard D (2002) Cell Mol Life Sci 59:1640
110. Pearl LH, Prodromou C (2001) Adv Protein Chem 59:157
111. Kang J, Kim T, Ko YG, Rho SB, Park SG, Kim MJ, Kwon HJ, Kim S (2000) J Biol Chem 275:31682
112. Pearl LH, Prodromou C (2000) Curr Opin Struct Biol 10:46
113. Pearl LH, Prodromou C (2006) Annu Rev Biochem 75:271
114. Jackson SE (2013) Top Curr Chem 328:155

Top Curr Chem (2014) 344: 145–166
DOI: 10.1007/128_2013_476
© Springer-Verlag Berlin Heidelberg 2013
Published online: 19 December 2013

Extracellular Activities of Aminoacyl-tRNA Synthetases: New Mediators for Cell–Cell Communication

Sung Hwa Son, Min Chul Park, and Sunghoon Kim

Abstract Over the last decade, many reports have discussed aminoacyl-tRNA synthetases (ARSs) in extracellular space. Now that so many of them are known to be secreted with distinct activities in the broad range of target cells including endothelial, various immune cells, and fibroblasts, they need to be classified as a new family of extracellular signal mediators. In this chapter the identity of the secreted ARSs, receptors, and their physiological and pathological implications will be described.

Keywords Aminoacyl-tRNA synthetase · Cytokine · Intercellular communication · Receptor · Signaling

Contents

1 Introduction 146
2 Secreted tRNA Synthetases and Related Activities 147
   2.1 Tryptophany-tRNA Synthetase 149
   2.2 Tyrosyl-tRNA Synthetase 149
   2.3 Glycyl-tRNA Synthetase (GRS) 151
   2.4 Lysyl-tRNA Synthetase (KRS) 152
   2.5 Histidyl-tRNA Synthetase/Asparaginyl-tRNA Synthetase/Seryl-tRNA Synthetase 152
   2.6 Aminoacyl-tRNA Synthetase-Interacting Multifunctional Protein 1/MSCp43 152
3 Working Mechanism of ARSs-Derived Cytokine 155
   3.1 Adhesion Protein-Mediated Mechanism 155
   3.2 Cytokine Receptor-Mediated Mechanism 157
4 Conclusion 158
References 159

S.H. Son, M.C. Park (✉), and S. Kim
Medicinal Bioconvergence Research Center, Graduate School of Convergence Science and Technology, College of Pharmacy, Seoul National University, Seoul 151-742, South Korea
e-mail: parkmin2@snu.ac.kr

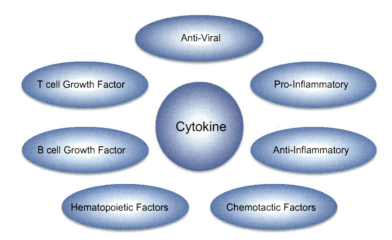

Fig. 1 Diagram showing different activities of classical cytokines in the human body. Cytokines are secreted in response to various stimuli to modulate the behavior of target cells. They have complex network and dynamic system involved in the control of numerous biological responses as above

1 Introduction

Cytokines are a unique class of intercellular regulatory proteins that are secreted into the extracellular environment and act on neighbor cells. Depending on their target cell and distance, cytokines can be categorized as autocrine (working to the same cell type as the secreting cell), paracrine (working to a different type of cell, but in the immediate vicinity of the secreting cell), or endocrine (working with the cells at a distance) [1]. Cytokines play a crucial role in immunologic homeostasis including the regulation of innate and adaptive immune responses and proliferation, apoptosis, migration, and differentiation of immune cells (Fig. 1) [2–4].

Cytokines are particularly associated with inflammation-related disorders including infection, allergy, autoimmune diseases, and cancer. For example, tumor necrosis factor (TNF)-α mediates the development processes in tumorigenesis by activating immune responses. In the earlier stages of tumorigenesis, TNF-α has an anti-tumor effect by activating M1 macrophage [5]. Conversely, in the later stages, it helps tumor metastasis by reorganizing tumor microenvironments [6, 7]. Due to their association with many diseases, cytokines, their functional receptors, and related signal pathways have been considered and investigated as promising candidates as therapeutic agents or targets [8]. Patients' cytokine profiles can provide information about disease [9].

The expression of classical cytokines is carefully regulated. Basal expression levels of cytokines are low under normal condition but in stressful conditions they are stimulated and secreted by a variety of cell types [10, 11]. Different cytokines can perform similar functions on target cells. Since classical cytokines are not sufficient to mediate complex cell–cell communications, additional proteins might be recruited to fill the gap for the intercellular communications that are not covered by classical cytokines. In the last few decades an increasing number of proteins whose functions are mainly intracellular have been found to be secreted and to work like cytokines. These proteins were originally known for intracellular fundamental maintenance of cells [12, 13]. The secretion of different aminoacyl-tRNA synthetases (ARSs) has been reported in many different cells [14–17], suggesting the physiological implications of these enzymes as a new extracellular signaling family. Autoantibodies against different ARSs have also been detected in autoimmune diseases such as polymyositis, dermatomyositis, and cancer [18, 19]. Interestingly, autoimmune disease patients that show anti-synthetase syndrome share a common selection of clinical features [20], suggesting a pathological association of the ARS secretion.

2 Secreted tRNA Synthetases and Related Activities

Multi-functional proteins generally control their activities by translocation or modification [21, 22]. Since all ARSs catalytically execute the first step of protein synthesis, namely, the ligation of amino acids to the cognate tRNAs that is essential for cell viability, they are ubiquitously expressed in all cell types to maintain protein synthesis and cell viability. However, in conditions that may interfere with their canonical functions, some ARSs are secreted from intact cells and work outside the cells like cytokines [23–29]. Translation is down-regulated when immune and endothelial cell are stimulated [30, 31]. During this time, ARSs can be secreted quickly from these cells and influence the neighboring cells, reorganizing the microenvironment [16]. The secreted ARSs work as the full-length proteins or are converted to active ligands after proteolytic cleavage or alternative splicing (Table 1).

Although cytokines are generally not expressed constitutively, some cytokines such as TGF-β, IL-1β, and IL-18, called procytokines, are produced constitutively as precursors in the cells. These cytokines are cleaved into mature bioactive forms by proteases in order to exhibit the cytokine effect on target cells. Since the active domain is sterically hindered in the intact form, only the truncated form can be secreted, the truncation process being essential for their extracellular activity. For example, IL-1 family cytokines such as IL-1β and IL-18 are produced constitutively as the pre-forms that are cleaved into active form by active caspase-1 [32, 33]. ARSs are ubiquitously expressed but their extracellular activities appear to be expressed as full-length and are also differentiated by the formation of different forms that are generated by alternative splicing or proteolytic cleavage.

Table 1 List of secreted human ARSs

ARSs	Active form (MW – kDa)	Target cells	Receptor	Effects	Reference
WRS	Mini-WRS, T1 and T2 (46,43,40)	Endothelial cell	VE-cadherin	Angiostatic effect	[23, 36–38, 42, 45–48]
YRS	Mini-YRS (20)	Leukocyte	CXCR1/2	Induction Immune response	[24, 120]
		Endothelial cell		Angiogenic effect	[51–54, 121]
	C-terminal YRS (40)	Leukocyte		Induction immune response (chemokine)	[25, 56]
GRS	Full length (74)	Cancer cell	K-cadherin	Induction apoptosis	[26, 61, 114, 115]
KRS	Full length (70)	Macrophage/monocyte		Pro-inflammatory effect	[27, 62]
HRS/NRS	Full length (80, 63)	Lymphocyte Stimulated monocyte iDCs	CCR5	Induction immune response (chemokine)	[28, 68–71]
SRS	Full length (60)	CCR3-transfected cell	CCR3	Induction cell migration	[28]
AIMP1	EMAP II (22)	Macrophage/monocyte		Induction immune response (chemokine)	[72, 88]
		Endothelial cell	ATP synthase, Integrin a5b1	Induction apoptosis	[75, 78, 80, 81]
	Full length (35)	Macrophage/monocyte	CD23	Pro-inflammatory effect	[92–94, 96, 97, 101, 104]
		Endothelial cell		Bi-phasic effect (Anti-/Proangiogenic effect)	[83, 101, 104]
		Fibroblast		Induction proliferation	[104, 105]
		Endothelial cell		Glucagon-like effect	[29]

2.1 Tryptophany-tRNA Synthetase

In normal cells, tryptophany-tRNA synthetase (WRS) exists in its full-length form and a truncated form, designated mini-WRS, in which the N-terminal 47 amino acids are deleted through an alternative splicing process [34, 35]. Production of the full-length and mini-WRS is stimulated by IFN-γ that is also responsible for the production of other angiostatic factors such as IP1 and MIG [36–38]. Two tandem promoters, both of which are responsive to IFN-γ, mediate the production of these two variants [39]. WRS and mini-WRS, which have the catalytic activity, are secreted from various cells, such as monocyte, keratinocytes, epithelial cells, and fibroblasts [34, 40], into the microenvironment of endothelial cells. However, only mini-WRS exhibits a potent anti-angiogenic effect in vivo and in vitro [23, 39, 41]. In addition to the alternative splicing process, there are two more truncated forms of WRS, named T1 and T2. After the full-length WRS is secreted into extracellular space, it is cleaved into T1 or T2 by leukocyte elastase or plasmin, which are critical proteases in angiogenesis [23, 42]. T1 and T2 are the variants with the N-terminal 70- and 93-amino acid deletions, respectively. While the two proteolytic forms (T1 and T2) possess angiostatic cytokine activity, T2 is catalytically inactive (Fig. 2a) [41]. T2 has no effect on pre-formed blood vessels, but arrests growth of developing vessels, suggesting its potential as an anti-cancer therapeutic agent. Since WRS is an essential component of the translation process, its secretion should be tightly regulated. To control WRS secretion, annexin II and S100A10, which have been implicated in endosome organization and trafficking [43], form ternary complexes with WRS in intracellular space [42].

In higher eukaryotes, ARSs have additional domains incorporated to the conserved catalytic domains [44]. The N-terminal appendix of WRS containing the WHEP domain regulates its cytokine effect. The ligand activity of the secreted WRS is dependent on its interaction with Trp residue exposed out from the receptor, VE-cadherin [45]. The WHEP domain sterically masks the catalytic domain to inhibit the approach of other molecules while maintaining the catalytic domain accessible to small substrates, Trp and AMP for catalysis. Since the WHEP domain hinders interaction between VE-cadherin and the catalytic domain of WRS [46–48], the full-length WRS cannot work as the ligand to VE-cadherin. It is to be seen whether the full-length WRS would have a cytokine activity distinct from the mini-WRS.

2.2 Tyrosyl-tRNA Synthetase

Higher eukaryotic tyrosyl-tRNA synthetase (YRS) has cytokine specific motif ELR (Glu-Leu-Arg) and endothelial-monocyte activating polypeptide-2 (EMAP II) domains in its N-terminal and C-terminal domains, respectively (Fig. 2b) [49]. Although the full-length YRS is secreted from leukocytes and endothelial

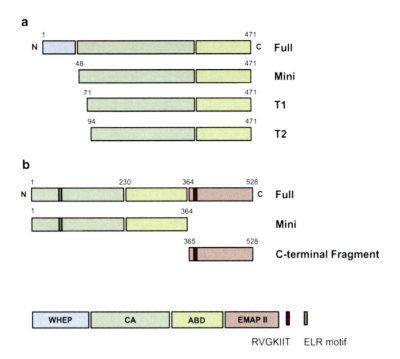

Fig. 2 Schematic representation for the arrangements of the functional domains of WRS and YRS. Arrangement of functional domains and active cytokine fragments of WRS and YRS are illustrated in (**a**) and (**b**), respectively. (**a**) Mini-, T1, and T2-WRSs lack the N-terminal 47, 70, and 98 amino acids, respectively. (**b**) YRS is divided to the two distinct cytokines. Mini-YRS represents the N-terminal 1–364 aa fragment. *WHEP* helix-turn-helix motif, *CA* catalytic domain, *ABD* anticodon-binding domain, *EMAP II* endothelial-monocyte activating polypeptide II, *RVGKIIT* Arg-Val-Gly-Lys-Ile-Ile-Thr heptapeptide sequence, *ELR motif* Glu-Leu-Arg peptide sequence

cells by apoptotic signals, it is subjected to proteolytic cleavage to generate two cytokine-like peptides, namely the N-terminal fragment (designated as mini-YRS) and the C-terminal EMAP II fragment by leukocyte elastase after secretion [24]. A mammalian YRS has an ELR (Glu-Leu-Arg) motif in the N-terminal region, which is a conserved sequence in a variety of chemokines and is critical in chemotactic and angiogenic activity [24, 50]. Due to this motif embedded in mini-YRS, it works with IL-8-like cytokine activity, which induces migration of polymorphonuclear (PMN) cells and stimulates angiogenesis [51, 52]. Mutation of this motif abolishes the migration of immune cells and proliferation of endothelial cells [53]. When the YRS ELR motif was added to yeast, which lacks the ELR motif and cytokine activity, it exhibited cytokine activity [54]. This demonstrates that YRS has acquired the cytokine activity by adopting the ELR motif during evolution.

Extracellular Activities of Aminoacyl-tRNA Synthetases: New Mediators for... 151

The C-terminal fragment of human YRS, which is conserved only in higher eukaryotes, shares 51% sequence homology with human EMAP II [25]. YRS also exhibits potent leukocyte and monocyte chemotaxis along with stimulating production of TNF-α, myeloperoxidase, and tissue factor [55]. The EMAP II-like C-terminal fragment harbors a motif of "RVGKIIT" sequence that is similar to the heptapeptides found in many eukaryotic proteins containing EMAP II-like domains. This motif is essential for the chemotaxis of EMAP II and C-terminal fragment of human YRS [56]. However, *C. elegans* MRS and *S. cerevisiae* Arc1p also harbor similar peptide motifs but lack the cytokine activity [57], suggesting that this motif itself is not sufficient for the activity.

The protein structure of human YRS explains why the full length is not active in the cytokine activity. The surface electrostatic potentials of the ELR motif and heptapeptide are positively and negatively charged, respectively. Since these motifs are located in proximity in the conformation, these motifs are sterically hindered in the full-length YRS and exposed only when the protein is cleaved [53]. A gain-of-function mutant can be made when a conserved tyrosine (Y341) that tethers the ELR motif is mutated to induce a subtle opening. In this case, the full-length YRS mutant exhibited angiogenic activity, further supporting the notion that exposure of the active motifs is critical and the truncation process is needed for the cytokine activity [58].

2.3 Glycyl-tRNA Synthetase (GRS)

Anti-GRS antibodies are observed not only in autoimmune diseases [59, 60] but also in breast cancer patients [18], suggesting the presence of GRS in extracellular space and its pathophysiological implications. The secretion of GRS was found to be induced by various apoptotic stresses from macrophages [26]. In particular, the activity of the secreted GRS from macrophages is intriguing, especially in the control of tumorigenesis. In response to the death signal such as Fas ligand, GRS is secreted from macrophages and induces cancer cell death via its interaction with K-cadherin (CDH6) that belongs to the cadherin superfamily. Upon binding to GRS, K-cadherin releases protein phosphatase 2A [26], which then dephosphorylates the activated ERK that is required for cancer cell survival [61]. Thus, GRS works as an anti-tumor peptide only to the subset of cancer cells that expresses K-cadherin and the activated Ras-ERK pathway. It is not clear at this point why macrophages have recruited GRS as a tool to fight against tumorigenesis. Perhaps it can respond rapidly to the challenges of cancer cells since it is constitutively expressed and is thus suitable as a mechanism for the first defense against cancer initiation. With specific anti-cancer activity, GRS has a great potential as a novel anti-cancer therapeutic agent.

2.4 Lysyl-tRNA Synthetase (KRS)

Anti-KRS antibodies are associated with autoimmune diseases such as polymyositis and dermatomyositis [62]. KRS interacts with syntenin-1 that controls intracellular trafficking and the vesicular secretion pathway [63, 64]. The association of KRS with anti-synthetase syndrome and interaction with syntenin-1 suggest its potential capability of secretion. In fact, KRS was shown to be secreted from cancer cells that are challenged with TNF-α [27]. The secreted KRS binds to macrophages and monocytes to provoke two mitogen-activated protein kinases (MAPK), ERK and p38MAPK, to activate immune cells. The secreted KRS induces pro-inflammatory responses such as cell migration and TNF-α production. Cancer cells can use the immune system for their survival [65–67] and the secreted KRS can change the microenvironment to the condition that is more suitable for survival and/or metastasis of cancer cells.

2.5 Histidyl-tRNA Synthetase/Asparaginyl-tRNA Synthetase/Seryl-tRNA Synthetase

Autoantibodies against histidyl-tRNA synthetase (HRS), asparaginyl-tRNA synthetase (NRS), GRS, AlaRS, IRS, and TRS have also been found in autoimmune patients. Several ARSs have been tested to determine whether they induce leukocyte migration. HRS and NRS function as chemokines that attract CD4$^+$ and CD8$^+$ lymphocytes, IL-2 activated monocytes, and immature dendritic cells (iDCs). Seryl-tRNA synthetase (SRS) induces migration of chemokine receptor 3 (CCR3)-transfected cells. Other autoantigenic ARSs may also provoke innate and adaptive immune responses by attracting mononuclear cells and finally inducing autoimmune diseases [28, 68, 69]. The N-terminal WHEP domain of HRS is reported as the antigenic region, which is recognized by autoantibodies [70, 71], and appears to be responsible for the chemotactic effect.

2.6 Aminoacyl-tRNA Synthetase-Interacting Multifunctional Protein 1/MSCp43

A polypeptide was discovered in the culture medium of activated fibrosarcoma cells and named EMAP II since it induced tissue factor (TF) on endothelial cell and enhanced pro-inflammatory activity on monocyte [72]. EMAP II was identified to be equivalent to the C-terminal fragment (147–312 residues) of Aminoacyl-tRNA Synthetase-Interacting Multifunctional Protein 1 (AIMP1)/MSCp43, which is one component of the multisynthetase complex (MSC) [73]. The full-length AIMP1 is also secreted and has extracellular signaling activities that are distinct depending on

Extracellular Activities of Aminoacyl-tRNA Synthetases: New Mediators for... 153

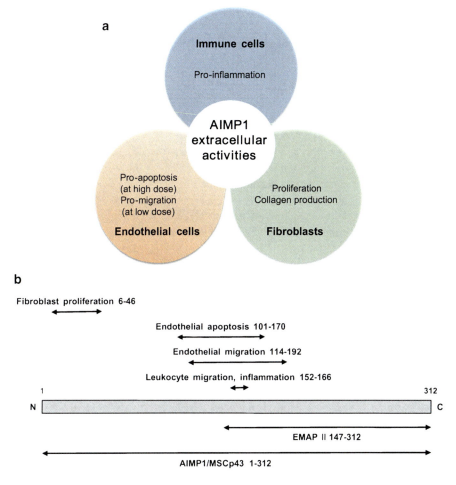

Fig. 3 Extracellular activities of AIMP1 and the structural distribution of the different activities. (**a**) AIMP1 exerts distinct activities depending on its target cells, can activate macrophages and monocytes and promote pro-inflammatory cytokine secretion, can promote migration of endothelial cells at low concentration but induces cell death at high concentration, and enhances proliferation of fibroblast cells. (**b**) The peptide regions for different activities of AIMP1 are schematically represented

target cells (Fig. 3a) [25, 74]. EMAP II is secreted from tumor cell lines and up-regulates TF on ECs [75]. TF is a cell-surface receptor that induces thrombin formations, and is mainly increased on ECs by cytokines such as IL-1β and TNF-α [76, 77]. Since this effect promotes local thrombohemorrhage and intense inflammation, EMAP II-treated ECs deliver TNF-α more easily [77]. EMAP II also induces apoptosis of endothelial cells by up-regulating Fas-associated signaling molecules and by down-regulating Bcl-2 [78]. In addition to its effect on ECs, the TNF receptor 1 on tumor cells is increased by EMAP II treatment [79]. Due to these effects, EMAP II induces apoptosis in cancer cells. In tumor mice models, EMAP II treatment with TNF-α suppresses TNF-α insensitive tumor growth without systemic toxicity [75, 80, 81]. When EMAP II over-expressed tumor cells are treated with TNF-α, tumor size is also decreased in a dose-dependent manner [82]. AIMP1 has multi-faceted properties in angiogenesis. While AIMP1 at low dose activates ERK, resulting in the induction and activation of matrix metalloproteinase 9 and promotion of migration of ECs, a high dose of AIMP1 induces apoptosis of ECs through the activation of Jun C-terminal kinase that mediates apoptosis of endothelial cells [83]. The dose-dependent biphasic activity of AIMP1 can finely control the complicated angiogenesis process depending on physiological context. Angiogenesis interruption is a critical anti-tumor therapeutic point because angiogenesis is associated with cancer development and metastasis [84].

In the immune system, EMAP II and AIMP1 exert chemokine and pro-inflammatory activity for leukocytes and macrophages. Immune cells including macrophages and monocytes rapidly remove cells undergoing apoptosis. When cells undergo stresses such as apoptosis and hypoxia, caspase 7, an apoptosis-related cysteine protease, cleaves AIMP1 to EMAP II [85–87]. EMAP II is then released from apoptotic cells and recruits monocytes and macrophage in apoptotic regions to scavenge damaged cells [88]. The chemotactic active domain of EMAP II appears to reside in the 14 amino acid peptide region from 152 to 166 amino acid in the full-length sequence (Fig. 3b) [89]. The structure of human EMAP II has been resolved by X-ray crystallography [82]. It contains three-stranded β-sheet with an α-helix and this domain is structurally homologous to other chemokines such as RANTES and neutrophil-activating peptide-2. The peptide sequence responsible for the cytokine activity mostly lies in the β1 strand. When EMAP II was locally treated in mice footpads, vascular congestion and leukostasis were observed in the footpads. Although expression of EMAP II is limited in central nervous system (CNS), it is up-regulated in autoimmune disorder lesions in microglial cells. Since microglial cells are a key element of CNS immune surveillance [90, 91], EMAP II may act as an important chemokine in inflammatory responses.

The secreted AIMP1 activates macrophages and monocytes inducing pro-inflammatory cytokine secretion such as TNF-α, IL-8, and IL-1β through ERK, p38 MAPK, and NF-κB signaling [92, 93]. In addition to these functions, AIMP1 promotes the cell adhesion activity of monocytes through induction of critical cell–cell adhesion molecule intercellular adhesion molecules 1 (ICAM-1) [94]. Because ICAM-1 is regulated via PI3K-ERK and p38 MAPK signaling [95], AIMP1 treatment enhances ICAM-1 expression levels in monocytes [96]. Since a high level of AIMP1 is observed in atherosclerotic lesions and TNF-α is considered as a pathogenic mediator in immune disease, its involvement in inflammation-related diseases is suggested. AIMP1 is also known to have an important role in cell-mediated immune response. When macrophages are stimulated by AIMP1, they induce and secrete IL-12 through NF-κB, which regulates cell-mediated Th1 immune response. Furthermore, aberrant IFN-γ expression is shown in CD4$^+$ T cells, co-cultured with AIMP1-activated macrophages [97]. AIMP1-treated murine bone marrow-derived dendritic cells (DCs) exhibit enhanced expression of cell surface markers such as CD40, CD86, and MHC class II and also induce IL-12. Due to these phenomena, AIMP1-treated DCs induce the activation of antigen-specific Th1 cells through enhanced antigen-presenting capabilities [98]. Because AIMP1 induces a Th1 immune response, which exhibits anti-tumor activity and TNF-α, AIMP1 treatment was shown to induce cytotoxic activities against tumor cells in mice models [99–101].

Wound healing is a complicated process that includes inflammation, proliferation, and maturation. Before fibroblast proliferation, inflammation is induced to protect the host and deliver nutrients [102, 103]. AIMP1 is used to connect the inflammation and proliferation steps. AIMP1 is secreted from TNF-α stimulated macrophages in wound regions connecting inflammation to proliferation. Interestingly, it activates ERK signaling in fibroblasts inducing proliferation and collagen synthesis. The N-terminal region of AIMP1, especially the 6–46 amino acid peptide, appears to be important and sufficient for fibroblast proliferation (Fig. 3b) [104]. In a mouse wound-healing model, AIMP1-treated mice exhibit increased fibroblast proliferation and collagen production, leading to enhanced wound healing [105].

AIMP1 is also enriched in pancreatic cells and is secreted by hypoglycemic stimulation [29]. When AIMP1 is secreted in the blood stream it exerts hormonal effects for glucose homeostasis. Namely, AIMP1 secreted from pancreatic α cells stimulates glucagon secretion and restores glucose levels in the blood by releasing liver-stored glucose. Also, AIMP1 induces lipolysis of triglyceride in fat tissue. All of these functions collectively appear to help to maintain glucose homeostasis.

3 Working Mechanism of ARSs-Derived Cytokine

Cytokines exert their signaling activities via binding to their cognate receptors that are expressed on the surface of target cells. After the cytokine binds to the receptor, the activated receptor induces a corresponding signal cascade, eliciting physiological responses [106, 107]. Although the functional receptors for all the secreted ARSs have not been identified, there are some cases in which the receptors have been suggested.

3.1 Adhesion Protein-Mediated Mechanism

Based on the structure of their extracellular domains, cytokine receptors can be divided into groups such as Type I and II, which are members of the Ig superfamily, TNF receptors, and seven transmembrane receptors [108]. Among these types, cadherins, which are the subfamily of seven transmembrane receptors, are calcium-dependent glycoproteins. They mediate cell–cell adhesion and maintain cell polarity through homo- or heterodimerization [109]. The secreted WRS and

GRS appear to use VE-cadherin and K-cadherin as their functional receptors, respectively [26, 45]. Integrins mediate the attachment between cells and extracellular matrix (ECM), and are involved in the regulation of cell motility and proliferation [110]. EMAP II was shown to interfere with the interaction of ECM and integrin alpha 5 beta 1 by competitive inhibition [111]. These results imply that at least some secreting ARSs are involved in the control of cell–cell adhesion.

3.1.1 WRS–VE-Cadherin Interaction

The angiostatic activity of the N-terminal truncated WRS appears to be mediated via its binding to vascular endothelial cadherin (VE-cadherin) that is a calcium-dependent adhesion molecule and specifically expressed in endothelial cells (Fig. 4a) [45]. VE-cadherin mediates endothelial adhesion through dimerization formation that regulates vascular morphology and survival [112]. WRS interacts with VE-cadherin via its catalytic active site [48]. VE-cadherin has two tryptophan residues at positions 2 and 4 in the N-terminus, which are the key determinants for homophilic interaction between neighboring endothelial cells [113]. Since the truncated WRS can recognize these tryptophan residues with its opened amino acid binding pocket, it inhibits VE-cadherin dimerization. This interferes with endothelial cell survival and organization inducing apoptosis. Through this mechanism, the truncated WRS inhibits endothelial cell migration and survival.

3.1.2 GRS–K-Cadherin Interaction

The secreted GRS also uses a cadherin protein, K-cadherin (CDH6), as the functional receptor that is highly expressed in fetal kidney, brain, and cancers (Fig. 4b) [114, 115]. In contrast to WRS, the interaction between GRS and K-cadherin triggers intercellular changes [26]. It releases protein phosphatase 2A (PP2A), which is a member of the phosphatase family and is bound to the intracellular domain of K-cadherin. The PP2A released dephosphorylates ERK that is activated by the upstream Ras [116]. Since some cancer cells rely on ERK signaling for survival [61], GRS-induced dephosphorylation of ERK can lead to the death of susceptible cancer cells. Using this activity, GRS can be used as anti-cancer therapeutics against cancer cells with K-cadherin expression and the activated ERK pathway.

3.1.3 EMAP II–Integrin Alpha 5 Beta 1

In endothelial cells the cell interaction with ECM is important to mediate cell migration and proliferation. Integrins, heterodimeric transmembrane glycoproteins, predominantly regulate this process [117]. Interestingly, EMAP II directly interferes with cell adhesion to fibronectin via its direct interaction with integrin alpha 5. This competitive inhibition disrupts endothelial cell adhesion causing

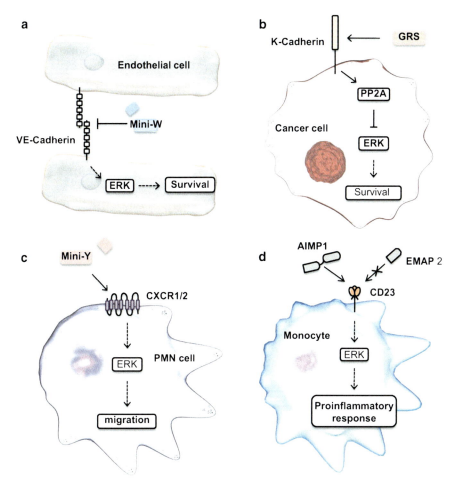

Fig. 4 The working mechanisms of secreted human ARSs. (**a**) Mini-WRS inhibits endothelial cell migration and survival by blocking cell adhesion mediated by VE-cadherin dimerization. (**b**) Mini-YRS binds to IL-8 type A receptor, CXCR1/2, to induce endothelial cell migration. (**c**) GRS binds to K-cadherin (CDH6) that then releases PP2A to inhibit ERK signaling by dephosphorylation of ERK. (**d**) AIMP1 binds to CD23 and provokes MAPKs signal pathway to activate immune cells

apoptosis [111], suggesting integrin alpha 5 as a potential functional receptor for angiostatic activity of EMAPII.

3.2 Cytokine Receptor-Mediated Mechanism

Some secreted ARSs appear to share the known cytokine receptors. Since the ELR motif, which is found in CXC chemokines [50], is embedded in YRS, it can interact with CXCR1/2 like other CXC chemokines [24]. CD23 is the receptor for immunoglobulin E (IgE) and implicated in cancer and inflammation [118]. It has recently been shown to serve as a functional receptor for AIMP1, but not EMAP II, in macrophages and monocytes [119]. This is the first receptor that distinguishes AIMP1 and EMAPII, implying that the secretion of the two forms may have unique physiological causes.

3.2.1 YRS-CXCR1/2

The CXC chemokines contain conserved ELR motifs that induce the migration of immune cells and have an angiogenic effect via CXC receptor (CXCR) (Fig. 4c). Mini-YRS, an ELR-positive chemokine, also binds to one kind of CXCR, the IL-8 type A receptors (CXCR1/2) [24]. High concentrations of mini-YRS dimerize and can't bind with the receptor, but low concentrations of mini-YRS can act as a functional agonist to CXCR1/2. This indicates the possibility that the dissociation of quaternary structures is critical for cytokine activity [120]. Mini-YRS induces endothelial cell migration through VEGFR2 and its angiogenic activity depends on the ELR motif. The secreted mini-YRS binds to vascular endothelial cells and activates angiogenic signals including vascular endothelial growth factor receptor-2 (VEGFR2). This stimulation leads to VEGF-independent phosphorylation of the receptor. Transactivation requires an intact ELR motif in the mini-YRS [121].

3.2.2 AIMP1–CD23

CD23, which is a low affinity receptor for immunoglobulin E (IgE), binds to ligands such as IgE, CD21, and CD11b, mediating allergic and inflammation processes [122, 123]. Although CD23 is highly expressed in macrophage/monocyte, it was not considered as a key molecule in macrophage/monocyte-derived inflammation until AIMP1 was identified as a ligand for CD23 [119]. AIMP1, but not EMAP II, activates the ERK pathway via CD23 binding, thereby inducing proinflammatory responses (Fig. 4d). The central domain of AIMP1 (101–192 amino acids) was identified as binding site and this domain also contains the inflammation-related domain (152–166 amino acids). Since EMAP II does not possess the CD23-binding domain, it does not use CD23 as functional receptor. This is the first evidence showing the distinct activity of AIMP1 that distinguishes it from its C-terminal domain, EMAPII.

4 Conclusion

There are increasing numbers of reports on the extracellular secretion of proteins whose functions were intracellular when they were first discovered. They include heat shock proteins, ferritin, and HMGB1 [12, 124, 125]. Now that many ARSs have been found in extracellular space with complex activities, these enzymes should be considered as a new family of secreting ligands with unique physiological implications. Although it is unclear at this moment why cells have recruited protein synthesis enzymes for cell signaling, one common property amongst ARSs is that all of them are constitutively expressed in all types of cells due to their being essential for protein synthesis. Thus, they can always be available to respond promptly to the urgent situation that may disrupt system homeostasis. In addition, the capability of diverse molecular interactions renders ARSs suitable as extracellular ligands to bind various receptors. More data is needed to see whether the secreted ARSs would share any common receptor family with other classical extracellular ligands or be a unique receptor family. Even though, there are receptors that have been found to interact with ARS, there is no novel receptor that is correlated specifically to only secreted ARSs. This implies that the secreted ARSs might overlap the functions of classical cytokines. For this overlap the secreted ARSs can modulate the normal activities of the classical cytokines through the competition or augmentation for the receptor binding. Whatever their functions, the secreted ARSs appear to play roles in the early stage before the classical cytokines work. The diverse extracellular activities of the secreted ARSs can also be explored as novel therapeutic agents and they are expected to constitute a new biological resource for future medicine. It is yet to be determined whether only a subset of ARSs are secreted or all ARSs secrete under appropriate conditions. Although there are a lot of questions that need to be answered, it seems clear that ARSs would play critical roles to maintain the balance of our body system through complex roles in extracellular space as well as through their intracellular canonical and non-canonical activities.

References

1. Foster D, Parrish-Novak J, Fox B, Xu W (2004) Cytokine-receptor pairing: accelerating discovery of cytokine function. Nat Rev Drug Discov 3(2):160–170
2. Kolls JK, McCray PB Jr, Chan YR (2008) Cytokine-mediated regulation of antimicrobial proteins. Nat Rev Immunol 8:829–835
3. Sims JE, Smith DE (2010) The IL-1 family: regulators of immunity. Nat Rev Immunol 10:89–102
4. Tracey KJ (2009) Reflex control of immunity. Nat Rev Immunol 9:418–428
5. Wang CY, Mayo MW, Baldwin AS Jr (1996) TNF- and cancer therapy-induced apoptosis: potentiation by inhibition of NF-kappaB. Science 274:784–787
6. Karin M, Cao Y, Greten FR, Li ZW (2002) NF-kappaB in cancer: from innocent bystander to major culprit. Nat Rev Cancer 2:301–310
7. Coussens LM, Werb Z (2002) Inflammation and cancer. Nature 420:860–867
8. Gutterman JU (1994) Cytokine therapeutics: lessons from interferon alpha. Proc Natl Acad Sci USA 91:1198–1205
9. Dvorak HF (2002) Vascular permeability factor/vascular endothelial growth factor: a critical cytokine in tumor angiogenesis and a potential target for diagnosis and therapy. J Clin Oncol 20:4368–4380

10. Hanada T, Yoshimura A (2002) Regulation of cytokine signaling and inflammation. Cytokine Growth Factor Rev 13:413–421
11. Mantovani A, Bussolino F, Dejana E (1992) Cytokine regulation of endothelial cell function. FASEB J 6:2591–2599
12. Gardella S, Andrei C, Ferrera D, Lotti LV, Torrisi MR, Bianchi ME, Rubartelli A (2002) The nuclear protein HMGB1 is secreted by monocytes via a non-classical, vesicle-mediated secretory pathway. EMBO Rep 3:995–1001
13. Srivastava P (2002) Roles of heat-shock proteins in innate and adaptive immunity. Nat Rev Immunol 2:185–194
14. Park SG, Ewalt KL, Kim S (2005) Functional expansion of aminoacyl-tRNA synthetases and their interacting factors: new perspectives on housekeepers. Trends Biochem Sci 30:569–574
15. Ewalt KL, Schimmel P (2002) Activation of angiogenic signaling pathways by two human tRNA synthetases. Biochemistry 41:13344–13349
16. Guo M, Yang XL, Schimmel P (2010) New functions of aminoacyl-tRNA synthetases beyond translation. Nat Rev Mol Cell Biol 11:668–674
17. Park SG, Schimmel P, Kim S (2008) Aminoacyl tRNA synthetases and their connections to disease. Proc Natl Acad Sci USA 105:11043–11049
18. Mun J, Kim YH, Yu J, Bae J, Noh DY, Yu MH, Lee C (2010) A proteomic approach based on multiple parallel separation for the unambiguous identification of an antibody cognate antigen. Electrophoresis 31:3428–3436
19. Hirakata M, Suwa A, Nagai S, Kron MA, Trieu EP, Mimori T, Akizuki M, Targoff IN (1999) Anti-KS: identification of autoantibodies to asparaginyl-transfer RNA synthetase associated with interstitial lung disease. J Immunol 162:2315–2320
20. Mammen AL (2011) Autoimmune myopathies: autoantibodies, phenotypes and pathogenesis. Nat Rev Neurol 7:343–354
21. Yang XJ (2005) Multisite protein modification and intramolecular signaling. Oncogene 24:1653–1662
22. Glomset JA, Farnsworth CC (1994) Role of protein modification reactions in programming interactions between ras-related GTPases and cell membranes. Annu Rev Cell Biol 10:181–205
23. Wakasugi K, Slike BM, Hood J, Otani A, Ewalt KL, Friedlander M, Cheresh DA, Schimmel P (2002) A human aminoacyl-tRNA synthetase as a regulator of angiogenesis. Proc Natl Acad Sci USA 99:173–177
24. Wakasugi K, Schimmel P (1999) Two distinct cytokines released from a human aminoacyl-tRNA synthetase. Science 284:147–151
25. van Horssen R, Eggermont AM, ten Hagen TL (2006) Endothelial monocyte-activating polypeptide-II and its functions in (patho)physiological processes. Cytokine Growth Factor Rev 17:339–348
26. Park MC, Kang T, Jin D, Han JM, Kim SB, Park YJ, Cho K, Park YW, Guo M, He W, Yang XL, Schimmel P, Kim S (2012) Secreted human glycyl-tRNA synthetase implicated in defense against ERK-activated tumorigenesis. Proc Natl Acad Sci USA 109:E640–E647
27. Park SG, Kim HJ, Min YH, Choi EC, Shin YK, Park BJ, Lee SW, Kim S (2005) Human lysyl-tRNA synthetase is secreted to trigger proinflammatory response. Proc Natl Acad Sci USA 102:6356–6361
28. Howard OM, Dong HF, Yang D, Raben N, Nagaraju K, Rosen A, Casciola-Rosen L, Hartlein M, Kron M, Yang D, Yiadom K, Dwivedi S, Plotz PH, Oppenheim JJ (2002) Histidyl-tRNA synthetase and asparaginyl-tRNA synthetase, autoantigens in myositis, activate chemokine receptors on T lymphocytes and immature dendritic cells. J Exp Med 196:781–791
29. Park SG, Kang YS, Kim JY, Lee CS, Ko YG, Lee WJ, Lee KU, Yeom YI, Kim S (2006) Hormonal activity of AIMP1/p43 for glucose homeostasis. Proc Natl Acad Sci USA 103:14913–14918

30. Deane JA, Fruman DA (2004) Phosphoinositide 3-kinase: diverse roles in immune cell activation. Annu Rev Immunol 22:563–598
31. Inoki K, Li Y, Zhu T, Wu J, Guan KL (2002) TSC2 is phosphorylated and inhibited by Akt and suppresses mTOR signalling. Nat Cell Biol 4:648–657
32. Martinon F, Mayor A, Tschopp J (2009) The inflammasomes: guardians of the body. Annu Rev Immunol 27:229–265
33. van de Veerdonk FL, Netea MG, Dinarello CA, Joosten LA (2011) Inflammasome activation and IL-1beta and IL-18 processing during infection. Trends Immunol 32:110–116
34. Tolstrup AB, Bejder A, Fleckner J, Justesen J (1995) Transcriptional regulation of the interferon-gamma-inducible tryptophanyl-tRNA synthetase includes alternative splicing. J Biol Chem 270:397–403
35. Turpaev KT, Zakhariev VM, Sokolova IV, Narovlyansky AN, Amchenkova AM, Justesen J, Frolova LY (1996) Alternative processing of the tryptophanyl-tRNA synthetase mRNA from interferon-treated human cells. Eur J Biochem 240:732–737
36. Kaplan G, Luster AD, Hancock G, Cohn ZA (1987) The expression of a gamma interferon-induced protein (IP-10) in delayed immune responses in human skin. J Exp Med 166:1098–1108
37. Farber JM (1993) HuMig: a new human member of the chemokine family of cytokines. Biochem Biophys Res Commun 192:223–230
38. Shaw AC, Rossel Larsen M, Roepstorff P, Justesen J, Christiansen G, Birkelund S (1999) Mapping and identification of interferon gamma-regulated HeLa cell proteins separated by immobilized pH gradient two-dimensional gel electrophoresis. Electrophoresis 20:984–993
39. Liu J, Shue E, Ewalt KL, Schimmel P (2004) A new gamma-interferon-inducible promoter and splice variants of an anti-angiogenic human tRNA synthetase. Nucleic Acids Res 32:719–727
40. Fleckner J, Martensen PM, Tolstrup AB, Kjeldgaard NO, Justesen J (1995) Differential regulation of the human, interferon inducible tryptophanyl-tRNA synthetase by various cytokines in cell lines. Cytokine 7:70–77
41. Otani A, Slike BM, Dorrell MI, Hood J, Kinder K, Ewalt KL, Cheresh D, Schimmel P, Friedlander M (2002) A fragment of human TrpRS as a potent antagonist of ocular angiogenesis. Proc Natl Acad Sci USA 99:178–183
42. Kapoor M, Zhou Q, Otero F, Myers CA, Bates A, Belani R, Liu J, Luo JK, Tzima E, Zhang DE, Yang XL, Schimmel P (2008) Evidence for annexin II-S100A10 complex and plasmin in mobilization of cytokine activity of human TrpRS. J Biol Chem 283:2070–2077
43. Zobiack N, Rescher U, Ludwig C, Zeuschner D, Gerke V (2003) The annexin 2/S100A10 complex controls the distribution of transferrin receptor-containing recycling endosomes. Mol Biol Cell 14:4896–4908
44. Guo M, Schimmel P, Yang XL (2010) Functional expansion of human tRNA synthetases achieved by structural inventions. FEBS Lett 584:434–442
45. Tzima E, Reader JS, Irani-Tehrani M, Ewalt KL, Schwartz MA, Schimmel P (2005) VE-cadherin links tRNA synthetase cytokine to anti-angiogenic function. J Biol Chem 280:2405–2408
46. Yang XL, Otero FJ, Skene RJ, McRee DE, Schimmel P, Ribas de Pouplana L (2003) Crystal structures that suggest late development of genetic code components for differentiating aromatic side chains. Proc Natl Acad Sci USA 100:15376–15380
47. Yang XL, Guo M, Kapoor M, Ewalt KL, Otero FJ, Skene RJ, McRee DE, Schimmel P (2007) Functional and crystal structure analysis of active site adaptations of a potent anti-angiogenic human tRNA synthetase. Structure 15:793–805
48. Zhou Q, Kapoor M, Guo M, Belani R, Xu X, Kiosses WB, Hanan M, Park C, Armour E, Do MH, Nangle LA, Schimmel P, Yang XL (2010) Orthogonal use of a human tRNA synthetase active site to achieve multifunctionality. Nat Struct Mol Biol 17:57–61

49. Renault L, Kerjan P, Pasqualato S, Menetrey J, Robinson JC, Kawaguchi S, Vassylyev DG, Yokoyama S, Mirande M, Cherfils J (2001) Structure of the EMAPII domain of human aminoacyl-tRNA synthetase complex reveals evolutionary dimer mimicry. EMBO J 20:570–578
50. Strieter RM, Polverini PJ, Kunkel SL, Arenberg DA, Burdick MD, Kasper J, Dzuiba J, Van Damme J, Walz A, Marriott D et al (1995) The functional role of the ELR motif in CXC chemokine-mediated angiogenesis. J Biol Chem 270:27348–27357
51. Wakasugi K, Slike BM, Hood J, Ewalt KL, Cheresh DA, Schimmel P (2002) Induction of angiogenesis by a fragment of human tyrosyl-tRNA synthetase. J Biol Chem 277:20124–20126
52. Cheng G, Zhang H, Yang X, Tzima E, Ewalt KL, Schimmel P, Faber JE (2008) Effect of mini-tyrosyl-tRNA synthetase on ischemic angiogenesis, leukocyte recruitment, and vascular permeability. Am J Physiol Regul Integr Comp Physiol 295:R1138–R1146
53. Yang XL, Skene RJ, McRee DE, Schimmel P (2002) Crystal structure of a human aminoacyl-tRNA synthetase cytokine. Proc Natl Acad Sci USA 99:15369–15374
54. Liu J, Yang XL, Ewalt KL, Schimmel P (2002) Mutational switching of a yeast tRNA synthetase into a mammalian-like synthetase cytokine. Biochemistry 41:14232–14237
55. Berger AC, Tang G, Alexander HR, Libutti SK (2000) Endothelial monocyte-activating polypeptide II, a tumor-derived cytokine that plays an important role in inflammation, apoptosis, and angiogenesis. J Immunother 23:519–527
56. Wakasugi K, Schimmel P (1999) Highly differentiated motifs responsible for two cytokine activities of a split human tRNA synthetase. J Biol Chem 274:23155–23159
57. Kleeman TA, Wei D, Simpson KL, First EA (1997) Human tyrosyl-tRNA synthetase shares amino acid sequence homology with a putative cytokine. J Biol Chem 272:14420–14425
58. Yang XL, Kapoor M, Otero FJ, Slike BM, Tsuruta H, Frausto R, Bates A, Ewalt KL, Cheresh DA, Schimmel P (2007) Gain-of-function mutational activation of human tRNA synthetase procytokine. Chem Biol 14:1323–1333
59. Targoff IN, Trieu EP, Plotz PH, Miller FW (1992) Antibodies to glycyl-transfer RNA synthetase in patients with myositis and interstitial lung disease. Arthritis Rheum 35:821–830
60. Targoff IN (1990) Autoantibodies to aminoacyl-transfer RNA synthetases for isoleucine and glycine. Two additional synthetases are antigenic in myositis. J Immunol 144:1737–1743
61. Anjum R, Blenis J (2008) The RSK family of kinases: emerging roles in cellular signalling. Nat Rev Mol Cell Biol 9:747–758
62. Gelpi C, Kanterewicz E, Gratacos J, Targoff IN, Rodriguez-Sanchez JL (1996) Coexistence of two antisynthetases in a patient with the antisynthetase syndrome. Arthritis Rheum 39:692–697
63. Meerschaert K, Remue E, De Ganck A, Staes A, Boucherie C, Gevaert K, Vandekerckhove J, Kleiman L, Gettemans J (2008) The tandem PDZ protein syntenin interacts with the aminoacyl tRNA synthetase complex in a lysyl-tRNA synthetase-dependent manner. J Proteome Res 7:4962–4973
64. Baietti MF, Zhang Z, Mortier E, Melchior A, Degeest G, Geeraerts A, Ivarsson Y, Depoortere F, Coomans C, Vermeiren E, Zimmermann P, David G (2012) Syndecan-syntenin-ALIX regulates the biogenesis of exosomes. Nat Cell Biol 14:677–685
65. Mantovani A, Allavena P, Sica A, Balkwill F (2008) Cancer-related inflammation. Nature 454:436–444
66. Mumm JB, Oft M (2008) Cytokine-based transformation of immune surveillance into tumor-promoting inflammation. Oncogene 27:5913–5919
67. Dunn GP, Bruce AT, Ikeda H, Old LJ, Schreiber RD (2002) Cancer immunoediting: from immunosurveillance to tumor escape. Nat Immunol 3:991–998
68. Wong FS, Karttunen J, Dumont C, Wen L, Visintin I, Pilip IM, Shastri N, Pamer EG, Janeway CA Jr (1999) Identification of an MHC class I-restricted autoantigen in type 1 diabetes by screening an organ-specific cDNA library. Nat Med 5:1026–1031

69. Soejima M, Kang EH, Gu X, Katsumata Y, Clemens PR, Ascherman DP (2011) Role of innate immunity in a murine model of histidyl-transfer RNA synthetase (Jo-1)-mediated myositis. Arthritis Rheum 63:479–487
70. Raben N, Nichols R, Dohlman J, McPhie P, Sridhar V, Hyde C, Leff R, Plotz P (1994) A motif in human histidyl-tRNA synthetase which is shared among several aminoacyl-tRNA synthetases is a coiled-coil that is essential for enzymatic activity and contains the major autoantigenic epitope. J Biol Chem 269:24277–24283
71. Xu Z, Wei Z, Zhou JJ, Ye F, Lo WS, Wang F, Lau CF, Wu J, Nangle LA, Chiang KP, Yang XL, Zhang M, Schimmel P (2012) Internally deleted human tRNA synthetase suggests evolutionary pressure for repurposing. Structure 20:1470–1477
72. Kao J, Ryan J, Brett G, Chen J, Shen H, Fan YG, Godman G, Familletti PC, Wang F, Pan YC et al (1992) Endothelial monocyte-activating polypeptide II. A novel tumor-derived polypeptide that activates host-response mechanisms. J Biol Chem 267:20239–20247
73. Shalak V, Kaminska M, Mitnacht-Kraus R, Vandenabeele P, Clauss M, Mirande M (2001) The EMAPII cytokine is released from the mammalian multisynthetase complex after cleavage of its p43/proEMAPII component. J Biol Chem 276:23769–23776
74. Lee SW, Cho BH, Park SG, Kim S (2004) Aminoacyl-tRNA synthetase complexes: beyond translation. J Cell Sci 117:3725–3734
75. Wu PC, Alexander HR, Huang J, Hwu P, Gnant M, Berger AC, Turner E, Wilson O, Libutti SK (1999) In vivo sensitivity of human melanoma to tumor necrosis factor (TNF)-alpha is determined by tumor production of the novel cytokine endothelial-monocyte activating polypeptide II (EMAPII). Cancer Res 59:205–212
76. Nawroth PP, Handley DA, Esmon CT, Stern DM (1986) Interleukin 1 induces endothelial cell procoagulant while suppressing cell-surface anticoagulant activity. Proc Natl Acad Sci USA 83:3460–3464
77. Bevilacqua MP, Pober JS, Majeau GR, Fiers W, Cotran RS, Gimbrone MA Jr (1986) Recombinant tumor necrosis factor induces procoagulant activity in cultured human vascular endothelium: characterization and comparison with the actions of interleukin 1. Proc Natl Acad Sci USA 83:4533–4537
78. Berger AC, Alexander HR, Wu PC, Tang G, Gnant MF, Mixon A, Turner ES, Libutti SK (2000) Tumour necrosis factor receptor I (p55) is upregulated on endothelial cells by exposure to the tumour-derived cytokine endothelial monocyte- activating polypeptide II (EMAP-II). Cytokine 12:992–1000
79. Berger AC, Alexander HR, Tang G, Wu PS, Hewitt SM, Turner E, Kruger E, Figg WD, Grove A, Kohn E, Stern D, Libutti SK (2000) Endothelial monocyte activating polypeptide II induces endothelial cell apoptosis and may inhibit tumor angiogenesis. Microvasc Res 60:70–80
80. Gnant MF, Berger AC, Huang J, Puhlmann M, Wu PC, Merino MJ, Bartlett DL, Alexander HR Jr, Libutti SK (1999) Sensitization of tumor necrosis factor alpha-resistant human melanoma by tumor-specific in vivo transfer of the gene encoding endothelial monocyte-activating polypeptide II using recombinant vaccinia virus. Cancer Res 59:4668–4674
81. Marvin MR, Libutti SK, Kayton M, Kao J, Hayward J, Grikscheit T, Fan Y, Brett J, Weinberg A, Nowygrod R, LoGerfo P, Feind C, Hansen KS, Schwartz M, Stern D, Chabot J (1996) A novel tumor-derived mediator that sensitizes cytokine-resistant tumors to tumor necrosis factor. J Surg Res 63:248–255
82. Kim Y, Shin J, Li R, Cheong C, Kim K, Kim S (2000) A novel anti-tumor cytokine contains an RNA binding motif present in aminoacyl-tRNA synthetases. J Biol Chem 275:27062–27068
83. Park SG, Kang YS, Ahn YH, Lee SH, Kim KR, Kim KW, Koh GY, Ko YG, Kim S (2002) Dose-dependent biphasic activity of tRNA synthetase-associating factor, p43, in angiogenesis. J Biol Chem 277:45243–45248
84. Fidler IJ, Ellis LM (1994) The implications of angiogenesis for the biology and therapy of cancer metastasis. Cell 79:185–188

85. Knies UE, Behrensdorf HA, Mitchell CA, Deutsch U, Risau W, Drexler HC, Clauss M (1998) Regulation of endothelial monocyte-activating polypeptide II release by apoptosis. Proc Natl Acad Sci USA 95:12322–12327

86. Barnett G, Jakobsen AM, Tas M, Rice K, Carmichael J, Murray JC (2000) Prostate adenocarcinoma cells release the novel proinflammatory polypeptide EMAP-II in response to stress. Cancer Res 60:2850–2857

87. Behrensdorf HA, van de Craen M, Knies UE, Vandenabeele P, Clauss M (2000) The endothelial monocyte-activating polypeptide II (EMAP II) is a substrate for caspase-7. FEBS Lett 466:143–147

88. Kao J, Fan YG, Haehnel I, Brett J, Greenberg S, Clauss M, Kayton M, Houck K, Kisiel W, Seljelid R et al (1994) A peptide derived from the amino terminus of endothelial-monocyte-activating polypeptide II modulates mononuclear and polymorphonuclear leukocyte functions, defines an apparently novel cellular interaction site, and induces an acute inflammatory response. J Biol Chem 269:9774–9782

89. Kao J, Houck K, Fan Y, Haehnel I, Libutti SK, Kayton ML, Grikscheit T, Chabot J, Nowygrod R, Greenberg S et al (1994) Characterization of a novel tumor-derived cytokine. Endothelial-monocyte activating polypeptide II. J Biol Chem 269:25106–25119

90. Schluesener HJ, Seid K, Zhao Y, Meyermann R (1997) Localization of endothelial-monocyte-activating polypeptide II (EMAP II), a novel proinflammatory cytokine, to lesions of experimental autoimmune encephalomyelitis, neuritis and uveitis: expression by monocytes and activated microglial cells. Glia 20:365–372

91. Mueller CA, Schluesener HJ, Conrad S, Meyermann R, Schwab JM (2003) Lesional expression of a proinflammatory and antiangiogenic cytokine EMAP II confined to endothelium and microglia/macrophages during secondary damage following experimental traumatic brain injury. J Neuroimmunol 135:1–9

92. Ko YG, Park H, Kim T, Lee JW, Park SG, Seol W, Kim JE, Lee WH, Kim SH, Park JE, Kim S (2001) A cofactor of tRNA synthetase, p43, is secreted to up-regulate proinflammatory genes. J Biol Chem 276:23028–23033

93. Park H, Park SG, Kim J, Ko YG, Kim S (2002) Signaling pathways for TNF production induced by human aminoacyl-tRNA synthetase-associating factor, p43. Cytokine 20:148–153

94. Elsner J, Sach M, Knopf HP, Norgauer J, Kapp A, Schollmeyer P, Dobos GJ (1995) Synthesis and surface expression of ICAM-1 in polymorphonuclear neutrophilic leukocytes in normal subjects and during inflammatory disease. Immunobiology 193:456–464

95. Capodici C, Hanft S, Feoktistov M, Pillinger MH (1998) Phosphatidylinositol 3-kinase mediates chemoattractant-stimulated, CD11b/CD18-dependent cell–cell adhesion of human neutrophils: evidence for an ERK-independent pathway. J Immunol 160:1901–1909

96. Park H, Park SG, Lee JW, Kim T, Kim G, Ko YG, Kim S (2002) Monocyte cell adhesion induced by a human aminoacyl-tRNA synthetase-associated factor, p43: identification of the related adhesion molecules and signal pathways. J Leukoc Biol 71:223–230

97. Kim E, Kim SH, Kim S, Kim TS (2006) The novel cytokine p43 induces IL-12 production in macrophages via NF-kappaB activation, leading to enhanced IFN-gamma production in CD4 + T cells. J Immunol 176:256–264

98. Kim E, Kim SH, Kim S, Cho D, Kim TS (2008) AIMP1/p43 protein induces the maturation of bone marrow-derived dendritic cells with T helper type 1-polarizing ability. J Immunol 180:2894–2902

99. Kim TS, Lee BC, Kim E, Cho D, Cohen EP (2008) Gene transfer of AIMP1 and B7.1 into epitope-loaded, fibroblasts induces tumor-specific CTL immunity, and prolongs the survival period of tumor-bearing mice. Vaccine 26:5928–5934

100. Lee BC, O'Sullivan I, Kim E, Park SG, Hwang SY, Cho D, Kim TS (2009) A DNA adjuvant encoding a fusion protein between anti-CD3 single-chain Fv and AIMP1 enhances T helper type 1 cell-mediated immune responses in antigen-sensitized mice. Immunology 126:84–91

101. Han JM, Myung H, Kim S (2010) Antitumor activity and pharmacokinetic properties of ARS-interacting multi-functional protein 1 (AIMP1/p43). Cancer Lett 287:157–164
102. Diegelmann RF, Evans MC (2004) Wound healing: an overview of acute, fibrotic and delayed healing. Front Biosci 9:283–289
103. Shaw TJ, Martin P (2009) Wound repair at a glance. J Cell Sci 122:3209–3213
104. Han JM, Park SG, Lee Y, Kim S (2006) Structural separation of different extracellular activities in aminoacyl-tRNA synthetase-interacting multi-functional protein, p43/AIMP1. Biochem Biophys Res Commun 342:113–118
105. Park SG, Shin H, Shin YK, Lee Y, Choi EC, Park BJ, Kim S (2005) The novel cytokine p43 stimulates dermal fibroblast proliferation and wound repair. Am J Pathol 166:387–398
106. Ihle JN, Kerr IM (1995) Jaks and stats in signaling by the cytokine receptor superfamily. Trends Genet 11:69–74
107. Kotenko SV, Pestka S (2000) Jak-Stat signal transduction pathway through the eyes of cytokine class II receptor complexes. Oncogene 19:2557–2565
108. Bazan JF (1990) Structural design and molecular evolution of a cytokine receptor superfamily. Proc Natl Acad Sci USA 87:6934–6938
109. Takeichi M (1991) Cadherin cell adhesion receptors as a morphogenetic regulator. Science 251:1451–1455
110. Giancotti FG, Ruoslahti E (1999) Integrin signaling. Science 285:1028–1032
111. Schwarz MA, Zheng H, Liu J, Corbett S, Schwarz RE (2005) Endothelial-monocyte activating polypeptide II alters fibronectin based endothelial cell adhesion and matrix assembly via alpha5 beta1 integrin. Exp Cell Res 311:229–239
112. Carmeliet P, Lampugnani MG, Moons L, Breviario F, Compernolle V, Bono F, Balconi G, Spagnuolo R, Oosthuyse B, Dewerchin M, Zanetti A, Angellilo A, Mattot V, Nuyens D, Lutgens E, Clotman F, de Ruiter MC, Gittenberger-de Groot A, Poelmann R, Lupu F, Herbert JM, Collen D, Dejana E (1999) Targeted deficiency or cytosolic truncation of the VE-cadherin gene in mice impairs VEGF-mediated endothelial survival and angiogenesis. Cell 98:147–157
113. Patel SD, Ciatto C, Chen CP, Bahna F, Rajebhosale M, Arkus N, Schieren I, Jessell TM, Honig B, Price SR, Shapiro L (2006) Type II cadherin ectodomain structures: implications for classical cadherin specificity. Cell 124:1255–1268
114. Shimoyama Y, Gotoh M, Terasaki T, Kitajima M, Hirohashi S (1995) Isolation and sequence analysis of human cadherin-6 complementary DNA for the full coding sequence and its expression in human carcinoma cells. Cancer Res 55:2206–2211
115. Osterhout JA, Josten N, Yamada J, Pan F, Wu SW, Nguyen PL, Panagiotakos G, Inoue YU, Egusa SF, Volgyi B, Inoue T, Bloomfield SA, Barres BA, Berson DM, Feldheim DA, Huberman AD (2011) Cadherin-6 mediates axon-target matching in a non-image-forming visual circuit. Neuron 71:632–639
116. Crespo P, Xu N, Simonds WF, Gutkind JS (1994) Ras-dependent activation of MAP kinase pathway mediated by G-protein beta gamma subunits. Nature 369:418–420
117. Sauer FG, Futterer K, Pinkner JS, Dodson KW, Hultgren SJ, Waksman G (1999) Structural basis of chaperone function and pilus biogenesis. Science 285:1058–1061
118. Linderoth J, Jerkeman M, Cavallin-Stahl E, Kvaloy S, Torlakovic E (2003) Immunohistochemical expression of CD23 and CD40 may identify prognostically favorable subgroups of diffuse large B-cell lymphoma: a Nordic lymphoma group study. Clin Cancer Res 9:722–728
119. Kwon HS, Park MC, Kim DG, Cho K, Park YW, Han JM, Kim S (2012) Identification of CD23 as a functional receptor for the proinflammatory cytokine AIMP1/p43. J Cell Sci 125:4620–4629
120. Vo MN, Yang XL, Schimmel P (2011) Dissociating quaternary structure regulates cell-signaling functions of a secreted human tRNA synthetase. J Biol Chem 286:11563–11568
121. Greenberg Y, King M, Kiosses WB, Ewalt K, Yang X, Schimmel P, Reader JS, Tzima E (2008) The novel fragment of tyrosyl tRNA synthetase, mini-TyrRS, is secreted to induce an angiogenic response in endothelial cells. FASEB J 22:1597–1605

122. Heyman B (2000) Regulation of antibody responses via antibodies, complement, and Fc receptors. Annu Rev Immunol 18:709–737
123. Delespesse G, Sarfati M, Wu CY, Fournier S, Letellier M (1992) The low-affinity receptor for IgE. Immunol Rev 125:77–97
124. Pockley AG (2003) Heat shock proteins as regulators of the immune response. Lancet 362:469–476
125. Jacobs A, Worwood M (1975) Ferritin in serum. Clinical and biochemical implications. N Engl J Med 292:951–956

Top Curr Chem (2014) 344: 167–188
DOI: 10.1007/128_2013_422
© Springer-Verlag Berlin Heidelberg 2013
Published online: 28 March 2013

Non-catalytic Regulation of Gene Expression by Aminoacyl-tRNA Synthetases

Peng Yao, Kiran Poruri, Susan A. Martinis, and Paul L. Fox

Abstract Aminoacyl-tRNA synthetases (AARSs) are a group of essential and ubiquitous "house-keeping" enzymes responsible for charging corresponding amino acids to their cognate transfer RNAs (tRNAs) and providing the correct substrates for high-fidelity protein synthesis. During the last three decades, wide-ranging biochemical and genetic studies have revealed non-catalytic regulatory functions of multiple AARSs in biological processes including gene transcription, mRNA translation, and mitochondrial RNA splicing, and in diverse species from bacteria through yeasts to vertebrates. Remarkably, ongoing exploration of non-canonical functions of AARSs has shown that they contribute importantly to control of inflammation, angiogenesis, immune response, and tumorigenesis, among other critical physiopathological processes. In this chapter we consider the non-canonical functions of AARSs in regulating gene expression by mechanisms not directly related to their enzymatic activities, namely, at the levels of mRNA production, processing, and translation. The scope of AARS-mediated gene regulation ranges from negative autoregulation of single AARS genes to gene-selective control, and ultimately to global gene regulation. Clearly, AARSs have evolved these auxiliary regulatory functions that optimize the survival and well-being of the organism, possibly with more complex regulatory mechanisms associated with more complex organisms. In the first section on transcriptional control, we introduce the roles of autoregulation by *Escherichia coli* AlaRS, transcriptional activation by human LysRS, and transcriptional inhibition by vertebrate SerRS. In the second section on translational control, we recapitulate the roles of GluProRS in translation repression at the initiation step, auto-inhibition of *E. coli thrS* mRNA translation

---

P. Yao and P.L. Fox (✉)

Department of Cellular and Molecular Medicine, Lerner Research Institute, Cleveland Clinic, 9500 Euclid Avenue, Cleveland, OH 44195, USA
e-mail: foxp@ccf.org

K. Poruri and S.A. Martinis
Department of Biochemistry, University of Illinois at Urbana-Champaign, Urbana, IL 61801, USA
e-mail: martinis@life.illinois.edu

168   P. Yao et al.

by ThrRS, and global translational arrest by phosphorylated human MetRS. Finally, in the third section, we describe the RNA splicing activities of mitochondrial TyrRS and LeuRS in *Neurospora* and yeasts, respectively.

Keywords Gene expression · Noncanonical function · Splicing · Transcriptional control · Translational control · tRNA synthetases

Contents

1 Introduction 168
2 Regulatory Functions of AARS in Transcriptional Control 170
   2.1 *E. coli* AlaRS Binds a Specific DNA Sequence and Regulates Its Own
      Transcription 170
   2.2 Phosphorylated LysRS in Transcriptional Regulation in the Immune Response of
      Human Mast Cells 171
   2.3 Transcription Repression by SerRS Contributes to Vascular Development in
      Zebrafish 171
3 Regulatory Functions of AARS in Translational Control 172
   3.1 Translation Control for Autoregulation of Bacterial ThrRS 172
   3.2 Function of GluProRS in GAIT-Mediated, Transcript-Selective Translational
      Silencing 174
   3.3 Human MetRS Regulates Global Translation and Tumor Suppressor Activity of
      AIMP3 178
4 RNA Splicing Mediated by AARS 178
   4.1 Group I Intron Splicing by *N. crassa* Mitochondrial TyrRS 178
   4.2 Group I Intron Splicing by Yeast Mitochondrial LeuRS 179
5 Concluding Remarks 180
References 181

1 Introduction

Aminoacyl-tRNA synthetases (AARS) are ancient house-keeping enzymes that are ubiquitous in the three domains of life and catalyze the ligation of amino acids to cognate tRNAs [1, 2]. By implementing highly specific aminoacylation reactions, they are uniquely responsible for deciphering the genetic code. In prokaryotes there are up to 22 distinct AARSs, 1 for each standard amino acid and also for pyrrolysine and selenocysteine aminoacylation [3]. Eukaryotes have two sets of AARSs, 20 cytoplasmic enzymes and 20 nuclear-encoded mitochondrial enzymes. In plants and some species of parasites such as *Plasmodium*, a third set of nuclear-encoded AARSs are targeted to the chloroplast and apicoplasts respectively [4, 5].

Synthetases contain catalytic and tRNA anticodon recognition sites in separate domains. They segregate into two structurally distinct classes [6, 7]. Ten Class I enzymes have a Rossmann fold that comprises the aminoacylation active site, bind the minor groove of the tRNA acceptor stem, and typically aminoacylate the 2′-hydroxyl position of the terminal ribose. In contrast, the aminoacylation cores of the Class II enzymes have an antiparallel β-sheet, bind the major groove of the acceptor stem, and aminoacylate the terminal ribose at its $3'$-hydroxyl group. Class I and II AARSs can be further grouped into subclasses that exhibit additional structural similarities and that typically recognize related amino acid substrates. Some AARSs contain an editing domain for removing erroneously aminoacylated products to maintain translation fidelity [8, 9].

As the translation machinery components responsible for the first committed reaction in protein synthesis, AARSs modulate global protein biosynthesis by catalyzing aminoacylation. In addition to their essential role, many AARSs exhibit noncanonical functions, unrelated to their primary function in protein synthesis [10]. These moonlighting activities include regulation of gene expression, signal transduction, cell migration, tumorigenesis, angiogenesis, and inflammation [11, 12]. Indeed, members of the translation apparatus could be considered as "hub proteins", establishing connectivity between nodes in protein networks that control key cell functions [13].

Several vertebrate cytoplasmic AARSs contain extra peptide domains, usually appended to the N- or C-terminus of the catalytic core [14, 15]. The domains include glutathione-S-transferase-like (GST-like) domains, endothelial monocyte-activating polypeptide II (EMAP II)-like domains, and WHEP-TRS domains. The latter were named after three AARSs containing these structures: Trp(W)RS, His(H)RS, and GluPro(EP)RS. They are also present in GlyRS and MetRS, but not in any other non-AARS proteins. The functions of these appended domains are typically not associated with the function of the enzymatic core, nor are they present in bacterial or archeal homologues. Rather, they are involved in non-canonical regulatory functions of AARSs that are unrelated to tRNA aminoacylation [16].

In vertebrate cells, nine of the AARS activities within eight polypeptides (including one bifunctional enzyme GluProRS), and three AARS-interacting multifunctional proteins (AIMPs), AIMP1, AIMP2, and AIMP3, reside in the cytosol in a 1.5 mDa tRNA multisynthetase complex (MSC) [17, 18]. Appended domains in several eukaryotic AARSs might contribute to assembly or stability of the MSC as well as high efficiency of protein synthesis by channeling charged tRNAs to the ribosome [19].

In eukaryotes, certain AARSs are translocated or shuttled between the cytoplasm and nucleus [20–22]. The nuclear AARSs can be involved in transcriptional control. In the first section we describe the role of AlaRS, LysRS, and SerRS in non-canonical transcriptional regulation. In contrast, in the cytoplasm, AARSs are localized near the translational machinery and the translating mRNA, and thus are well-positioned to exert translational control. In the second section on translational control we describe the role of GluProRS in translation regulation at the levels of initiation, auto-inhibition of *E. coli thrS* mRNA translation by ThrRS, and global translational arrest by human MetRS. In the third section we describe the RNA splicing activity of mitochondrial TyrRS and LeuRS in *Neurospora* and yeast, respectively.

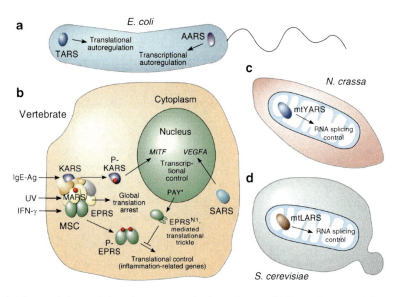

Fig. 1 Non-catalytic regulation of gene expression by aminoacyl-tRNA synthetases. (**a**) Transcriptional control of AlaRS and translational control of ThrRS for autoregulation in *E. coli*. (**b**) Transcriptional and translational control mechanisms by AARSs in human cells; phosphorylated LysRS regulates gene transcription in mast cells in the immune response; SerRS drives transcriptional repression and controls vascular development in zebrafish and possibly in humans; phosphorylated MetRS down-regulates global protein translation; GluProRS mediates IFN-γ-activated transcript-selective translational silencing of inflammation-related genes and EPRS[N1] maintains a translational "trickle." (**c**) RNA splicing mediated by *Neurospora crassa* mitochondrial TyrRS. (**d**) RNA splicing mediated by yeast mitochondrial LeuRS

2 Regulatory Functions of AARS in Transcriptional Control

2.1 E. coli *AlaRS Binds a Specific DNA Sequence and Regulates Its Own Transcription*

An anti-termination mechanism driving read-through of the transcription terminator in the mRNA leader regulates most AARS genes in Gram-positive bacteria such as *Bacillus subtilis* [23, 24]. Alternatively, in one of the earliest examples of a non-canonical function of an AARS, *E. coli* AlaRS represses transcription of its own gene by binding to palindromic sequences flanking the start site of the gene (Fig. 1a) [25, 26]. The sequence centers are 19 base pairs apart corresponding to 2 turns of a DNA B helix, which can be spanned by AlaRS. Transcription repression is dramatically enhanced by elevated concentrations of the cognate amino acid alanine. The K_m for alanine in the standard amino acid-dependent ATP-PP$_i$ exchange reaction is comparable to the concentration of alanine required to inhibit AlaRS transcription. Hence, the AlaRS binding site for alanine for aminoacylation is hypothesized to be similar to that which confers alanine dependence to transcriptional control. The amino acid effect is caused by direct interaction of the ligand with the synthetase, which in turn promotes tighter binding to the DNA region and consequently enhances control. PheRS also binds DNA but its function remains unknown [27]. The interaction is structural rather than sequence-specific and involves a loop in the double-stranded DNA [28]. Interestingly, the crystal structure of the α-subunit of human cytosolic PheRS suggests that three DNA/dsRNA-binding domains at the N-terminus might be involved in DNA-binding and in regulation of intracellular functions [29].

2.2 Phosphorylated LysRS in Transcriptional Regulation in the Immune Response of Human Mast Cells

LysRS produces diadenosine tetraphosphate (Ap(4)A) in vitro [30, 31] and is involved in transcriptional control of inflammatory gene expression in human mast cells (Fig. 1b) [32, 33]. When human mast cells are exposed to IgE-Ag (immuno-globulin E and antigen), LysRS is phosphorylated on Ser[207] in a MAPK-dependent manner and released from the MSC, and then translocates into the nucleus. Nuclear LysRS forms a complex with microphthalmia-associated transcription factor (MITF, a basic helix–loop–helix leucine zipper transcription factor involved in development and function of mast cells in allergic reactions) and its repressor, histidine triad nucleotide-binding protein-1 (Hint-1, it can hydrolyze substrates such as AMP-morpholidate and AMP-NH2 through interactions with these substrates via a histi-dine triad motif), which is released from the complex by binding to Ap(4)A, enabling MITF to transcribe its select target genes. Silencing LysRS led to reduced Ap(4)A production in immunologically activated cells, which resulted in diminished expression of MITF-inducible genes. Specific LysRS Ser[207] phosphorylation regulates Ap(4)A production in immunologically stimulated mast cells, thus implicating LysRS as a key activator in gene regulation. The full details of the mechanism are described in Chap. 7.

2.3 Transcription Repression by SerRS Contributes to Vascular Development in Zebrafish

Vascular patterning in vertebrates is established and maintained through a highly complex and dynamic process. Vascular endothelial growth factor-A (VEGF-A), a potent hypoxia-inducible angiogenic cytokine, is essential for blood vessel formation during development and plays important roles in the establishment of stereotypical vascular patterning in vertebrates [34, 35]. In two separate genetic screens in zebrafish, mutations in the SerRS gene (SARS) have been shown to affect vascular development and maintenance [36–38]. Multiple mutations at the SARS gene locus cause pronounced dilatation of the aortic arch vessels and aberrant branching of cranial and intersegmental vessels [36, 37]. Elevated vegfa mRNA in SARS mutants suggests that SerRS might regulate vegfa transcription and affect vascular development. Rescue of the abnormal vascular branching phenotype of *sars* mutants by forced overexpression of an enzymatically inactive form of SerRS T429A implies that an aminoacylation-independent, non-canonical function of SerRS may contribute to the vascular development in the zebrafish [36]. Moreover, the aberrant vascular sprouting in *sars* mutants was effectively inhibited through knockdown of *vegf receptor* or *vegfa* genes, suggesting that SerRS may function in modulating VEGF-A-mediated signal transduction [36].

In a more generalized scenario, excessive vessel branch-points after *SARS* knockdown in human umbilical vein endothelial cells, and rescued vascular branching in *SARS* mutant zebrafish by injections of highly homologous human *SARS* mRNA, suggest that SerRS function in vascular development is conserved between zebrafish and humans [37]. The UNE-S (Unique-S) domain appended to SerRS at the C-terminus contains a nuclear localization signal (NLS), and was identified as essential for shuttling the enzyme into the nucleus and possibly controlling *vegfa* mRNA transcription for normal vascular development (Fig. 1b) [39]. Intriguingly, the SerRS UNE-S domain appears only in species that harbor closed circulatory systems. Consistent with these observations, *SARS* mutants lacking the NLS or its activity, exhibit vasculature abnormalities. A crystal structure-based compensatory mutation released the buried NLS and restored normal vasculature [39]. Whether SerRS directly binds to the *vegfa* gene promoter or other DNA regions to regulate its transcription remains an open question. Full details of this mechanism are given in Chap. 5.

3 Regulatory Functions of AARS in Translational Control

Translational control of gene expression provides numerous advantages over other regulatory modes, including rapid responsiveness, regulatable intracellular localization, non-destruction of template mRNA, and coordinated regulation of transcript ensembles. Control can be either global, generally directed by post-translational modification of essential translation factors, or transcript-selective. Transcript-selective translation control is most often driven by the specific interaction of factor(s) with the 5′ or 3′ untranslated region (UTR), thereby influencing initiation, elongation, or termination of mRNA translation by ribosomes. Translational control generally involves negative regulation, but occasionally is positive. In this section we will describe auto-regulatory, transcript-selective, and global control of mRNA translation by several AARSs.

3.1 Translation Control for Autoregulation of Bacterial ThrRS

Regulation of protein synthesis at the translational level allows rapid adaptation to changes in environmental conditions. In prokaryotes, the synthesis of RNA-binding proteins is often regulated by competition between the natural substrate of the protein and its corresponding mRNA (e.g., rRNA and its binding ribosomal protein, or tRNA and its cognate AARS). These two RNA segments can be similar in primary sequence and secondary structure. *E. coli* ThrRS negatively regulates the expression of its own gene (*ThrS*) at the translational level (Fig. 1a) [40]. ThrRS binds to the operator, positioned in the leader of its own mRNA, and inhibits translation initiation by competing with binding of the 30S ribosomal subunit [41]. The operator is composed of four domains. Domain 1 is single-stranded and contains the Shine–Dalgarno sequence and initiation codon. Domains 2 and 4 are two-stem loop structures that carry sequence analogies with the anticodon loop of tRNA^{Thr} and are linked by single-stranded domain 3 [42]. The tRNA^{Thr} anticodon-like sequences in these stem-loops are important, as the first base of the sequence can be changed without major effect, while the second and third bases are crucial for the control activity similar to the anticodon loop in tRNA^{Thr} [43].

In the aminoacylation complex, the homodimeric enzyme ThrRS binds to two tRNAs and the anticodon loop of tRNA^{Thr} is recognized by C-terminal domain of ThrRS and the acceptor arm bound by catalytic and N-terminal domains [44]. In the translation control complex, ThrRS binds to two stem-loop domains (domains 2 and 4) of a single operator, which mimic the tRNA^{Thr} anticodon arms [42, 45, 46]. Recognition by ThrRS of either the operator or tRNA is governed principally by similar specific interactions between the C-terminal domain of the synthetase and the anticodon loop of the tRNA, or the analogous sequences in the apical loops of operator domains 2 and 4. Unlike tRNA^{Thr}, the operator lacks an acceptor arm. The loss is compensated by the existence of domain 4 and internal loop in domain 2. This is confirmed by the finding that changes in amino acids that interact with the acceptor arm of the tRNA do not affect control. Point mutations within domain 2 which alter the internal loop structure strongly affect ThrRS recognition and translation control. Presumably, the position of domain 2, which is closer to the catalytic domain than the anticodon stem of tRNA^{Thr}, combined with the binding of domain 4 to the other subunit of ThrRS, are able to compensate for the lack of an equivalent of the tRNA acceptor arm. Thus, the affinity of ThrRS for the entire operator is similar to that for the cognate tRNA [42, 46, 47]. Biochemical, genetic and structural experiments indicate that ThrRS competes with the ribosome for binding to *thrS* mRNA, binding to distinct intermingled domains of the operator [41, 48]. The favored model of inhibition is that a domain of the synthetase sterically hinders ribosome binding. The N-terminal domain of ThrRS is optimally placed to block the binding of the ribosome, most likely by obstructing the mRNA from folding around the neck of the 30S subunit. Moreover, the distance between the ribosome and ThrRS-binding domains must be restricted to permit competition [49].

Translational regulation of the *thrS* gene has only been reported in *E. coli*; however, the feedback mechanism might also be present in γ-proteobacteria. Having two RNAs competing for the same binding site and restrained by the binding requirements of ThrRS, the operator most likely has evolved to converge on the tRNA structure. Mimicry has important functional implications when similar structures allow for competition in binding to a common partner. Due to the resemblance between the tRNA^{Thr} and the operator in *thrS* mRNA 5'UTR, the operator acts as a competitive inhibitor of tRNA aminoacylation. When tRNA concentration is high, ThrRS works exclusively as an enzyme that is saturated by its substrate other than a repressor. When tRNA concentration is low, ThrRS is free to bind to its own mRNA and suppress its translation. Thus, *thrS* mRNA translation is stringently regulated by two layers of competition mechanisms (the competition between tRNA and the operator for ThrRS binding and that between ribosome and ThrRS for mRNA binding) in a coordinated manner dependent on cell growth rate [23].

There is much to be learned about the structural and functional details of the RNA–protein interactions described here. Moreover, the combination of modern genetic and biochemical, as well as genomic, proteomic, and structural approaches will greatly facilitate future discovery of new regulatory pathways involving RNA–protein interactions in both prokaryotes and eukaryotes.

3.2 Function of GluProRS in GAIT-Mediated, Transcript-Selective Translational Silencing

In prokaryotes, the translation or stability of a specific mRNA is generally under the control of a single regulator. However, in eukaryotes, certain mRNAs contain multiple *cis*-acting sequences, each recognized by a different RNA-binding protein, possibly regulating their localization and/or translation. Moreover, the regulator is often a multi-component complex rather than a single protein. The presence of similar structural or sequence elements in an ensemble of mRNAs facilitates coordinated regulation as a posttranscriptional "operon" or "regulon." The mean length of human 3'UTRs is about fivefold greater than that of 5'UTRs, indicating the expanded potential for motifs, structural elements, and binding sites for *trans*-acting factors that exert transcript-selective translational control [50]. Moreover, the 3'UTR, unlike the 5'UTR, is not constrained by the requirement to permit ribosome scanning, and thus highly stable and complex structural elements are permitted. Here, we briefly review the principal discoveries of EPRS-directed 3'UTR-dependent translational control of inflammation-related gene expression reported during the last decade, emphasizing the novel aspects of the regulatory mechanism and its potential pathophysiological significance (Fig. 1b).

Vertebrate GluProRS (EPRS) is the only bifunctional AARS; it catalyzes the ligation of Glu and Pro amino acids to their cognate tRNAs [51, 52]. It is localized almost entirely in the MSC [53]. EPRS is a large, 172-kDa protein consisting of a short GST-like domain at the N-terminus followed by the GluRS domain, a linker consisting of three tandem WHEP-TRS domains, and finally the C-terminal ProRS domain [54, 55]. All five AARSs bearing WHEP-TRS domains have non-canonical functions unrelated to tRNA synthetase activity [14, 16, 56].

3.2.1 Discovery of the IFN-γ-Inducible, GluProRS-Bearing GAIT Complex

Macrophages function in innate and adaptive immunity in vertebrates by phagocytosing cellular debris and pathogens and by stimulating lymphocytes and other immune cells to respond to pathogens. Macrophages are activated by cytokines to produce and secrete proteins and small molecules to eliminate invading microorganisms to provide defense against infection or injury. Ceruloplasmin (Cp) is an acute-phase plasma protein made by liver hepatocytes and monocytes/ macrophages. The protein serves important roles in iron homeostasis and inflammation. The pro-inflammatory function of macrophage-derived Cp is uncertain, but roles in bactericidal activity, defense against oxidant stress, and lipoprotein oxidation have been reported [57, 58]. Overexpression and accumulation of injurious agents can be detrimental to host tissues and organisms, contributing to chronic inflammatory disease. *Cp* mRNA is undetectable in human monocytic cells, but treatment with interferon (IFN)-γ causes a robust induction of *Cp* mRNA and protein within 2–4 h of treatment [59]. Despite the continued presence of abundant *Cp* mRNA for at least 24 h, synthesis of Cp protein stops abruptly, and almost completely, about 12–16 h after IFN-γ treatment [60]. Silencing of Cp translation is highly selective since global protein synthesis is unaffected by IFN-γ.

Translational silencing of *Cp* mRNA requires a structural element (termed IFN-gamma-activated inhibitor of translation, or GAIT element) in its 3′UTR [61]. Deletion and mutation analysis indicates that the GAIT element is a 29-nucleotide, bipartite stem-loop with sequence and structural features that distinguish it from other defined translational control elements [61]. By proteomic and genetic methods, four protein constituents of the GAIT complex have been identified: GluProRS, NS1-associated protein-1 (NSAP1), glyceraldehyde 3-phosphate dehydrogenase (GAPDH), and ribosomal protein L13a (Fig. 1b) [56, 62]. GAIT complex assembly occurs in two distinct stages. During the first 2–4 h, a pre-GAIT complex, which does not bind the GAIT element, is formed from GluProRS and NSAP1. During the second stage, approximately 12 h later, L13a joins GAPDH and the pre-GAIT complex to form the functional, quaternary GAIT complex that binds the 3′UTR GAIT element of *Cp* mRNA and silences its translation [56]. The GAIT pathway might have evolved to restrict the expression of injurious proteins, thereby participating in the "resolution of inflammation" process [56].

3.2.2 GluProRS WHEP Domain Directs Binding of GAIT Element-Bearing Transcripts

The MSC is the GluProRS donor since its loss from the complex coincides temporally and quantitatively with its appearance in the pre-GAIT complex. GluProRS is the unique GAIT complex component that binds GAIT elements in the target mRNA 3′UTR (Fig. 1b). The upstream pair of the three WHEP-TRS domains (R1 and R2) in the GluProRS linker is the GAIT element-binding site [63].

NSAP1 functions as a negative regulator of target mRNA binding. NSAP1 binds the downstream pair of the three WHEP-TRS domains (R2 and R3) in the GluProRS linker, and competitively inhibits binding to the GAIT element. Binding of L13a and GAPDH to the pre-GAIT complex causes a conformational change that permits the interaction of the holo-complex with the GAIT element, despite the presence of NSAP1 [63].

3.2.3 Phosphorylation-Dependent GluProRS Translocation to GAIT Complex Induces Translational Silencing

In human monocyte/macrophage cells, IFN-γ activates sequential phosphorylation of Ser^{886} and Ser^{999} of GluProRS in the noncatalytic linker connecting the synthetase cores and initiates assembly of the GAIT complex [64, 65]. IFN-γ activates cyclin-dependent kinase 5 (Cdk5) through induction of its regulatory protein Cdk5R1 (p35) to phosphorylate Ser^{886}, and Cdk5/p35 induces phosphorylation of Ser^{999} via a distinct AGC kinase. Phosphorylation of both sites is required for GluProRS release from the parental MSC [65]. Ser^{886} phosphorylation is required for binding NSAP1 to form the pre-GAIT complex. The function of the pre-GAIT complex is still unknown; possibly the complex binds to a different set of target mRNAs and has its own regulatory function distinct from mature GAIT complex. Ser^{999} phosphorylation directs the formation of the active GAIT complex that binds initiation factor eIF4G, prevents the recruitment of pre-initiation complex, and thus switches off translation. Despite the phosphorylation and release of about half of the GluProRS from its parent complexes, global protein synthesis is unaltered. This finding has been generalized in a "depot hypothesis" which asserts that macromolecular assemblies, in addition to serving their primary purpose coordinating complex tasks, also function as depots for stimulus-dependent release of regulatory proteins [66].

3.2.4 Potential Pathophysiological Significance of the GAIT System

The physiological role of GAIT complex-mediated translational silencing remains unclear. The abundance of the four GAIT complex components, and the near-stoichiometric release of two proteins from their parent complexes, indicates that the GAIT complex is likely to be in marked excess of Cp mRNA. About 55 candidate human transcripts containing putative GAIT elements in their 3'UTR have been identified by application of a "pattern-matching" algorithm against a human 3'UTR database [67]. The candidates are consistent with a pathway in which GAIT-mediated translational silencing contributes to the resolution of inflammation, particularly the local inflammation due to cytokine activation of infiltrating or resident macrophages. Among the candidates, vascular endothelial growth factor A (VEGF-A), death-associated protein kinase (DAPK), and zipper-interacting protein kinase (ZIPK) have been experimentally validated as authentic GAIT targets [67, 68]. Translation state and pathway analyses revealed several chemokine ligands and receptors as GAIT targets that were validated experimentally [69].

Together these results indicate that the GAIT system regulates expression of a posttranscriptional "operon" of functionally related genes [70]. Moreover, the GAIT pathway has been shown to be functionally conserved in mice and humans [71, 72]. EPRS with linked catalytic domains is found in the primitive cnidarian *Nematostella vectensis*, indicating it appeared early during animal evolution [15]. The *N. vectensis EPRS* gene is alternatively spliced in the linker region to yield three isoforms with one or two WHEP domains in each. Remarkably, the 2-WHEP domain linker binds human *Cp* GAIT element with high affinity, possibly indicating an early evolution of a GAIT-like regulatory system.

The GAIT system is itself subject to regulation by environmental signals. When myeloid cells are simultaneously subjected to the inhibitory activity of IFN-γ and stimulatory activity of hypoxia, high-level VEGF-A expression continues unabated, indicating that hypoxia prevails [71]. Immediately adjacent to the GAIT element in the *vegfa* 3'UTR there is a CA-rich element (CARE) that is a binding site for heterogenous nuclear ribonucleoprotein (hnRNP) L. In response to hypoxia, hnRNP L binding to the CARE induces a conformational change in the 3'UTR that prevents GAIT complex binding and translational silencing. Analogous to metabolite-dependent bacterial riboswitches, the protein-dependent *vegfa* RNA switch exemplifies a new class of vertebrate regulatory systems in which combinatorial utilization of nearby RNA elements can integrate disparate input signals to produce the appropriate output.

The GAIT system is also subject to pathological dysregulation. Treatment of human peripheral blood monocytes (PBMs) with oxidatively modified low density lipoprotein (oxLDL), a pathophysiological agent present in human atherosclerotic lesions and thought to contribute to the progression of the condition [73], completely suppresses GAIT pathway activity, and enhances expression of VEGF-A and other GAIT targets [74]. OxLDL induces *S*-nitrosylation of GAPDH, thereby suppressing its chaperone-like binding to L13a and permitting its proteasomal degradation. In the absence of L13a, the GAIT system is dysfunctional and synthesis of inflammation-related proteins continues unabated. Genetic defects in GAIT constituents, or GAIT system dysfunction caused by environmental stress, e.g., oxLDL, may prolong or increase inflammation and contribute to progression of chronic inflammatory disorders including atherosclerosis.

3.2.5 Coding Region Polyadenylation Generates a Truncated GluProRS that Maintains a "Translational Trickle" of VEGF-A Expression

Differential GAIT complex activation by increasing levels of IFN-γ revealed an "uncoupling" of VEGF-A protein and *vegfa* mRNA expression; expression of VEGF-A and other GAIT target genes was reduced to a low, constant rate independent of mRNA expression [75, 76]. A C-terminal truncated form of GluProRS was discovered, termed GluProRS[N1] (or EPRS[N1]) which contains the ERS domain followed by the upstream pair of WHEP domains (Fig. 1b). This truncation allowed binding of GAIT element-bearing mRNA, but not other GAIT complex proteins, thus acting as a dominant-negative inhibitor of the GAIT complex. *EPRS[N1]* mRNA is generated by a polyadenylation-directed conversion of a Tyr codon in the GluProRS coding sequence to a stop codon (PAY*). By shielding a small, constant amount of GAIT target transcripts from translational repression, constitutive, low-level expression of GluProRS[N1] imposes a robust "translational trickle" of target protein expression. This process is postulated to maintain VEGF-A expression, and that of other pro-inflammatory proteins, at basal levels required for tissue health and organismal advantage.

3.3 Human MetRS Regulates Global Translation and Tumor Suppressor Activity of AIMP3

Human cytoplasmic MetRS plays an essential role in translation initiation by transferring Met to initiator $tRNA_i^{Met}$. MetRS is a component of the MSC and provides a docking site for AIMP3. AIMP3 is a potent tumor suppressor that translocates to the nucleus for DNA repair. Dissociation of AIMP3 from MetRS is induced by phosphorylation of MetRS at Ser^{662} by GCN2 (general control nonrepressed-2), and by reduced affinity for AIMP3 following UV irradiation [77]. This phosphorylated MetRS exhibited significantly reduced catalytic activity due to diminished $tRNA_i^{Met}$ binding and led to down-regulation of global translation (Fig. 1b). Thus, MetRS, with $eIF2\alpha$, co-regulates GCN2-mediated translational inhibition, and by release of AIMP3 for nuclear translocation couples translation inhibition and DNA repair following DNA damage. UV-stress-induced MetRS phosphorylation, and consequent global repression of translation, mirrors to some extent the IFN-γ-induced GluProRS phosphorylation and transcript-selective translational silencing program. More mechanistic details are included in Chap. 9.

4 RNA Splicing Mediated by AARS

Group I introns are ribozymes that excise themselves and self-splice in vitro [78]. Although autocatalytic, these natural ribozymes are aided by proteins to facilitate proper RNA folding and catalytic efficiency [79]. Splicing partners include two nuclear-encoded mitochondrial AARS, LeuRS (NAM2p [80]) and TyrRS (CYT-18p [81]), for mitochondrial group I RNA splicing in yeast and *N. crassa*, respectively.

4.1 Group I Intron Splicing by N. crassa Mitochondrial TyrRS

Only a few TyrRSs have adapted to aid RNA splicing and require three unique insertions that are essential for its secondary role [82]. This contrasts with bacterial and yeast mitochondrial TyrRSs which lack these inserts and do not splice [83].

The bifunctional *N. crassa* mitochondrial TyrRS promotes splicing of several group I introns including the LSU, ND1, *cob*-I2, and T4 *td* bacteriophage (Fig. 1c) [84–86]. It has been proposed that CYT-18p substitutes for the function of the RNA P5abc domain that is missing in these particular group I introns [87, 88].

RNA footprinting and X-ray crystallography suggested that CYT-18p binds conserved class II tRNA-like structures within the stacked P4-P6 helix of the group I intron that mimics the D-anticodon stacked helices and extra-long variable loop [82, 89–92]. X-Ray structures showed that the intron RNA binds across the homodimer to the nucleotide binding fold opposite CYT-18p's aminoacylation active site [88]. Thus, the intron site of binding is distinct from the tRNA binding site. A small insertion in the nucleotide-binding fold (called insertion II) that isn't found in non-splicing TyrRSs also recognizes the intron [83, 93, 94]. In addition, the enzyme depends on its tRNA binding domain as well as the unique α-helical N-terminal extension (called H0) to bind the intron [92]. Although the CYT-18p C-terminal domain contributes to binding group I introns, it is essential only for splicing the *N. crassa* mt LSU intron [95, 96]. Biochemical analyses support the fact that CYT-18p stabilizes the catalytically active group I intron structure [86, 89, 90, 97]. Initially, the CYT-18p homodimer binds a single intron [98] via its central P4-P6 domain to promote accurate folding. It then makes contact with the P3-P9 domain to align these two RNA domains to form the intron's active site [82, 99].

4.2 Group I Intron Splicing by Yeast Mitochondrial LeuRS

Yeast mitochondrial LeuRS is essential for excision of the self-splicing *cob* bI4 and *cox1a* aI4α group I introns (Fig. 1d) [80, 100–102]. These two related introns are also dependent on a second splicing partner, the bI4 intron-encoded bI4 maturase, which functions in conjunction with LeuRS [103]. The LeuRS and bI4 maturase protein co-factors bind directly and independently to the bI4 intron to support RNA splicing. Each can independently stimulate splicing, albeit ribozyme activity is more efficient in a ternary complex with both protein splicing factors [104].

Unlike CYT18p [83, 93], yeast mitochondrial LeuRS lacks obvious idiosyncratic domains or insertions that confer splicing. Indeed, LeuRSs from diverse origins that lack group I introns, such as *M. tuberculosis, E. coli* and human mitochondria, substitute as splicing factors for the yeast enzyme [105, 106]. It was thus hypothesized that LeuRS elements that are conserved for aminoacylation were also responsible for RNA splicing. Subsequently, the LeuRS-specific CP1 domain that is responsible for amino acid editing was also shown to act as a splicing factor, even when independent of the full-length protein [106, 107]. In addition, mutation of a single site within the LeuRS CP1 domain inactivated its splicing activity [107].

The LeuRS CP1 domain is linked to the canonical aminoacylation core via two β-strands at its N- and C-terminus [108, 109]. A conserved WIG' tripeptide at the end of the N-terminal β-strand is critical for LeuRS splicing activity [102]. Mutation of the conserved tryptophan to a cysteine (W238C) abolishes splicing, while mutation of the nearby glycine to a serine (G240S) suppresses splicing defects of the bI4 maturase splicing factor [102, 110]. Although the isolated LeuRS CP1 domain core can independently support RNA splicing, the conserved WIG sequence is distal to the CP1 domain and more closely associated with the aminoacylation core. In addition, the β-strands can undergo significant conformation changes during cycling of the LeuRS aminoacylation and editing complexes [111]. Thus, it is likely that the core of the enzyme as well as a LeuRS-specific C-terminal domain that binds the corner of $tRNA^{Leu}$ for aminoacylation and editing [112] contribute to RNA splicing in an auxiliary manner to the CP1 domain.

5 Concluding Remarks

In addition to its function as an essential component of the protein synthesis machinery, certain AARSs have evolved non-catalytic regulatory functions that modulate gene expression at multiple levels ranging from mRNA transcription to RNA splicing and ultimately protein translation. Lower eukaryotes, have capitalized upon AARS to facilitate gene processing by interacting with critical introns to facilitate their excision. For example, in yeast and *Neurospora* the AARS target mitochondrial introns in non-AARS genes, including essential ribosomal and respiratory genes, and might represent a novel regulatory pathway for mitochondrial gene expression.

Protein-mediated gene regulation at the transcriptional or translational levels can be classified by two distinguishable events: (1) the binding of the regulator to its operator, and (2) the consequent effect of binding on gene expression. In simpler and smaller organisms such as prokaryotes, AARS can exhibit self-balancing activities to control their own expression such as *E. coli* AlaRS binding to *alars* mRNA to regulate its transcription [26]. AARSs have evolved relatively straightforward auxiliary functions in autoregulation of expression, possibly driven by the highly compact genome in bacteria. In this case, the two essential regulatory events are implemented by two separate domains from a single enzyme. In the case of *E. coli* ThrRS, the C-terminal domain binds to the operator in its own mRNA and the N-terminal domain blocks ribosome recruitment, thereby restricting translation [48]. In the absence of microRNA-based regulatory networks as found in higher organisms, we propose that such self-regulatory mechanisms are required for maintaining optimal expression of AARS in bacteria, enhancing their adaptability and survival.

During evolution of complex organisms such as vertebrates, AARS accrued new domains and new cellular functions that regulate expression of stress-responsive genes to adapt to environmental changes. Intriguingly, many of these functions involve the immune system, including macrophages and mast cells. In these cases, multiple proteins, generally in complexes, can distribute the responsibility for distinct regulatory events, for example, GluProRS and L13a in the GAIT complex are responsible for target mRNA binding and inhibition of translation initiation, respectively [56, 113]. Alternatively, the regulatory mode of gene-selective transcriptional control by human LysRS differs from that of EPRS in transcript-selective translational control. LysRS responds to environmental stimuli and binds to transcription factor MITF as an activator through upregulation of target gene transcription under the allergic response [114]. As a common feature, both GluProRS and LysRS reside in the MSC, and are released upon stress activation. We have proposed that the mammalian MSC, in addition to improving efficiency of protein synthesis, functions as a "depot" that releases specific AARS (and AIMPs) to perform equally specific non-catalytic regulatory functions [66]. In the future we might find that many, if not all, of the MSC components might be released under a broad spectrum of physiopathological conditions.

Acknowledgements This work was supported in part by National Institutes of Health grants P01 HL029582, P01 HL076491, R01 GM086430, and R01 DK083359 to P.L.F., and National Science Foundation grant MCB 0843611 to S.A.M. P.Y. was supported by a Postdoctoral Fellowship from the American Heart Association, Great Rivers Affiliate.

References

1. Ibba M, Soll D (2000) Aminoacyl-tRNA synthesis. Annu Rev Biochem 69:617–650
2. Ribas de Pouplana L, Schimmel P (2001) Aminoacyl-tRNA synthetases: potential markers of genetic code development. Trends Biochem Sci 26:591–596
3. Yuan J, O'Donoghue P, Ambrogelly A, Gundllapalli S, Sherrer RL, Palioura S, Simonovic M, Soll D (2010) Distinct genetic code expansion strategies for selenocysteine and pyrrolysine are reflected in different aminoacyl-tRNA formation systems. FEBS Lett 584:342–349
4. Brandao MM, Silva-Filho MC (2011) Evolutionary history of Arabidopsis thaliana aminoacyl-tRNA synthetase dual-targeted proteins. Mol Biol Evol 28:79–85
5. Jackson KE, Habib S, Frugier M, Hoen R, Khan S, Pham JS, Ribas de Pouplana L, Royo M, Santos MA, Sharma A, Ralph SA (2011) Protein translation in Plasmodium parasites. Trends Parasitol 27:467–476
6. Eriani G, Delarue M, Poch O, Gangloff J, Moras D (1990) Partition of tRNA synthetases into two classes based on mutually exclusive sets of sequence motifs. Nature 347:203–206
7. Cusack S, Berthet-Colominas C, Hartlein M, Nassar N, Leberman R (1990) A second class of synthetase structure revealed by X-ray analysis of *Escherichia coli* seryl-tRNA synthetase at 2.5 A. Nature 347:249–255
8. Ling J, Reynolds N, Ibba M (2009) Aminoacyl-tRNA synthesis and translational quality control. Annu Rev Microbiol 63:61–78
9. Martinis SA, Boniecki MT (2010) The balance between pre- and post-transfer editing in tRNA synthetases. FEBS Lett 584:455–459
10. Martinis SA, Plateau P, Cavarelli J, Florentz C (1999) Aminoacyl-tRNA synthetases: a family of expanding functions. Mittelwihr, France, October 10–15, 1999. EMBO J 18:4591–4596
11. Park SG, Schimmel P, Kim S (2008) Aminoacyl tRNA synthetases and their connections to disease. Proc Natl Acad Sci U S A 105:11043–11049
12. Kim S, You S, Hwang D (2011) Aminoacyl-tRNA synthetases and tumorigenesis: more than housekeeping. Nat Rev Cancer 11:708–718

13. Barabasi AL, Oltvai ZN (2004) Network biology: understanding the cell's functional organization. Nat Rev Genet 5:101–113
14. Shiba K (2002) Intron positions delineate the evolutionary path of a pervasively appended peptide in five human aminoacyl-tRNA synthetases. J Mol Evol 55:727–733
15. Ray PS, Sullivan JC, Jia J, Francis J, Finnerty JR, Fox PL (2011) Evolution of function of a fused metazoan tRNA synthetase. Mol Biol Evol 28:437–447
16. Guo M, Yang XL, Schimmel P (2010) New functions of aminoacyl-tRNA synthetases beyond translation. Nat Rev Mol Cell Biol 11:668–674
17. Rho SB, Kim MJ, Lee JS, Seol W, Motegi H, Kim S, Shiba K (1999) Genetic dissection of protein-protein interactions in multi-tRNA synthetase complex. Proc Natl Acad Sci U S A 96:4488–4493
18. Robinson JC, Kerjan P, Mirande M (2000) Macromolecular assemblage of aminoacyl-tRNA synthetases: quantitative analysis of protein-protein interactions and mechanism of complex assembly. J Mol Biol 304:983–994
19. Kyriacou SV, Deutscher MP (2008) An important role for the multienzyme aminoacyl-tRNA synthetase complex in mammalian translation and cell growth. Mol Cell 29:419–427
20. Nathanson L, Deutscher MP (2000) Active aminoacyl-tRNA synthetases are present in nuclei as a high molecular weight multienzyme complex. J Biol Chem 275:31559–31562
21. Sajish M, Zhou Q, Kishi S, Valdez DM Jr, Kapoor M, Guo M, Lee S, Kim S, Yang XL, Schimmel P (2012) Trp-tRNA synthetase bridges DNA-PKcs to PARP-1 to link IFN-gamma and p53 signaling. Nat Chem Biol 8:547–554
22. Fu G, Xu T, Shi Y, Wei N, Yang XL (2012) tRNA-controlled nuclear import of a human tRNA synthetase. J Biol Chem 287:9330–9334
23. Putzer H, Grunberg-Manago M, Springer M (1995) Bacterial aminoacyl-tRNA synthetases: genes and regulation of expression. In: Söll D, RajBhandary UL (eds) tRNA: structure, biosynthesis, and function. American Society for Microbiology, Washington, DC, pp 293–333
24. Grundy FJ, Henkin TM (1993) tRNA as a positive regulator of transcription antitermination in *B. subtilis*. Cell 74:475–482
25. Putney SD, Royal NJ, Neuman de Vegvar H, Herlihy WC, Biemann K, Schimmel P (1981) Primary structure of a large aminoacyl-tRNA synthetase. Science 213:1497–1501
26. Putney SD, Schimmel P (1981) An aminoacyl tRNA synthetase binds to a specific DNA sequence and regulates its gene transcription. Nature 291:632–635
27. Lechler A, Kreutzer R (1998) The phenylalanyl-tRNA synthetase specifically binds DNA. J Mol Biol 278:897–901
28. Dou X, Limmer S, Kreutzer R (2001) DNA-binding of phenylalanyl-tRNA synthetase is accompanied by loop formation of the double-stranded DNA. J Mol Biol 305:451–458
29. Finarov I, Moor N, Kessler N, Klipcan L, Safro MG (2010) Structure of human cytosolic phenylalanyl-tRNA synthetase: evidence for kingdom-specific design of the active sites and tRNA binding patterns. Structure 18:343–353
30. Hilderman RH, Ortwerth BJ (1987) A preferential role for lysyl-tRNA4 in the synthesis of diadenosine 5′,5′′′-P1, P4-tetraphosphate by an arginyl-tRNA synthetase-lysyl-tRNA synthetase complex from rat liver. Biochemistry 26:1586–1591
31. Brevet A, Plateau P, Cirakoglu B, Pailliez JP, Blanquet S (1982) Zinc-dependent synthesis of 5′,5′-diadenosine tetraphosphate by sheep liver lysyl- and phenylalanyl-tRNA synthetases. J Biol Chem 257:14613–14615
32. Yannay-Cohen N, Carmi-Levy I, Kay G, Yang CM, Han JM, Kemeny DM, Kim S, Nechushtan H, Razin E (2009) LysRS serves as a key signaling molecule in the immune response by regulating gene expression. Mol Cell 34:603–611
33. Lee YN, Nechushtan H, Figov N, Razin E (2004) The function of lysyl-tRNA synthetase and Ap4A as signaling regulators of MITF activity in FcεRI-activated mast cells. Immunity 20:145–151
34. Tammela T, Enholm B, Alitalo K, Paavonen K (2005) The biology of vascular endothelial growth factors. Cardiovasc Res 65:550–563

35. Coultas L, Chawengsaksophak K, Rossant J (2005) Endothelial cells and VEGF in vascular development. Nature 438:937–945
36. Fukui H, Hanaoka R, Kawahara A (2009) Noncanonical activity of seryl-tRNA synthetase is involved in vascular development. Circ Res 104:1253–1259
37. Herzog W, Muller K, Huisken J, Stainier DY (2009) Genetic evidence for a noncanonical function of seryl-tRNA synthetase in vascular development. Circ Res 104:1260–1266
38. Jin SW, Herzog W, Santoro MM, Mitchell TS, Frantsve J, Jungblut B, Beis D, Scott IC, D'Amico LA, Ober EA, Verkade H, Field HA, Chi NC, Wehman AM, Baier H, Stainier DY (2007) A transgene-assisted genetic screen identifies essential regulators of vascular development in vertebrate embryos. Dev Biol 307:29–42
39. Xu X, Shi Y, Zhang HM, Swindell EC, Marshall AG, Guo M, Kishi S, Yang XL (2012) Unique domain appended to vertebrate tRNA synthetase is essential for vascular development. Nat Commun 3:681
40. Springer M, Plumbridge JA, Butler JS, Graffe M, Dondon J, Mayaux JF, Fayat G, Lestienne P, Blanquet S, Grunberg-Manago M (1985) Autogenous control of *Escherichia coli* threonyl-tRNA synthetase expression in vivo. J Mol Biol 185:93–104
41. Moine H, Romby P, Springer M, Grunberg-Manago M, Ebel JP, Ehresmann B, Ehresmann C (1990) *Escherichia coli* threonyl-tRNA synthetase and tRNA^{Thr} modulate the binding of the ribosome to the translational initiation site of the thrS mRNA. J Mol Biol 216:299–310
42. Romby P, Caillet J, Ebel C, Sacerdot C, Graffe M, Eyermann F, Brunel C, Moine H, Ehresmann C, Ehresmann B, Springer M (1996) The expression of *E. coli* threonyl-tRNA synthetase is regulated at the translational level by symmetrical operator-repressor interactions. EMBO J 15:5976–5987
43. Romby P, Brunel C, Caillet J, Springer M, Grunberg-Manago M, Westhof E, Ehresmann C, Ehresmann B (1992) Molecular mimicry in translational control of *E. coli* threonyl-tRNA synthetase gene. Competitive inhibition in tRNA aminoacylation and operator-repressor recognition switch using tRNA identity rules. Nucleic Acids Res 20:5633–5640
44. Sankaranarayanan R, Dock-Bregeon AC, Romby P, Caillet J, Springer M, Rees B, Ehresmann C, Ehresmann B, Moras D (1999) The structure of threonyl-tRNA synthetase-tRNA^{Thr} complex enlightens its repressor activity and reveals an essential zinc ion in the active site. Cell 97:371–381
45. Torres-Larios A, Dock-Bregeon AC, Romby P, Rees B, Sankaranarayanan R, Caillet J, Springer M, Ehresmann C, Ehresmann B, Moras D (2002) Structural basis of translational control by *Escherichia coli* threonyl tRNA synthetase. Nat Struct Biol 9:343–347
46. Caillet J, Nogueira T, Masquida B, Winter F, Graffe M, Dock-Bregeon AC, Torres-Larios A, Sankaranarayanan R, Westhof E, Ehresmann B, Ehresmann C, Romby P, Springer M (2003) The modular structure of *Escherichia coli* threonyl-tRNA synthetase as both an enzyme and a regulator of gene expression. Mol Microbiol 47:961–974
47. Brunel C, Caillet J, Lesage P, Graffe M, Dondon J, Moine H, Romby P, Ehresmann C, Ehresmann B, Grunberg-Manago M et al (1992) Domains of the *Escherichia coli* threonyl-tRNA synthetase translational operator and their relation to threonine tRNA isoacceptors. J Mol Biol 227:621–634
48. Jenner L, Romby P, Rees B, Schulze-Briese C, Springer M, Ehresmann C, Ehresmann B, Moras D, Yusupova G, Yusupov M (2005) Translational operator of mRNA on the ribosome: how repressor proteins exclude ribosome binding. Science 308:120–123
49. Springer M, Graffe M, Butler JS, Grunberg-Manago M (1986) Genetic definition of the translational operator of the threonine-tRNA ligase gene in *Escherichia coli*. Proc Natl Acad Sci U S A 83:4384–4388
50. Mazumder B, Seshadri V, Fox PL (2003) Translational control by the 3′-UTR: the ends specify the means. Trends Biochem Sci 28:91–98
51. Cerini C, Kerjan P, Astier M, Gratecos D, Mirande M, Semeriva M (1991) A component of the multisynthetase complex is a multifunctional aminoacyl-tRNA synthetase. EMBO J 10:4267–4277

52. Kaiser E, Hu B, Becher S, Eberhard D, Schray B, Baack M, Hameister H, Knippers R (1994) The human EPRS locus (formerly the QARS locus): a gene encoding a class I and a class II aminoacyl-tRNA synthetase. Genomics 19:280–290
53. Han JM, Kim JY, Kim S (2003) Molecular network and functional implications of macromolecular tRNA synthetase complex. Biochem Biophys Res Commun 303:985–993
54. Cahuzac B, Berthonneau E, Birlirakis N, Guittet E, Mirande M (2000) A recurrent RNA-binding domain is appended to eukaryotic aminoacyl-tRNA synthetases. EMBO J 19:445–452
55. Jeong EJ, Hwang GS, Kim KH, Kim MJ, Kim S, Kim KS (2000) Structural analysis of multifunctional peptide motifs in human bifunctional tRNA synthetase: identification of RNA-binding residues and functional implications for tandem repeats. Biochemistry 39:15775–15782
56. Sampath P, Mazumder B, Seshadri V, Gerber CA, Chavatte L, Kinter M, Ting SM, Dignam JD, Kim S, Driscoll DM, Fox PL (2004) Noncanonical function of glutamyl-prolyl-tRNA synthetase: gene-specific silencing of translation. Cell 119:195–208
57. Musci G, Polticelli F, Calabrese L (1999) Structure/function relationships in ceruloplasmin. Adv Exp Med Biol 448:175–182
58. Ehrenwald E, Chisolm GM, Fox PL (1994) Intact human ceruloplasmin oxidatively modifies low density lipoprotein. J Clin Invest 93:1493–1501
59. Mazumder B, Mukhopadhyay CK, Prok A, Cathcart MK, Fox PL (1997) Induction of ceruloplasmin synthesis by IFN-γ in human monocytic cells. J Immunol 159:1938–1944
60. Mazumder B, Fox PL (1999) Delayed translational silencing of ceruloplasmin transcript in gamma interferon-activated U937 monocytic cells: role of the 3′ untranslated region. Mol Cell Biol 19:6898–6905
61. Sampath P, Mazumder B, Seshadri V, Fox PL (2003) Transcript-selective translational silencing by gamma interferon is directed by a novel structural element in the ceruloplasmin mRNA 3′ untranslated region. Mol Cell Biol 23:1509–1519
62. Mazumder B, Sampath P, Seshadri V, Maitra RK, DiCorleto PE, Fox PL (2003) Regulated release of L13a from the 60S ribosomal subunit as a mechanism of transcript-specific translational control. Cell 115:187–198
63. Jia J, Arif A, Ray PS, Fox PL (2008) WHEP domains direct noncanonical function of glutamyl-prolyl tRNA synthetase in translational control of gene expression. Mol Cell 29:679–690
64. Arif A, Jia J, Mukhopadhyay R, Willard B, Kinter M, Fox PL (2009) Two-site phosphorylation of EPRS coordinates multimodal regulation of noncanonical translational control activity. Mol Cell 35:164–180
65. Arif A, Jia J, Moodt RA, DiCorleto PE, Fox PL (2011) Phosphorylation of glutamyl-prolyl tRNA synthetase by cyclin-dependent kinase 5 dictates transcript-selective translational control. Proc Natl Acad Sci U S A 108:1415–1420
66. Ray PS, Arif A, Fox PL (2007) Macromolecular complexes as depots for releasable regulatory proteins. Trends Biochem Sci 32:158–164
67. Ray PS, Fox PL (2007) A post-transcriptional pathway represses monocyte VEGF-A expression and angiogenic activity. EMBO J 26:3360–3372
68. Mukhopadhyay R, Ray PS, Arif A, Brady AK, Kinter M, Fox PL (2008) DAPK-ZIPK-L13a axis constitutes a negative-feedback module regulating inflammatory gene expression. Mol Cell 32:371–382
69. Vyas K, Chaudhuri S, Leaman DW, Komar AA, Musiyenko A, Barik S, Mazumder B (2009) Genome-wide polysome profiling reveals an inflammation-responsive posttranscriptional operon in gamma interferon-activated monocytes. Mol Cell Biol 29:458–470
70. Keene JD, Tenenbaum SA (2002) Eukaryotic mRNPs may represent posttranscriptional operons. Mol Cell 9:1161–1167
71. Ray PS, Jia J, Yao P, Majumder M, Hatzoglou M, Fox PL (2009) A stress-responsive RNA switch regulates VEGFA expression. Nature 457:915–919

72. Arif A, Chatterjee P, Moodt RA, Fox PL (2012) Heterotrimeric GAIT complex drives transcript-selective translation inhibition in murine macrophages. Mol Cell Biol 32:5046–5055
73. Stocker R, Keaney JF Jr (2004) Role of oxidative modifications in atherosclerosis. Physiol Rev 84:1381–1478
74. Jia J, Arif A, Willard B, Smith JD, Stuehr DJ, Hazen SL, Fox PL (2012) Protection of extraribosomal RPL13a by GAPDH and dysregulation by S-nitrosylation. Mol Cell 47:656–663
75. Yao P, Potdar AA, Arif A, Ray PS, Mukhopadhyay R, Willard B, Xu Y, Yan J, Saidel GM, Fox PL (2012) Coding region polyadenylation generates a truncated tRNA synthetase that counters translation repression. Cell 149:88–100
76. Yao P, Fox PL (2012) A truncated tRNA synthetase directs a "translational trickle" of gene expression. Cell Cycle 11:1868–1869
77. Kwon NH, Kang T, Lee JY, Kim HH, Kim HR, Hong J, Oh YS, Han JM, Ku MJ, Lee SY, Kim S (2011) Dual role of methionyl-tRNA synthetase in the regulation of translation and tumor suppressor activity of aminoacyl-tRNA synthetase-interacting multifunctional protein-3. Proc Natl Acad Sci U S A 108:19635–19640
78. Cech TR (1990) Self-splicing of group I introns. Annu Rev Biochem 59:543–568
79. Lambowitz AM, Perlman PS (1990) Involvement of aminoacyl-tRNA synthetases and other proteins in group I and group II intron splicing. Trends Biochem Sci 15:440–444
80. Labouesse M (1990) The yeast mitochondrial leucyl-tRNA synthetase is a splicing factor for the excision of several group I introns. Mol Gen Genet 224:209–221
81. Akins RA, Lambowitz AM (1987) A protein required for splicing group I introns in *Neurospora* mitochondria is mitochondrial tyrosyl-tRNA synthetase or a derivative thereof. Cell 50:331–345
82. Myers CA, Wallweber GJ, Rennard R, Kemel Y, Caprara MG, Mohr G, Lambowitz AM (1996) A tyrosyl-tRNA synthetase suppresses structural defects in the two major helical domains of the group I intron catalytic core. J Mol Biol 262:87–104
83. Kamper U, Kuck U, Cherniack AD, Lambowitz AM (1992) The mitochondrial tyrosyl-tRNA synthetase of *Podospora anserina* is a bifunctional enzyme active in protein synthesis and RNA splicing. Mol Cell Biol 12:499–511
84. Mannella CA, Collins RA, Green MR, Lambowitz AM (1979) Defective splicing of mitochondrial rRNA in cytochrome-deficient nuclear mutants of *Neurospora crassa*. Proc Natl Acad Sci U S A 76:2635–2639
85. Collins RA, Lambowitz AM (1985) RNA splicing in Neurospora mitochondria. Defective splicing of mitochondrial mRNA precursors in the nuclear mutant cyt18-1. J Mol Biol 184:413–428
86. Mohr G, Zhang A, Gianelos JA, Belfort M, Lambowitz AM (1992) The *neurospora* CYT-18 protein suppresses defects in the phage T4 td intron by stabilizing the catalytically active structure of the intron core. Cell 69:483–494
87. Mohr G, Caprara MG, Guo Q, Lambowitz AM (1994) A tyrosyl-tRNA synthetase can function similarly to an RNA structure in the *Tetrahymena* ribozyme. Nature 370:147–150
88. Paukstelis PJ, Coon R, Madabusi L, Nowakowski J, Monzingo A, Robertus J, Lambowitz AM (2005) A tyrosyl-tRNA synthetase adapted to function in group I intron splicing by acquiring a new RNA binding surface. Mol Cell 17:417–428
89. Caprara MG, Lehnert V, Lambowitz AM, Westhof E (1996) A tyrosyl-tRNA synthetase recognizes a conserved tRNA-like structural motif in the group I intron catalytic core. Cell 87:1135–1145
90. Caprara MG, Mohr G, Lambowitz AM (1996) A tyrosyl-tRNA synthetase protein induces tertiary folding of the group I intron catalytic core. J Mol Biol 257:512–531
91. Webb AE, Rose MA, Westhof E, Weeks KM (2001) Protein-dependent transition states for ribonucleoprotein assembly. J Mol Biol 309:1087–1100

92. Myers CA, Kuhla B, Cusack S, Lambowitz AM (2002) tRNA-like recognition of group I introns by a tyrosyl-tRNA synthetase. Proc Natl Acad Sci U S A 99:2630–2635
93. Cherniack AD, Garriga G, Kittle JD Jr, Akins RA, Lambowitz AM (1990) Function of *Neurospora* mitochondrial tyrosyl-tRNA synthetase in RNA splicing requires an idiosyncratic domain not found in other synthetases. Cell 62:745–755
94. Kittle JD Jr, Mohr G, Gianelos JA, Wang H, Lambowitz AM (1991) The Neurospora mitochondrial tyrosyl-tRNA synthetase is sufficient for group I intron splicing in vitro and uses the carboxy-terminal tRNA-binding domain along with other regions. Genes Dev 5:1009–1021
95. Chen X, Mohr G, Lambowitz AM (2004) The *Neurospora crassa* CYT-18 protein C-terminal RNA-binding domain helps stabilize interdomain tertiary interactions in group I introns. RNA 10:634–644
96. Mohr G, Rennard R, Cherniack AD, Stryker J, Lambowitz AM (2001) Function of the *Neurospora crassa* mitochondrial tyrosyl-tRNA synthetase in RNA splicing. Role of the idiosyncratic N-terminal extension and different modes of interaction with different group I introns. J Mol Biol 307:75–92
97. Guo Q, Lambowitz AM (1992) A tyrosyl-tRNA synthetase binds specifically to the group I intron catalytic core. Genes Dev 6:1357–1372
98. Saldanha RJ, Patel SS, Surendran R, Lee JC, Lambowitz AM (1995) Involvement of Neurospora mitochondrial tyrosyl-tRNA synthetase in RNA splicing. A new method for purifying the protein and characterization of physical and enzymatic properties pertinent to splicing. Biochemistry 34:1275–1287
99. Saldanha R, Ellington A, Lambowitz AM (1996) Analysis of the CYT-18 protein binding site at the junction of stacked helices in a group I intron RNA by quantitative binding assays and in vitro selection. J Mol Biol 261:23–42
100. De La Salle H, Jacq C, Slonimski PP (1982) Critical sequences within mitochondrial introns: pleiotropic mRNA maturase and cis-dominant signals of the box intron controlling reductase and oxidase. Cell 28:721–732
101. Labouesse M, Netter P, Schroeder R (1984) Molecular basis of the 'box effect', a maturase deficiency leading to the absence of splicing of two introns located in two split genes of yeast mitochondrial DNA. Eur J Biochem 144:85–93
102. Li GY, Becam AM, Slonimski PP, Herbert CJ (1996) In vitro mutagenesis of the mitochondrial leucyl tRNA synthetase of *Saccharomyces cerevisiae* shows that the suppressor activity of the mutant proteins is related to the splicing function of the wild-type protein. Mol Gen Genet 252:667–675
103. Boniecki MT, Rho SB, Tukalo M, Hsu JL, Romero EP, Martinis SA (2009) Leucyl-tRNA synthetase-dependent and -independent activation of a group I intron. J Biol Chem 284:26243–26250
104. Rho SB, Martinis SA (2000) The bI4 group I intron binds directly to both its protein splicing partners, a tRNA synthetase and maturase, to facilitate RNA splicing activity. RNA 6:1882–1894
105. Houman F, Rho SB, Zhang J, Shen X, Wang CC, Schimmel P, Martinis SA (2000) A prokaryote and human tRNA synthetase provide an essential RNA splicing function in yeast mitochondria. Proc Natl Acad Sci U S A 97:13743–13748
106. Sarkar J, Poruri K, Boniecki MT, McTavish KK, Martinis SA (2012) Yeast mitochondrial leucyl-tRNA synthetase CP1 domain has functionally diverged to accommodate RNA splicing at expense of hydrolytic editing. J Biol Chem 287:14772–14781
107. Rho SB, Lincecum TL Jr, Martinis SA (2002) An inserted region of leucyl-tRNA synthetase plays a critical role in group I intron splicing. EMBO J 21:6874–6881
108. Cusack S, Yaremchuk A, Tukalo M (2000) The 2 A crystal structure of leucyl-tRNA synthetase and its complex with a leucyl-adenylate analogue. EMBO J 19:2351–2361

109. Tukalo M, Yaremchuk A, Fukunaga R, Yokoyama S, Cusack S (2005) The crystal structure of leucyl-tRNA synthetase complexed with tRNALeu in the post-transfer-editing conformation. Nat Struct Mol Biol 12:923–930
110. Labouesse M, Herbert CJ, Dujardin G, Slonimski PP (1987) Three suppressor mutations which cure a mitochondrial RNA maturase deficiency occur at the same codon in the open reading frame of the nuclear NAM2 gene. EMBO J 6:713–721
111. Palencia A, Crepin T, Vu MT, Lincecum TL Jr, Martinis SA, Cusack S (2012) Structural dynamics of the aminoacylation and proofreading functional cycle of bacterial leucyl-tRNA synthetase. Nat Struct Mol Biol 19:677–684
112. Hsu JL, Rho SB, Vannella KM, Martinis SA (2006) Functional divergence of a unique C-terminal domain of leucyl-tRNA synthetase to accommodate its splicing and aminoacylation roles. J Biol Chem 281:23075–23082
113. Kapasi P, Chaudhuri S, Vyas K, Baus D, Komar AA, Fox PL, Merrick WC, Mazumder B (2007) L13a blocks 48S assembly: role of a general initiation factor in mRNA-specific translational control. Mol Cell 25:113–126
114. Ofir-Birin Y, Fang P, Bennett SP, Zhang HM, Wang J, Rachmin I, Shapiro R, Song J, Dagan A, Pozo J, Kim S, Marshall AG, Schimmel P, Yang XL, Nechushtan H, Razin E, Guo M (2012) Structural switch of lysyl-tRNA synthetase between translation and transcription. Mol Cell 49:30–42

Top Curr Chem (2014) 344: 189–206
DOI: 10.1007/128_2013_426
© Springer-Verlag Berlin Heidelberg 2013
Published online: 28 March 2013

Amino-Acyl tRNA Synthetases Generate Dinucleotide Polyphosphates as Second Messengers: Functional Implications

Sagi Tshori, Ehud Razin, and Hovav Nechushtan

Abstract In this chapter we describe aminoacyl-tRNA synthetase (aaRS) production of dinucleotide polyphosphate in response to stimuli, their interaction with various signaling pathways, and the role of diadenosine tetraphosphate and diadenosine triphosphate as second messengers. The primary role of aaRS is to mediate aminoacylation of cognate tRNAs, thereby providing a central role for the decoding of genetic code during protein translation. However, recent studies suggest that during evolution, "moonlighting" or non-canonical roles were acquired through incorporation of additional domains, leading to regulation by aaRSs of a spectrum of important biological processes, including cell cycle control, tissue differentiation, cellular chemotaxis, and inflammation. In addition to aminoacylation of tRNA, most aaRSs can also produce dinucleotide polyphosphates in a variety of physiological conditions. The dinucleotide polyphosphates produced by aaRS are biologically active both extra- and intra-cellularly, and seem to function as important signaling molecules. Recent findings established the role of dinucleotide polyphosphates as second messengers.

Keywords Alarmones · Ap_3A · Ap_4A · Ap_4A hydrolase · Dinucleotide polyphosphate · Fragile histidine triad protein · Lysyl tRNA synthetase · Microphthalmia transcription factor · Second messenger

---

S. Tshori
Department of Nuclear Medicine, Hadassah Hebrew University Medical Center, Jerusalem, Israel

E. Razin (✉)
Department of Biochemistry and Molecular Biology, Institute for Medical Research Israel-Canada, Hebrew University Hadassah Medical School, Jerusalem, Israel
e-mail: ehudr@ekmd.huji.ac.il

H. Nechushtan
Department of Oncology, Hadassah Hebrew University Medical Center, Jerusalem, Israel

Contents

1 Introduction 190
2 Second Messengers 191
    2.1   What Is a Second Messenger? 191
    2.2   Identifying New Second Messengers 192
    2.3   Non-kinase Second-Messenger Signaling 193
    2.4   Diadenosine Oligophosphates as Signaling Molecules 193
    2.5   Diadenosine Oligophosphates Are Synthesized by aaRSs 194
    2.6   Diadenosine Oligophosphates Are Degraded by Hydrolysis 195
    2.7 Ap_n A Binding Proteins 196
    2.8   Physiological Actions of Diadenosine Oligophosphates 197
    2.9   Diadenosine Oligophosphates Levels Are Regulated
          by Physiological Stimuli 197
    2.10 Ap_4A as a Second Messenger: The Case of LysRS-MITF Pathway 198
    2.11 Ap_4A as a Second Messenger in Other Systems 200
    2.12 Ap_3A Directs Fhit Activity 201
3 Summary 202
References 202

Abbreviations

aaRS   Aminoacyl tRNA synthetase
Ap_3A Di-adenosine tri-phosphate
Ap_4A Di-adenosine tetra-phosphate
Ap_nN Di-adenosine polynucleotide
cAMP   Cyclic adenosine monophosphate
Fhit   Fragile histidine triad protein
Hint-1   Histidine triad nucleotide-binding protein 1
LysRS   Lysyl tRNA synthetase
MITF   Microphthalmia transcription factor
MSC   Multisynthetase complex
Np_nN Dinucleotides polyphosphates
TrpRS   Tryptophenyl tRNA synthetase

1 Introduction

Enzymes are expected to be efficient, specific, and highly evolved to serve certain catalytic functions. In recent years, however, there has been a growing appreciation that this picture is oversimplified. Many enzymes are catalytically promiscuous as they are capable of catalyzing secondary reactions at an active site that was thought to be specific to the catalysis of a particular primary reaction.

In addition to their main role, several aminoacyl tRNA synthetases (aaRSs) can synthesize dinucleotide polyphosphates (Np_nNs) both in prokaryotes and in eukaryotes by transference of adenylate from the enzyme's aminoacyladenylate intermediate complex to an acceptor nucleotide.

These dinucleotides polyphosphates are biologically active, exerting a wide range of cellular responses. These compounds have emerged as putative intracellular signaling molecules implicated in the regulation of essential cellular processes, but until recently there was insufficient data to support the notion that diadenosine polyphosphates (Ap_nA) are second messengers. With the accumulation of more data it has become apparent that indeed extracellular signals cause increased synthesis of Ap_nAs by aaRS, and that these Ap_nAs act as second messengers, activating effector proteins.

2 Second Messengers

In 1959, Earl Sutherland discovered a small heat-stable molecule that mediated the effects of the hormone epinephrine on extracts of liver cells. This molecule, cyclic AMP, was produced inside cells after hormone treatment and was able to activate the same glycolytic enzymes activated by extracellular epinephrine. These experiments established the concept of the intracellular second messenger [1].

Since that time many intracellular second messengers have been identified. The four prototypical second messengers include cyclic adenosine monophosphate (cAMP), cyclic guanosine monophosphate (cGMP), diacylglycerol (DAG), and calcium. Investigation of the mechanisms by which these second messengers mediate the effects of various hormones, growth factors, and neurotransmitters identified a common theme: these molecules function as activators of dedicated protein kinases that, in turn, activate downstream kinase cascades. This process subsequently leads to the phosphorylation of target proteins, which mediates change in cellular physiology (Fig. 1) [2].

2.1 What Is a Second Messenger?

Second messengers are defined as "small molecules that are synthesized in the cell in response to extracellular first messengers, and diffuse through the cytoplasm to mediate their effects" [3]. Intercellular signaling by first messenger molecules produces second messenger molecules that in turn activate principal effector proteins such as protein kinase A (PKA) and protein kinase C (PKC) [2]. Second messengers amplify the strength of the initial signaling event of the extracellular ligand.

Synthesis and degradation of second messengers are regulated by a number of enzymes expressed in mammalian cells. Enzymes that synthesize second messengers include, but are not limited to, adenylyl cyclase (AC) for synthesis of cyclic AMP (cAMP) and guanylyl cyclase for synthesis of cyclic GMP (cGMP). Second messengers are degraded by specific enzymes, including cyclic nucleotide phosphodiesterase (PDE) for hydrolysis of cAMP and cGMP, and phospholipase C (PLC) for hydrolysis of phosphatidylinositol 4,5-bisphosphate to inositol 1,4,5-trisphosphate (IP3) and diacylglycerol (DAG) [3].

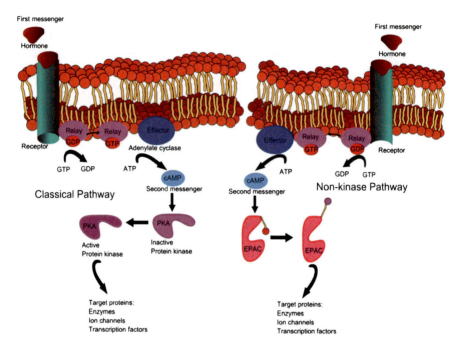

Fig. 1 Classical and non-kinase signal transduction pathways. Extracellular ligands (e.g., hormones, neurotransmitters, and growth factors) bind plasma membrane receptors. G protein relay activates an effector protein, such as adenylate cyclase, creating a second messenger. In the classical pathway, a protein kinase (PKA) is activated, resulting in the activation of various cellular processes. In the non-kinase pathway, EPAC proteins relay the signal without the involvement of secondary effector protein kinases

2.2 *Identifying New Second Messengers*

After cyclic AMP was discovered by Sutherland and Rall [1], which led to the concept of second messengers described above, Sutherland set out to determine which hormones exert which of their effects by this mechanism. Sutherland decided to embark on a series of studies of different biological systems, and formulated a set of criteria [4]. With some adaptations, these criteria are still used to identify new second messenger molecules [5, 6]. These criteria are as follows:

1. The messenger should mimic the effect of the extracellular stimulus when applied intracellularly.
2. Enzymatic machinery should be present to synthesize and metabolize the prospective messenger.
3. The messenger's intracellular levels should change in response to the extracellular stimulus.
4. Antagonists of the action of the messenger should block the effects of the stimulus whose effects are proposed to be tranduced by the messenger.
5. Specific intracellular targets should be present for the candidate second messenger.

2.3 Non-kinase Second-Messenger Signaling

After half a century of second messenger research, one would have expected that few surprises would remain; however, that is not the case. For instance, many experimental observations have hinted at the existence of PKA-independent mechanisms of cAMP action. Most of the physiological effects of cAMP can be ascribed to the action of one or more members of the PKA family, which have distinct tissue-specific patterns of expression. However, cAMP exerts physiological effects that are not readily explained by PKA's action or by the other known cAMP receptors. In 1998 the discovery of a new family of cAMP receptor/effector proteins suggested a solution to this puzzle (Fig. 1) [7]. These proteins, designated EPAC (exchange protein directly activated by cAMP), cAMP-GEF (cAMP regulated guanine nucleotide exchange factor), CalDAG-GEF (calcium and diacylglycerol regulated guanine nucleotide exchange factor), and RasGRP (ras guanine nucleotide releasing protein), can mediate some of the physiological effects of the second messengers in a protein-kinase-independent fashion. These proteins are all exchange factors for Ras family GTPases that operate in pathways that run in parallel to the classic kinase-dependent pathways. Since the identification of the EPAC/cAMP-GEF and RasGRP/CalDAG-GEF proteins, more than 300 articles have cited the initial reports, reflecting an emerging recognition that the physiological effects of second messengers are not only a consequence of their interactions with protein kinase effectors. Thus, in the analysis of any physiological response to a second messenger molecule, the first question should be whether the effect is kinase dependent, kinase independent, or a combination of the two mechanisms [2].

2.4 Diadenosine Oligophosphates as Signaling Molecules

Ap_nAs are made up of two adenosine moieties joined in $5'–5'$ linkage by a chain ranging from 2 to 6 phosphodiester links. These compounds were discovered by Zamecnik and his co-workers in the mid-1960s [8]. In prokaryotes, heat shock and oxidative stress were found to cause Ap_nA accumulation [9, 10], and for this reason these molecules were considered as pleiotropically acting "alarmones" [11], which are intracellular signaling molecules that are produced due to harsh environmental factors, and can convey sophisticated messages [12]. Studies showed that Ap_nA binds to and inhibits oxidative stress-related proteins [13, 14]. Later, further studies demonstrated a ubiquitous occurrence of Ap_nA in the whole spectrum of organisms from bacteria to higher eukaryotes [15, 16].

In multicellular organisms Ap_nAs have been shown to be physiologically active. Many diverse tissues or cells respond to Ap_nA (e.g., heart, hippocampus, sperm, hepatocytes, neutrophils, and pancreatic cells), indicating the involvement of the Ap_nA family in widespread biochemical events, not restricted to any specialised cell type or tissue (reviewed in [15]). Another important feature of Ap_nAs' effects is the varied responses observed in various cell types, following administration or addition of Ap_nA in animal models and cell culture respectively (reviewed in [15]).

2.5 *Diadenosine Oligophosphates Are Synthesized by aaRSs*

In addition to the main role of aaRSs in the production of aminoacyl-tRNAs, several aaRSs can synthesize dinucleotide polyphosphates.

aaRS proteins react differently in the presence of their cognate tRNAs and other cellular factors such as ATP. In terms of formation of second messengers relevant to this review, one such Ap_nA family member, diadenosine $5',5'''-P_1,P_4$-tetraphosphate (Ap_4A), is an aaRS- synthesized second messenger with proven historical and physiological implications. The synthesis of Ap_4A was first detected in an in vitro system consisting of ATP, Mg^{2+}, L-lysine, and purified *Escherichia coli (E. coli)* lysyl-tRNA synthetase [8]. It was later shown that Ap_4A can be synthesized in vitro by several other aaRS proteins, all using aminoacyladenylate as an obligatory intermediate [11].

Later studies confirmed the physiological relevence of this observation by showing that, in vivo, Ap_4A is synthesized by a side reaction during aminoacyl-tRNA synthesis [Eq. (1)] [17, 18]. This process involves the attack of the enzyme-bound aminoacyl adenylate intermediate (I) by the pyrophosphate group of ATP, resulting in formation of Ap_4A (IIb) rather than aminoacyl-tRNA (IIa). Nucleophilic attack by other nucleotides results in formation of a variety of dinucleoside polyphosphates. However, Ap_4A is the predominant species due to the high concentration of ATP in cells [19].

Equation 1. Ap_nN synthesis by a side reaction of aaRS, usually in the absence of cognate tRNA, through aminoacyl intermediate

$$aaRS + ATP + aa \rightarrow aaRS \bullet aa \sim AMP + PPi \qquad \text{(I)}$$

$$aaRS \bullet aa \sim AMP + tRNA \rightarrow aa \sim tRNA + aaRS + AMP \qquad \text{(IIa)}$$

$$aaRS \bullet aa \sim AMP + NTP/NDP \rightarrow aaRS + AMP + Ap_nN \qquad \text{(IIb)}$$

However, a number of aaRS proteins rely on different mechanisms of actions, and therefore provide interesting considerations in terms of the role of aaRS in homeostasis, as opposed to production of second messengers important for intra- or extracellular signaling. Reactions catalyzed by the aaRS proteins arginyl-tRNA synthetase (ArgRS) and glutaminyl-tRNA synthetase (GlnRS) are unique, since both ArgRS and GlnRS are unable to produce Ap_4A, which is related to the fact that both of these enzymes belong to the group of aaRSs that can only form aminoacyl adenylate in the presence of the cognate tRNA. They are unusual insofar as the presence of cognate tRNA in most cases inhibits Ap_4A synthesis.

In contrast, glycinyl-tRNA synthetase (GlyRS) is responsible for the most robust production of Ap_4A by aaRS family members. It uses direct ATP condensation to synthesize Ap_4A and this unique mechanism is probably conserved for GlyRSs in eukaryotes and archaea [20]. It is intriguing to speculate that GlyRS is responsible for maintaining the basal level of cellular Ap_4A. In this scenerio, whereby GlyRS may be linked to Ap_4A homeostasis, LysRS could play an alternative role in regulating Ap_4A concentration in response to selective stimuli or specific physiological conditions.

Along with ATP, aaRSs can also utilize any nucleoside oligophosphate (NDP, NTP, ppGpp, etc.) as substrates, at least in vitro, resulting in the formation of a family of adenylated dinucleoside oligophosphates (AP_nNs) (reviewed in [10, 11]). Although most aaRSs are able to catalyze the synthesis of Np_nNs and/or p_nN, it is not clear whether such enzymes are responsible for the majority of the Np_nNs that accumulate in vivo. In the cases of lysyl-tRNA synthetase (LysRS), phenylalanyl-tRNA synthetase (PheRS), methioninyl-tRNA synthetase (MetRS), and valinyl-tRNA synthetase (ValRS), it has been observed that the intracellular levels of Ap_4N increased in transformed $E.$ $coli$ strains that overproduced these enzymes [16].

Tryptophanyl-tRNA synthetase (TrpRS) is another interesting exception, as it is unable to catalyze in vitro formation of Ap_4A in contrast to some other aminoacyl-tRNA synthetases. However, in the presence of L-tryptophan, ATP-Mg^{2+}, and ADP, the enzyme catalyzes the Ap_3A synthesis via adenylate intermediate [21]. TrpRS was shown in vivo to catalyze the synthesis of Ap_3A, but not Ap_4A, in human monocytes (J96 cells) and myeloid leukaemia HL60 cells incubated with interferon [22].

2.6 Diadenosine Oligophosphates Are Degraded by Hydrolysis

Ap_nA molecules can be non-specifically degraded by a variety of phosphodiesterases and nucleotidases, such as Ap_4A phosphorylase [23, 24] in $S.$ $cerevisiae$ or in cyanobacteria $Anabaena$ $flos$-$aquae$ [25]. However, specific Np_nN hydrolases occur in both prokaryotes and eukaryotes [26]. These include hydrolases belonging to either the histidine triad (HIT) superfamily or to the nucleoside diphosphate linked moiety X (Nudix) superfamily.

The human fragile HIT (Fhit) protein is a dinucleoside $5',5'''$-P1,P3-triphosphate hydrolase that cleaves Ap_3A to AMP and ADP [27]. HIT proteins share a conserved His-X-His-X-His-X-X sequence motif, where X is a hydrophobic amino acid residue. The histidine residues of the HIT sequence are critical for the Ap_3A hydrolase activity of Fhit. It can also hydrolase Ap_4A, but with a threefold higher K_M.

In most higher eukaryotes, Ap_4A hydrolase metabolizes Ap_4A molecules by asymmetrical hydrolysis, resulting in the formation of ATP and AMP. Ap_4A hydrolase is encoded by the NudT2 gene [28]. This gene encodes a member of the MutT family of nucleotide pyrophosphatases, a subset of the larger Nudix hydrolase family. The gene product possesses a modification of the MutT sequence motif found in certain nucleotide pyrophosphatases. Ap_4A hydrolase has some cleaving activity for Ap_5A and Ap_6A, but it does not cleave Ap_3A.

Ap$_5$A and Ap$_6$A are preferentially cleaved by NudT3 (YOR163w) and NudT11 (Aps1) gene products [29], which also cleave a β phosphate from the diphosphate groups in diphosphoinositol pentakisphosphate (PP-InsP$_5$) and bisdiphosphoinositol tetrakisphosphate ([PP]$_2$-InsP$_4$). The major reaction products are ADP and p$_4$A from Ap$_6$A, and either ADP and ATP or AMP and p$_4$A from Ap$_5$A. The MutT/Nudix motif, which appears in a number of proteins from across the phylogenetic spectrum, is characterized by the following (or closely related) sequence: GX5EX7REUXEEXGU, where U is usually I, L, or V [30].

This motif represents one general solution to the challenge of regulating the levels of metabolic intermediates that either act as cellular signals or can be deleterious to cell function [30]. Thus, by analogy to the "housekeeping" or maintenance genes, those coding for the nudix hydrolases could be considered "housecleaning" genes whose function is to cleanse the cell of potentially deleterious endogenous metabolites and to modulate the accumulation of intermediates in biochemical pathways.

2.7 Ap$_n$ A Binding Proteins

Ap$_n$A binding proteins comprise a diverse class of proteins which facilitate the linking and controlling of various cellular processes. These include hemoglobin, glycogen phosphorylase, glyceraldehyde-3-phosphate dehydrogenase, DNA polymerase K associated protein, hormones, and hormone-like factors. Ap$_n$A binding proteins can be categorized into three main groups:

1. Receptors located at the cell surface – P2 purinergic receptors [31], brain membrane receptor [32].
2. Enzymes utilizing Ap$_n$A as substrates – Fhit [33], diadenosine tetraphosphate hydrolase [34].
3. Other enzymes that bind Ap$_n$A but do not metabolize them – Ap$_4$A binding protein [35], heat shock protein ClpB, histidine triad nucleotide-binding protein 1 (Hint-1) [36].

As most signal transducing molecules perform their functions through protein–protein interactions, it is interesting to consider how Ap$_n$A interactions with such proteins mediate their biological activities, and through what molecular mechanisms. If the signal transduction hypothesis for the Ap$_n$A family is correct, the structural dissimilarities between the various members of the Ap$_n$A family caused by differences in the length of their oligophosphate chain may result in the different patterns of binding proteins for each Ap$_n$A molecule. If, in contrast, proteins mediating biological effects bind Ap$_n$A nonselectively with regards to phosphate chain length, an alternative mechanism for signal transduction is that the binding protein assumes different protein conformational states as a consequence of its binding to the Ap$_n$A molecule, or by charge differences induced by different Ap$_n$A molecules.

2.8 Physiological Actions of Diadenosine Oligophosphates

A wealth of data now suggest that the whole set of Ap_nAs, with "n" signifying 2–6 phosphates, possess biological activity, and thus can be considered a new class of signaling molecules used by eukaryotic cells to regulate many housekeeping and specialized functions [15]. For instance, dinucleotide polyphosphates have been shown to play roles in both extracellular and intracellular physiology: a variety of tissues and cells respond to extracellular application of Ap_nA, including the heart [37], hippocampus [38], hepatocytes [39], neutrophils [40], and pancreatic cells [41]. Ap_nAs are also involved in the physiology of various systems, including inhibition of sperm motility, decrease of coronary perfusion pressure, and prevention of aggregation of human platelets induced by ADP (reviewed in [15]), indicating the involvement of the Ap_nA family in myriad biochemical events not restricted to any specialized cell type or tissue.

Less is known about the intracellular role of Ap_nAs. Many studies indicate that Ap_nAs are involved in various intracellular processes. These include nuclear functions such as stimulation of DNA synthesis, mitogenic activity, and activation of gene transcription and membrane processes such as induction of Ca^{2+} oscillations and release and inhibition of K ATP channels.

However, it should be noted that some of these initial reports were performed before methods to measure and control intracellular Ap_nA levels were optimized, and therefore some of the results are inconsistent and so should be treated with caution. For example, although initial reports indicated a role for Ap_4A in cell proliferation [42], a later study in HeLa D98/AH2 and L929 cells could not support the hypothesis that changes in Ap_4A levels regulate cellular proliferation [43]. However, further studies supported the earlier findings, suggesting that NUDT2 (Ap_4A hydrolase) is an estrogen repressed gene and is also induced by HER2 pathways in breast carcinoma cells, and that NUDT2 promotes proliferation of breast carcinoma cells and is a potent prognostic factor in human breast carcinomas [44].

We should, therefore, be careful not to overinterpret the physiological roles attributed to Ap_4A, especially in the absence of coroborating data from recent years.

2.9 Diadenosine Oligophosphates Levels Are Regulated by Physiological Stimuli

Ap_nA levels are regulated by a variety of important physiological triggers. One example is a study on the effects of interferons on Ap_3A levels. Interferons are well known to cause inhibition of cell proliferation. One study reported that, after incubating interferon γ or α with human monocyte J96 cells and human myeloid leukemia HL60 cells, Ap_3A concentration considerably increased, in parallel with accumulation of tryptophanyl-tRNA synthetase (TrpRS). The Ap_3A/Ap_4A ratio in HL60 cells was reported to increase threefold [22].

However, there is also evidence that in proliferating cells, Ap_4A acts as a second messenger of mitogenic induction similar to cAMP in hormone induction [42]. After mitogenic stimulation of G1-arrested mouse 3T3 cells and baby hamster kidney fibroblasts, the Ap_4A pool gradually increased 1,000-fold during progression through the G1 phase, reaching maximum Ap_4A concentrations in the S phase [42]. Other studies reported that Ap_4A binds to mammalian DNA polymerase α [45] and stimulates DNA replication [46], indicating that Ap_4A could play a role as a DNA inducer.

Interestingly, the Ap_3A/Ap_4A ratio shows distinct polar behavior in apoptotic vs differentiating HL60 cells. One study showed that induction of cell differentiation by phorbol ester in the promyelocytic cell line HL60 resulted in Ap_3A accumulation [47]. In contrast, when HL60 cells were induced to undergo apoptosis by VP16, an inhibitor of DNA topoisomerase II, a striking inversion of the Ap_3A/Ap_4A ratio was reported. This inversion is due to elevation of Ap_4A concentration in apoptotic cells and drop of Ap_3A level. The opposite effect was observed when HL60 cells underwent differentiation: the Ap_4A concentration remained constant whereas the Ap_3A level was greatly elevated [47]. These data indicate that Ap_4A and Ap_3A act as physiological antagonists in the determination of the cellular status, with Ap_4A inducing apoptosis, and Ap_3A acting as a co-inductor of differentiation. In both cases the mechanism of signal transduction remains unknown [48].

Another example of Ap_nA regulation by physiological signals has been observed in murine pancreatic β cells, where glucose at concentrations inducing insulin release were reported to produce a 30- to 70-fold increase in Ap_3A and Ap_4A levels [23].

Ap_nA levels can also be controlled by regulation of Ap_4A hydrolase. One study showed that in mast cells, immunological challenge triggers both importin β dependent translocation of Ap_4A hydrolase to the nucleus, and de-phosphorylation of Ap_4A hydrolase. Decrease of its hydrolytic activity results in increased Ap_4A levels [49].

2.10 Ap_4A as a Second Messenger: The Case of LysRS-MITF Pathway

Ap_4A, which is produced by LysRS, provides an example of a second messenger that meets Sutherland's criteria, as discussed earlier in this review. LysRS can translocate to the nucleus where it forms a multiprotein complex with microphthalmia transcription factor (MITF) and Hint-1 or with upstream stimulatory factor 2 (USF2) and Hint-1 [36, 50]. LysRS nuclear localization occurs in response to immunological challenge, and is mediated by MAPK-dependent phosphorylation of LysRS serine 207, which releases it from the multisynthetase complex, allowing it to translocate to the nucleus and regulate gene expression [51].

LysRS has long been known to produce diadenosine tetraphosphate (Ap_4A), but the link to transcriptional regulation in eukaryotic cells has only recently been established. This regulation occurs through Ap_4A-induced dissociation of Hint from the respective transcription factor, allowing for transcription of MITF- or USF2-responsive genes (Fig. 2) [50].

As mentioned, it was speculated as early as 1984 that Ap_4A acts as a second messenger of mitogenic induction similar to the action of cAMP in hormone induction [42]. If Ap_4A is indeed a second messenger, it should fulfill Sutherland's criteria [5, 6]. Is this indeed the case? The latest findings should be taken into account:

1. *The messenger should mimic the effect of the extracellular stimulus when applied intracellularly.* Upon RBL activation by immunoglobulin E (IgE) and antigen (IgE-Ag), Ap_4A accumulates and MITF target expression is induced. The introduction of Ap_4A by cold shock into RBL cells results in the activation of MITF target gene expression, such as granzyme B and tryptophane hydroxylase, which closely follows the effect of immunological activation of these cells by IgE-Ag [36].
2. *Enzymatic machinery should be present to synthesize and metabolize the prospective messenger.* As described earlier, Ap_4A can be synthesized by LysRS. It is known that several enzymes are able to hydrolyze Ap_4A, but there is evidence that the "Nudix" type 2 gene product, Ap_4A hydrolase, is responsible for Ap_4A degradation following the immunological activation of mast cells [6]. Thus enzymatic machinery is present in the cell that can both synthesize and degrade Ap_4A, and this mechanism responds to physiological stimuli.
3. *The messenger's intracellular levels should change in response to the extracellular stimulus.* Upon engagement of the surface high affinity Fc receptors (FcεRI) in mast cells, Ap_4A is synthesized by LysRS in close proximity to a multiprotein complex of Hint, MITF, and LysRS [36]. Ap_4A levels peak 15 min after stimulation, followed by a sharp drop and a return to the basal level within 60 min [6].
4. *Antagonists of the action of the messenger should block the effects of the stimulus whose effects are proposed to be tranduced by the messenger.* Currently there are no antagonists for Ap_4A. However, silencing of LysRS or transfection of S207A mutated LysRS decrease Ap_4A production and abrogate the expression of MITF target genes, such as tryptophan hydroxylase [51].
5. *Specific intracellular targets should be present for the candidate messenger.* Hint-1 is a HIT protein that can bind AMP [52] and GMP [53], and it is known to catalyze AMP-NH$_4$ [54]. Hint-1 binds and inhibits the transcription factors MITF and USF2 [36, 50]. Ap_4A can bind to Hint-1 but is not degraded by it. However, Ap_4A binding to Hint-1 causes its release from both MITF and USF2 [36, 50], thereby abolishing the inhibitory effect of Hint-1, and activating both MITF target genes (tryptophan hydroxylase, granzyme B, myosin light chain 1a) and USF2 target gene telomerase reverse transcriptase (TERT).

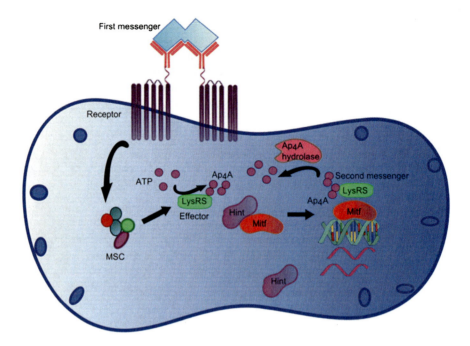

Fig. 2 LysRS-MITF signaling pathway. Extracellular ligand, such as antigen, binds plasma membrane high affinity Fc receptor in mast cells. LysRS is released from the MSC, and starts producing Ap$_4$A, detaching Hint from MITF. MITF is free to bind DNA and activate target genes. Finally, Ap$_4$A hydrolase degrades Ap$_4$A, and Hint can bind and inhibit MITF

Thus, Ap$_4$A indeed fulfills all of Sutherland's criteria, and is therefore a second messenger.

2.11 Ap$_4$A as a Second Messenger in Other Systems

Second messengers are usually involved in multiple signal transduction pathways. For example, in *Pseudomonas fluorescens*, loss of the ApaH function and subsequent accumulation of Ap$_4$A promote biofilm formation by two mechanisms [19]:

Ap$_4$A inhibits Pho regulon, resulting in decreased levels of RapA, and the subsequent increases in the level of cyclic diguanylate (c-di-GMP). Higher levels of c-di-GMP result in increased biofilm formation by promoting localization of the adhesin LapA to the cell surface. Furthermore, Ap$_4$A promotes expression of LapA and its transporter, LapEBC, which contributes to increased biofilm formation.

2.12 Ap₃A Directs Fhit Activity

Ap_3A has been found to be another dinucleotide polyphosphate with important second messenger activities, linking interferon signaling to the fragile histidine triad (Fhit) gene, a tumor-suppressor gene present on the short arm of chromosome 3. Many cancer cell lines and primary cancers exhibiting hemizygous or homozygous deletions with end points within the FHIT gene. It is well known that overexpression of Fhit results in apoptosis in Fhit-deficient cancer cells, and recent studies have demonstrated a role for Fhit in responses to damage induced by ultraviolet C (UVC) light. Furthermore, after oxidative stress, Fhit-positive cells produce higher levels of ROS, while Fhit-deficient cancer cells show only a mild response to oxidative stress, producing less ROS, and are more resistant to chemotherapy-induced cell death (reviewed in [55]).

Fhit tumor suppressor activity was initially believed to be related to its Ap_3A hydrolytic activity, but further studies disproved this theory. For instance, one study reported that the H96N mutation of Fhit, which decreases Fhit hydrolase activity by several orders of magnitude, did not affect its tumor suppressor activity [56]. Another study constructed a series of mutations in the Fhit gene, designed to increase K_M and/or to decrease K_{cat} [33] of the Fhit protein. However, apoptotic activity correlated exclusively with K_M. It was, therefore, assumed that the pro-apoptotic function of Fhit depends on formation of an enzyme–substrate complex, which is limited by substrate binding, and is unrelated to substrate hydrolysis. Recent data indicate that Fhit–2 Ap_3A complex represents the active form of Fhit. This active form exposes an extensively phosphorylated surface with two adenosine moieties in a deep cavity lined with charged amino acids [57]. The negatively supercharged surface might favor Fhit binding to secondary effector molecules.

Fhit interacts with several signal transduction pathways [58]. Fhit directly interacts with the C-terminus of β-catenin and represses transcription of target genes such as cyclin D1, MMP-14, and survivin [59]. Multiple key molecules in the Ras/Rho GTPase molecular switch, including Ran, Rab, and Rac, were negatively regulated by Fhit in non-small cell lung carcinoma (NSCLC) [60]. The finding of Ras/Rho GTPases as direct cellular targets of Fhit may provide a biological support for an earlier crystallographic study that suggested that the Fhit–substrate complex is an active signaling molecule mediating its tumor-suppression activity and functionally analogous to the Ras-GTP complexes [57]. Furthermore, loss of endogenous Fhit expression caused increased Akt activity in vitro and in vivo, which leads to the inhibition of apoptosis. The results of the study indicated that the Fhit Y114 residue plays a critical role in Fhit-induced apoptosis through inactivation of the PI3K-Akt-survivin signal pathway [61].

As mentioned above, in myeloid leukemia HL60 cells treated with interferon γ or α, the Ap_3A concentration considerably increases in parallel with accumulation of TrpRS [22]. The competitive inhibitors of TrpRS adenylate formation, tryptamine and β-indolilpyruvic acid, considerably decrease Ap_3A accumulation. Thus, Ap_3A accumulation results from TrpRS activity.

IFN-γ is known to upregulate TrpRS expression [62, 63]. Under normal conditions, little TrpRS could be detected in the nucleus. However, upon IFN-γ treatment, the presence of TrpRS in the nucleus, along with its overexpression in the cytoplasm, was detected in a variety of cell lines. This observation demonstrated that IFN-γ-mediated upregulation and translocation of TrpRS into the nucleus was not specific to a particular cell line [64].

Thus the following model can be suggested. Interferons induce the expression and translocation of TrpRS to the nucleus. Ap_3A is synthesized by TrpRS and binds to Fhit, forming an active Fhit–Ap_3A complex. The active complex interacts with mediators of several signal transduction pathways, including β catenin and PI3K-Akt pathways, resulting in the regulation of specific intracellular targets. Ap_3A can then be degraded by Fhit, terminating the signal.

3 Summary

tRNA synthetases are an ancient class of proteins, perhaps the first to develop sites for binding specific amino acids. Thus they were in an ideal position to develop new functions needed, at least in part, for the development of the complexity of organisms, whereby tRNA synthetases became connected to signaling pathways mediating angiogenesis [65], immune response [66], inflammation [67], and apoptosis [68].

These expanded functions often involve direct interaction partners. For example, human tyrosyl-tRNA synthetase interacts with chemokine receptor CXCR1 to induce cell migration [65] and human glutaminyl-tRNA synthetase interacts with ASK1 to regulate apoptosis. However, functional expansion can also be achieved indirectly via reaction products of aaRSs. Namely, the majority of aaRSs have the capacity to catalyze a side reaction to form Ap_nA in the absence of cognate tRNA. These reactions of aaRS are the best known sources of Ap_nA in vivo.

In this chapter we have presented the concept that Ap_nA family members produced by aaRSs form a new class of second messenger molecules used by eukaryotic cells to regulate many intracellular processes. This class of molecules deserves further detailed research. We would not be surprised if more physiological functions will be attributed to the Ap_nA family.

References

1. Rall TW, Sutherland EW (1958) Formation of a cyclic adenine ribonucleotide by tissue particles. J Biol Chem 232:1065–1076
2. Springett GM, Kawasaki H, Spriggs DR (2004) Non-kinase second-messenger signaling: new pathways with new promise. Bioessays 26:730–738

Amino-Acyl tRNA Synthetases Generate Dinucleotide Polyphosphates as Second... 203

3. Bornfeldt KE (2006) A single second messenger: several possible cellular responses depending on distinct subcellular pools. Circ Res 99:790–792
4. Sutherland EW (1972) Studies on the mechanism of hormone action. Science 177:401–408
5. Bolander FF (2004) Molecular endocrinology, 3rd edn. Elsevier Academic, Amsterdam
6. Carmi-Levy I, Yannay-Cohen N, Kay G, Razin E, Nechushtan H (2008) Diadenosine tetraphosphate hydrolase is part of the transcriptional regulation network in immunologically activated mast cells. Mol Cell Biol 28:5777–5784
7. de Rooij J, Zwartkruis FJ, Verheijen MH, Cool RH, Nijman SM, Wittinghofer A, Bos JL (1998) Epac is a Rap1 guanine-nucleotide-exchange factor directly activated by cyclic AMP. Nature 396:474–477
8. Zamecnik PC, Stephenson ML, Janeway CM, Randerath K (1966) Enzymatic synthesis of diadenosine tetraphosphate and diadenosine triphosphate with a purified lysyl-sRNA synthetase. Biochem Biophys Res Commun 24:91–97
9. Lee PC, Bochner BR, Ames BN (1983) AppppA, heat-shock stress, and cell oxidation. Proc Natl Acad Sci U S A 80:7496–7500
10. Lee PC, Bochner BR, Ames BN (1983) Diadenosine 5′,5‴-P1,P4-tetraphosphate and related adenylylated nucleotides in Salmonella typhimurium. J Biol Chem 258:6827–6834
11. Varshavsky A (1983) Diadenosine 5′,5‴-P1,P4-tetraphosphate: a pleiotropically acting alarmone? Cell 34:711–712
12. Huang R, Li M, Gregory RL (2011) Bacterial interactions in dental biofilm. Virulence 2:435–444
13. Baker JC, Jacobson MK (1986) Alteration of adenyl dinucleotide metabolism by environmental stress. Proc Natl Acad Sci U S A 83:2350–2352
14. Johnstone DB, Farr SB (1991) AppppA binds to several proteins in Escherichia coli, including the heat shock and oxidative stress proteins DnaK, GroEL, E89, C45 and C40. EMBO J 10:3897–3904
15. Kisselev LL, Justesen J, Wolfson AD, Frolova LY (1998) Diadenosine oligophosphates (Ap(n)A), a novel class of signalling molecules? FEBS Lett 427:157–163
16. Fraga H, Fontes R (2011) Enzymatic synthesis of mono and dinucleoside polyphosphates. Biochim Biophys Acta 1810:1195–1204
17. Brevet A, Chen J, Leveque F, Plateau P, Blanquet S (1989) In vivo synthesis of adenylylated bis(5′-nucleosidyl) tetraphosphates (Ap4N) by Escherichia coli aminoacyl-tRNA synthetases. Proc Natl Acad Sci U S A 86:8275–8279
18. Plateau P, Blanquet S (1982) Zinc-dependent synthesis of various dinucleoside 5′,5‴-P1, P3-tri- or 5″,5‴-P1,P4-tetraphosphates by Escherichia coli lysyl-tRNA synthetase. Biochemistry 21:5273–5279
19. Monds RD, Newell PD, Wagner JC, Schwartzman JA, Lu W, Rabinowitz JD, O'Toole GA (2010) Di-adenosine tetraphosphate (Ap4A) metabolism impacts biofilm formation by Pseudomonas fluorescens via modulation of c-di-GMP-dependent pathways. J Bacteriol 192: 3011–3023
20. Guo RT, Chong YE, Guo M, Yang XL (2009) Crystal structures and biochemical analyses suggest a unique mechanism and role for human glycyl-tRNA synthetase in Ap4A homeostasis. J Biol Chem 284:28968–28976
21. Merkulova T, Kovaleva G, Kisselev L (1994) P1,P3-bis(5′-adenosyl)triphosphate (Ap3A) as a substrate and a product of mammalian tryptophanyl-tRNA synthetase. FEBS Lett 350: 287–290
22. Vartanian A, Narovlyansky A, Amchenkova A, Turpaev K, Kisselev L (1996) Interferons induce accumulation of diadenosine triphosphate (Ap3A) in human cultured cells. FEBS Lett 381:32–34
23. Guranowski A, Blanquet S (1985) Phosphorolytic cleavage of diadenosine 5′,5‴-P1, P4-tetraphosphate. Properties of homogeneous diadenosine 5′,5‴-P1,P4-tetraphosphate alpha, beta-phosphorylase from Saccharomyces cerevisiae. J Biol Chem 260:3542–3547

24. Plateau P, Fromant M, Schmitter JM, Blanquet S (1990) Catabolism of bis(5′-nucleosidyl) tetraphosphates in Saccharomyces cerevisiae. J Bacteriol 172:6892–6899
25. McLennan AG, Mayers E, Adams DG (1996) Anabaena flos-aquae and other cyanobacteria possess diadenosine 5′,5‴-P1,P4-tetraphosphate (Ap4A) phosphorylase activity. Biochem J 320(Pt 3):795–800
26. Guranowski A (2000) Specific and nonspecific enzymes involved in the catabolism of mononucleoside and dinucleoside polyphosphates. Pharmacol Ther 87:117–139
27. Barnes LD, Garrison PN, Siprashvili Z, Guranowski A, Robinson AK, Ingram SW, Croce CM, Ohta M, Huebner K (1996) Fhit, a putative tumor suppressor in humans, is a dinucleoside 5′,5‴-P1,P3-triphosphate hydrolase. Biochemistry 35:11529–11535
28. Thorne NM, Hankin S, Wilkinson MC, Nunez C, Barraclough R, McLennan AG (1995) Human diadenosine 5′,5‴-P1,P4-tetraphosphate pyrophosphohydrolase is a member of the MutT family of nucleotide pyrophosphatases. Biochem J 311(Pt 3):717–721
29. Safrany ST, Ingram SW, Cartwright JL, Falck JR, McLennan AG, Barnes LD, Shears SB (1999) The diadenosine hexaphosphate hydrolases from Schizosaccharomyces pombe and Saccharomyces cerevisiae are homologues of the human diphosphoinositol polyphosphate phosphohydrolase. Overlapping substrate specificities in a MutT-type protein. J Biol Chem 274:21735–21740
30. Bessman MJ, Frick DN, O'Handley SF (1996) The MutT proteins or "Nudix" hydrolases, a family of versatile, widely distributed, "housecleaning" enzymes. J Biol Chem 271:25059–25062
31. Pintor J, King BF, Ziganshin AU, Miras-Portugal MT, Burnstock G (1996) Diadenosine polyphosphate-activated inward and outward currents in follicular oocytes of Xenopus laevis. Life Sci 59:PL179–184
32. Hilderman RH, Martin M, Zimmerman JK, Pivorun EB (1991) Identification of a unique membrane receptor for adenosine 5′,5‴-P1,P4-tetraphosphate. J Biol Chem 266:6915–6918
33. Trapasso F, Krakowiak A, Cesari R, Arkles J, Yendamuri S, Ishii H, Vecchione A, Kuroki T, Bieganowski P, Pace HC, Huebner K, Croce CM, Brenner C (2003) Designed FHIT alleles establish that Fhit-induced apoptosis in cancer cells is limited by substrate binding. Proc Natl Acad Sci U S A 100:1592–1597
34. Feussner K, Guranowski A, Kostka S, Wasternack C (1996) Diadenosine 5′,5‴-P1, P4-tetraphosphate (Ap4A) hydrolase from tomato (Lycopersicon esculentum cv. Lukullus) – purification, biochemical properties and behaviour during stress. Z Naturforsch C 51:477–486
35. Zourgui L, Baltz D, Baltz T, Oukerro F, Tarrago-Litvak L (1988) Purification, immunological and biochemical characterization of Ap4A binding protein from Xenopus laevis oocytes. Nucleic Acids Res 16:2913–2929
36. Lee YN, Nechushtan H, Figov N, Razin E (2004) The function of lysyl-tRNA synthetase and Ap4A as signaling regulators of MITF activity in FcεRI-activated mast cells. Immunity 20:145–151
37. Jovanovic A, Zhang S, Alekseev AE, Terzic A (1996) Diadenosine polyphosphate-induced inhibition of cardiac KATP channels: operative state-dependent regulation by a nucleoside diphosphate. Pflugers Arch 431:800–802
38. Klishin A, Lozovaya N, Pintor J, Miras-Portugal MT, Krishtal O (1994) Possible functional role of diadenosine polyphosphates: negative feedback for excitation in hippocampus. Neuroscience 58:235–236
39. Green AK, Cobbold PH, Dixon CJ (1995) Cytosolic free Ca^{2+} oscillations induced by diadenosine 5′,5‴-P1,P3-triphosphate and diadenosine 5′,5‴-P1,P4-tetraphosphate in single rat hepatocytes are indistinguishable from those induced by ADP and ATP respectively. Biochem J 310(Pt 2):629–635
40. Gasmi L, McLennan AG, Edwards SW (1996) Neutrophil apoptosis is delayed by the diadenosine polyphosphates, Ap5A and Ap6A: synergism with granulocyte-macrophage colony-stimulating factor. Br J Haematol 95:637–639
41. Ripoll C, Martin F, Manuel Rovira J, Pintor J, Miras-Portugal MT, Soria B (1996) Diadenosine polyphosphates. A novel class of glucose-induced intracellular messengers in the pancreatic beta-cell. Diabetes 45:1431–1434

42. Weinmann-Dorsch C, Hedl A, Grummt I, Albert W, Ferdinand FJ, Friis RR, Pierron G, Moll W, Grummt F (1984) Drastic rise of intracellular adenosine(5′)tetraphospho(5′)adenosine correlates with onset of DNA synthesis in eukaryotic cells. Eur J Biochem 138:179–185

43. Perret J, Hepburn A, Cochaux P, Van Sande J, Dumont JE (1990) Diadenosine 5′,5‴-P1, P4-tetraphosphate (AP4A) levels under various proliferative and cytotoxic conditions in several mammalian cell types. Cell Signal 2:57–65

44. Oka K, Suzuki T, Onodera Y, Miki Y, Takagi K, Nagasaki S, Akahira J, Ishida T, Watanabe M, Hirakawa H, Ohuchi N, Sasano H (2011) Nudix-type motif 2 in human breast carcinoma: a potent prognostic factor associated with cell proliferation. Int J Cancer 128: 1770–1782

45. Rapaport E, Zamecnik PC, Baril EF (1981) HeLa cell DNA polymerase alpha is tightly associated with tryptophanyl-tRNA synthetase and diadenosine 5′,5‴-P1,P4-tetraphosphate binding activities. Proc Natl Acad Sci U S A 78:838–842

46. Grummt F (1978) Diadenosine 5′,5‴-P1,P4-tetraphosphate triggers initiation of in vitro DNA replication in baby hamster kidney cells. Proc Natl Acad Sci U S A 75:371–375

47. Vartanian A, Prudovsky I, Suzuki H, Dal Pra I, Kisselev L (1997) Opposite effects of cell differentiation and apoptosis on Ap3A/Ap4A ratio in human cell cultures. FEBS Lett 415: 160–162

48. Vartanian A, Alexandrov I, Prudowski I, McLennan A, Kisselev L (1999) Ap4A induces apoptosis in human cultured cells. FEBS Lett 456:175–180

49. Carmi-Levy I, Motzik A, Ofir-Birin Y, Yagil Z, Yang CM, Kemeny DM, Han JM, Kim S, Kay G, Nechushtan H, Suzuki R, Rivera J, Razin E (2011) Importin beta plays an essential role in the regulation of the LysRS-Ap(4)A pathway in immunologically activated mast cells. Mol Cell Biol 31:2111–2121

50. Lee YN, Razin E (2005) Nonconventional involvement of LysRS in the molecular mechanism of USF2 transcriptional activity in FcepsilonRI-activated mast cells. Mol Cell Biol 25: 8904–8912

51. Yannay-Cohen N, Carmi-Levy I, Kay G, Yang CM, Han JM, Kemeny DM, Kim S, Nechushtan H, Razin E (2009) LysRS serves as a key signaling molecule in the immune response by regulating gene expression. Mol Cell 34:603–611

52. Lima CD, Klein MG, Hendrickson WA (1997) Structure-based analysis of catalysis and substrate definition in the HIT protein family. Science 278:286–290

53. Brenner C, Garrison P, Gilmour J, Peisach D, Ringe D, Petsko GA, Lowenstein JM (1997) Crystal structures of HINT demonstrate that histidine triad proteins are GalT-related nucleotide-binding proteins. Nat Struct Biol 4:231–238

54. Brenner C (2002) Hint, Fhit, and GalT: function, structure, evolution, and mechanism of three branches of the histidine triad superfamily of nucleotide hydrolases and transferases. Biochemistry 41:9003–9014

55. Pichiorri F, Palumbo T, Suh SS, Okamura H, Trapasso F, Ishii H, Huebner K, Croce CM (2008) Fhit tumor suppressor: guardian of the preneoplastic genome. Future Oncol 4:815–824

56. Siprashvili Z, Sozzi G, Barnes LD, McCue P, Robinson AK, Eryomin V, Sard L, Tagliabue E, Greco A, Fusetti L, Schwartz G, Pierotti MA, Croce CM, Huebner K (1997) Replacement of Fhit in cancer cells suppresses tumorigenicity. Proc Natl Acad Sci U S A 94:13771–13776

57. Pace HC, Garrison PN, Robinson AK, Barnes LD, Draganescu A, Rosler A, Blackburn GM, Siprashvili Z, Croce CM, Huebner K, Brenner C (1998) Genetic, biochemical, and crystallographic characterization of Fhit-substrate complexes as the active signaling form of Fhit. Proc Natl Acad Sci U S A 95:5484–5489

58. Wali A (2010) FHIT: doubts are clear now. Sci World J 10:1142–1151

59. Weiske J, Albring KF, Huber O (2007) The tumor suppressor Fhit acts as a repressor of beta-catenin transcriptional activity. Proc Natl Acad Sci U S A 104:20344–20349

60. Jayachandran G, Sazaki J, Nishizaki M, Xu K, Girard L, Minna JD, Roth JA, Ji L (2007) Fragile histidine triad-mediated tumor suppression of lung cancer by targeting multiple components of the Ras/Rho GTPase molecular switch. Cancer Res 67:10379–10388

61. Semba S, Trapasso F, Fabbri M, McCorkell KA, Volinia S, Druck T, Iliopoulos D, Pekarsky Y, Ishii H, Garrison PN, Barnes LD, Croce CM, Huebner K (2006) Fhit modulation of the Akt-survivin pathway in lung cancer cells: Fhit-tyrosine 114 (Y114) is essential. Oncogene 25:2860–2872
62. Bange FC, Flohr T, Buwitt U, Bottger EC (1992) An interferon-induced protein with release factor activity is a tryptophanyl-tRNA synthetase. FEBS Lett 300:162–166
63. Rubin BY, Anderson SL, Xing L, Powell RJ, Tate WP (1991) Interferon induces tryptophanyl-tRNA synthetase expression in human fibroblasts. J Biol Chem 266:24245–24248
64. Sajish M, Zhou Q, Kishi S, Valdez DM Jr, Kapoor M, Guo M, Lee S, Kim S, Yang XL, Schimmel P (2012) Trp-tRNA synthetase bridges DNA-PKcs to PARP-1 to link IFN-gamma and p53 signaling. Nat Chem Biol 8:547–554
65. Wakasugi K, Schimmel P (1999) Two distinct cytokines released from a human aminoacyl-tRNA synthetase. Science 284:147–151
66. Amsterdam A, Nissen RM, Sun Z, Swindell EC, Farrington S, Hopkins N (2004) Identification of 315 genes essential for early zebrafish development. Proc Natl Acad Sci USA 101: 12792–12797
67. Mukhopadhyay R, Jia J, Arif A, Ray PS, Fox PL (2009) The GAIT system: a gatekeeper of inflammatory gene expression. Trends Biochem Sci 34:324–331
68. Ko YG, Kim EY, Kim T, Park H, Park HS, Choi EJ, Kim S (2001) Glutamine-dependent antiapoptotic interaction of human glutaminyl-tRNA synthetase with apoptosis signal-regulating kinase 1. J Biol Chem 276:6030–6036

Top Curr Chem (2014) 344: 207–246
DOI: 10.1007/128_2013_455
© Springer-Verlag Berlin Heidelberg 2013
Published online: 2 July 2013

Association of Aminoacyl-tRNA Synthetases with Cancer

Doyeun Kim, Nam Hoon Kwon, and Sunghoon Kim

Abstract Although aminoacyl-tRNA synthetases (ARSs) and ARS-interacting multi-functional proteins (AIMPs) have long been recognized as housekeeping proteins, evidence indicating that they play a key role in regulating cancer is now accumulating. In this chapter we will review the conventional and non-conventional functions of ARSs and AIMPs with respect to carcinogenesis. First, we will address how ARSs and AIMPs are altered in terms of expression, mutation, splicing, and post-translational modifications. Second, the molecular mechanisms for ARSs' and AIMPs' involvement in the initiation, maintenance, and progress of carcinogenesis will be covered. Finally, we will introduce the development of therapeutic approaches that target ARSs and AIMPs with the goal of treating cancer.

Keywords Aminoacyl tRNA synthetase · ARS-interacting multi-functional protein · Cancer · Therapeutics

Contents

1	Introduction	208
2	Alterations of ARSs and AIMPs in Cancer	209
	2.1 Changes in Expression Level	209
	2.2 Mutations and Splicing	214
	2.3 Post-Translational Modifications	221
3	Molecular Mechanisms of ARSs and AIMPs in Cancer	223
	3.1 Aminoacyl-tRNA Synthetases	223
	3.2 ARS-Interacting Multi-functional Proteins	230

D. Kim, N.H. Kwon (✉), and S. Kim
Medicinal Bioconvergence Research Center, Graduate School of Convergence Science and Technology, College of Pharmacy, Seoul National University, Seoul 151-742, Korea
e-mail: doyeun.kim@snu.ac.kr; nanarom1@snu.ac.kr; sungkim@snu.ac.kr

4	Therapeutics	233
	4.1 Aminoacyl-tRNA Synthetases	233
	4.2 ARS-Interacting Multi-functional Proteins	234
5	Future Perspectives	235
References		236

1 Introduction

To form cancer cells, parental cells must overcome multiple barriers of innate responses [1, 2]. During the initial stages of carcinogenesis, differentiated dormant cells regain sustained proliferative signals, evade growth suppressors, and overcome replicative mortality. They deregulate cellular energetics, and induce angiogenesis in order to have a plentiful supply of energy. They acquire invasion and metastatic ability to dominate the whole organism. Genome instability or inflammation often contributes to the whole process of tumor progression.

ARSs (aminoacyl-tRNA synthetases) are fundamental enzymes that charge amino acids to their cognate tRNA, thereby providing the building blocks for translation. As part of translational machinery, ARSs are obviously involved in sustaining growth, although whether their role is "driver" or "passenger" is not resolved. Interestingly, ARSs in higher eukaryotes harbor certain additional domains or scaffolding proteins for complex formation that are not conserved in lower complex organisms, suggesting extra-translational functions have evolved with increasing organismal complexity [3]. Accordingly, non-conventional functions of ARSs and AIMPs (ARS-Interacting multi-functional proteins; AIMP1/p43, AIMP2/p38, and AIMP3/p18) in amino acid metabolism, inflammation, tumorigenesis, and angiogenesis in higher eukaryotes have been revealed during the last decades [4]. Furthermore, misregulation or altered expression of ARSs and AIMPs (collectively termed ARSN herein) are often linked to human diseases related to neurodegeneration, immune modulation, and cancer [4–7].

In tumor formation, ARSN plays roles in triggering or suppressing the hallmarks of cancer: Leucyl-tRNA synthetase (LRS) and methionyl-tRNA synthetase (MRS) can sustain proliferative signaling [8, 9]. LRS, MRS, arginyl-tRNA synthetase (RRS), aspartyl-tRNA synthetase (DRS), mitochondrial DRS, and mitochondrial asparaginyl-tRNA synthetase (NRS) can deregulate cellular energetics [8–11]. Secreted ARSN, glycyl-tRNA synthetase (GRS), lysyl-tRNA synthetase (KRS), threonyl-tRNA synthetase (TRS), tryptophanyl-tRNA synthetase (WRS), tyrosyl-tRNA synthetase (YRS), and AIMP1 can impact tumor-promoting inflammation, angiogenesis, and metastasis [12–15]. WRS, AIMP2, and AIMP3 can activate growth suppressors and induce cell death, revealing anti-tumorigenic functions [16–20]. AIMP2-DX2, the exon 2-deleted isoform of AIMP2, enables cancer cells to evade the suppressive effects of AIMP2, thereby sustaining cell proliferating signaling [21, 22]. Glutaminyl-tRNA synthetase (QRS) also sustains proliferative signaling in the presence of glutamine [23]. KRS promotes cancer cell migration [24] and glutamyl-proryl-tRNA synthetase (EPRS), seryl-tRNA synthetase (SRS), TRS, WRS, and YRS are considered as angiogenesis-regulating factors [5, 14, 15, 25–27].

2 Alterations of ARSs and AIMPs in Cancer

Cancer is a genetic disease, with accumulation of mutations in the genome underlying the development of malignant tumors. When the accumulation of mutations exceeds what the cell can regulate by homeostasis, certain molecular pathways malfunction. Consequently, what begins as focal functional defects leads to acquisition of cancer cell characteristics through molecular interactions and crosstalk of signaling cascades. In this section we discuss cases where ARSN is involved in different types of cancer. In some forms of cancer, ARSN itself is mutated, while in others regulation of ARSN expression levels or its function seems to play a role.

2.1 Changes in Expression Level

Increasingly, cancer tissues or cells are studied for genome-wide genetic changes in an unbiased manner using bioinformatics approaches. Such an unbiased assessment of the ARS's role in cancer was performed in silico [6]. Differential expression and copy number variations of 23 ARSN (20 ARSs and 3 AIMPs) were analyzed in ten different cancer groups whose data were strictly selected from the Gene Expression Omnibus database (GEO) [28] and ArrayExpress database [29]. In this approach, 3,501 cancer associated genes (CAGs) which showed strong cancer association and clinical indications were used to identify possible interaction partners of ARSN based on curated protein–protein interactions collected from public databases. Finally, 123 first neighbor CAGs which can directly associate with ARSN and 1,295 second neighbor CAGs that may bind the first neighbors were chosen and analyzed together with ARSN. As shown in Fig. 1a, the expression profiles of ARSN are similar to those of first and second neighbor CAGs, whereas the patterns are clearly distinguishable from those of non-CAGs that are not included in the US National Cancer Institute's cancer gene index. Copy number variations of ARSN, first and second neighbor CAGs and non-CAGs were also analyzed in nine cancer groups using the data obtained from CanGEM database [31]. Positive correlation in the copy number variations was also observed between ARSN and CAGs while little variation was shown in non-CAGs (Fig. 1b).

In addition to bioinformatic approaches such as these, experimental evidence supporting the quantifiable changes in ARS during carcinogenesis is being accumulated for some cancers and will be discussed below (Fig. 2).

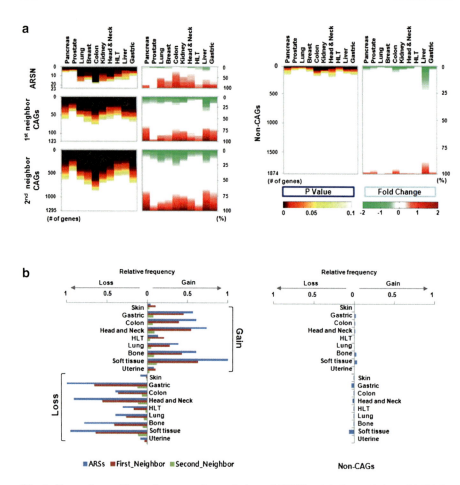

Fig. 1 Expression profiles and copy number variations of ARSN and their correlation with CAGs. (**a**) The gene expression profiles of ARSN were compared with 123 first neighbor CAGs and 1,295 second neighbor CAGs. CAGs are selected by curated protein–protein interaction analysis based on 11 public databases. Non-CAGs consist of 1,874 genes which are not included in the US National Cancer Institute's cancer gene index and have no significant cancer-associated expression profiles in more than 22 cancer data sets ($P > 0.5$). The gene expression data obtained from GEO and ArrayExpress contains 40 data sets which have more than 10 normal and cancer samples. Each cancer group has at least two data sets from independent studies. Differentially expressed genes between normal and cancer were identified using integrative statistical hypothesis testing ($P < 0.05$) for each data set in the each cancer group. For multiple data sets in individual cancer groups, Stouffer's methods [30] were used for the summarization of the P values for each gene revealing the combined P values of all the genes in the ten cancer groups. (**b**) The comparison of copy number variations between ARSN, CAGs, and non-CAGs. Data for the copy number variations in the nine cancer groups were obtained from the CanGEM database. The relative frequency of copy number variations is presented for ARSN, first neighbor CAGs, second neighbor CAGs, and non-CAGs, respectively. *HLT* hematopoietic and lymphoid tissue. These figures were originally published in Nature Reviews Cancer [6]

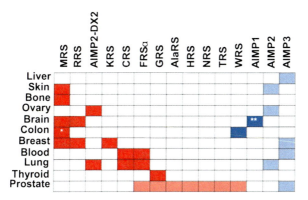

Fig. 2 Expressional variation of ARSN observed in cancer. Upregulation or downregulation of ARSN based on various cancer tissues analysis are shown in *red* or *blue color*, respectively. Results obtained from cancer cell lines or mice models were represented as *pale red* or *pale blue*. *Catalytic activity of MRS was increased; **Secretion of AIMP1 was decreased. See the text for more detail

2.1.1 Expression of ARSs in Cancer

LRS. Anabolic energy metabolism is a hallmark of fast growing cancer cells, and the mTOR (mammalian target of rapamycin) promotes anabolic cellular processes in response to growth factors and metabolic cues. Among signals for cell proliferation, the mTORC1 (mTOR complex 1) signaling pathway is implicated in the process of cancer development. mTORC1-regulating drugs, including rapamycin analogues, are undergoing clinical trials [32]. LRS was recently reported as a regulator of mTORC1 signaling via sensing its substrate leucine [8, 33]. Whether LRS-mTORC1 pathway is amplified or mutated for constitutive activity is as yet unknown. Interestingly, LRS was upregulated in lung cancer cells in comparison with normal cells. In addition, downregulation of LRS reduced migration and colony formation ability of cancer cells [34]. Upregulated expression of LRS in acute myeloid leukemia was also reported [35].

WRS. WRS was identified as a potential prognostic marker in colorectal cancer by the human protein atlas [36]. Upon further examination in a cohort of patients ($n = 320$) it was found that low expression of WRS correlated with more advanced tumor stage and an increased risk of lymph node metastasis and recurrence in colorectal cancer [37]. WRS is a well-known angiogenesis suppressor when its N-terminus is truncated by proteolysis or splicing [38, 39]. Moreover, nuclear WRS has been identified as a mediator in IFN-γ (interferon-γ)-dependent antiproliferative responses [40]. It is not surprising, therefore, that expression level of WRS is inversely proportional to tumor progression.

In a proteomic study using prostate cancer cells challenged with androgen, WRS and several ARSs were found to be significantly upregulated in cancer cells but not in untreated control cells [41]. In accordance with the androgen study, another proteomic approach also revealed that WRS was significantly upregulated when

HepG2 cells or rat primary hepatocyte were infected with hepatitis B virus [42]. Viral infection has also been found, through viral oncogenes, to perturb proliferative signals, leading to cervical and liver cancers. The results from these proteomic approaches suggest that WRS is transiently induced in response to various stimuli, although the distinct roles in these contexts require further study.

ARSs in Prostate Cancer. An increased response to hormones is one way for breast or prostate cancer cells to acquire an uncontrolled proliferative phenotype. ARSs have been reported to respond to androgen and, moreover, ARSs have also been reported to be overexpressed in the prostate cancer cell line LNCaP, as well as prostate cancer tissues [41]. At the protein level, semi-quantitative MudPIT ESI-ion trap MS/MS and quantitative iTRAQ MALDI-TOF MS/MS platforms showed that, in androgen-treated LNCaP cells, several ARSs were highly expressed, including Alanyl-tRNA synthetase (ARS), α subunit of phenylalanyl-tRNA synthetase (FRS), GRS, NRS, TRS, histidyl-tRNA synthetase (HRS), and WRS. The increased transcription and translation of these ARSs upon androgen stimuli were partially verified by microarray and Oncomine database analyses. Interestingly, the promoter regions of GRS and KRS were also enriched in sites mediating responses to androgen receptor binding in an androgen dependent manner. The exact function and role of overexpressed ARSs in prostate cancer are unclear although a pivotal role in protein biosynthesis has been suggested in these studies.

MicroRNA and ARSs. Recently, many researchers have focused on the connection between microRNAs (*miR*) and cancers [43]. Examples are *miR-15* and *miR-16*, about 68% of which are deleted or downregulated in B-cell chronic lymphocytic leukemia [44]. The reduced expression of these *miRs* was also observed in pituitary adenomas, showing an inverse correlation with arginyl-tRNA synthetase (RRS) expression as well as tumor size [45]. RRS mRNA has antisense complementary regions with 85% sequence homology to *miR-16*, suggesting RRS as a putative target of *miR-16* [44]. Moreover, reduced secretion of AIMP1, the binding partner of RRS in the multi-tRNA synthetase complex (MSC), was observed in the pituitary adenoma tissues with low expression of *miR-15* and *miR-16*. The relationship between *miR-15*, *miR-16*, RRS, and AIMP1 remains to be elucidated, but this finding suggests that expression level of RRS and the inverse secretion of AIMP1 may be related to the tumor growth.

Metabolism and ARSs. Growth signaling mediated by IGF (insulin-like growth factor) is implicated in the normal and cancerous growth of breast epithelial cells. Expression profiling of breast cancer cells overexpressing IGF1/2 revealed that half of genes with increased expression were related to amino acid metabolism, including RRS, MRS, and other amino acid transport proteins [10]. Expression of these genes may contribute to the increased rate of protein synthesis observed in proliferative cancer cells.

Other ARSs. Overexpression of KRS was reported in breast cancer, metastatic cell lines, and glioma-derived stem cells [13, 35, 46]. Additionally, the α-subunit of FRS is overexpressed in lung solid tumors and acute phase chronic myeloid leukemia [47, 48], and GRS is up-regulated in papillary thyroid carcinoma [49]. The increased aminoacylation activity of MRS has been reported in human colon cancer [50], and overexpression of MRS was also reported in several types of cancers, such as malignant fibrous histiocytomas, lipoma, osteosarcomas, malignant gliomas, and glioblastomas [51–54].

2.1.2 Expression of AIMPs in Cancer

AIMPs. AIMP1, AIMP2, and AIMP3 are scaffold proteins for the MSC. They are, however, functionally versatile and play important roles in other locations as well as in the MSC [5]. Analysis of hundreds of gastric and colorectal cancers showed a significant decrease of all three proteins, suggesting a tumor suppressive role for the AIMPs [6]. In the same context, low expression of AIMP1 was observed in pituitary adenomas [45].

AIMP2 and Its Splicing Variant AIMP2-DX2. AIMP2 is located in the MSC under normal conditions. It is released from the MSC under certain conditions, such as TGF-β (transforming growth factor-β) and TNF-α (tumor necrosis factor-α) signaling and DNA damage stress to induce growth arrest and apoptosis [18–20, 22, 55]. AIMP2 heterozygous mice become highly susceptible to various types of tumors including skin cancer carcinogenesis [20]. AIMP2 homozygous mice show neonatal lethality due to lung dysfunction which is caused by overproliferation of lung epithelial cells [18].

AIMP2-DX2 is a splicing variant of AIMP2 but with exon 2 deleted in the transcript. Increased expression of AIMP2-DX2 was observed in tumorigenic lung cancer cell lines as well as in lung cancer patients' samples. AIMP2-DX2 hampers the tumor suppressive effect of AIMP2 by competing with AIMP2 for binding to target proteins [21, 22]. This property means that AIMP2-DX2 works like a dominant negative mutant against the pro-apoptotic interaction of AIMP2.

As mentioned above, AIMP2 acts as a potent tumor suppressor in response to different stimuli, whereas its splicing variant AIMP2-DX2 is an antagonist of AIMP2. Therefore it is expected that an abnormal ratio of expression of AIMP2-DX2 to AIMP2 might be observed in cancers. In fact, an increased expression ratio of AIMP2-DX2 to AIMP2 was reported in lung cancer patients and drug-resistant epithelial ovarian cancer patients [21, 22].

AIMP3. AIMP3, which shows strong interaction to MRS in the MSC, is released upon DNA damage and oncogenic stress and translocated into nucleus [16, 17]. AIMP3 associates with ATM/ATR (ataxia telangiectasia mutated/ATM and Rad3-related) and induces DNA damage response by activating p53. AIMP3 is a well-known tumor suppressor, and it was reported that mice with a single allelic deletion of AIMP3 suffered from breast adenocarcinoma, a sarcoma of unknown origin, adenocarcinoma in seminal vesicles, hepatocarcinoma, and lymphoma [16]. Single allelic loss of AIMP3 could not activate p53 in response to oncogenic stress such as Ras or Myc and caused abnormalities in chromosomal structure [17]. Cell senescence is recognized as a protective barrier to neoplastic expansion [2], and AIMP3-overexpressing mice develop progeroid [56]. These data suggest that AIMP3 is a controller of cell fate at the demarcation between tumorigenesis and senescence, and this relationship should be further studied.

2.2 Mutations and Splicing

In addition to its changes in translation, mutations of ARSN or alterations in the mRNA transcript levels were reported in various cancers. Mutations ranged from point mutations to chromosomal deletions. How these mutations affect the function of ARSN remains to be elucidated (Table 1).

2.2.1 Small Scale Mutations

Point Mutation. There are several reports of ARS mutations in cancer tissues. Several point mutations in AIMP3 were found in chronic myeloid leukemia patients [57], and these mutations affected the association between AIMP3 and ATM, resulting in modulation of p53 activation, supporting the role of AIMP3 in the initiation or progression of human leukemias.

Mutations in the promoter region (at position -318 to -291) of mitochondrial isoleucyl-tRNA synthetase (IRS) were observed in 59% of nonpolyposis colorectal cancer and Turcot syndrome [58]. It is not clear whether these mutations affect the expression level of mitochondrial IRS, but still suggests its relationship to the progression of nonpolyposis colorectal cancer and Turcot syndrome.

L-Asparaginase is an anticancer drug used in treating several cancers including acute lymphoblastic leukemia (ALL) [67]. Unlike normal cells, in ALL, leukemic cells are unable to synthesize the non-essential amino acid asparagine; thus they depend on circulating asparagine. Asparaginase catalyzes L-asparagine producing aspartic acid and ammonia. This deprives leukemic cells of circulating asparagine, which leads to cell death. Polymorphism of DRS, mitochondrial DRS and mitochondrial NRS were observed in ALL patients and B-lymphoblastoid cell lines, and among them a few single nucleotide polymorphisms in mitochondrial NRS contributed significantly to asparaginase sensitivity [11]. It was supposed that intracellular amino acid metabolism and protein synthesis related to aspartate and asparagine pathways might be altered by these polymorphisms, rescuing cells from the effects of asparaginase.

Recently a somatic frameshift mutation in MRS exon 3 was reported in colorectal cancer and gastric cancer [59]. This mutation in the T9 repeat sequence results in a premature stop of translation with loss of the entire catalytic domain of MRS. The second allele remains intact, and therefore the catalytic function of MRS in translation should not be affected by the mutant allele. This mutant may affect the non-canonical function of MRS and further study could reveal the importance of its function in cancer.

Association of Aminoacyl-tRNA Synthetases with Cancer

Table 1 Mutations of ARSN found in cancers

Gene	Mutations		Cancer	References
AIMP3	Point mutation	T35A, S40A, E76A, T80P, S87A, R144A, V106A	Chronic myeloid leukemia	Kim et al. [57]
IRS2		A and/or TA deletion at (A)10(TA)9 repeat in promoter region	Nonpolyposis colorectal cancer and Turcot syndrome	Miyaki et al. [58]
DRS		G/T (SNP ID rs3768998), T/C (SNP ID rs7587285), T/C (SNP ID rs11893318), T/C (SNP ID rs2322725), A/C (SNP ID rs2278683)	Acute lymphoblastic leukemia and B-lymphoblastoid cell lines	Chen et al. [11]
DRS2		A/G (SNP ID rs2068871), A/G (SNP ID rs16846526), C/T (SNP ID rs2759328), C/T (SNP ID rs941988), C/T (SNP ID rs2227589)	Acute lymphoblastic leukemia and B-lymphoblastoid cell lines	Chen et al. [11]
NRS2		G/A (SNP ID rs11237537)	Acute lymphoblastic leukemia and B-lymphoblastoid cell lines	Chen et al. [11]
MRS	Frame shift mutation	c.212delT resulting in production of 102 amino-acid-long-peptide	Colorectal cancer and gastric cancer	Park et al. [59]
EPRS	PAY*	Introduction of UAA stop codon at Y864	Monocytes	Yao et al. [60]
CRS and *ALK*	Chromosome rearrangement	t(2;11;2)(p23;p15; q31)	Inflammatory myofibroblastic tumor	Debelenko et al. [61]
MRS	Chromosome amplification	12q13-q15 locus	Malignant fibrous histiocytomas, sarcomas, malignant gliomas and glioblastomas	Forus et al. [51], Nilbert et al. [52], Palmer et al. [53], Reifenberger et al. [54]

(continued)

216   D. Kim et al.

Table 1 (continued)

Gene	Mutations		Cancer	References
CRS	Deletion	11p15.5-p15.4 locus	Wilms tumor and embryonal rhab-domyosarcoma, adrenocortical carcinoma, and lung, ovarian, and breast cancers	Hu et al. [62], Reid et al. [63], Xu et al. [64], Zhao and Bepler [65]
LRS2		3p21.3 locus	Nasopharyngeal carcinoma	Zhou et al. [66]

IRS2 mitochondrial IRS, *DRS2* mitochondrial DRS, *NRS2* mitochondrial NRS, *LRS2* mitochondrial LRS, *PAY** polyadenylation-directed conversion of a Tyr codon in the EPRS-coding sequence to a stop codon

2.2.2 Large Scale Mutations

Chromosomal Rearrangement. Chromosomal translocation is a well-known hallmark of cancer, where abnormal mutations are introduced into dormant oncogenes. Anaplastic lymphoma kinase (ALK), a receptor tyrosine kinase, is implicated in various cancers including lymphomas and lung cancers, where it is translocated and fused to other proteins [68]. In one example a fusion between cysteinyl-tRNAsynthetase (CRS) and ALK was identified in inflammatory myofibroblastic tumors [61], where the in-frame chimeric protein preserved the functional catalytic domain of ALK at the C terminus. In cancers, uncontrolled receptor tyrosine kinases deliver cell proliferative signals in the absence of growth stimuli, and therefore misregulation of ALK activity may play a causative role in inflammatory myofibroblastic tumor. However, the role of CRS in the context of this chromosomal rearrangement is an open question, and studies focused on the CRS might provide insights to tumorigenesis.

Malignant fibrous histiocytomas, sarcomas, and malignant gliomas and glioblastomas have an amplification of the chromosome 12q13 locus where MRS and CHOP (C/EBP homologous protein transcription factor) reside [51–54]. This amplification likely results in overexpression of MRS and CHOP, which may promote tumor progression. CHOP and MRS genes share a 56 bp tail to tail complementary sequence which reveals an in vivo interaction between the two transcripts through each of their 3′ untranslated regions (3′ UTR) [69]. The association between CHOP and MRS mRNAs could increase the stability of CHOP transcript, enabling a growth advantage for tumors [70]. The functional link between the expression of MRS and CHOP can also be found in myxoid and round cell liposarcomas. Myxoid liposarcomas have a characteristic chromosomal translocation (t12, 16)(q13 p11), which results in production of a fusion protein, TLS-CHOP (fusion between fus-like protein and CHOP). TLS-CHOP also contains a 3′ UTR sequence complementary to that of MRS and has been shown to transform NIH3T3 fibroblasts and nude mice [70]. It is likely that the regulation of MRS expression is altered by the chromosomal translocation in myxoid liposarcomas and the increase of MRS transcript enhances the stability of the TLS-CHOP transcript.

Deletion. Loss of CRS by chromosomal deletion has been observed in several cancers including Wilms tumor and embryonal rhabdomyosarcoma [62, 63]. The 11p15.5 chromosomal location harboring CRS is an important region for tumor-suppressor activity, with loss of heterozygosity in the above cancers as well as in adrenocortical carcinoma, and lung, ovarian, and breast cancers [64, 65]. This region includes genes such as RRM1, GOK, Nup98, hNAP2, p57KIP2, KVLQT1, TAPA-1, ASCL2, as well as several novel genes. Some or all of these genes may be involved in the malignancy caused by deletion of this region. However, the exact mechanism for how these genes relate to tumor-suppressive effects has not yet been fully examined.

In 95–100% of nasopharyngeal carcinomas, chromosomal region 3p21.3, where mitochondrial LRS resides, is frequently deleted [66]. Zhou et al. intensively examined the impact of chromosomal deletion on mitochondrial LRS expression and modification. They found that 78% of patients' samples showed down-regulation or no expression of mitochondrial LRS. In addition, 28% of specimens had homozygous deletions of mitochondrial LRS and 64% had hyper-methylation of mitochondrial LRS. These results imply that inactivation of mitochondrial LRS by both genetic and epigenetic mechanisms may be an important event in the carcinogenesis of nasopharyngeal carcinoma.

Polyadenylation-Mediated Deletion Mutation. EPRS is released from the MSC by IFN-γ and with other proteins forms the GAIT (IFN-γ activated inhibitor of translation) complex, which blocks translation of specific target mRNA [71] (Fig. 3). A truncated N-terminal fragment of EPRS, EPRS^{N1}, was recently identified [60]. EPRS^{N1} works as a GAIT component and protects the target mRNA from the inhibitory GAIT complex, thereby maintaining basal transcription of target genes. Interestingly, C-terminal truncation of EPRS^{N1} is not caused by genetic mutation, but is generated by polyadenylation-directed conversion of a Tyr codon to a stop codon. Sequencing data revealed that the UAU codon for Y864 was replaced by a UAA stop codon which was followed by a poly(A) tail without an intervening 3′ UTR. This replacement resulted in deletion of the last 12 amino acids of R2, R3, and PRS (Fig. 3a). It remains unclear whether in cancer the expression of EPRS and EPRS^{N1} is controlled in an opposing manner, and how the regulatory mechanism is perturbed. It is expected that there would be a positive correlation between expression of EPRS^{N1} and tumor malignancy, since GAIT controls the translation of VEGF-A (vascular endothelial growth factor-A) which is an important signal mediator in angiogenesis.

2.2.3 Splicing

For several ARSN, including AIMP2, AIMP2-DX2, EPRS, and WRS, splicing is involved in cancer regulation.

AIMP2-DX2, the splicing variant of AIMP2 (Fig. 4), has been shown to be induced in cancer cells and human cancer patients. The proportion of AIMP2-DX2 compared with full length AIMP2 was increased depending on lung cancer stage,

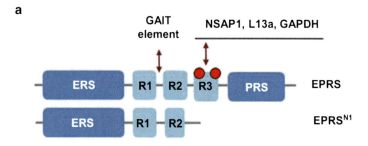

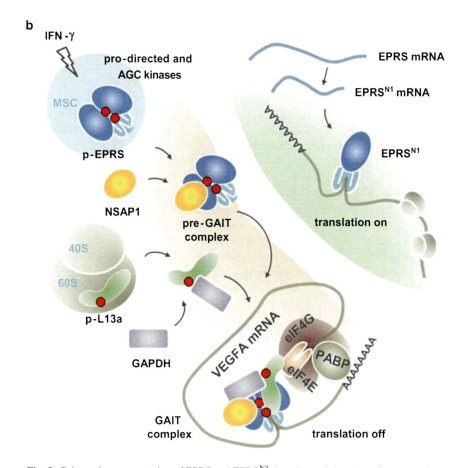

Fig. 3 Schematic representation of EPRS and EPRS[N1] domains and the roles of two proteins on the control of VEGFA transcription. (**a**) Human EPRS contains ERS, PRS, and three tandem-repeats (R1, R2, and R3) linking the ERS and PRS. Polyadenylation-mediated stop codon is introduced in the transcript of EPRS[N1] atY864 residue and results in partial deletion of R2

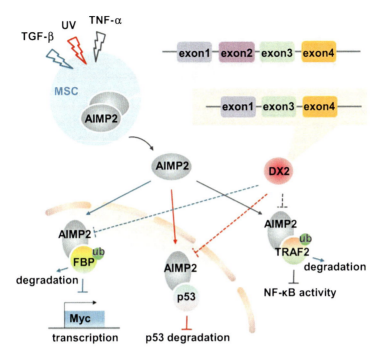

Fig. 4 AIMP2 competes with AIMP2-DX2 under TGF-β, TNF-α, and DNA damage signaling. AIMP2 consists of four exons and, among them, exon 2 is deleted in AIMP2-DX2 (DX2) by alternative splicing. Upon TGF-β and TNF-α stimuli, AIMP2 released from the MSC binds FBP and TRAF2, respectively, and induces the ubiquitin-mediated degradation of FBP and TRAF2 to induce apoptosis. When a DNA damage signal such as UV irradiation is encountered, AIMP2 interacts with p53 and protects p53 from MDM2-mediated degradation. AIMP2-DX2 competitively inhibits the pro-apoptotic interaction of AIMP2 by associating with FBP, TRAF2, and p53

Fig. 3 (continued) C-terminus and whole deletions of R3 and PRS. The R1 and R2 interface is required for its binding with GAIT element in the target mRNA. R3 harbors the phosphorylation residues (*red dots*) and is critical for binding with other protein components of GAIT complex. EPRS^{N1}, therefore, can interact with target mRNA without forming GAIT complex. (**b**) Upon IFN-γ stimulation, EPRS is phosphorylated by pro-directed kinases and AGC kinases (protein kinase A, G, and C families) at S886 and S999. These phosphorylation events (depicted by *red dots*) elicit EPRS release from the MSC and formation of GAIT complex for the translation control of VEGFA mRNA. Phosphorylated EPRS (p-EPRS) interacts with NSAP1 and forms an inactive, pre-GAIT complex. Later, L13a is phosphorylated and released from the 60S ribosomal subunit. Phosphorylated L13a (p-L13a) and GAPDH join the pre-GAIT complex to form the functional GAIT complex. GAIT complex binds the GAIT element in the 3′UTR of VEGFA transcript circularized by simultaneous interaction with PABP, eIF4G and the poly(A) tail. Interaction between p-L13a and eIF4G in the translation initiation complex blocks the recruitment of the 43S ribosomal complex subunit and repress translation. EPRS^{N1} does not inhibit the translation of target mRNA, and mediates the basal translation of VEGFA

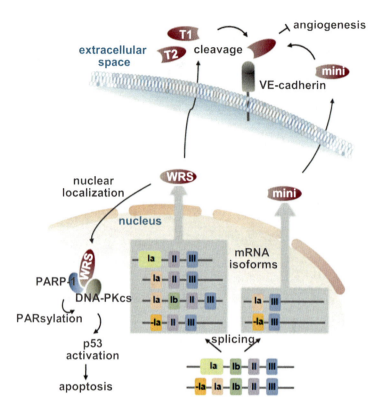

Fig. 5 Isoforms of WRS and their roles in the nuclear and extracellular spaces. The full-length WRS and mini WRS (mini) can be generated by alternative splicing. N-terminal truncated T1 and T2 isoforms are produced by proteolytic cleavage of the full-length WRS in the extracellular space. T1, T2, and mini WRS bind to VE-cadherin and induce anti-angiogenesis. Upon IFN-γ signaling, the full-length WRS is also translocated into nucleus and delivers apoptotic signal via DNA-PKcs-mediated p53 activation

and it showed a positive correlation with patient survival rate [21]. Increased splicing of AIMP2-DX2 in cancer cells is regulated by an oncogenic splicing factor, SF2/ASF (splicing factor 2/alternative splicing factor), which enhances the expression of oncogenic proteins through splicing regulation [72]. The A152G mutation was identified in exon 2 of AIMP2 from stably transformed WI-26 cells that had been treated with benzo[a]pyrene-7,8-dihydrodiol-9,10-epoxide (BPDE) to induce expression of AIMP2-DX2 [21]. This mutation increased binding of the AIMP2 mRNA transcript to SF2/ASF, implying that this site is important for the generation of AIMP2-DX2 transcript via SF2/ASF-mediated splicing.

WRS can be generated either as the full length protein, or "mini" WRS fragments in which the N-terminal 47 amino acids are removed by alternative splicing [38, 73, 74] (Fig. 5). Promoters for transcription of both the full-length WRS and mini WRS are responsive to IFN-γ, IFN-α, and interferon regulatory factor 1. Both the full-length WRS and mini WRS are catalytically active in translation and are secreted into the extracellular environment of vascular endothelial cells, but only mini WRS possesses cytokine activity and exhibits anti-angiogenic effects. To date, over five splicing variants have been identified [38, 39, 74, 75]. These isoforms are generated by different combinatorial splicing of different mRNA transcripts with the exons–Ia, Ia, Ib, II, and III. Almost all the WRS mRNA variants were detected in vivo, and result in the expression of two protein forms, the full-length WRS and mini WRS. Regulation of WRS splicing is not fully understood [39, 74] and will require more detailed analysis. Recently, the splicing signature of WRS was investigated in hypoxic pancreatic cancer cells [76], and the distinct pattern of the splicing signature suggests complex regulation of WRS in terms of transcript levels.

2.3 Post-Translational Modifications

2.3.1 Cleavage

Among the secreted ARSN, WRS, YRS, and AIMP1 are found as digested forms. Secreted WRS is processed in the extracellular space by elastase or plasmin to generate truncated forms of WRS, T1 and T2, where the N-terminal 70- and 93-amino acids are absent, respectively [38, 77] (Fig. 5). Both T1 and T2 play roles in cancer microenvironments, and have angiostatic activity. Interestingly, the full-length WRS does not have this angiostatic effect because the N-terminus of WRS sterically hinders the interaction between WRS and its receptor, VE-cadherin [78]. It implies that proteolytic digestion of WRS which removes the N-terminal domain WHEP, which became attached to WRS during evolution [3, 79], is critically required for the inhibition of angiogenesis by WRS.

YRS is a well-known cytokine which is involved in the pro-inflammatory response and angiogenesis, and its pro-angiogenic activity is only revealed after it is cleaved into fragments [80]. Secreted YRS is processed by elastase to form an N-terminal fragment (known as mini YRS) and acts like a cytokine inducing migration of polymorphonuclear cells and endothelial cells. Mini YRS has an ELR motif in the N-terminus, which mediates its interactions with the receptor CXCR1/2, and induces transactivation of VEGFR2 (VEGF receptor 2), leading to endothelial migration and tumor angiogenesis [81, 82]. In the structure of the full-length YRS, the ELR motif is masked by the C-terminal domain, and therefore mini YRS, but not the full-length YRS, lacks the angiogenic activity. The C-terminal fragment of YRS is structurally similar to EMAPII (endothelial monocyte-activating polypeptide II) and exhibits EMAP II-like activity, stimulating secretion of potent leukocyte and monocyte chemokines and pro-inflammatory cytokines [83].

Whereas full-length AIMP1 is secreted and works as a functional cytokine, in response to apoptosis, AIMP1 can be processed by caspase-7 at the amino acid motif ASTD, and the C-terminal fragment produced by this cleavage is EMAP II

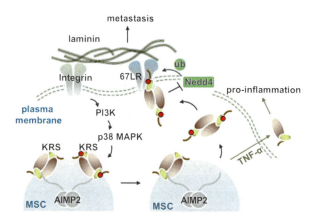

Fig. 6 KRS promotes cancer cell migration. KRS is phosphorylated by p38 MAPK under laminin stimuli, and moves to plasma membrane. Membranous KRS interacts with 67LR and increases the stability of 67LR by blocking Nedd4-mediated ubiquitination. KRS is also secreted to extracellular space upon TNF-α stimuli and induces TNF-α secretion from immune cells

[84–86]. The C-terminal fragment of AIMP1 has been well studied in the context of proinflammatory cytokines involved in angiogenesis; however, full-length AIMP1 demonstrates higher potency in inflammation as well as angiogenesis [87]. The C-terminal peptide of AIMP1 has been detected in the culture medium of stimulated murine fibrosarcoma cells and prostate adenocarcinoma cells [86, 88].

2.3.2 Phosphorylation

Whereas WRS and YRS directly affect angiogenesis, EPRS is indirectly involved in this process via suppression of VEGF-A translation [71]. To perform this function, EPRS is released from the MSC by IFN-γ-dependent sequential phosphorylation of S886 and S999 [89] (Fig. 3). These modifications are also required for GAIT complex formation, which controls VEFG-A gene silencing.

AIMP3 is also translocated into nucleus, which activates the tumor suppressor protein p53 [16]. AIMP3 associates with MRS in the MSC, and, upon DNA damage, GCN2 (general control non-repressed 2) is activated, resulting in phosphorylation of the S662 residue of MRS [90]. Phosphorylation at S662 induces conformational change of MRS, which releases AIMP3. AIMP3 appears to be further modified after release from the MSC, although the mechanism by which it occurs is unclear.

AIMP2 binds to p53 and directly protects it from ubiquitin-mediated degradation induced by DNA damage stress [55]. AIMP2 is phosphorylated by Jun N-terminal kinase (JNK) after UV irradiation, resulting in nuclear translocation. The exact site of phosphorylation was not identified.

KRS modification is critically required for the enhancement of cancer cell migration [24] (Fig. 6). Upon laminin signal, KRS is phosphorylated by p38 MAPK (mitogen-activated protein kinase) at the T52 residue which induces its release from the MSC. Free KRS moves to plasma membrane and regulates the stability of 67 kDa laminin receptor (67LR) to induce laminin-dependent cancer metastasis.

3 Molecular Mechanisms of ARSs and AIMPs in Cancer

During the last few decades, the way that ARSN regulates growth, proliferation, or cell death has been the subject of active investigation. Surprisingly, it was found that ARSN functions in multiple signaling pathways beyond its canonical role in aminoacylation. In this section we will overview the molecular mechanisms for ARSN's activities in cancer cells (Fig. 7).

3.1 Aminoacyl-tRNA Synthetases

LRS. mTORC1 is upregulated in various cancers [91, 92] and therefore targeting mTOR-related signaling is an ongoing strategy in the development of anticancer therapeutics. It is known that activation of mTORC1 is dependent on amino acid concentration; however, the amino acid sensor that directly couples intracellular amino acid-mediated signaling to mTORC1 has been unknown for a long time. Recently it was suggested that LRS, which uses leucine for catalytic aminoacylation, acts as a leucine sensor for mTORC1 activation [8] (Figs. 7 and 8). Activation of mTORC1 in response to amino acids requires the involvement of Rag GTPase, the mediator of amino acid signaling for mTORC1 [93, 94]. LRS directly binds to RagD GTPase in an amino acid-dependent manner, and acts as GTPase-activating protein (GAP) for RagD GTPase [8]. The C-terminal region of LRS has a putative GAP motif which is found in several ADP-ribosylation factor-GAP (Arf-GAP) proteins. Alanine substitution for H844 and R845 in this LRS motif resulted in loss of GAP activity. Rag GTPase works as a heterodimer, and LRS showed higher affinity for RagB/RagD rather than RagA/RagD. A similar mechanism was found to act in yeast, suggesting the importance of LRS as a conserved mechanism for the control of mTOR activity in eukaryotes [33].

LRS binding to RagD GTPase is dependent on the C-terminal domain of LRS; however, its catalytic domain is also required for sensing leucine to activate mTORC1. It suggests that LRS-mediated mTORC1 activation can be blocked by leucine analogues and that compounds targeting protein–protein interaction may also be applied for drug development to treat cancers whose proliferation is mediated by LRS-mTORC1 signaling. Increased expression levels of LRS in cancer patients have been reported [34]. However, whether LRS misregulation leads to uncontrolled activation of mTORC1 in cancer patients remains to be investigated.

KRS. KRS is secreted from cancer cells upon induction by TNF-α, and induces proinflammatory signaling in immune cells [13] (Figs. 6 and 7). A receptor for KRS has not yet been identified, but downstream signaling includes ERK (extracellular signal-regulated kinase) and p38 MAPKs and G protein, Gαi. KRS increases TNF-α secretion which can provide a cancer-supporting microenvironment, suggesting that cancers may use KRS for survival or metastasis by exerting paracrine activity.

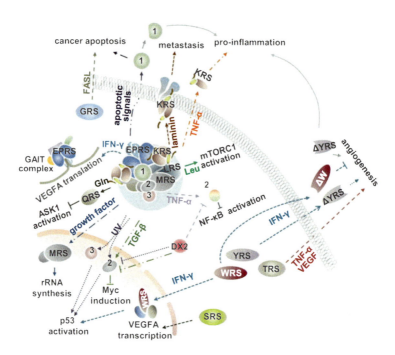

Fig. 7 Non-classical functions of ARSN and their underlying mechanisms in cancer. IFN-γ induces release of EPRS from the MSC and assembly of the GAIT complex to silence VEGFA translation. WRS and YRS also respond to IFN-γ, and relocalize to play roles in angiogenesis and cell apoptosis. AIMP2 (*2*) and AIMP3 (*3*) is released from the MSC by UV irradiation and induce p53-mediated DNA damage response. KRS is translocated to plasma membrane or extracellular space and increases cell migration or pro-inflammatory responses, respectively. LRS triggers mTORC1 activation in response to leucine and regulates cell proliferation and autophagy, and MRS enhances rRNA synthesis in nucleolus under proliferative condition. QRS inhibits ASK1 activation in the presence of glutamine. GRS and AIMP1 (*1*) are secreted from immune cells upon apoptotic signal and act as cytokines to induce cancer cell apoptosis. AIMP2-DX2 inhibits the tumor-suppressive roles of AIMP2 in TGF-β, TNF-α, and DNA damage stress. TRS enhances angiogenesis in extracellular space and SRS directs transcriptional repression of VEGFA during vascular development. *ΔW* truncated WRS, *ΔYRS* truncated YRS. See details in the text

KRS also regulates immune response by translocating into nucleus and interacting with the transcriptional regulator MITF (microphthalmia associated transcription factor), MITF repressor Hint-1, and Ap4A, the catalytic product of KRS [95, 96]. A phosphorylation switch at the S207 residue regulates the transition between its translational and transcriptional roles [97]. More studies need to be performed to determine the role of KRS on the immune response in the context of carcinogenesis.

High expression of 67LR in lymphomas and breast, lung, ovary, colon, and prostate carcinomas has been observed [98–102]. 67LR expression is positively correlated with cancer progression and malignancy, and therefore it is considered a marker for metastasis in various cancers. Recently, involvement of KRS in laminin-induced cell migration was reported [24] (Fig. 6). KRS binds to and stabilizes 67LR

Fig. 8 LRS senses leucine and activates mTORC1 signaling pathway. LRS binds GTP form of RagD GTPase (RagD) under leucine-rich circumstance, and acts as GAP to convert GTP to GDP. mTORC1 is activated by GTP-RagB/GDP-RagD and induces cell proliferation and translation

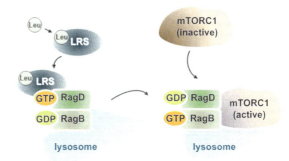

in the plasma membrane to enhance laminin-dependent migration. P38 MAPK phosphorylation at T52 residue, which is activated by laminin signaling, triggers the translocation of KRS from the MSC to the plasma membrane. Membranous KRS binds to 67LR and protects 67LR from Nedd4-mediated degradation. As shown in the structure of KRS S207D, the mimetic of p-S207 KRS [97], phosphorylation at T52 is expected to open the structure of KRS. Overexpression of KRS transcripts was reported in breast cancer, metastatic cell lines and glioma derived stem cells [13, 35, 46]. Study of the relationship between KRS and 67LR is ongoing for drug development to treat cancer metastasis.

YRS. YRS is secreted from leukocytes in response to apoptotic signals, and then cleaved into two fragments by the leukocyte elastase. C-terminal fragments of YRS show homology to human EMAP II, and induce chemotaxis of monocytes as well as angiogenesis of endothelial cells, by inducing TNF-α, myeloperoxidase, and tissue factor [103, 104] (Fig. 7).

Mini YRS, the N-terminal fragment of YRS, contains an ELR motif on a surface helix (α5), which is typically found in angiogenic CXC chemokines. Through the ELR motif, YRS interacts with IL-8 type A receptors CXCR1/2, and exhibits IL-8-like cytokine activity, promoting cell migration and angiogenesis. Dimerization of mini YRS occurs at high concentrations of mini YRS. Unlike monomeric forms, the dimer acts as a non-functional antagonist to CXCR1/2 [81].

Mini YRS induces endothelial cell migration through transactivation of VEGFR2, which is mediated by complex formation with the CXCR receptor [81, 82] (Fig. 7). This stimulation and transactivation requires the intact ELR motif and leads to VEGF-independent phosphorylation of the receptor.

The reason why mini YRS, but not the full-length protein, harbors angiogenic activity lies in its unique protein structure [105]. The ELR motif is predicted to be sterically hidden by the EMAP II domain within the full-length protein, and exposed only when the protein is cleaved. The reason why the Y341A mutation of YRS is a gain-of-function mutation is that the mutant protein has a subtle open structure which exposes the ELR, thus imparting angiogenic activity in vitro and in vivo in the absence of proteolysis. Substitutions of either the E or L of the ELR motif ablated cytokine activity of YRS without showing any critical effects on aminoacylation. On the other hand, R93Q substitution of the ELR affected both cytokine activity and aminoacylation. It seems that cytokine function of YRS has been developed independent of catalytic activity, as R93, but not E91 and L92, is conserved in almost all eukaryotes.

WRS. The major functions of WRS are its anti-angiogenic and pro-apoptotic activities, exerted in extracellular space and nucleus, respectively. To act as an angiostatic cytokine, WRS must be processed by proteolytic digestion or pre-mRNA splicing, to generate T1 and T2 or mini WRS, respectively [38, 73, 74, 77] (Figs. 5 and 7).

WRS pre-mRNA transcripts are alternatively spliced to generate several different transcripts in response to IFN-γ and IFN-α. The final end products are the full-length WRS and mini WRS, in which the 47 N-terminal amino acids are truncated. According to Turpaev et al., WRS is under the control of two promoters with transcription initiation sites P1 and P2 [74]. Using these two promoters, four mRNA isoforms are generated, using different combinations of exons, such as exon IA–exon II–exon III (isoform 1), Δexon IA–exon II–exon III (isoform 2), exon IA–exon IB–exon II–exon III (isoform 3), and exon IA–exon III (isoform 4). Liu et al. discovered additional isoforms that possess yet another exon, yielding combinations of exon Ia–exon II–exon III or exon Ia–exon III [39]. Since the translational starting site is located in exon II and exon III, isoforms containing exon II generate the full-length WRS, while isoforms without exon II produce mini WRS.

The secreted full-length WRS, in the extracellular space, is cleaved by elastase or plasmin to yield the T1 and T2 isoforms, which lack the N-terminal 70- and 93-amino acids, respectively. Both isoforms possess angiostatic activity [106, 107]. However, whereas T2 is catalytically inactive, its angiostatic activity is more potent than those of the full-length WRS and T1.

WRS secretion is tightly regulated. Not a component of the MSC, it forms a ternary complex with annexin II and S100A10 in the intracellular environment, which regulates exocytosis in endothelial cells. When cells are stimulated by IFN-γ, a small fraction of total WRS is dissociated from the annexin II-S100A10 complex and secreted [106]. WRS secretion was not inhibited by brefeldin A or A23187, the inhibitors for Golgi-involving classical secretion pathway and calcium-dependent exocytosis, respectively. It suggests that WRS may use a non-classical secretion pathway.

The receptor of WRS T2 was identified as vascular endothelial cadherin (VE-cadherin) [77]. VE-cadherin, a cellular adhesion protein, is selectively expressed at the intercellular junctions between endothelial cells and has a critical role in angiogenesis and vascular permeability. VE-cadherin contains two conserved tryptophan residues at positions 2 and 4 of the N-terminal extracellular cadherin domain 1 (EC1), which is critical for the dimerization of VE-cadherin [108]. These tryptophan residues are recognized by the catalytically active site of WRS T2. The interaction between EC1 of VE-cadherin and T2 of WRS prevents the activation of VEGFR and subsequent ERK-mediated signaling, which suppresses endothelial cell migration and proliferation [109]. It has been proposed that the full-length WRS lacks angiostatic activity, even though possessing an intact catalytic active site. It seemed that the N-terminus of human WRS would sterically hinder the interaction between EC1 of VE-cadherin and WRS [109]. Therefore mini WRS, T1, and T2, but not the full-length WRS, possess angiostatic activity, suggesting an evolutionary mechanism for VE-cadherin regulation in multicellular organisms.

In addition to its role as a cytokine, WRS also plays a pro-apoptotic function in the nucleus [40]. IFN-γ signaling can induce expression of WRS and mediate its translocation to the nucleus. Without nuclear WRS, DNA-PKcs (catalytic subunit of DNA-dependent protein kinase) is indirectly linked to PARP-1 (poly [ADP-ribose] polymerase 1) with Ku70/80 acting as a bridge between them. Ku70/80 binding to both DNA-PKcs and PARP-1 orients the C-terminal domain of PARP-1 for phosphorylation by DNA-PKcs. In this state, PARP-1 cannot PARsylate DNA-PKcs. Upon IFN-γ stimuli, nuclear WRS replaces Ku70/80 and binds to both DNA-PKcs and PARP-1. The WHEP domain of WRS bridges the C-terminal kinase domain of DNA-PKcs to the N-terminal domain of PARP-1 and consequently stimulates the PARsylation of DNA-PKcs by PARP-1. PARsylated DNA-PKcs then phosphorylates p53, thereby inducing p53-driven apoptosis and senescence.

Interestingly, active site occupancy has proven critical for the function of nuclear WRS. WRS generates Trp-AMP as an intermediate aminoacyl-adenylate product during the aminoacylation step. When the active site in WRS is occupied with Trp-AMP, the WHEP appears to fold back toward the active sites. In the absence of Trp-AMP, WHEP appears to be opened and available for interacting with DNA-PKcs and PARP-1.

Because WRS links IFN-γ signaling to p53 activation, which is implicated in other anticancer processes, including anti-proliferative and anti-angiogenic effects, it is worth noting that WRS is downregulated in colorectal cancer and metastasis [36].

TRS. Auto-antibodies against TRS have been detected in patients with polymyositis and dermatomyositis [110], implying that TRS can be secreted into extracellular space from apoptotic cells. Recently the role of secreted TRS in angiogenesis has also been reported [26]. TRS was secreted from human vascular endothelial cells exposed to TNF-α or VEGF (Fig. 7). Secreted TRS worked as an autocrine or paracrine factor, and stimulated the migration and tube formation of vascular endothelial cells. The pro-angiogenic effect of TRS appears to require either catalytic activity or the unoccupied active site of TRS, in that inhibition of TRS catalytic activity suppressed the angiogenic function of TRS.

EPRS. Halofuginone is a herbal medicine used to treat various diseases including cancers and inflammatory diseases. The recent discovery that halofuginone inhibits EPRS activity suggests a possible role of EPRS in pathogenesis [111]. EPRS specifically inhibits the translation of VEGF-A mRNA, which is an important signal mediator in angiogenesis, via the GAIT complex. EPRS forms the GAIT complex with NSAP1 (NS1-associated protein 1, also known as heterogenous nuclear ribonucleoprotein Q1), GAPDH (glyceraldehyde-3-phosphate dehydrogenase), and ribosomal protein L13a [112, 113] (Fig. 3).

GAIT complex formation is sequential [25]. First, 2 h after IFN-γ treatment, IFN-γ induces phosphorylation of EPRS at S886 and S999, which induces its dissociation from the MSC and interaction with NSAP1, resulting in inactive, pre-GAIT complex formation. Then, 12–16 h post-IFN-γ treatment, L13a is phosphorylated at S77 and released from the 60S ribosomal subunit. Phosphorylated L13a associates with GAPDH, and joins the pre-GAIT complex to form the functional GAIT complex. The complex then binds GAIT element in the 3' UTR of target transcripts, such as VEFG-A mRNA. The target mRNA is circularized by simultaneous interactions of PABP (poly(A)-binding protein)-eIF4G (eukaryotic translational initiation factor 4G) complex and the poly(A) tail. The interaction of phosphorylated L13a with eIF4G blocks the recruitment of the eIF3-containing 43S ribosomal complex subunit, resulting in inhibition of translation.

Post-transcriptional control via the GAIT complex is a complex regulatory system: EPRS can bind the target mRNA without phosphorylation but NSAP1 interacts only with p-EPRS. The pre-GAIT complex cannot interact with target mRNA because NSAP1 hinders the binding between p-EPRS and mRNA. Binding of p-EPRS to target mRNA is fully achieved by joining of both L13a and GAPDH, which restores EPRS binding to GAIT element RNA even in the presence of NSAP1 [114]. In addition, translational inhibition of VEFG-A requires the presence of p-L13a, suggesting that functional GAIT complex needs only these four proteins with their modification.

Several chemokine ligand and receptor mRNAs have been analyzed as possible targets for the GAIT complex [25]. The results of this analysis suggested that EPRS might be a key regulator of inflammation. The GAIT complex mediates translational control through a delayed silencing mechanism; therefore it can prevent the excess accumulation of inflammatory proteins that can be detrimental to the host.

SRS. A recent study has indicated that nuclear SRS also controls transcription of VEGF-A [27] (Fig. 7). In the study, an insertional null mutation in SRS in zebrafish caused abnormalities in vessel formation, which was rescued by wild type SRS. The NLS (nuclear localization signal) and UNE-S (unique-S) domain in SRS appear to play critical roles for normal vascular developments. The mechanism by which SRS controls VEGF-A expression is unclear; however, the relationship between SRS mutations and vascular development and maintenance has been demonstrated in the zebrafish model [115, 116], suggesting the possibility of SRS involvement in neo-angiogenesis in cancer.

GRS. GRS is secreted from macrophages stimulated by cancer-cell secreted Fas ligand [12] (Fig. 7). Secreted GRS induces cancer cell death via binding to the CDH6 expressed on the surface of cancer cells. Interestingly, only cancer cells with hyper-activated ERK respond to the GRS apoptotic signal, because the CDH6-mediated GRS stimuli induces a signal cascade which activates PP2A (protein phosphatase 2), and in turn reduces ERK phosphorylation.

MRS. Ribosome biogenesis is related to cancer in that cell growth and proliferation is associated with an increase in the rate of ribosome production [117]. During tumorigenesis, the tight regulation between extracellular signaling and ribosome biogenesis is disrupted. Deregulation of Pol I (RNA polymerase I), which transcribes rRNA in the nucleolus, can contribute to tumorigenesis [117], and the increase in rRNA synthesis, the major component of ribosome, is adversely correlated with the prognosis of cancer patients [118].

MRS plays a role in rRNA synthesis [9] (Fig. 7). Although its precise function in rRNA synthesis is still unknown, modification or deregulation of MRS causing its accumulation in the nucleus may be involved in the upregulation of rRNA synthesis, which can accelerate tumorigenesis.

In addition, MRS strikingly increases misaminoacylation in response to oxidative stress [119]. Misaminoacylation caused by MRS is thought to play a role as a defense mechanism in protecting cells from anti-oxidative stress, because methionine can acts as a ROS (reactive oxygen species) scavenger [120]. However, the long-term existence of protein with misincorporated methionine residues may induce unfolded protein response which is implicated in tumorigenesis [121]. In fact, down-regulation of methionine sulfoxide reductase A, a regenerator of methionine from its oxidized form, has been shown to increase cell proliferation, extracellular matrix degradation, and the aggressiveness of breast cancer cells in vitro and in vivo [121]. The molecular and biochemical mechanisms of MRS misaminoacylation, and how misaminoacylation is controlled in and contributes to cancer needs more investigation.

AlaRS. Translational fidelity is required for the precise control in development and homeostasis of complex organisms. Malfunction of translational fidelity leads to the production of non-functional or unstable proteins that need clearing, and compromised protein degradation causes formation of protein aggregates linked to numerous neuronal disease conditions including Alzheimer's and Parkinson's diseases, and amyotrophic lateral sclerosis [122]. In animal models harboring an AlaRS mutation which causes misacylation of serine to $tRNA^{Ala}$, protein aggregation in the Purkinje cells of the cerebellum led to neuronal cell death [123]. To remove misacylated Ser-$tRNA^{Ala}$, organisms use AlaXp, the separate homologue of the editing domain of AlaRS [124, 125]. As described above for MRS, misaminoacylation and the consequent unfolded protein response can lead to tumorigenesis. Although as of yet there is no evidence linking cancer to misaminoacylation caused by AlaRS or the expression of an AlaRS mutant, the possibility of its role in tumorigenesis with regard to misaminoacylation is still open.

QRS. ASK1 (apoptosis signal-regulating kinase 1) mediates apoptosis in response to various oncogenic stresses [126]. Upon Fas ligand stimuli, ASK1 activates JNK in glutamine-deprived cells and induce cell death. QRS can bind the catalytic domain of ASK1 through its catalytic domain in the presence of glutamine and inactivate ASK1 [23]. The association between QRS and ASK1 is enhanced by glutamine but disappears by addition of Fas ligand. This results show that QRS can sustain cancer cell proliferation in a glutamine-dependent manner by suppressing proapoptotic role of ASK1. It also suggests the importance of the relationship between ARSN and substrates for the control of cancer survival.

3.2 ARS-Interacting Multi-functional Proteins

AIMPs serve as scaffolding proteins which comprise the MSC. Three components, AIMP1, AIMP2, and AIMP3, interact with complex-forming ARSs to form the stable MSC, and to support the fidelity and efficiency of translation [127, 128]. Recent studies have shown that, in addition to their activities in protein translation, various stimuli trigger the dissociation of the AIMPs from the MSC, and enable them move to other intracellular or extracellular locations [3, 4, 6]. Outside of the MSC, AIMPs partner with other proteins to play critical roles in various forms of cancer.

AIMP1. Inflammatory signals are generally considered to eradicate tumor cells; however they can also provide cells with pro-oncogenic microenvironments enriched with proliferative cytokines. Based on this contradictory effect, inflammation can promote cancer through various mechanisms [2]. AIMP1 was initially found as a pro-EMAP II, a polypeptide with cytokine activity. AIMP1 secretion is induced by stress signaling pathways, and acts as a cytokine on various target cells such as immune cells, endothelial cells, and fibroblasts [129] (Fig. 7). In endothelial cells, AIMP1 exhibits dose-dependent biphasic activity, promoting migration and apoptosis of endothelial cells at low and high doses, respectively [87]. At higher doses, AIMP1 enhances phosphorylation of JNK and sequentially induces activation of caspase-3 to stimulate apoptosis of endothelial cells, which are linked to tumor angiogenesis [87]. In addition, extracellular AIMP1 is internalized through lipid rafts and then interacts with cytoskeletal proteins critical to biological processes such as cell polarity, adhesion, and migration. AIMP1-mediated modulation of cellular architecture maintenance and remodeling through its interactions with cytoskeletal proteins seems to be involved in its angiostatic effects reported to inhibit tumor growth [130, 131].

AIMP1 also stimulates the secretion of TNF-α and IL-8 from monocytes or macrophages in both an MAPK-and NF-kB-dependent manner. In antigen-sensitized mice, AIMP1 enhances production of IFN-γ, TNF-α, and T cell activation to develop Th1 (type 1 helper T cell) response [132]. In a xenograft mouse model, AIMP1-treated mice exhibited induction of serum TNF-α and IL-1b levels, along with reduced tumor size [133]. AIMP1 appears to invoke multiple signaling pathways in the production of TNF-α [134]. Among them, the ERK1/2 MAPK signaling pathway is conserved [135].

Recently, CD23, a known receptor for IgE, was identified as a receptor for secreted AIMP1 [136]. This study showed that downregulation of CD23 reduced the activation of ERK1/2 and AIMP1-dependent cytokine secretion in primary human monocytes. Another study showed that AIMP1 triggers secretion of TNF-α from macrophages, and induces fibroblast proliferation and collagen synthesis around an artificial wound region in dorsal skin of mice in an ERK-dependent manner [135]. These effects revealed a rapid wound healing in *AIMP1* wild type mice compared with *AIMP1* knockout mice.

AIMP2. Loss of AIMP2 in mice was shown to cause neonatal lethality due to the failure of lung differentiation [18]. At a molecular level, AIMP2 binds to and promotes ubiquitin-mediated degradation of FBP (FUSE-binding protein), a transcriptional activator of Myc, which is important for lung cell differentiation. Therefore AIMP2 deficiency appears to cause neonatal lethality of mice through uncontrolled hyper-proliferation of lung cells (Fig. 4).

AIMP2 mediates pro-apoptotic signaling upon induction of various stimuli. The main function of AIMP2 is to stabilize target proteins by binding to them. Under TGF-β signaling, highly expressed AIMP2 is translocated into the nucleus, whereupon it binds to FBP and increases ubiquitination and degradation of FBP [18]. AIMP2-dependent FBP degradation was inhibited by addition of 26S proteasome inhibitor, suggesting that AIMP2 facilitated the ubiquitination of FBP. Under DNA damage stress like UV, AIMP2 appears to be phosphorylated by JNK, moves into nucleus, and associates with p53. Further studies showed that the interaction between AIMP2 and p53 competitively impaired the binding of MDM2 to p53 and protected p53 from MDM2-dependent degradation [55].

In the same context, AIMP2 was reported to be released from the MSC under TNF-α stimuli, whereupon it binds to TRAF2 (TNF receptor associated factor 2). TRAF2 plays a pivotal role in the TNF-α-related chemo-resistance in ovarian cancer [22, 137]. AIMP2 stimulates the ubiquitin-mediated degradation of TRAF2 by facilitating the association of c-IAP1 (cellular inhibitor of apoptosis 1), the E3 ubiquitin ligase, with TRAF2. As a result, AIMP2 promotes the TNF-α-dependent pro-apoptotic effects. Interestingly, the tumor suppressive role of AIMP2 in the TNF-α signaling pathway was observed in chemo-resistant ovarian cancer cells in which p53 was functionally inactive [22]. This study suggests that AIMP2 induces apoptosis through multiple signaling pathways and can still be effective even when one of the related routes has been disrupted during tumorigenesis.

AIMP2-DX2. Splicing regulates gene function at a transcriptional level. AIMP2-DX2 has been reported to be highly elevated in tumor cells, and its expression to be positively correlated with cancer stage and survival of patients [21]. AIMP2-DX2 was also involved in the activation of NF-kB, which plays a role in the chemo-resistance of ovarian cancer [22, 137]. The main function of AIMP2-DX2 is to compete with the full-length AIMP2 for association with binding partners. It contributes to tumorigenesis by blocking the pro-apoptotic function of the full-length AIMP2 (Figs. 4 and 7).

Based on the function of AIMP2, AIMP2-DX2 hinders the interaction of AIMP2 with FBP (unpublished data), p53, and TRAF2 under the stimuli of TGF-β, DNA damage stress, and TNF-α, respectively. Based on domain mapping of AIMP2 for its interaction with p53 [55], the residues critical for its binding to p53 were amino acids 162–225, which span the interface between exon 3 and exon 4 peptides. This region is still present in AIMP2-DX2, supporting studies showing that AIMP2-DX2 binds to the target proteins and thereby competitively hinders the ability of AIMP2 to interact with target proteins.

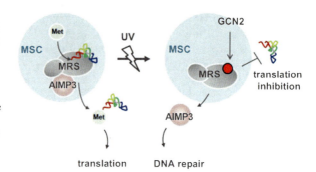

Fig. 9 Tumor suppressor AIMP3 is activated by MRS modification. UV irradiation triggers GCN2-mediated MRS phosphorylation (*red dot*) and induces conformational change in MRS to release AIMP3. Free AIMP3 enhances DNA damage response through its interaction with ATM/ATR

AIMP3. AIMP3 heterozygosity increased the rate of spontaneous tumors in aged mouse [16]. A decreased level of AIMP3 has also been demonstrated in cancer cell lines and patient tissues. At the molecular level, AIMP3 is released from the MSC by DNA damage stress such as UV irradiation. Levels of uncharged tRNA are increased under UV irradiation, which activates GCN2 kinase which in turn suppresses protein synthesis by phosphorylating eIF2α (eukaryotic initiation factor subunit α) at S51 residue [138]. GCN2 has also been reported to phosphorylate MRS at S662 residue, and this modification induces a conformational change in MRS which releases bound AIMP3 [90] (Figs. 7 and 9).

MRS is a strong binding partner of AIMP3 in the MSC [127, 128]. In normal condition, AIMP3 is located in the MSC interacting with MRS. MRS and AIMP3 associate with eIF2γ together and AIMP3 transfers methionine-charged initiator tRNA (Met-tRNA$_i^{Met}$), but not elongator tRNA, from MRS to eIF2γ-containing initiation complex. By doing so, AIMP3 seems to be involved in the accurate and efficient control of translation initiation in the MSC. This study also shows that AIMP3 has the ability to bind specific nucleic acids like Met-tRNA$_i^{Met}$, suggesting the possible involvement of AIMP3 in the direct recognition of damaged DNA.

It seems that further modification of AIMP3 such as phosphorylation is required for its nuclear translocation to induce DNA damage response [16], suggesting that fine control of AIMP3 is required for regulating cell homeostasis. It is, therefore, not surprising that deregulation of AIMP3 expression causes spontaneous tumor or senescence [16, 56]. AIMP3 also responds to genotoxic stress or the oncogenes Ras or Myc and prevents cell transformation by inducing p53 [17]. In transformed AIMP3 heterozygous cells, cell division and chromosomal structure were severely affected [17]. Senescence is a homeostatic mechanism to prevent continuous proliferation of cancer cells. Overexpression of AIMP3 induces cellular senescence. Mice overexpressing AIMP3 exhibited progeroid phenotype [56] due to the proteasome dependent degradation of mature lamin A. Whether crosstalk between the molecular pathways of p53 activation and lamin A degradation can control tumorigenesis remains to be studied.

4 Therapeutics

4.1 Aminoacyl-tRNA Synthetases

For many years, antibiotics have been developed by targeting ARSs in pathogenic microorganisms [139, 140]. Mupirocin is the representative commercial inhibitor of eubacterial and archaeal IRS used for the treatment of Gram-positive pathogens [141]. Tavaborole (formerly AN2690) is an antifungal agent that targets the LRS editing site, and is used in clinical trials for onychomycosis, with positive results from the first of two planned Phase 3 studies (www.anacor.com).

Recent studies have suggested that targeting human ARSs may provide new strategies for developing therapies for human diseases. Halofuginone is a derivative of febrifugine, the active principal of the Chinese herb changshan (*Dichroafebrifuga*) which is used as an anti-malarial treatment in Chinese traditional medicine [142]. Halofuginone targets prolyl-tRNA synthetase (PRS) in EPRS and is in clinical trials for the treatment of cancer and fibrosis [111, 143]. Halofuginone acts as a prolyladenylate-mimetic and inhibits aminoacylation activity of PRS. Unlike other inhibitors of ARSs, which mimic aminoacyl-adenylates and out-compete the substrate ATP, halofuginone-mediated PRS inhibition requires ATP. By associating with ATP as well as PRS, halofuginone occupies two active sites of PRS: the proline binding pocket and the binding pocket for the tRNA interaction [143]. As a result, halofuginone-mediated PRS inhibition triggers amino acid starvation response, inhibits the differentiation of Th17 cells, and suppresses fibrosis [111]. The unique mechanism of action of halofuginone suggests that such a strategy of dual site inhibition should be tested for other ARSs. In addition, the amino acid starvation response might represent another target pathway for developing anticancer and anti-inflammatory therapeutics.

Borrelidin, a nitrile-containing macrolide, is isolated from Streptomyces [144]. It was first used in the treatment of infectious diseases such as malaria because it suppresses translation by inhibiting the catalytic activity of bacterial TRS. Borrelidin inhibits eukaryotic TRS as well, and by doing so it increases the levels of uncharged tRNAs and activates GCN2 kinase-mediated stress response, and ultimately it induces translation inhibition and cell apoptosis [145]. Borrelidin also seems to target secreted TRS for inhibiting angiogenesis [26]. Treating human endothelial cells and chorioallantoic membrane with TRS triggers tube formation and angiogenesis, respectively. The borrelidin derivative BC194 significantly suppresses TRS-mediated endothelial cell proliferation and vessel formation. Based on these studies, both intracellular and extracellular TRS are targets of borrelidin, and the catalytic activity of TRS appears to be related to cancer cell proliferation and angiogenesis. The mechanism of TRS and mode of action of borrelidin warrants further study. BC194 showed reduced cytotoxicity and the same levels of anti-angiogenic properties compared to borrelidin [146, 147]. These results suggest that TRS could be a useful target for anti-metastasis therapeutics. Furthermore, TRS might mediate progression of ALL and metastasis of osteosarcoma and melanoma, based on studies showing that borrelidin inhibited growth of these cell lines and mouse model [145, 148, 149].

WRS T2 has potent angiostatic activity, and is in pre-clinical testing for the inhibition of neovasculization [38, 73]. In addition, it is headed to initiate clinical trials for the treatment of wet age-related macular degeneration or acute myocardial infarction [150] (http://www.pacificu.edu/optometry/ce/courses/20591/armdpg4. cfm; http://www.eyecyte.com/clinical.htm). aTyrPharma (www.atyrpharma.com) and Pangu BioPharma have been founded to develop drugs targeting secreted ARSN, and they are developing several products in therapeutic areas such as blood, immune, and metabolic disorders.

Other ARSs also represent potential therapeutic targets. These include a mimetic peptide that mimics the apoptotic activity of GRS, synthetic compounds which inhibit the pro-metastatic interaction between KRS and 67LR, and leucine derivatives or compounds which prevent the interaction between LRS and RagD GTPase to block mTORC1 signaling. In addition, Neomics Co., Ltd. (http://www.neomics.com) is developing therapeutics against ARSN targets to develop therapies for immune disease and cancer.

4.2 ARS-Interacting Multi-functional Proteins

AIMP1's role as an antitumor cytokine was directly tested in the xenograft model of stomach cancer [133]. A single intravenous injection of AIMP1 (up to 10 mg/kg) showed significant reduction of tumor volume and weight within a week. However, the stability of AIMP1 is so low that a strategy to prolong its effects needs to be considered. The antitumor effect of AIMP1 has also been demonstrated using immune cell therapy. Tumor-bearing mice immunized with fibroblasts that were genetically modified to harbor the ovalbumin epitope and AIMP1 showed prolonged survival compared to the control group expressing the ovalbumin epitope only [151]. The higher immune response was explained in part by CD8-mediated cytotoxicity of cancer cells. Fusion of anti-CD3 single-chain Fv and AIMP1 enhanced the Th-mediated immune response, supporting the use of AIMP1 to boost antitumor immunity [132]. AIMP1-targeting strategies will require further optimization but these approaches demonstrate the most promising application of AIMP1 for anticancer therapeutics.

AIMP2-DX2 is another attractive target for anticancer drug development, and can employ small molecules, specific siRNAs or viral vectors to inhibit the expression of AIMP2-DX2. For example, suppression of AIMP2-DX2 using lentivirus-mediated shRNA in lung cancer cells inhibited cell growth by blocking glucose uptake and growth factor mediated signaling pathways [152]. A trans-splicing ribozyme was used to induce selective death of AIMP2-DX2 expressing lung cancer cells [153]. The ribozyme specifically targets the AIMP2-DX2 transcript, and delivers the suicide gene into the cancer cells, resulting in growth retardation. Due to the so-far limited application of genetic manipulation for human therapies, multiple approaches to perturb the tumorigenic function of AIMP2-DX2 should be investigated.

5 Future Perspectives

ARSN has long been recognized as housekeeping proteins. Due to an accumulation of new evidence, these are now being accepted as regulators whose functions are critically related to induction, maintenance and progress of human disease, especially cancer. ARSN consists of 23 proteins and has as many as 123 cancer-related binding partners, CAGs, suggesting the potential pathophysiological implications of ARSN in tumorigenesis [6]. However, much work needs to be done to uncover exhaustively their role in cancer.

ARSN consists of multifunctional proteins and their expression, modification, and localization are exquisitely regulated. ARSN maintains cellular homeostasis under physiological conditions. However, upon various signals and stimuli, it is mobilized to provide an "emergency response" by controlling key regulators like tumor suppressors and oncogenes (Fig. 7). Because research has focused on revealing the function of individual ARSN, most published data suggest that ARSN works as individual entities. However, several ARSN have been shown to be activated under the same circumstances: AIMP2 and AIMP3 are activated by UV; EPRS, SRS, WRS, and YRS respond to IFN-γ; EPRS, SRS, TRS, WRS, and YRS may affect angiogenesis; and GRS and KRS may play roles in adversely affecting tumor microenvironment. Considering the delicacy and accuracy of the mechanisms controlling living organisms, ARSN may communicate and act systemically and cooperatively. Given the availability of new technologies and devices, and evidence for how genomics and proteomics has widened the understanding of biology and brought to bear a pool of new drug target candidates, it is time to use a systems-based approach to understand the networking within ARSN as well as between ARSN and CAGs. Whereas cancer-specific markers, novel modifications, and variations in genes and proteins have been the focus of much investigation in recent years, little progress in these areas has been made for ARSN research. "Omics" approaches using more focused and well-designed strategies to target selected gene pools will be helpful for identifying more meaningful candidates of ARSN in the context of cancer.

Recently, important ARS substrates and intermediate or end products have come to prominence. These include amino acids, tRNA, adenosine polyphosphate such as Ap4A, and aminoacyl-adenylates, which can control cell proliferation, translation, transcription, and apoptosis [23, 40, 154–157]. The regulation of these factors and their functional relationship with ARSN will be a promising theme with regard to cancer. The involvement of ARSN in amino acid-related cancer metabolism has also been reported [10]. QRS blocks ASK1-mediated apoptotic signal with glutamine supplemented [23]. In addition, LRS-driven mTORC1 activation suggests that mTORC1-mediated metabolism regulation is critical for the effect of anticancer agents and anticancer drug discovery [158, 159].

Currently translational components are emerging as key regulators in tumorigenesis [160]. Introduction of human LRS to treat MELAS syndrome, a type of neurodegenerative disorder, and the consequent correction of causative tRNA mutant effect provided a step toward the use of human ARSs in new therapeutics for other disease indications [7, 161]. For cancer research, halofuginone-PRS and borrelidin-TRS provide a bright prospect for the possibility of ARS as novel drug targets. Importantly, these studies suggest that inhibiting specific functions of human ARSs, even their catalytic activity, can have clinical benefits without significant cytotoxicity. We believe the direction of ARSN-targeting strategies will be to find inhibitors or mimetics through two paths: one is to modulate ARSN's enzymatic activity and the other to target its interactions with other important binding partner proteins.

References

1. Hanahan D, Weinberg RA (2000) The hallmarks of cancer. Cell 100(1):57–70
2. Hanahan D, Weinberg RA (2011) Hallmarks of cancer: the next generation. Cell 144(5):646–674. doi:10.1016/j.cell.2011.02.013
3. Guo M, Yang XL, Schimmel P (2010) New functions of aminoacyl-tRNA synthetases beyond translation. Nat Rev Mol Cell Biol 11(9):668–674. doi:10.1038/nrm2956
4. Min G, Schimmel P (2013) Essential nontranslational functions of tRNA synthetases. Nat Chem Biol 9(3):145–153. doi:10.1038/nchembio.1158
5. Park SG, Schimmel P, Kim S (2008) Aminoacyl tRNA synthetases and their connections to disease. Proc Natl Acad Sci USA 105(32):11043–11049. doi:10.1073/pnas.0802862105
6. Kim S, You S, Hwang D (2011) Aminoacyl-tRNA synthetases and tumorigenesis: more than housekeeping. Nat Rev Cancer 11(10):708–718. doi:10.1038/nrc3124
7. Yao P, Fox PL (2013) Aminoacyl-tRNA synthetases in medicine and disease. EMBO Mol Med. doi:10.1002/emmm.201100626
8. Han JM, Jeong SJ, Park MC, Kim G, Kwon NH, Kim HK, Ha SH, Ryu SH, Kim S (2012) Leucyl-tRNA synthetase is an intracellular leucine sensor for the mTORC1-signaling pathway. Cell 149(2):410–424. doi:10.1016/j.cell.2012.02.044
9. Ko YG, Kang YS, Kim EK, Park SG, Kim S (2000) Nucleolar localization of human methionyl-tRNA synthetase and its role in ribosomal RNA synthesis. J Cell Biol 149(3):567–574
10. Pacher M, Seewald MJ, Mikula M, Oehler S, Mogg M, Vinatzer U, Eger A, Schweifer N, Varecka R, Sommergruber W, Mikulits W, Schreiber M (2007) Impact of constitutive IGF1/IGF2 stimulation on the transcriptional program of human breast cancer cells. Carcinogenesis 28(1):49–59. doi:10.1093/carcin/bgl091
11. Chen SH, Yang W, Fan Y, Stocco G, Crews KR, Yang JJ, Paugh SW, Pui CH, Evans WE, Relling MV (2011) A genome-wide approach identifies that the aspartate metabolism pathway contributes to asparaginase sensitivity. Leukemia 25(1):66–74. doi:10.1038/leu.2010.256
12. Park MC, Kang T, Jin D, Han JM, Kim SB, Park YJ, Cho K, Park YW, Guo M, He W, Yang XL, Schimmel P, Kim S (2012) Secreted human glycyl-tRNA synthetase implicated in defense against ERK-activated tumorigenesis. Proc Natl Acad Sci USA 109(11):E640–E647. doi:10.1073/pnas.1200194109
13. Park SG, Kim HJ, Min YH, Choi EC, Shin YK, Park BJ, Lee SW, Kim S (2005) Human lysyl-tRNA synthetase is secreted to trigger proinflammatory response. Proc Natl Acad Sci USA 102(18):6356–6361. doi:10.1073/pnas.0500226102
14. Tzima E, Schimmel P (2006) Inhibition of tumor angiogenesis by a natural fragment of a tRNA synthetase. Trends Biochem Sci 31(1):7–10. doi:10.1016/j.tibs.2005.11.002

15. Ivakhno SS, Kornelyuk AI (2004) Cytokine-like activities of some aminoacyl-tRNA synthetases and auxiliary p43 cofactor of aminoacylation reaction and their role in oncogenesis. Exp Oncol 26(4):250–255
16. Park BJ, Kang JW, Lee SW, Choi SJ, Shin YK, Ahn YH, Choi YH, Choi D, Lee KS, Kim S (2005) The haploinsufficient tumor suppressor p18 upregulates p53 via interactions with ATM/ATR. Cell 120(2):209–221. doi:10.1016/j.cell.2004.11.054
17. Park BJ, Oh YS, Park SY, Choi SJ, Rudolph C, Schlegelberger B, Kim S (2006) AIMP3 haploinsufficiency disrupts oncogene-induced p53 activation and genomic stability. Cancer Res 66(14):6913–6918. doi:10.1158/0008-5472.CAN-05-3740
18. Kim MJ, Park BJ, Kang YS, Kim HJ, Park JH, Kang JW, Lee SW, Han JM, Lee HW, Kim S (2003) Downregulation of FUSE-binding protein and c-myc by tRNA synthetase cofactor p38 is required for lung cell differentiation. Nat Genet 34(3):330–336. doi:10.1038/ng1182
19. Choi JW, Kim DG, Park MC, Um JY, Han JM, Park SG, Choi EC, Kim S (2009) AIMP2 promotes TNFalpha-dependent apoptosis via ubiquitin-mediated degradation of TRAF2. J Cell Sci 122(Pt 15):2710–2715. doi:10.1242/jcs.049767
20. Choi JW, Um JY, Kundu JK, Surh YJ, Kim S (2009) Multidirectional tumor-suppressive activity of AIMP2/p38 and the enhanced susceptibility of AIMP2 heterozygous mice to carcinogenesis. Carcinogenesis 30(9):1638–1644. doi:10.1093/carcin/bgp170
21. Choi JW, Kim DG, Lee AE, Kim HR, Lee JY, Kwon NH, Shin YK, Hwang SK, Chang SH, Cho MH, Choi YL, Kim J, Oh SH, Kim B, Kim SY, Jeon HS, Park JY, Kang HP, Park BJ, Han JM, Kim S (2011) Cancer-associated splicing variant of tumor suppressor AIMP2/p38: pathological implication in tumorigenesis. PLoS Genet 7(3):e1001351. doi:10.1371/journal.pgen.1001351
22. Choi JW, Lee JW, Kim JK, Jeon HK, Choi JJ, Kim DG, Kim BG, Nam DH, Kim HJ, Yun SH, Kim S (2012) Splicing variant of AIMP2 as an effective target against chemoresistant ovarian cancer. J Mol Cell Biol 4(3):164–173. doi:10.1093/jmcb/mjs018
23. Ko YG, Kim EY, Kim T, Park H, Park HS, Choi EJ, Kim S (2001) Glutamine-dependent antiapoptotic interaction of human glutaminyl-tRNA synthetase with apoptosis signal-regulating kinase 1. J Biol Chem 276(8):6030–6036. doi:10.1074/jbc.M006189200
24. Kim DG, Choi JW, Lee JY, Kim H, Oh YS, Lee JW, Tak YK, Song JM, Razin E, Yun SH, Kim S (2012) Interaction of two translational components, lysyl-tRNA synthetase and p40/37LRP, in plasma membrane promotes laminin-dependent cell migration. FASEB J 26 (10):4142–4159. doi:10.1096/fj.12-207639
25. Mukhopadhyay R, Jia J, Arif A, Ray PS, Fox PL (2009) The GAIT system: a gatekeeper of inflammatory gene expression. Trends Biochem Sci 34(7):324–331. doi:10.1016/j.tibs.2009.03.004
26. Williams TF, Mirando AC, Wilkinson B, Francklyn CS, Lounsbury KM (2013) Secreted threonyl-tRNA synthetase stimulates endothelial cell migration and angiogenesis. Sci Rep 3:1317. doi:10.1038/srep01317
27. Xu X, Shi Y, Zhang HM, Swindell EC, Marshall AG, Guo M, Kishi S, Yang XL (2012) Unique domain appended to vertebrate tRNA synthetase is essential for vascular development. Nat Commun 3:681. doi:10.1038/ncomms1686
28. Barrett T, Troup DB, Wilhite SE, Ledoux P, Rudnev D, Evangelista C, Kim IF, Soboleva A, Tomashevsky M, Marshall KA, Phillippy KH, Sherman PM, Muertter RN, Edgar R (2009) NCBI GEO: archive for high-throughput functional genomic data. Nucleic Acids Res 37 (Database issue):D885–D890. doi:10.1093/nar/gkn764
29. Parkinson H, Kapushesky M, Shojatalab M, Abeygunawardena N, Coulson R, Farne A, Holloway E, Kolesnykov N, Lilja P, Lukk M, Mani R, Rayner T, Sharma A, William E, Sarkans U, Brazma A (2007) ArrayExpress – a public database of microarray experiments and gene expression profiles. Nucleic Acids Res 35(Database issue):D747–D750. doi:10.1093/nar/gkl995

30. Hwang D, Rust AG, Ramsey S, Smith JJ, Leslie DM, Weston AD, de Atauri P, Aitchison JD, Hood L, Siegel AF, Bolouri H (2005) A data integration methodology for systems biology. Proc Natl Acad Sci USA 102(48):17296–17301. doi:10.1073/pnas.0508647102
31. Scheinin I, Myllykangas S, Borze I, Bohling T, Knuutila S, Saharinen J (2008) CanGEM: mining gene copy number changes in cancer. Nucleic Acids Res 36(Database issue): D830–D835. doi:10.1093/nar/gkm802
32. Plas DR, Thomas G (2009) Tubers and tumors: rapamycin therapy for benign and malignant tumors. Curr Opin Cell Biol 21(2):230–236. doi:10.1016/j.ceb.2008.12.013
33. Bonfils G, Jaquenoud M, Bontron S, Ostrowicz C, Ungermann C, De Virgilio C (2012) Leucyl-tRNA synthetase controls TORC1 via the EGO complex. Mol Cell 46(1):105–110. doi:10.1016/j.molcel.2012.02.009
34. Shin SH, Kim HS, Jung SH, Xu HD, Jeong YB, Chung YJ (2008) Implication of leucyl-tRNA synthetase 1 (LARS1) over-expression in growth and migration of lung cancer cells detected by siRNA targeted knock-down analysis. Exp Mol Med 40(2):229–236
35. Lukk M, Kapushesky M, Nikkila J, Parkinson H, Goncalves A, Huber W, Ukkonen E, Brazma A (2010) A global map of human gene expression. Nat Biotechnol 28(4):322–324. doi:10.1038/nbt0410-322
36. Uhlen M, Bjorling E, Agaton C, Szigyarto CA, Amini B, Andersen E, Andersson AC, Angelidou P, Asplund A, Asplund C, Berglund L, Bergstrom K, Brumer H, Cerjan D, Ekstrom M, Elobeid A, Eriksson C, Fagerberg L, Falk R, Fall J, Forsberg M, Bjorklund MG, Gumbel K, Halimi A, Hallin I, Hamsten C, Hansson M, Hedhammar M, Hercules G, Kampf C, Larsson K, Lindskog M, Lodewyckx W, Lund J, Lundeberg J, Magnusson K, Malm E, Nilsson P, Odling J, Oksvold P, Olsson I, Oster E, Ottosson J, Paavilainen L, Persson A, Rimini R, Rockberg J, Runeson M, Sivertsson A, Skollermo A, Steen J, Stenvall M, Sterky F, Stromberg S, Sundberg M, Tegel H, Tourle S, Wahlund E, Walden A, Wan J, Wernerus H, Westberg J, Wester K, Wrethagen U, Xu LL, Hober S, Ponten F (2005) A human protein atlas for normal and cancer tissues based on antibody proteomics. Mol Cell Proteomics 4(12):1920–1932. doi:10.1074/mcp.M500279-MCP200
37. Ghanipour A, Jirstrom K, Ponten F, Glimelius B, Pahlman L, Birgisson H (2009) The prognostic significance of tryptophanyl-tRNA synthetase in colorectal cancer. Cancer Epidemiol Biomarkers Prev 18(11):2949–2956. doi:10.1158/1055-9965.EPI-09-0456
38. Otani A, Slike BM, Dorrell MI, Hood J, Kinder K, Ewalt KL, Cheresh D, Schimmel P, Friedlander M (2002) A fragment of human TrpRS as a potent antagonist of ocular angiogenesis. Proc Natl Acad Sci USA 99(1):178–183. doi:10.1073/pnas.012601899
39. Liu J, Shue E, Ewalt KL, Schimmel P (2004) A new gamma-interferon-inducible promoter and splice variants of an anti-angiogenic human tRNA synthetase. Nucleic Acids Res 32(2):719–727. doi:10.1093/nar/gkh240
40. Sajish M, Zhou Q, Kishi S, Valdez DM Jr, Kapoor M, Guo M, Lee S, Kim S, Yang XL, Schimmel P (2012) Trp-tRNA synthetase bridges DNA-PKcs to PARP-1 to link IFN-gamma and p53 signaling. Nat Chem Biol 8(6):547–554. doi:10.1038/nchembio.937
41. Vellaichamy A, Sreekumar A, Strahler JR, Rajendiran T, Yu J, Varambally S, Li Y, Omenn GS, Chinnaiyan AM, Nesvizhskii AI (2009) Proteomic interrogation of androgen action in prostate cancer cells reveals roles of aminoacyl tRNA synthetases. PLoS One 4(9):e7075. doi:10.1371/journal.pone.0007075
42. Zhang J, Niu D, Sui J, Ching CB, Chen WN (2009) Protein profile in hepatitis B virus replicating rat primary hepatocytes and HepG2 cells by iTRAQ-coupled 2-D LC-MS/MS analysis: insights on liver angiogenesis. Proteomics 9(10):2836–2845. doi:10.1002/pmic.200800911
43. Calin GA, Croce CM (2006) MicroRNA signatures in human cancers. Nat Rev Cancer 6(11):857–866. doi:10.1038/nrc1997
44. Calin GA, Dumitru CD, Shimizu M, Bichi R, Zupo S, Noch E, Aldler H, Rattan S, Keating M, Rai K, Rassenti L, Kipps T, Negrini M, Bullrich F, Croce CM (2002) Frequent deletions and

down-regulation of micro-RNA genes miR15 and miR16 at 13q14 in chronic lymphocytic leukemia. Proc Natl Acad Sci USA 99(24):15524–15529. doi:10.1073/pnas.242606799

45. Bottoni A, Piccin D, Tagliati F, Luchin A, Zatelli MC, degli Uberti EC (2005) miR-15a and miR-16-1 down-regulation in pituitary adenomas. J Cell Physiol 204(1):280–285. doi:10.1002/jcp.20282

46. Sun L, Hui AM, Su Q, Vortmeyer A, Kotliarov Y, Pastorino S, Passaniti A, Menon J, Walling J, Bailey R, Rosenblum M, Mikkelsen T, Fine HA (2006) Neuronal and glioma-derived stem cell factor induces angiogenesis within the brain. Cancer Cell 9(4):287–300. doi:10.1016/j.ccr.2006.03.003

47. Sen S, Zhou H, Ripmaster T, Hittelman WN, Schimmel P, White RA (1997) Expression of a gene encoding a tRNA synthetase-like protein is enhanced in tumorigenic human myeloid leukemia cells and is cell cycle stage- and differentiation-dependent. Proc Natl Acad Sci USA 94(12):6164–6169

48. Lieber M, Smith B, Szakal A, Nelson-Rees W, Todaro G (1976) A continuous tumor-cell line from a human lung carcinoma with properties of type II alveolar epithelial cells. Int J Cancer 17(1):62–70

49. Wasenius VM, Hemmer S, Kettunen E, Knuutila S, Franssila K, Joensuu H (2003) Hepatocyte growth factor receptor, matrix metalloproteinase-11, tissue inhibitor of metalloproteinase-1, and fibronectin are up-regulated in papillary thyroid carcinoma: a cDNA and tissue microarray study. Clin Cancer Res 9(1):68–75

50. Kushner JP, Boll D, Quagliana J, Dickman S (1976) Elevated methionine-tRNA synthetase activity in human colon cancer. Proc Soc Exp Biol Med 153(2):273–276

51. Forus A, Florenes VA, Maelandsmo GM, Fodstad O, Myklebost O (1994) The protooncogene CHOP/GADD153, involved in growth arrest and DNA damage response, is amplified in a subset of human sarcomas. Cancer Genet Cytogenet 78(2):165–171

52. Nilbert M, Rydholm A, Mitelman F, Meltzer PS, Mandahl N (1995) Characterization of the 12q13-15 amplicon in soft tissue tumors. Cancer Genet Cytogenet 83(1):32–36

53. Palmer JL, Masui S, Pritchard S, Kalousek DK, Sorensen PH (1997) Cytogenetic and molecular genetic analysis of a pediatric pleomorphic sarcoma reveals similarities to adult malignant fibrous histiocytoma. Cancer Genet Cytogenet 95(2):141–147

54. Reifenberger G, Ichimura K, Reifenberger J, Elkahloun AG, Meltzer PS, Collins VP (1996) Refined mapping of 12q13-q15 amplicons in human malignant gliomas suggests CDK4/SAS and MDM2 as independent amplification targets. Cancer Res 56(22):5141–5145

55. Han JM, Park BJ, Park SG, Oh YS, Choi SJ, Lee SW, Hwang SK, Chang SH, Cho MH, Kim S (2008) AIMP2/p38, the scaffold for the multi-tRNA synthetase complex, responds to genotoxic stresses via p53. Proc Natl Acad Sci USA 105(32):11206–11211. doi:10.1073/pnas.0800297105

56. Oh YS, Kim DG, Kim G, Choi EC, Kennedy BK, Suh Y, Park BJ, Kim S (2010) Downregulation of lamin A by tumor suppressor AIMP3/p18 leads to a progeroid phenotype in mice. Aging Cell 9(5):810–822. doi:10.1111/j.1474-9726.2010.00614.x

57. Kim KJ, Park MC, Choi SJ, Oh YS, Choi EC, Cho HJ, Kim MH, Kim SH, Kim DW, Kim S, Kang BS (2008) Determination of three-dimensional structure and residues of the novel tumor suppressor AIMP3/p18 required for the interaction with ATM. J Biol Chem 283(20):14032–14040. doi:10.1074/jbc.M800859200

58. Miyaki M, Iijima T, Shiba K, Aki T, Kita Y, Yasuno M, Mori T, Kuroki T, Iwama T (2001) Alterations of repeated sequences in 5′ upstream and coding regions in colorectal tumors from patients with hereditary nonpolyposis colorectal cancer and Turcot syndrome. Oncogene 20(37):5215–5218. doi:10.1038/sj.onc.1204578

59. Park SW, Kim SS, Yoo NJ, Lee SH (2010) Frameshift mutation of MARS gene encoding an aminoacyl-tRNA synthetase in gastric and colorectal carcinomas with microsatellite instability. Gut Liver 4(3):430–431. doi:10.5009/gnl.2010.4.3.430

60. Yao P, Potdar AA, Arif A, Ray PS, Mukhopadhyay R, Willard B, Xu Y, Yan J, Saidel GM, Fox PL (2012) Coding region polyadenylation generates a truncated tRNA synthetase that counters translation repression. Cell 149(1):88–100. doi:10.1016/j.cell.2012.02.018

61. Debelenko LV, Arthur DC, Pack SD, Helman LJ, Schrump DS, Tsokos M (2003) Identification of CARS-ALK fusion in primary and metastatic lesions of an inflammatory myofibroblastic tumor. Lab Invest 83(9):1255–1265

62. Hu RJ, Lee MP, Connors TD, Johnson LA, Burn TC, Su K, Landes GM, Feinberg AP (1997) A 2.5-Mb transcript map of a tumor-suppressing subchromosomal transferable fragment from 11p15.5, and isolation and sequence analysis of three novel genes. Genomics 46(1):9–17. doi:10.1006/geno.1997.4981

63. Reid LH, Davies C, Cooper PR, Crider-Miller SJ, Sait SN, Nowak NJ, Evans G, Stanbridge EJ, deJong P, Shows TB, Weissman BE, Higgins MJ (1997) A 1-Mb physical map and PAC contig of the imprinted domain in 11p15.5 that contains TAPA1 and the BWSCR1/WT2 region. Genomics 43(3):366–375. doi:10.1006/geno.1997.4826

64. Xu XL, Wu LC, Du F, Davis A, Peyton M, Tomizawa Y, Maitra A, Tomlinson G, Gazdar AF, Weissman BE, Bowcock AM, Baer R, Minna JD (2001) Inactivation of human SRBC, located within the 11p15.5-p15.4 tumor suppressor region, in breast and lung cancers. Cancer Res 61(21):7943–7949

65. Zhao B, Bepler G (2001) Transcript map and complete genomic sequence for the 310 kb region of minimal allele loss on chromosome segment 11p15.5 in non-small-cell lung cancer. Oncogene 20(56):8154–8164. doi:10.1038/sj.onc.1205027

66. Zhou W, Feng X, Li H, Wang L, Zhu B, Liu W, Zhao M, Yao K, Ren C (2009) Inactivation of LARS2, located at the commonly deleted region 3p21.3, by both epigenetic and genetic mechanisms in nasopharyngeal carcinoma. Acta Biochim Biophys Sin 41(1):54–62

67. Pui CH, Jeha S (2007) New therapeutic strategies for the treatment of acute lymphoblastic leukaemia. Nat Rev Drug Discov 6(2):149–165. doi:10.1038/nrd2240

68. Palmer RH, Vernersson E, Grabbe C, Hallberg B (2009) Anaplastic lymphoma kinase: signalling in development and disease. Biochem J 420(3):345–361. doi:10.1042/BJ20090387

69. Lee SW, Kang YS, Kim S (2006) Multifunctional proteins in tumorigenesis: aminoacyl-tRNA synthetases and translational components. Curr Proteomics 3(4):233–247

70. Ubeda M, Schmitt-Ney M, Ferrer J, Habener JF (1999) CHOP/GADD153 and methionyl-tRNA synthetase (MetRS) genes overlap in a conserved region that controls mRNA stability. Biochem Biophys Res Commun 262(1):31–38. doi:10.1006/bbrc.1999.1140

71. Ray PS, Fox PL (2007) A post-transcriptional pathway represses monocyte VEGF-A expression and angiogenic activity. EMBO J 26(14):3360–3372. doi:10.1038/sj.emboj.7601774

72. Karni R, de Stanchina E, Lowe SW, Sinha R, Mu D, Krainer AR (2007) The gene encoding the splicing factor SF2/ASF is a proto-oncogene. Nat Struct Mol Biol 14(3):185–193. doi:10.1038/nsmb1209

73. Wakasugi K, Slike BM, Hood J, Otani A, Ewalt KL, Friedlander M, Cheresh DA, Schimmel P (2002) A human aminoacyl-tRNA synthetase as a regulator of angiogenesis. Proc Natl Acad Sci USA 99(1):173–177. doi:10.1073/pnas.012602099

74. Turpaev KT, Zakhariev VM, Sokolova IV, Narovlyansky AN, Amchenkova AM, Justesen J, Frolova LY (1996) Alternative processing of the tryptophanyl-tRNA synthetase mRNA from interferon-treated human cells. Eur J Biochem (FEBS) 240(3):732–737

75. Tolstrup AB, Bejder A, Fleckner J, Justesen J (1995) Transcriptional regulation of the interferon-gamma-inducible tryptophanyl-tRNA synthetase includes alternative splicing. J Biol Chem 270(1):397–403

76. Paley EL, Paley DE, Merkulova-Rainon T, Subbarayan PR (2011) Hypoxia signature of splice forms of tryptophanyl-tRNA synthetase marks pancreatic cancer cells with distinct metastatic abilities. Pancreas 40(7):1043–1056. doi:10.1097/MPA.0b013e318222e635

77. Tzima E, Reader JS, Irani-Tehrani M, Ewalt KL, Schwartz MA, Schimmel P (2005) VE-cadherin links tRNA synthetase cytokine to anti-angiogenic function. J Biol Chem 280(4):2405–2408. doi:10.1074/jbc.C400431200

78. Yang XL, Guo M, Kapoor M, Ewalt KL, Otero FJ, Skene RJ, McRee DE, Schimmel P (2007) Functional and crystal structure analysis of active site adaptations of a potent anti-angiogenic human tRNA synthetase. Structure 15(7):793–805. doi:10.1016/j.str.2007.05.009

79. Guo M, Schimmel P, Yang XL (2010) Functional expansion of human tRNA synthetases achieved by structural inventions. FEBS Lett 584(2):434–442. doi:10.1016/j.febslet. 2009.11.064

80. Wakasugi K, Schimmel P (1999) Highly differentiated motifs responsible for two cytokine activities of a split human tRNA synthetase. J Biol Chem 274(33):23155–23159

81. Vo MN, Yang XL, Schimmel P (2011) Dissociating quaternary structure regulates cell-signaling functions of a secreted human tRNA synthetase. J Biol Chem 286 (13):11563–11568. doi:10.1074/jbc.C110.213876

82. Zeng R, Chen YC, Zeng Z, Liu XX, Liu R, Qiang O, Li X (2012) Inhibition of mini-TyrRS-induced angiogenesis response in endothelial cells by VE-cadherin-dependent mini-TrpRS. Heart Vessels 27(2):193–201. doi:10.1007/s00380-011-0137-1

83. van Horssen R, Eggermont AM, ten Hagen TL (2006) Endothelial monocyte-activating polypeptide-II and its functions in (patho)physiological processes. Cytokine Growth Factor Rev 17(5):339–348. doi:10.1016/j.cytogfr.2006.08.001

84. Behrensdorf HA, van de Craen M, Knies UE, Vandenabeele P, Clauss M (2000) The endothelial monocyte-activating polypeptide II (EMAP II) is a substrate for caspase-7. FEBS Lett 466(1):143–147

85. Knies UE, Behrensdorf HA, Mitchell CA, Deutsch U, Risau W, Drexler HC, Clauss M (1998) Regulation of endothelial monocyte-activating polypeptide II release by apoptosis. Proc Natl Acad Sci USA 95(21):12322–12327

86. Barnett G, Jakobsen AM, Tas M, Rice K, Carmichael J, Murray JC (2000) Prostate adeno-carcinoma cells release the novel proinflammatory polypeptide EMAP-II in response to stress. Cancer Res 60(11):2850–2857

87. Park SG, Kang YS, Ahn YH, Lee SH, Kim KR, Kim KW, Koh GY, Ko YG, Kim S (2002) Dose-dependent biphasic activity of tRNA synthetase-associating factor, p43, in angiogene-sis. J Biol Chem 277(47):45243–45248. doi:10.1074/jbc.M207934200

88. Tas MP, Murray JC (1996) Endothelial-monocyte-activating polypeptide II. Int J Biochem Cell Biol 28(8):837–841

89. Arif A, Jia J, Mukhopadhyay R, Willard B, Kinter M, Fox PL (2009) Two-site phosphoryla-tion of EPRS coordinates multimodal regulation of noncanonical translational control activ-ity. Mol Cell 35(2):164–180. doi:10.1016/j.molcel.2009.05.028

90. Kwon NH, Kang T, Lee JY, Kim HH, Kim HR, Hong J, Oh YS, Han JM, Ku MJ, Lee SY, Kim S (2011) Dual role of methionyl-tRNA synthetase in the regulation of translation and tumor suppressor activity of aminoacyl-tRNA synthetase-interacting multifunctional protein-3. Proc Natl Acad Sci USA 108(49):19635–19640. doi:10.1073/pnas.1103922108

91. Hsieh AC, Liu Y, Edlind MP, Ingolia NT, Janes MR, Sher A, Shi EY, Stumpf CR, Christensen C, Bonham MJ, Wang S, Ren P, Martin M, Jessen K, Feldman ME, Weissman JS, Shokat KM, Rommel C, Ruggero D (2012) The translational landscape of mTOR signalling steers cancer initiation and metastasis. Nature 485(7396):55–61. doi:10.1038/nature10912

92. Guertin DA, Sabatini DM (2005) An expanding role for mTOR in cancer. Trends Mol Med 11(8):353–361. doi:10.1016/j.molmed.2005.06.007

93. Sancak Y, Peterson TR, Shaul YD, Lindquist RA, Thoreen CC, Bar-Peled L, Sabatini DM (2008) The rag GTPases bind raptor and mediate amino acid signaling to mTORC1. Science 320(5882):1496–1501. doi:10.1126/science.1157535

94. Kim E, Goraksha-Hicks P, Li L, Neufeld TP, Guan KL (2008) Regulation of TORC1 by Rag GTPases in nutrient response. Nat Cell Biol 10(8):935–945. doi:10.1038/ncb1753

95. Lee YN, Nechushtan H, Figov N, Razin E (2004) The function of lysyl-tRNA synthetase and Ap4A as signaling regulators of MITF activity in FcepsilonRI-activated mast cells. Immunity 20(2):145–151

96. Yannay-Cohen N, Carmi-Levy I, Kay G, Yang CM, Han JM, Kemeny DM, Kim S, Nechushtan H, Razin E (2009) LysRS serves as a key signaling molecule in the immune response by regulating gene expression. Mol Cell 34(5):603–611. doi:10.1016/j.molcel.2009.05.019

97. Ofir-Birin Y, Fang P, Bennett SP, Zhang HM, Wang J, Rachmin I, Shapiro R, Song J, Dagan A, Pozo J, Kim S, Marshall AG, Schimmel P, Yang XL, Nechushtan H, Razin E, Guo M (2013) Structural switch of lysyl-tRNA synthetase between translation and transcription. Mol Cell 49(1):30–42. doi:10.1016/j.molcel.2012.10.010

98. Berno V, Porrini D, Castiglioni F, Campiglio M, Casalini P, Pupa SM, Balsari A, Menard S, Tagliabue E (2005) The 67 kDa laminin receptor increases tumor aggressiveness by remodeling laminin-1. Endocr Relat Cancer 12(2):393–406. doi:10.1677/erc.1.00870

99. van den Brule FA, Buicu C, Berchuck A, Bast RC, Deprez M, Liu FT, Cooper DN, Pieters C, Sobel ME, Castronovo V (1996) Expression of the 67-kD laminin receptor, galectin-1, and galectin-3 in advanced human uterine adenocarcinoma. Hum Pathol 27(11):1185–1191

100. Sanjuan X, Fernandez PL, Miquel R, Munoz J, Castronovo V, Menard S, Palacin A, Cardesa A, Campo E (1996) Overexpression of the 67-kD laminin receptor correlates with tumour progression in human colorectal carcinoma. J Pathol 179(4):376–380. doi:10.1002/(SICI)1096-9896(199608)179:4<376::AID-PATH591>3.0.CO;2-V

101. Fontanini G, Vignati S, Chine S, Lucchi M, Mussi A, Angeletti CA, Menard S, Castronovo V, Bevilacqua G (1997) 67-Kilodalton laminin receptor expression correlates with worse prognostic indicators in non-small cell lung carcinomas. Clin Cancer Res 3(2):227–231

102. Liu L, Sun L, Zhao P, Yao L, Jin H, Liang S, Wang Y, Zhang D, Pang Y, Shi Y, Chai N, Zhang H (2010) Hypoxia promotes metastasis in human gastric cancer by up-regulating the 67-kDa laminin receptor. Cancer Sci 101(7):1653–1660. doi:10.1111/j.1349-7006.2010.01592.x

103. Wakasugi K, Slike BM, Hood J, Ewalt KL, Cheresh DA, Schimmel P (2002) Induction of angiogenesis by a fragment of human tyrosyl-tRNA synthetase. J Biol Chem 277(23):20124–20126. doi:10.1074/jbc.C200126200

104. Wakasugi K, Schimmel P (1999) Two distinct cytokines released from a human aminoacyl-tRNA synthetase. Science 284(5411):147–151

105. Kapoor M, Otero FJ, Slike BM, Ewalt KL, Yang XL (2009) Mutational separation of aminoacylation and cytokine activities of human tyrosyl-tRNA synthetase. Chem Biol 16(5):531–539. doi:10.1016/j.chembiol.2009.03.006

106. Kapoor M, Zhou Q, Otero F, Myers CA, Bates A, Belani R, Liu J, Luo JK, Tzima E, Zhang DE, Yang XL, Schimmel P (2008) Evidence for annexin II-S100A10 complex and plasmin in mobilization of cytokine activity of human TrpRS. J Biol Chem 283(4):2070–2077. doi:10.1074/jbc.M706028200

107. Yu Y, Liu Y, Shen N, Xu X, Xu F, Jia J, Jin Y, Arnold E, Ding J (2004) Crystal structure of human tryptophanyl-tRNA synthetase catalytic fragment: insights into substrate recognition, tRNA binding, and angiogenesis activity. J Biol Chem 279(9):8378–8388. doi:10.1074/jbc.M311284200

108. Patel SD, Ciatto C, Chen CP, Bahna F, Rajebhosale M, Arkus N, Schieren I, Jessell TM, Honig B, Price SR, Shapiro L (2006) Type II cadherin ectodomain structures: implications for classical cadherin specificity. Cell 124(6):1255–1268. doi:10.1016/j.cell.2005.12.046

109. Zhou Q, Kapoor M, Guo M, Belani R, Xu X, Kiosses WB, Hanan M, Park C, Armour E, Do MH, Nangle LA, Schimmel P, Yang XL (2010) Orthogonal use of a human tRNA synthetase active site to achieve multifunctionality. Nat Struct Mol Biol 17(1):57–61. doi:10.1038/nsmb.1706

110. Labirua A, Lundberg IE (2010) Interstitial lung disease and idiopathic inflammatory myopathies: progress and pitfalls. Curr Opin Rheumatol 22(6):633–638. doi:10.1097/BOR.0b013e32833f1970

111. Keller TL, Zocco D, Sundrud MS, Hendrick M, Edenius M, Yum J, Kim YJ, Lee HK, Cortese JF, Wirth DF, Dignam JD, Rao A, Yeo CY, Mazitschek R, Whitman M (2012) Halofuginone

and other febrifugine derivatives inhibit prolyl-tRNA synthetase. Nat Chem Biol 8(3):311–317. doi:10.1038/nchembio.790

112. Mazumder B, Sampath P, Seshadri V, Maitra RK, DiCorleto PE, Fox PL (2003) Regulated release of L13a from the 60S ribosomal subunit as a mechanism of transcript-specific translational control. Cell 115(2):187–198

113. Sampath P, Mazumder B, Seshadri V, Gerber CA, Chavatte L, Kinter M, Ting SM, Dignam JD, Kim S, Driscoll DM, Fox PL (2004) Noncanonical function of glutamyl-prolyl-tRNA synthetase: gene-specific silencing of translation. Cell 119(2):195–208. doi:10.1016/j.cell.2004.09.030

114. Jia J, Arif A, Ray PS, Fox PL (2008) WHEP domains direct noncanonical function of glutamyl-prolyl tRNA synthetase in translational control of gene expression. Mol Cell 29 (6):679–690. doi:10.1016/j.molcel.2008.01.010

115. Fukui H, Hanaoka R, Kawahara A (2009) Noncanonical activity of seryl-tRNA synthetase is involved in vascular development. Circ Res 104(11):1253–1259. doi:10.1161/CIRCRESAHA.108.191189

116. Herzog W, Muller K, Huisken J, Stainier DY (2009) Genetic evidence for a noncanonical function of seryl-tRNA synthetase in vascular development. Circ Res 104(11):1260–1266. doi:10.1161/CIRCRESAHA.108.191718

117. Ruggero D, Pandolfi PP (2003) Does the ribosome translate cancer? Nat Rev Cancer 3(3):179–192. doi:10.1038/nrc1015

118. Drygin D, Lin A, Bliesath J, Ho CB, O'Brien SE, Proffitt C, Omori M, Haddach M, Schwaebe MK, Siddiqui-Jain A, Streiner N, Quin JE, Sanij E, Bywater MJ, Hannan RD, Ryckman D, Anderes K, Rice WG (2011) Targeting RNA polymerase I with an oral small molecule CX-5461 inhibits ribosomal RNA synthesis and solid tumor growth. Cancer Res 71(4):1418–1430. doi:10.1158/0008-5472.CAN-10-1728

119. Netzer N, Goodenbour JM, David A, Dittmar KA, Jones RB, Schneider JR, Boone D, Eves EM, Rosner MR, Gibbs JS, Embry A, Dolan B, Das S, Hickman HD, Berglund P, Bennink JR, Yewdell JW, Pan T (2009) Innate immune and chemically triggered oxidative stress modifies translational fidelity. Nature 462(7272):522–526. doi:10.1038/nature08576

120. De Luca A, Sanna F, Sallese M, Ruggiero C, Grossi M, Sacchetta P, Rossi C, De Laurenzi V, Di Ilio C, Favaloro B (2010) Methionine sulfoxide reductase A down-regulation in human breast cancer cells results in a more aggressive phenotype. Proc Natl Acad Sci USA 107(43):18628–18633. doi:10.1073/pnas.1010171107

121. Ma Y, Hendershot LM (2004) The role of the unfolded protein response in tumour development: friend or foe? Nat Rev Cancer 4(12):966–977. doi:10.1038/nrc1505

122. Irvine GB, El-Agnaf OM, Shankar GM, Walsh DM (2008) Protein aggregation in the brain: the molecular basis for Alzheimer's and Parkinson's diseases. Mol Med 14(7–8):451–464. doi:10.2119/2007-00100.Irvine

123. Lee JW, Beebe K, Nangle LA, Jang J, Longo-Guess CM, Cook SA, Davisson MT, Sundberg JP, Schimmel P, Ackerman SL (2006) Editing-defective tRNA synthetase causes protein misfolding and neurodegeneration. Nature 443(7107):50–55. doi:10.1038/nature05096

124. Beebe K, Mock M, Merriman E, Schimmel P (2008) Distinct domains of tRNA synthetase recognize the same base pair. Nature 451(7174):90–93. doi:10.1038/nature06454

125. Ahel I, Korencic D, Ibba M, Soll D (2003) Trans-editing of mischarged tRNAs. Proc Natl Acad Sci USA 100(26):15422–15427. doi:10.1073/pnas.2136934100

126. Chen Z, Seimiya H, Naito M, Mashima T, Kizaki A, Dan S, Imaizumi M, Ichijo H, Miyazono K, Tsuruo T (1999) ASK1 mediates apoptotic cell death induced by genotoxic stress. Oncogene 18(1):173–180. doi:10.1038/sj.onc.1202276

127. Kang T, Kwon NH, Lee JY, Park MC, Kang E, Kim HH, Kang TJ, Kim S (2012) AIMP3/p18 controls translational initiation by mediating the delivery of charged initiator tRNA to initiation complex. J Mol Biol 423(4):475–481. doi:10.1016/j.jmb.2012.07.020

128. Han JM, Lee MJ, Park SG, Lee SH, Razin E, Choi EC, Kim S (2006) Hierarchical network between the components of the multi-tRNA synthetase complex: implications for complex formation. J Biol Chem 281(50):38663–38667. doi:10.1074/jbc.M605211200

129. Lee YS, Han JM, Son SH, Choi JW, Jeon EJ, Bae SC, Park YI, Kim S (2008) AIMP1/p43 downregulates TGF-beta signaling via stabilization of smurf2. Biochem Biophys Res Commun 371(3):395–400. doi:10.1016/j.bbrc.2008.04.099

130. Yi JS, Lee JY, Chi SG, Kim JH, Park SG, Kim S, Ko YG (2005) Aminoacyl-tRNA synthetase-interacting multi-functional protein, p43, is imported to endothelial cells via lipid rafts. J Cell Biochem 96(6):1286–1295. doi:10.1002/jcb.20632

131. Jackson VC, Dewilde S, Albo AG, Lis K, Corpillo D, Canepa B (2011) The activity of aminoacyl-tRNA synthetase-interacting multi-functional protein 1 (AIMP1) on endothelial cells is mediated by the assembly of a cytoskeletal protein complex. J Cell Biochem 112(7):1857–1868. doi:10.1002/jcb.23104

132. Lee BC, O'Sullivan I, Kim E, Park SG, Hwang SY, Cho D, Kim TS (2009) A DNA adjuvant encoding a fusion protein between anti-CD3 single-chain Fv and AIMP1 enhances T helper type 1 cell-mediated immune responses in antigen-sensitized mice. Immunology 126(1):84–91. doi:10.1111/j.1365-2567.2008.02880.x

133. Han JM, Myung H, Kim S (2010) Antitumor activity and pharmacokinetic properties of ARS-interacting multi-functional protein 1 (AIMP1/p43). Cancer Lett 287(2):157–164. doi:10.1016/j.canlet.2009.06.005

134. Park H, Park SG, Kim J, Ko YG, Kim S (2002) Signaling pathways for TNF production induced by human aminoacyl-tRNA synthetase-associating factor, p43. Cytokine 20(4):148–153

135. Park SG, Shin H, Shin YK, Lee Y, Choi EC, Park BJ, Kim S (2005) The novel cytokine p43 stimulates dermal fibroblast proliferation and wound repair. Am J Pathol 166(2):387–398. doi:10.1016/S0002-9440(10)62262-6

136. Kwon HS, Park MC, Kim DG, Cho K, Park YW, Han JM, Kim S (2012) Identification of CD23 as a functional receptor for the proinflammatory cytokine AIMP1/p43. J Cell Sci 125(Pt 19):4620–4629. doi:10.1242/jcs.108209

137. Janssens S, Tinel A, Lippens S, Tschopp J (2005) PIDD mediates NF-kappaB activation in response to DNA damage. Cell 123(6):1079–1092. doi:10.1016/j.cell.2005.09.036

138. Deng J, Harding HP, Raught B, Gingras AC, Berlanga JJ, Scheuner D, Kaufman RJ, Ron D, Sonenberg N (2002) Activation of GCN2 in UV-irradiated cells inhibits translation. Curr Biol 12(15):1279–1286

139. Schimmel P, Tao J, Hill J (1998) Aminoacyl tRNA synthetases as targets for new anti-infectives. FASEB J 12(15):1599–1609

140. Hurdle JG, O'Neill AJ, Chopra I (2005) Prospects for aminoacyl-tRNA synthetase inhibitors as new antimicrobial agents. Antimicrob Agents Chemother 49(12):4821–4833. doi:10.1128/AAC.49.12.4821-4833.2005

141. Silvian LF, Wang J, Steitz TA (1999) Insights into editing from an ile-tRNA synthetase structure with tRNAile and mupirocin. Science 285(5430):1074–1077

142. Jiang S, Zeng Q, Gettayacamin M, Tungtaeng A, Wannaying S, Lim A, Hansukjariya P, Okunji CO, Zhu S, Fang D (2005) Antimalarial activities and therapeutic properties of febrifugine analogs. Antimicrob Agents Chemother 49(3):1169–1176. doi:10.1128/AAC.49.3.1169-1176.2005

143. Zhou H, Sun L, Yang XL, Schimmel P (2013) ATP-directed capture of bioactive herbal-based medicine on human tRNA synthetase. Nature 494(7435):121–124. doi:10.1038/nature11774

144. Kawamura T, Liu D, Towle MJ, Kageyama R, Tsukahara N, Wakabayashi T, Littlefield BA (2003) Anti-angiogenesis effects of borrelidin are mediated through distinct pathways: threonyl-tRNA synthetase and caspases are independently involved in suppression of proliferation and induction of apoptosis in endothelial cells. J Antibiot 56(8):709–715

145. Habibi D, Ogloff N, Jalili RB, Yost A, Weng AP, Ghahary A, Ong CJ (2012) Borrelidin, a small molecule nitrile-containing macrolide inhibitor of threonyl-tRNA synthetase, is a potent inducer of apoptosis in acute lymphoblastic leukemia. Invest New Drugs 30(4):1361–1370. doi:10.1007/s10637-011-9700-y

146. Moss SJ, Carletti I, Olano C, Sheridan RM, Ward M, Math V, Nur EAM, Brana AF, Zhang MQ, Leadlay PF, Mendez C, Salas JA, Wilkinson B (2006) Biosynthesis of the angiogenesis inhibitor borrelidin: directed biosynthesis of novel analogues. Chem Commun (Camb) (22):2341–2343. doi:10.1039/b602931k
147. Wilkinson B, Gregory MA, Moss SJ, Carletti I, Sheridan RM, Kaja A, Ward M, Olano C, Mendez C, Salas JA, Leadlay PF, vanGinckel R, Zhang MQ (2006) Separation of anti-angiogenic and cytotoxic activities of borrelidin by modification at the C17 side chain. Bioorg Med Chem Lett 16(22):5814–5817. doi:10.1016/j.bmcl.2006.08.073
148. Harisi R, Kenessey I, Olah JN, Timar F, Babo I, Pogany G, Paku S, Jeney A (2009) Differential inhibition of single and cluster type tumor cell migration. Anticancer Res 29(8):2981–2985
149. Wakabayashi T, Kageyama R, Naruse N, Tsukahara N, Funahashi Y, Kitoh K, Watanabe Y (1997) Borrelidin is an angiogenesis inhibitor; disruption of angiogenic capillary vessels in a rat aorta matrix culture model. J Antibiot 50(8):671–676
150. Zeng R, Chen YC, Zeng Z, Liu WQ, Liu XX, Liu R, Qiang O, Li X (2010) Different angiogenesis effect of mini-TyrRS/mini-TrpRS by systemic administration of modified siRNAs in rats with acute myocardial infarction. Heart Vessels 25(4):324–332. doi:10.1007/s00380-009-1200-z
151. Kim TS, Lee BC, Kim E, Cho D, Cohen EP (2008) Gene transfer of AIMP1 and B7.1 into epitope-loaded, fibroblasts induces tumor-specific CTL immunity, and prolongs the survival period of tumor-bearing mice. Vaccine 26(47):5928–5934. doi:10.1016/j.vaccine.2008. 08.051
152. Chang SH, Chung YS, Hwang SK, Kwon JT, Minai-Tehrani A, Kim S, Park SB, Kim YS, Cho MH (2012) Lentiviral vector-mediated shRNA against AIMP2-DX2 suppresses lung cancer cell growth through blocking glucose uptake. Mol Cells 33(6):553–562. doi:10.1007/s10059-012-2269-2
153. Won YS, Lee SW (2012) Selective regression of cancer cells expressing a splicing variant of AIMP2 through targeted RNA replacement by trans-splicing ribozyme. J Biotechnol 158 (1–2):44–49. doi:10.1016/j.jbiotec.2012.01.006
154. Thompson DM, Parker R (2009) Stressing out over tRNA cleavage. Cell 138(2):215–219. doi:10.1016/j.cell.2009.07.001
155. Mei Y, Yong J, Liu H, Shi Y, Meinkoth J, Dreyfuss G, Yang X (2010) tRNA binds to cytochrome c and inhibits caspase activation. Mol Cell 37(5):668–678. doi:10.1016/j.molcel.2010.01.023
156. Nechushtan H, Kim S, Kay G, Razin E (2009) Chapter 1: the physiological role of lysyl tRNA synthetase in the immune system. Adv Immunol 103:1–27. doi:10.1016/S0065-2776(09) 03001-6
157. Jonker R, Engelen MP, Deutz NE (2012) Role of specific dietary amino acids in clinical conditions. Br J Nutr 108(Suppl 2):S139–S148. doi:10.1017/S0007114512002358
158. Townsend KN, Hughson LR, Schlie K, Poon VI, Westerback A, Lum JJ (2012) Autophagy inhibition in cancer therapy: metabolic considerations for antitumor immunity. Immunol Rev 249(1):176–194. doi:10.1111/j.1600-065X.2012.01141.x
159. Liu Q, Thoreen C, Wang J, Sabatini D, Gray NS (2009) mTOR mediated anti-cancer drug discovery. Drug Discov Today Ther Strateg 6(2):47–55. doi:10.1016/j.ddstr.2009.12.001
160. Silvera D, Formenti SC, Schneider RJ (2010) Translational control in cancer. Nat Rev Cancer 10(4):254–266. doi:10.1038/nrc2824
161. Li R, Guan MX (2010) Human mitochondrial leucyl-tRNA synthetase corrects mitochondrial dysfunctions due to the tRNALeu(UUR) A3243G mutation, associated with mitochondrial encephalomyopathy, lactic acidosis, and stroke-like symptoms and diabetes. Mol Cell Biol 30(9):2147–2154. doi:10.1128/MCB.01614-09

Pathogenic Implications of Human Mitochondrial Aminoacyl-tRNA Synthetases

Hagen Schwenzer, Joffrey Zoll, Catherine Florentz, and Marie Sissler

Abstract Mitochondria are considered as the powerhouse of eukaryotic cells. They host several central metabolic processes fueling the oxidative phosphorylation pathway (OXPHOS) that produces ATP from its precursors ADP and inorganic phosphate Pi (PPi). The respiratory chain complexes responsible for the OXPHOS pathway are formed from complementary sets of protein subunits encoded by the nuclear genome and the mitochondrial genome, respectively. The expression of the mitochondrial genome requires a specific and fully active translation machinery from which aminoacyl-tRNA synthetases (aaRSs) are key actors. Whilst the macromolecules involved in mammalian mitochondrial translation have been under investigation for many years, there has been an explosion of interest in human mitochondrial aaRSs (mt-aaRSs) since the discovery of a large (and growing) number of mutations in these genes that are linked to a variety of neurodegenerative disorders. Herein we will review the present knowledge on mt-aaRSs in terms of their biogenesis, their connection to mitochondrial respiration, i.e., the respiratory chain (RC) complexes, and to the mitochondrial translation machinery. The pathology-related mutations detected so far are described, with special attention given to their impact on mt-aaRSs biogenesis, functioning, and/or subsequent

---

Note: Rigorously, amino acid conversion of a given mutation should be preceded by the letter "p." to indicate that the protein level is considered. For example, the 172C > G nucleotide change engenders the p.R58G mutation in *DARS2* (referencing the gene) or mt-AspRS (referencing the protein). For sake of simplicity, the "p." is omitted throughout the chapter.

H. Schwenzer, C. Florentz, and M. Sissler (✉)
Architecture et Réactivité de l'ARN, CNRS, Université de Strasbourg, IBMC, 15 rue René Descartes, 67084 Strasbourg Cedex, France
e-mail: H.Schwenzer@ibmc-cnrs.unistra.fr; C.Florentz@ibmc-cnrs.unistra.fr; M.Sissler@ibmc-cnrs.unistra.fr

J. Zoll
EA 3072, Université de Strasbourg, Faculté de médecine, 4 rue Kirschleger, 67085 Strasbourg, France
e-mail: Joffrey.zoll@unistra.fr

activities. The collected data to date shed light on the diverse routes that are linking primary molecular possible impact of a mutation to its phenotypic expression. It is envisioned that a variety of mechanisms, inside and outside the translation machinery, would play a role on the heterogeneous manifestations of mitochondrial disorders.

Keywords Aminoacyl-tRNA synthetase · Human mitochondrial disorders · Pathology-related mutations · Respiratory chain defects

Contents

1 Mt-aaRSs and Mitochondrial ATP Synthesis	249
1.1 Mitochondrial Respiratory Chain Complexes	249
1.2 The Human Mitochondrial Translation Machinery	252
1.3 Link Between Mitochondrial Translation, Mitochondrial Respiration, and Mitochondrial Disorders	253
1.4 Biochemical Analysis of Mitochondrial Respiration: A Potential Diagnostic Tool for the Detection of Mitochondrial Translation Defects	254
2 Mt-aaRSs in Mitochondrial Translation	257
2.1 Nuclear-Encoded aaRSs of Mitochondrial Location and Evolutionary Considerations	258
2.2 Mt-aaRSs are Imported Proteins	259
2.3 Structural Insights of mt-aaRSs	261
2.4 Some Functional Peculiarities	263
3 Mt-aaRSs in Human Disorders	264
3.1 Discovery of aaRS-Related Disorders	264
3.2 The Present-Day List of Pathology-Related Mutations Within mt-aaRS Encoding Genes	264
3.3 Compound Heterozygous vs. Homozygous States	274
4 Diverse Molecular Impacts	276
4.1 Impact of Mutations on mt-aaRSs Biogenesis	277
4.2 Impact of Mutations on mt-aaRSs Function	279
4.3 Impact of Mutations on Mitochondrial Translation and Activity of the Respiratory Chain Complexes	282
5 Outlook	284
References	286

Abbreviations

AaRS	Aminoacyl-tRNA synthetase (specificity is indicated by the name of the amino acid (abbreviated in a three-letter code) transferred to the cognate tRNA. As an example, AspRS stands for aspartyl-tRNA synthetase)
mt	Mitochondrial
MTS	Mitochondrial targeting sequence
RC	Respiratory chain

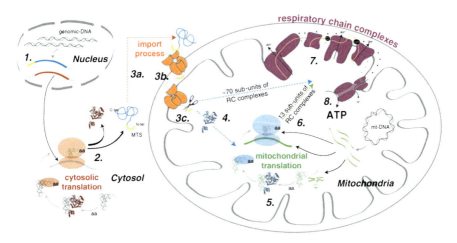

Fig. 1 From mitochondrial aminoacyl-tRNA synthetases to mitochondrial ATP synthesis. The route from the place of encoding of mt-aaRSs (the nucleus) to their place of biosynthesis (the cytosol) and their place of use (the mitochondria) is schematized. Mt-aaRSs biogenesis comprises mRNAs expression and processing (*1*), mt-aaRSs synthesis (*2*), import process into mitochondria (*3a* addressing; *3b* translocation; *3c* processing), and proper folding, oligomerization, and stability upon entry to mitochondria (*4*). Mt-aaRSs functioning includes amino acid activation, tRNA recognition, tRNA charging (*5*). Mt-aaRSs are devoted to the mitochondrial translation, and thus, to the synthesis of the 13 mt-DNA-encoded respiratory chain (RC) complexes (*6*), for which the activity (*7*) ultimately lead to ATP production (*8*). Of note, all other sub-units of the RC complexes (~70) are also imported from the cytosol

1 Mt-aaRSs and Mitochondrial ATP Synthesis

In order to facilitate the understanding of the connection between human mt-aaRSs and ATP synthesis, the route from their place of encoding (the nucleus) to their place of biogenesis (the cytosol) and their place of use (the mitochondria) is schematized in Fig. 1.

1.1 Mitochondrial Respiratory Chain Complexes

One of the most prominent functions of mitochondria is the production of cellular free energy in the form of ATP, in a process known as oxidative phosphorylation (OXPHOS). This process takes place in five large multi-subunit complexes (the respiratory chain complexes) that are located in the inner mitochondrial membrane (Fig. 2a). Complexes I to IV, accompanied by the mobile elements Coenzyme Q and cytochrome *c*, allow for the activity of Complex V, the ATP synthase. In mitochondria, the final oxidation of nutrients releases CO_2 (mainly through the Krebs cycle) concomitantly with the reduction of NAD^+ into

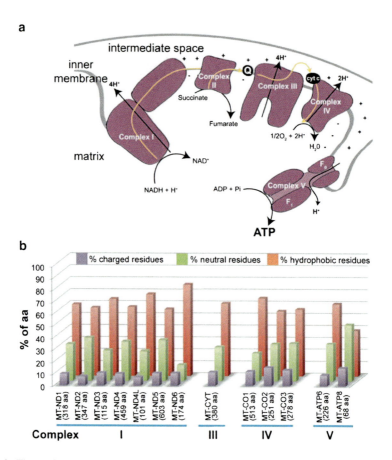

Fig. 2 The respiratory chain complexes. (**a**) Organization of the complexes along the mitochondrial inner membrane. Complex I: NADH: ubiquinone reductase; Complex II: Succinate-coenzyme Q reductase; Complex III: Coenzyme Q: cytochrome c oxidoreductase; Complex IV: Cytochrome c oxidase; Complex V: ATP synthase. Q stands for Coenzyme Q, and cyt c for cytochrome c. Sub-units composition of the five complexes in given is Table 1. (**b**) Composition of the 13 mt-DNA-encoded sub-units

NADH + H$^+$ and FAD into FADH$_2$. Oxidation of these hydrogen carriers involves the transfer of protons and electrons to the respiratory chain Complexes I and II respectively, followed by the channeling of these electrons through Complexes III and IV, where the electron is finally accepted by oxygen to form metabolic water. During electron transport, proton pumps in Complexes I, III, and IV become activated, leading to the expulsion of protons from the mitochondrial matrix to the mitochondrial intra-membrane space. The generated proton gradient (chemical potential) combined with the electron's movement (electrical potential) lead to return of protons to the matrix, thus activating ATP synthase by a proton-motive force. This enzyme binds ADP to inorganic phosphate and generates ATP. In summary, final oxidation of nutrients into CO$_2$ and H$_2$O takes place inside mitochondria and is directly correlated to oxygen consumption and ATP synthesis by the respiratory chain complexes [1].

Table 1 Name and composition of sub-units or the respiratory chain complexes

Complex	Name	Mitochondrial-encoded sub-units	Nuclear-encoded sub-units
I	*NADH:ubiquinone reductase*	7 (MT-ND1, MT-ND2, MT-ND3, MT-ND4, MT-ND4L, MT-ND5, MT-ND6)	38 (NDUFA1, NDUFA2, NDUFA3, NDUFA4, NDUFA5, NDUFA6, NDUFA7, NDUFA8, NDUFA9, NDUFA10, NDUFA11, NDUFA12, NDUFA13, NDUFB1, NDUFAB1, NDUFB2, NDUFB3, NDUFB4, NDUFB5, NDUFB6, NDUFB7, NDUFB8, NDUFB9, NDUFB10, NDUFB11, NDUFC1, NDUFC2, NDUFS1, NDUFS2, NDUFS3, NDUFS4, NDUFS5, NDUFS6, NDUFS7, NDUFS8, NDUFV1, NDUFV2, NDUFV3)
II	*Succinate: coenzyme Q reductase*	None	4 (SDHA, SDHB, SDHC, SDHD)
III	*Coenzyme Q: cytochrome c oxidoreductase*	1 (MT-CYB)	9 (CYC1, UQCR10, UQCR11, UQCRB, UQCRC1, UQCRC2, UQCRFS1, UQCRH, UQCRQ)
IV	*Cytochrome c oxidase*	3 (MT-CO1, MT-CO2, MT-CO3)	16 (COX4I1, COX4I2, COX5A, COX5B, COX6A1, COX6A2, COX6B1, COX6B2,COX6C, COX7A1, COX7A2, COX7B, COS7B2, COX7C, COX8A, COX8C)
V	*ATP Synthase*	2 (MT-ATP6, MT-ATP8)	17 (ATP5A1, ATP5B, ATP5C1, ATP5D, ATP5E, ATP5F1, ATP5G1, ATP5G2, ATP5G3, ATP5H, ATP5I, ATP5J, ATP5J2, ATP5L, ATP5L2, ATP5O, ATP5IF1)
	Mitochondrial respiratory chain complex assembly factors	None	28 (ATPAF1, ATPAF2, BCS1L, COA1, COA3, COA4, COA5, COA6, COX10, COX11, COX14, COX15, COX17, COX18, COX19, NDUFAF1, NDUFAF2, NDUFAF3, NDUFAF4, NDUFAF5, NDUFAF6, NDUFAF7, NUBPL, SCO1, SOC2, SDHAF1, SDHAF2, SURF1)

Respiratory chain complexes are large multi-protein complexes. Interestingly, the sets of proteins involved are of dual genetic origin. In humans, a total of 84 subunits and an additional 28 assembly factors are nuclear-encoded while 13 subunits are encoded by the mitochondrial genome (Table 1). These correspond to seven subunits of Complex I (NADH:ubiquinone reductase), one subunit of Complex III (Coenzyme Q: cytochrome *c* oxidoreductase), three subunits of Complex IV (cytochrome *c* oxidase), and two subunits of Complex V (ATP synthase). Complex II (succinate:coenzyme Q reductase) is the only complex formed exclusively by nuclear-encoded subunits. The size of the 13 mt-DNA-encoded proteins ranges from 68 amino acids (aa) (ATP8) to 603 aa (ND5) (Fig. 2b). The proteins are rather hydrophobic with 59.4% $\pm$ 8.5% aliphatic and aromatic residues, 29.8% $\pm$ 7.8% neutral residues, and 10.8% $\pm$ 2.6% charged residues. Leucine residues are present at the highest levels (14.4%); isoleucine, serine, and threonine are present at more than 7%, while some residues represent less than 3% (arginine, aspartate, cysteine, glutamine, glutamate, and lysine) of the protein compositions. It should be noted that a full and active set of mitochondrial translation components has been maintained for the synthesis of solely these 13 mt-DNA-encoded subunits.

1.2 The Human Mitochondrial Translation Machinery

The human mitochondrial genome is a circular double-stranded DNA of 16,569 bp [2]. This genome is tightly packed (with a single non-coding domain, the D-loop) and codes for the 13 respiratory chain subunits, in addition to 2 ribosomal RNAs (rRNAs) and 22 transfer RNAs (tRNAs). All three families of RNAs – mRNAs, rRNAs, and tRNAs – are processed from large primary transcripts according to the tRNA punctuation model [3]. The mitochondrial translation apparatus (Fig. 3) further involves a large number of proteins that are all nuclear-encoded, and are synthesized in the cytosol, before being imported into mitochondria for maturation. These include the full set of ribosomal proteins and ribosomal assembly proteins (translation initiation, elongation, termination factors) and tRNA maturation and modifying enzymes (enzymes cleaving the 5'- and 3'- ends of tRNAs primary transcripts, enzymes fixing the non-coded CCA 3'-end, enzymes of post-transcriptional modification). More than 100 proteins have been reported so far as being actors of the mitochondrial translation machinery [4–7] and will not be further discussed herein. Last but not least, a full set of nuclear-encoded aaRSs is required. These enzymes catalyze the specific esterification of tRNA 3'-ends with the corresponding amino acid so that the aminoacyl-tRNA (aa-tRNA) can be taken up by the translation factors and brought to the ribosome where the nascent protein is synthesized [8, 9]. The present knowledge of human mt-aaRSs will be discussed extensively below.

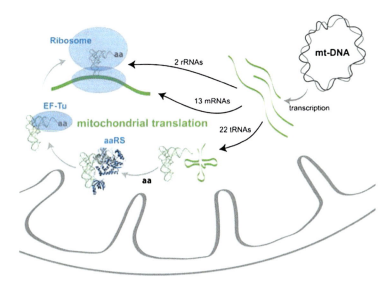

Fig. 3 The mitochondrial translation machinery. The human mt-DNA codes for 13 respiratory chain subunits (mRNAs), 2 ribosomal RNAs (rRNAs), and 22 transfer RNAs (tRNAs). All other requested proteins, such as, e.g., ribosomal proteins and ribosomal assembly proteins, translation initiation, elongation, and termination factors, tRNA maturation and modifying enzymes, mt-aaRSs, are encoded by the nucleus, synthesized in the cytosol, and imported into the mitochondria

1.3 Link Between Mitochondrial Translation, Mitochondrial Respiration, and Mitochondrial Disorders

As mentioned above, ATP synthesis by mitochondria is dependent on the coordinated expression of nuclear and mitochondrial genes. First, there is a need to coordinate the biogenesis of the respiratory chain complexes along the inner mitochondrial membrane so that partner proteins find each other to form the individual multiprotein complexes. Second, there is also a need to coordinate the setup and maintenance of mitochondrial translation machinery. Above all, this involves efficient partnerships between mt-DNA-encoded RNAs and nuclear-encoded proteins, especially between rRNAs and ribosomal proteins to form active ribosomes and between tRNAs and aaRSs to allow for accurate synthesis of aa-tRNAs. Accordingly, there are key links between the aminoacylation activity of aaRSs in charge of the synthesis of the 13 mt-DNA-encoded proteins, and the activity of respiratory chain complexes. It can be anticipated that any dysfunction of a single macromolecule of the translation machinery may have severe impacts on the activity of the respiratory chain complexes.

Mitochondrial disorders were defined as pathologies with aberrant oxidative phosphorylation (OXPHOS). Potential causes include an aberrant ROS production, elevation of $NADH/NAD^+$ ratio and lactate production, and/or ATP production deficiency. Defects were observed in a large variety of organs, and could manifest at any stage of life [10]. They were described in the late 1980s as exclusively related to mutations within the mt-DNA and thus maternally inherited. Additional disorders were subsequently associated with mutations within nuclear genes coding for proteins of mitochondrial location, and thus followed Mendelian inheritance. Mitochondrial disorders are nowadays classified according to the genetic origins of the involved-mutations. (1) The first category, actually the firstly reported, concerns mt-DNA-encoded RNAs. All the 22 mt-DNA-encoded tRNAs have been linked to pathology-related mutations. The most striking examples concern the tRNA^{Lys} and tRNA^{Leu}, currently described as "hot spots" for mutations, and correlated respectively with Myoclonus Epilepsy with Ragged Red Fibers (MERRF [11]) and Mitochondrial Encephalomyopathy with Lactic Acidosis and Stroke-like episodes (MELAS [12]). More than 230 mutations in tRNA genes are presently referenced in the 2012 MITOMAP Human Mitochondrial Genome Database (http://www.mitomap.org). Mt-DNA also codes for 2 rRNAs, with around 50 disease-related mutations described that are most frequently connected with aminoglycoside-induced deafness or non-syndromic sensorineural deafness (DEAF). (2) The second category concerns mt-DNA-encoded proteins. They are all sub-units of the respiratory chain complexes, making those complexes sensitive to mt-DNA mutations. As an example, mutations in ND5 (subunit 5 of complex I) can lead to MELAS, Leigh syndrome, MERRF, or Leber's Hereditary Optic Neuropathy (LHON) defects (reviewed in, e.g., [13]). (3) The third category, the most diverse, concerns nuclear-encoded proteins of mitochondrial location. On the one hand, mutations can affect proteins of the RC, which are directly contributing to OXPHOS (e.g., mutation in cytochrome oxidase subunits which are linked with Leigh syndrome, reviewed in [14]). On the other hand, mutations can affect proteins involved in the mt-DNA maintenance (e.g., DNA polymerase gamma, reviewed in [15]) and/or translation (e.g., mitochondrial elongation factor, reviewed in, e.g., [16]) interfering indirectly with OXPHOS. AaRSs of mitochondrial location belong to the last category. The recent discovery of mutations within mt-aaRS genes and the growing number of reported cases is opening a path to an emerging field of investigation (reviewed in, e.g., [6, 17]), and is the reason for a strong interest in the understanding of fundamental function of the mt-aaRSs and their implication in mitochondrial disorders.

Before reviewing in detail the molecular aspects of point mutations on mt-aaRSs properties, the various approaches available for the evaluation of respiratory chain complex activities used as tools either for diagnosis or for molecular investigation of this links between translation and respiration will be described.

1.4 Biochemical Analysis of Mitochondrial Respiration: A Potential Diagnostic Tool for the Detection of Mitochondrial Translation Defects

Because mitochondria provide much of the cellular energy, mitochondrial disorders preferentially affect tissues with high energy demands, including brain, muscle, heart, and endocrine systems. Consequently, mitochondrial defects play a central role in hereditary mitochondrial diseases, ischemia reperfusion injury, heart failure,

Pathogenic Implications of Human Mitochondrial Aminoacyl-tRNA Synthetases

Table 2 Some of the various experimental procedures used to evaluate the impact of a mutation on the different steps from mt-aaRS expression to mitochondrial ATP synthesis

Steps of mt-aaRSs life cycle		Methods
1	aaRS mRNAs expression/ processing	RT-PCR, qPCR, northern blot
2	aaRS synthesis/stability	Western blot, inhibition of cytosolic translation
3	aaRS import	GFP-fusion protein, immuno-cytochemistry, in vitro import assay, in vitro maturation assay
4	aaRS folding/oligomerization/ stability	Protein refolding, coexpression of diffently tagged proteins
5	amino acid activation	$[P^{32}]$/colorimetric-based ATP-PPi exchange assay
	tRNA recognition	Aminoacylation assay
	tRNA charging	Northern blot, aminoacylation assay
6	Synthesis of mt-DNA encoded RC sub-units	BN-PAGE, pulse-chase experiment
7/8	RC complex activity	Histochemical and immunohistochemical methods, polarography

Numbers on the left recall the steps displayed in Fig. 1

metabolic syndrome, neurodegenerative diseases, and cancer [18, 19]. There are remarkably diverse causes for mitochondrial disorders. These may be linked to the dual genetic systems encoding components of the respiratory chain complexes, to the need for mitochondrial translation machinery, but also to mechanisms required for the biosynthesis and maintenance of mt-DNA and to the biogenesis of the organelle itself. Moreover, each cell may contain hundreds to thousands of copies of the mitochondrial genome. The distribution of the affected tissues and the proportion of mutant to wild-type mt-DNA (termed heteroplasmy) lead to clinical manifestations, which are remarkably variable and heterogeneous. An additional breakthrough came from the later discovery of mutations within nuclear genes as the causes for similar diseases. Therefore, the establishment of a mitochondrial disorder diagnosis can be very difficult. It requires an evaluation of the family pedigree, in conjunction with a thorough assessment of the clinical, imaging, and muscle biopsy analysis [20]. Isolated OXPHOS deficiencies are generally caused by mutations in genes encoding subunits of the OXPHOS system. Combined deficiencies in the respiratory chain complexes may reflect the consequence of mutations in mt-DNA-encoded tRNAs or rRNAs, or are due to rearrangements or depletion of mt-DNA [21]. They may also reflect a dysfunction of the mitochondrial translation machinery.

Table 2 summarizes some of the various experimental procedures used to evaluate the impact of a mutation on the different steps from mt-aaRS expression to mitochondrial ATP synthesis. It includes the initial screening procedure of muscle biopsy analysis, i.e., histochemistry and immunohistochemistry. The activity of the five multiprotein enzymatic complexes can be assayed globally by measuring the mitochondrial inner membrane electrochemical potential, oxygen consumption, or ATP synthesis, or assayed individually by measurement of their enzymatic activities. This is performed by polarography. The structural integrity of the multiprotein complexes, evaluated by blue native gel electrophoresis is also increasingly used for diagnostics [22]. Finally, once the diagnosis is performed, it is possible to evaluate the impact of mutations on the rate of mitochondrial protein synthesis using chase experiments on specific recombinant cell-lines [23].

Several *histochemical and immunohistochemical methods* can be used as reliable morphological tools in order to visualize respiratory chain abnormalities on tissue sections. Classical histochemistry techniques allow visualizing succinate dehydrogenase (SDH) and cytochrome *c* oxidase (COX) activity. Indeed, the most informative histochemical impairment of mitochondria in skeletal muscle is ragged red fibers (RRF), observed on frozen sections traditionally with the modified Gomori trichrome method [24]. Since the accumulation of material other than mitochondria sometimes simulates a RRF appearance, the identification of deposits suspected of being mitochondrial proliferation should be confirmed by histochemical staining of oxidative enzymes as SDH and COX. Muscle from mitochondrial myopathy associated with mt-DNA mutations tends to show a mosaic expression of COX consisting of a variable number of COX-deficient and COX-positive fibers, and RRFs can be COX negative or COX positive. The mosaic pattern of COX expression could be considered as the histochemical signature of a heteroplasmic mt-DNA mutation affecting the expression of mt-DNA-encoded genes in skeletal muscle [25]. Immunohistochemical methods allow the visualization of the expression of several mt-DNA and nuclear-encoded subunits of the respiratory chain. An antibody against COX IV is routinely used as a probe for nuclear-encoded mitochondrial protein and an antibody against COX II as a probe for mt-DNA-encoded protein. However, any other combination of antibodies can also be used.

Polarography (spectrophotometric assays) can be applied to both tissue samples and cultured cells and is designed to assess the enzymatic activity of the individual OXPHOS Complexes I–V, along with the Krebs cycle enzyme citrate synthase as a mitochondrial control. Determining the enzymatic activities can be valuable in defining isolated or multicomplex disorders and may be relevant to the design of future molecular investigations [26]. Different assays have to be performed in order to analyze each complex separately. Assays for Complexes I, II, II + III, III, and IV are routinely performed when there is a suspicion of mitochondrial defects. The principle of polarographic approach is based on the fact that mitochondria require oxygen to produce ATP. The rate of oxygen consumption from isolated mitochondria or directly in skinned fibers is a useful and valuable technique in the research and evaluation of mitochondrial dysfunction and disease, because ADP-dependent oxygen consumption directly reflects OXPHOS efficiency [27–31]. In the presence of oxidizable substrates, freshly isolated mitochondria are introduced into the polarograph and oxygen consumption is measured in the presence of exogenously added ADP as well as several inhibitors. The typical parameters determined from mitochondrial polarography include state III rate (maximal mitochondrial respiration with ADP), state IV rate (basal mitochondrial respiration), and RCR (respiratory control ratio or state III rate/state IV rate). RCR is a good indicator of the integrity of the inner membrane of the isolated mitochondria and is sensitive for indicating OXPHOS defects. The ratio of ADP consumed/oxygen consumed during the experiment is a direct reflection of phosphorylation efficiency and can indicate abnormalities of the ATP synthase activity or uncoupling between the activities of Complexes I to IV and complex V (ATP synthase). It is possible to explore the respiratory parameters of skeletal muscle with permeabilized muscle fibers, thus skipping a mitochondria purification step [32]. Muscle fibers are permeabilized by saponin, allowing respiratory substrates and inhibitors to reach the mitochondria [31, 33].

Blue native polyacrylamide (or agarose) gel electrophoresis (BN-PAGE) allows for the isolation of intact respiratory chain complexes and analysis of their subunit content (reviewed in [34]). Briefly, after solubilization of mitochondria in the presence of dodecylmaltoside, large complexes are first separated by native gels electrophoresis on low percentage polyacrylamide or agarose gel. A second dimension electrophoresis is then performed under non-native conditions in the presence of SDS and β-mercaptoethanol using a 10% polyacrylamide gel. Individual subunits are detected by western blotting or mass spectrometry. The two-dimensional separation approach can also be adapted to perform in gel activity assays to address the dynamics of protein synthesis and complex assembly (in combination with pulse-chase labeling of proteins in cultured cells).

In conclusion, a variety of biochemical approaches are available to evaluate mitochondrial function. These form a powerful toolkit that can be used to diagnose the mitochondrial origin of a disorder. However, the gap is large between the dysfunction of the respiratory chain (as a whole, or as individual complexes) and the understanding of the dysfunction at the molecular level. Indeed, considering the translation rate of each of the 13 mt-DNA-encoded proteins as the sole outcome resulting from defects in tRNAs and/or aaRSs is too simplistic. The links between the mitochondrial translation of a given aaRS on the one hand, and ATP production on the other, involve a number of issues that need to be explored. Some of these will be discussed in what follows. One must also take into account the possibility of alternative functions of mt-aaRSs, outside their strict housekeeping role in translation.

2   Mt-aaRSs in Mitochondrial Translation

The housekeeping function of aaRSs is to provide aminoacylated-tRNAs (aa-tRNAs) for translation. Enzymes connect tRNAs with their corresponding amino acid in an efficient and specific way through a two-step reaction named aminoacylation. In the first step, the amino acid is activated by ATP into an aminoacyl-adenylate (followed by the release of PPi). In the second step, the activated amino acid is transferred onto the cognate tRNA, releasing AMP. The formation of the 20 canonical aa-tRNA species in human mitochondria requires the import of a complete set of mt-targeted aaRSs encoded by the nuclear genome. The faster evolution rate of mt-DNA than the nuclear genome [35, 36] leads to abnormal RNAs, shrunken in size and often lacking important signals. For instance, most mt-tRNAs have shortened sizes, miss crucial folding and recognition nucleotides as compared to "classical" tRNAs, and are more flexible [6, 7, 37, 38]. Mechanisms compensating for the degeneration of mt-tRNAs remain mainly unsolved and raise the question of molecular adaptation of partner proteins, especially mt-aaRSs. The structural and functional deciphering of the set of aaRSs of human mitochondrial location is at the early stages.

2.1 Nuclear-Encoded aaRSs of Mitochondrial Location and Evolutionary Considerations

The nuclear gene annotation of the human aaRSs of mitochondrial location was completed a decade ago [8]. Access to gene sequences highlighted that the set of aaRSs dedicated to translation in human mitochondria is mainly different from the set acting in the cytosolic translation (Fig. 4). This concerns 17 out of the 19 pairs of aaRS. The two exceptions concern the GlyRSs and LysRSs. GlyRSs are generated from two translation initiation sites on the same gene, leading to one enzyme with a mitochondrial targeting sequence (MTS), mt-GlyRS, and a second without, cytosolic GlyRS [39, 40]. With the LysRSs, an alternative mRNA splicing pathway allows for the insertion – or not – of the nucleotide sequence coding for the MTS, leading to two mature mt- and cytosolic LysRSs differing only by a few residues at their N-terminus [41]. No gene coding for mt-GlnRS, the 20th synthetase, has been found so far, leaving open the question about how glutaminylation in human mitochondria is performed. Among possible explanations, it is proposed that either the sequence of human mt-GlnRS has evolved so much that it has become unrelated to any of the known GlnRSs, or its function is fulfilled by an mt-addressed version of cyt-GlnRS (as proposed in yeast [42]), or that the synthesis of mt Gln-tRNA^{Gln} occurs via an indirect pathway (the transamidation pathway) involving misacylation of tRNA^{Gln} by a non-discriminative GluRS followed by Glu-amidation [43]. The existence of an indirect pathway in mitochondria was demonstrated in the cases of plants [44, 45], yeast [46], and more recently humans [47]. However, the coexistence of direct and indirect pathways for Gln-tRNA^{Gln} synthesis in yeast and mammalian mitochondria is still under consideration [48, 49].

Despite the conventional view of the endosymbiotic origin of mitochondria [50], the source of nuclear genes for mt-addressed aaRSs is diverse and not necessarily easy to trace back. Some of the mt-addressed aaRSs originate from the bacterial domain, but none specifically from the alpha-proteobacteria, although the alpha-proteobacterial contribution to the mitochondrial genome is well established. This favors the hypothesis that mt-aaRSs have been acquired by numerous post-endosymbiotic and/or lateral gene transfer events, from sources representative of all kingdoms of life [51]. The precise knowledge on the origin of all human mt-aaRS genes is necessary for the investigation of pathology-related mutations (see below) and will be of help in building up homology models in cases where crystallographic structures for mt-aaRS are not available. This global view on the origin of all human mt-aaRS genes will be established soon (Sissler et al., in preparation).

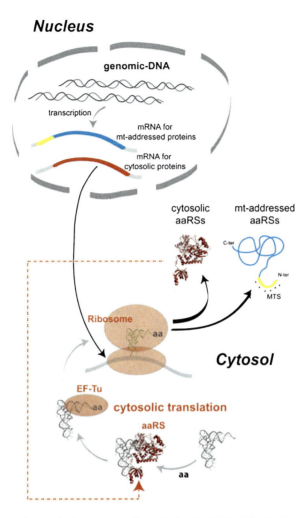

Fig. 4 Two sets of genes for human cytosolic and mitochondrial aaRSs. Achievement of human genome sequencing and gene annotation of the human aaRSs of mitochondrial location [8] reveal the presence of two distinct sets of nuclear genes for aaRSs (with two exceptions, see text). One set (in *red*) codes for the aaRSs of cytosolic location. The second set (in *blue*) codes for the aaRSs of mitochondrial location. The latter distinguishes by the presence of an encoded N-terminal mitochondrial targeting sequence (MTS, in *yellow*). The two sets of genes are translated via the cytosolic translation machinery. Sequences that are subsequently addressed to mitochondria are maintained unfolded in the cytosol, pending their entry into the organelle

2.2 Mt-aaRSs are Imported Proteins

Human mt-aaRSs are translated within the cytosol and subsequently imported into the mitochondria thanks to the presence of MTSs. As is the case for the vast majority of proteins of mitochondrial matrix and/or inner membrane location, these sequences are predicted to be located at the N-termini of precursor mt-aaRSs [8]. However, to date, no consensus sequences have been deciphered. An MTS typically consists of ~15–50 amino acids including numerous positively charged residues (e.g., lysines and arginines), and forms amphipathic alpha-helices [52]. The MTS first directs the precursor proteins to mitochondria where they are further translocated by the translocase complexes of the outer (TOM) and inner (TIM) membranes (reviewed in, e.g., [53–59]). Upon arrival in the matrix, MTS are proteolytically cleaved by the mitochondrial processing peptidase (MPP). This process was recently shown possibly to affect the half-life of the proteins [60]. Removal of the pre-sequence exposes new amino-termini of the imported proteins, which may contain a stabilizing or a destabilizing amino acid (bulky hydrophobic residues are typically destabilizing). The N-end rule indeed states that regulation of the proteolytic degradation is closely related to the N-terminal residue of proteins [61]. It has recently been proposed that two additional peptidases (that function subsequently to MPP) are implicated in protein stabilization by removing the newly exposed N-terminal destabilizing residue(s). The first is the intermediate cleaving peptidase Icp55, which removes a single amino acid. The second is the mitochondrial intermediate peptidase Oct1, which removes an octapeptide. Accordingly, the processing of imported proteins is closely connected to protein turnover and quality control (reviewed in [54]).

Cleavage (maturation) sites of the MTS for human mt-aaRSs are so far mostly defined according to theoretical predictions based on computer programs (e.g., Predotar http://urgi.versailles.inra.fr/predotar/predotar.html, MitoProt http://ihg.gsf.de/ihg/mitoprot.html, TargetP http://www.cbs.dtu.dk/services/TargetP/, and iPSORT http://ipsort.hgc.jp/). However, the expression of recombinant human mt-aaRS proteins, deprived of theoretically predicted MTSs, appears to be difficult due to low solubility and the tendency of proteins to aggregate [9, 62]. This is likely to indicate that many predictions may be incorrect. Indeed, inaccurate prediction of the cleavage site was previously shown to be responsible for suboptimal expression of human mt-LeuRS. Only the LeuRS variant deprived of its 39 N-terminal amino acids was sufficiently overexpressed in *Escherichia coli*, efficiently purified, and fully active, while the variant deprived of the predicted 21 amino acids remained insoluble [63, 64]. Along the same line, the re-design of the N-terminus of human mt-AspRS enhances expression, solubility, and crystallizability of the mitochondrial protein [65, 66]. Discrepancies are also apparent between the predicted cleavages sites, the starting amino acid of the recombinant proteins, and the first residue visible in the established crystallographic structures to date (reviewed in [65]).

None of the above-mentioned examples have the experimental exact cleavage points of the mature proteins been established so far. These examples however emphasized that the preparation of stable recombinant molecules would gain from optimized criteria to predict MTS cleavage sites. A more systematic effort to determine experimentally the precise N-terminus of mature mitochondrial proteins (as done for, e.g., the yeast mt proteome [60]) would be of help to determine unambiguously the sequence of a functional mt protein.

Pathogenic Implications of Human Mitochondrial Aminoacyl-tRNA Synthetases

However, this analysis remains difficult to perform experimentally (by sequencing or mass spectrometry) due to the minute amount of protein that can be isolated from human mitochondria and because of the risk of secondary proteolysis.

2.3 Structural Insights of mt-aaRSs

When considering their primary structures, all mt-aaRSs fall into the expected classes of the aaRSs as originally defined in [67, 68], with signature motifs being respectively HIGH and KMSK for class I enzymes, and motifs 1, 2, and 3 for class II enzymes. Striking divergences are, however, observed when considering their modular organizations. Modular design of the aaRSs is a result of a patchwork assembly of different functional modules during evolution. Minimal cores are the catalytic domain and the tRNA anticodon binding domain that are possibly surrounded by additional components for structural or functional purposes. Those are either remnants from early ancestors or structural inventions for functional expansion (e.g., [69]). The most striking observation concerns the divergence between the tetrameric cytosolic PheRS ($\alpha_2\beta_2$) and the monomeric (α) version found in human mitochondria [70], a situation also observed for PheRSs in yeast [71]. Human mt-PheRS possesses the minimum set of structural domains, making this enzyme the smallest exhibiting aminoacylation activity and the only class II monomeric synthetase [72]. Another striking observation is that two independent coding sequences have been found for mt-GluRS and mt-ProRS, as opposed to a single gene for both activities in the human cytosol leading to the bifunctional GluProRS [73].

Although the first crystallographic structure of a bacterial aaRS was published three decades ago (TyrRS from *Bacillus stearothermophilus* [74]), the first 3D structures of mt enzymes were only established recently: a bovine structure in 2005 (mt-SerRS [75]), the first human structure in 2007 (mt-TyrRS [76]), followed by two additional ones in 2012 (mt-PheRS [72] and mt-AspRS [66]) (Fig. 5). The time lag between resolving structures of bacterial aaRS and mammalian mt-aaRSs was mainly due to the difficulties involved in producing large amounts of stable mitochondrial proteins (defining the N-terminus leading to a soluble protein). Resolution of crystallographic structures and investigation of biophysical properties of mt-aaRSs reveal similarities, but also distinctive features, when compared to related prokaryotic homologs (the four mt-aaRSs for which crystallographic structures have been obtained are of prokaryotic origin [51]).

Structural idiosyncrasies (Fig. 5) concern, for instance, the very unique structural organization of mt-PheRS (monomer instead of heterotetramer [77]) or the presence of two distinctive insertions in mt-SerRS (an 8 aa amino-terminal "distal helix" and a carboxy-terminal "C-tail" composed of an over 40 Å long flexible loop stretching away from the body of the monomer [75]). More generally, it has been observed that, besides having similar architectures to prokaryotic homologs, mitochondrial enzymes are distinguished by more electropositive surface potentials (specifically along the

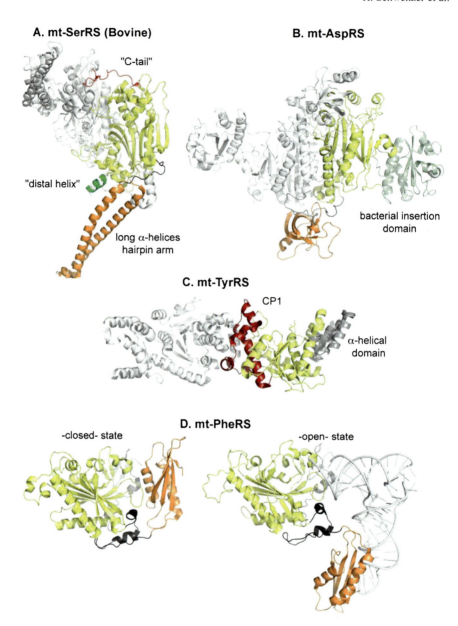

Fig. 5 Known crystallographic structures of mammalian mt-aaRSs. (**A**) Bovine mt-SerRS [75], where the specific "distal helix" and "C-tail" are emphasized in *green* and *red*, respectively. In addition, the bacterial-type N-terminal long α-helices hairpin arm is shown in *orange*. (**B**) Human mt-AspRS [66], where the bacterial insertion domain is highlighted in *light green*. (**C**) Human mt-TyrRS [76], where the CP1 and the α-helical domains are indicated in *red* and *gray*, respectively. Note that the S4-like domain is missing in this structure. (**D**) Human mt-PheRS in the –closed– state [77], and in the –open– state, complexed within *Thermus thermophilus* tRNA^{Phe} (in *white*) [72]. Binding of tRNA engenders a drastic conformational change of mt-PheRS through ~160° hinge-type

tRNA binding surface) or by enlarged grooves for tRNA accommodation. The latest aspect is visible for instance in mt-AspRS, where the angle formed by the bacterial-specific insertion domain and the catalytic domain is more opened by 26°C [66]. This can also be seen in mt-PheRS where the PheRS/tRNA^{Phe} complex formation was shown to be accompanied by a considerable rearrangement through an ~160° hinge-type rotation from a –closed– to an –open– state of the PheRS and the global repositioning of the anticodon binding domain upon tRNA binding [72, 77]. In addition, an alternative interaction network has been observed at the subunit interface of the dimeric mt-AspRS (weaker in terms of salt-bridges and hydrogen bonds). Biophysical investigations also demonstrated a thermal stability reduced by as much as 12°C for mt-AspRS, compared to *E. coli* AspRS [66]. Finally, a gain of plasticity is proposed for both the mt-TyrRS, where the KMSKS loop is rather remote from the active site, explaining the relative lacks of constraints in the structure [76], and the mt-AspRS, where unusual thermodynamic properties of tRNA binding are observed [66]. It has been suggested that the gain of plasticity may be a more general property of mt-aaRSs, as they have to deal with degenerated mt-tRNAs [66].

2.4 Some Functional Peculiarities

The bacterial origin (established for many of the mt-aaRSs) predicts that most of the mt-aaRSs should behave as prokaryotic aaRSs. However, this is not the case, raising interesting questions regarding the evolution of macromolecules of the mitochondrial translation machinery. The human mt-AspRS has been extensively studied along these lines. It shares 43% of identical residues (including residues specific for all AspRSs), the same modular organization (including the bacterial-type insertion and C-terminal extension domains), and the same architecture as *E. coli* AspRS, a representative bacterial homolog [66]. However, and despite the fact that the two enzymes are likely descendants from a common ancestor, numerous functional idiosyncrasies/discrepancies were reported. Indeed, the mt-AspRS exhibits a reduced catalytic efficiency [8, 9], requires a minimal set of recognition determinants within its cognate tRNA [78], displays a higher sensitivity to small substrate analogs [79], is able to cross aminoacylate bacterial tRNAs (while the bacterial enzyme unilaterally recognizes bacterial tRNAs [80, 81]), and shows an increased intrinsic plasticity when compared to its bacterial homolog [66].

It is proposed that all structural and functional peculiarities of the mt-aaRSs (exemplified here by the mt-AspRS) with respect to the bacterial homologs may represent an evolutionary process, allowing nuclear-encoded proteins to cooperate with degenerated organelle RNAs [66].

Fig. 5 (continued) rotation and the global repositioning of the anticodon binding domain. For all structures the catalytic core is in *yellow*, the anticodon binding domain in *orange*, and the hinge region in *black*. When appropriate, the second dimer is displayed in *light gray*

3 Mt-aaRSs in Human Disorders

As already evoked, a breakthrough took place in 2007 with the discovery of a first set of mutations present in the nuclear gene of an mt-aaRS, namely mt-AspRS [82]. This first discovery was followed very rapidly by the description of numerous additional mutations not only on the same gene but also on other mt-aaRS-coding genes, so that half of them are presently known to be affected [6, 17]. This discovery sheds light on a new family of nuclear genes involved in human disorders allowing the new naming of "mt-aaRS disorders."

3.1 Discovery of aaRS-Related Disorders

Mutations in *DARS2*, the nuclear gene coding for mt-AspRS, were first found in 2007 in patients with cerebral white matter abnormalities of unknown origin [82]. These abnormalities were part of a childhood-onset disorder called Leukoencephalopathy with Brain stem and Spinal cord involvement and Lactate elevation (LBSL). Since this first discovery, mutations in eight additional mt-aaRS-encoding genes have been reported (Table 3). They hit *RARS2* (patients with PontoCerebellar Hypoplasia type 6, PCH6), *YARS2* (Myopathy, Lactic Acidosis and Sideroblastic Anemia, MLASA syndrome), *SARS2* (HyperUricemia, Pulmonary hypertensions and Renal failure in infancy and Alkalosis, HUPRA syndrome), *HARS2* (Perrault Syndrome, PS), *AARS2* (Infantile Mitochondrial Cardiomyopathy), *MARS2* (Autosomal Recessive Spastic Ataxia with Leukoencephalopathy, ARSAL), *EARS2* (Early-onset Leuko-encephalopathy with Thalamus and Brainstem involvement and High Lactate), and *FARS2* (Infantile mitochondrial Alpers Encephalopathy). An observation among the numerous reported cases is that despite a dominant effect on brain and neuronal system, sporadic manifestations also occur in skeletal muscle, kidney, lung, and/or heart.

3.2 The Present-Day List of Pathology-Related Mutations Within mt-aaRS Encoding Genes

Nine mt-aaRS genes are currently known to harbor a total of 65 mutations, found in patients in 64 different genetic combinations (Table 4). All mutations can also be visualized on schematic representations of modular organizations of the proteins (Fig. 6). The most prominent affected gene is *DARS2*, located on chromosome I. It comprises 32,475 bp, codes for 17 exons and is translated into a 645 aa long mt-AspRS. To date, 28 different mutations are known: 8 nonsense mutations (frameshift, premature stop), 16 missense mutations (amino acid exchange), and 4 insertions/deletions have been described in 13 different reports. These mutations

Table 3 Human mt-aaRSs involved in mitochondrial disorders

Gene	Pathogenic manifestation	Tissue	First report
DARS2 Aspartyl-tRNA Synthetase	Leukoencephalopathy with Brain stem and Spinal cord involvement and Lactate elevation (LBSL)	Brain, spinal cord	[82]
RARS2 Arginyl-tRNA Synthetase	PontoCerebellar Hypoplasia type 6 (PCH6)	Brain	[101]
YARS2 Tyrosyl-tRNA Synthetase	Myopathy, Lactic Acidosis and Sideroblastic Anemia (MLASA syndrome)	Blood, skeletal muscle	[97]
SARS2 Seryl-tRNA Synthetase	HyperUricemia, Pulmonary hypertension and Renal failure in infancy and Alkalosis (HUPRA syndrome)	Kidney, lung	[108]
HARS2 Histidyl-tRNA Synthetase	Perrault Syndrome (PS)	Ovarian, sensorineural system	[90]
AARS2 Alanyl-tRNA Synthetase	Infantil Mitochondrial Cardiomyopathy	Heart	[110]
MARS2 Methionyl-tRNA synthetase	Autosomal Recessive Spastic Ataxia with Leukoencephalopathy (ARSAL)	Brain	[98]
EARS2 Glutamyl-tRNA Synthetase	Early-onset Leukoencephalopathy with Thalamus and Brain stem involvment and high Lactate	Brain	[89]
FARS2 Phenylalalnyl-tRNA Synthetase	Infantile mitochondrial Alpers Encephalopathy	Brain	[104]

Initial chronological reports connecting affected mt-aaRS genes with pathogenic manifestations, and affected tissues

Table 4 Pathology-related mutations on mitochondrial aaRS genes, transcripts, and proteins. Mutations are separated by half line spaces so that to illustrate those found in the two alleles of a single patient

AARS2 (VI)	Gene mutations (13,672 bp; gene ID: 57505)	Exons (22; NM_020745.3)	Proteins (985 aa; NP_065796.1)	Domains	Ref.
	1774C<T	13	R592W	CD	[110]
	464T>G	3	L155R	CD	
	1774C<T	13	R592W	CD	
	1774C<T	13	R592W	CD	
DARS2 (I)	Gene mutations (32,475 bp; gene ID: 55157)	Exons (17; NM_018122.4)	Proteins (645 aa; NP_060592.2)	Domains	Ref.
	228-20_21delTTinsC	3	R76SfsX5	ABD	[82]
	1876C>G	17	L626V	BED	
	228-20_-21delTTinsC	3	R76SfsX5	ABD	
	787C>T	9	R263X	CD	
	228-11C>G	3	R76SfsX5	ABD	
	536G>A	6	R179H	CD	
	228-20_-21delTTinsC	3	R76SfsX5	ABD	
	788G>A	9	R263Q	CD	
	228-20_-21delTTinsC	3	R76SfsX5	ABD	
		15	A522_K558del	CD	
	228-20_-21delTTinsC	3	R76SfsX5	ABD	
	492+2T>C	5	M134_K165del	ABD/HR	
	228-20_-21delTTinsC	3	R76SfsX5	ABD	
	455G>T	5	C152F	HR	
	228-20_-21delTTinsC	3	R76SfsX5	ABD	
	617-663del	7	F207CfsX24	CD	
	133A>G	2	S45G	ABD	
	228-15C>A	3	R76SfsX5	CD	

536G>A	R179H	CD
1273G>T	E425X	BID
1837C>T	L613F	BED
1876T>A	L626Q	BED
228-15C>G	R76SfsX5	ABD
1886A>G	Y629C	BED
228-16C>A	R76SfsX5	ABD
295-2A>G	A100_P132del	ABD
228-20_-21delTTinsC	R76SfsX5	CD
1679A>T	D560V	ABD
228-20_-21delTTinsC	R76SfsX5	ABD
550C>A	Q184K	CD
228-15C>A	R76SfsX5	ABD
396+2T>G	A100_P132del	ABD
228-20_-21delTTinsC	R76SfsX5	ABD
742C>A	Q248K	CD
228-20_-21delTTinsC	R76SfsX5	ABD
1272_1273GG>C	E424NfsX1	BID
228-20_-21delTTinsC	R76SfsX5	ABD
397-2A>G	M134_K165del	ABD/HR
228-20_-21delTTinsC	R76SfsX5	ABD
536G>A	R179H	CD
228-10C>A	R76SfsX5	ABD
492+2T>C	M134_K165del	ABD/HR

(continued)

Table 4 (continued)

DARS2 (I)	Gene mutations (32,475 bp; gene ID: 55157)	Exons (17; NM_018122.4)	Proteins (645 aa; NP_060592.2)	Domains	Ref.
	1345-17del13	14	C449_K521del	CD	[131]
	228-20_-21delTTinsC	3	R76SfsX5	ABD	
	228-16C>A	3	R76SfsX5	ABD	[85]
	716T>C	8	L239P	CD	
	228-20_21delTTinsC	3	R76SfsX5	ABD	[83]
	492+2T>C	5	M134_K165del	ABD/HR	
	228-20_21delTTinsC	3	R76SfsX5	ABD	
	455G>T	5	C152F	HR	
	228-16C>A	3	R76SfsX5	ABD	[84]
	745C>A	8	L249I	CD	
	228-16C>A	3	R76SfsX4	ABD	[99]
	?	?	?	?	
	228-16C>A	3	R76SfsX5	ABD	[94]
	228-16C>A	3	R76SfsX5	ABD	
	228-12C>G	3	R76SfsX5	CD	[112]
	228-24CinsT	3	R76SfsX6	CD	
	228-12C>A	3	R76SfsX5	ABD	[88]
	1069C>T	11	Q357X	CD	
	1825C>T	17	R609W	BED	[95]
	1825C>T	17	R609W	BED	
	228-20_21delTTinsC	3	R76SfsX5	ABD	[132]
	1395_1396delAA	14	T465TfsX7	CD	
	228-22T>A	3	R76SfsX5	ABD	[100]
	228-22T>A	3	R76SfsX5	ABD	

EARS2 (XVI)	Gene mutations (32,980 bp; gene ID: 124454)	Exons (9; NM_001083614.1)	Proteins (523 aa; NP_001077083.1)	Domains	Ref.
	172C>G	2	R58G	ABD	[102]
	406A>T	5	T136S	ABD	
	502A>G	4	R168G	CD	[89]
	1279_1280insTCC	7	T426_R427insL	ABD	
	322C>T	3	R108W	CD	
	322C>T	3	R108W	CD	
	1194C>G	6	Y398X	ABD	
	322C>T	3	R108W	CD	
	328G>A	3	G110S	CD	
	328G>A	3	G110S	CD	
	610G>A	4	G204S	CD	
	286G>A	2	E96K	CD	
	500G>A	4	C167Y	CD	
	322C>T	3	R108W	CD	
	949G>T	4	G317C	CD	
	164G>A	2	R55H	CD	
	670G>A	4	G224S	CD	
	1A>G	1	M1?	MTS	
	320G>A	3	R107H	CD	
	322C>T	3	R108W	CD	
	1547G>A	9	R516Q	CD	
	19A>T	1	R7X	MTS	
	322C>T	3	R108W	CD	

(continued)

Table 4 (continued)

EARS2 (XVI)	Gene mutations (32,980 bp; gene ID: 124454)	Exons (9; NM_001083614.1)	Proteins (523 aa; NP_001077083.1)	Domains	Ref.
	286G>A	2	E96K	CD	
	500G>A	4	C167Y	CD	
	193A>G	2	K65E	CD	[133]
	193A>G	2	K65E	CD	
FARS2 (VI)	Gene mutations (403,026 bp; gene ID: 10667)	Exons (6; NM_006567.3)	Proteins (451 aa; NP_006558.1)	Domains	Ref.
	986T>C	5	I329T	ABD	[104]
	1172A>T	6	D391V	CD	
	1275G>C	7	L425L	CD	
	1277C>T	7	S426F	CD	
	431A>G	3	Y144C	ABD	[134]
	431A>G	3	Y144C	ABD	
HARS2 (V)	Gene mutations (6,900 bp; gene ID: 234338)	Exons (13, NM_012208.2)	Proteins (506 aa; NP_036340.1)	Domains	Ref.
	1102G>T	10	V368L	ABD	[90]
	598C>G	6	L200V/L200_K211del	CD	
MARS2 (II)	Gene mutations (1,779 bp; gene ID: 92935)	Exons (1; NM_138395.3)	Proteins (593 aa, NP_612404)	Domains	Ref.
	681D268bpfs236X/wt				[98]
	Dup1/Dup1				
	Dup1/Dup2				
	Dup1/Dup-del				
	Yorube Insertion/wt				
RARS2 (VI)	Gene mutations (75,618 bp; gene ID: 57038)	Exons (20; NM_020320.3)	Proteins (562 aa, NP_612404)	Domains	Ref.
	1704A>G*	20	K568K	ABD	[101]
	872A>G*	10	K291R	CD	
	IVS2+5(A>G)*	2	L13RfsX15	MTS	
	35A>G	1	Q12R(Q12fsX25)	MTS	[87]
	1024A>G	12	M342V	CD	

	Gene mutations	Exons	Proteins	Domains	Ref.
SARS2 (XIX)	Gene mutations (15,127 bp; gene ID: 54938)	Exons (16; NM_017827.3)	Proteins (518 aa; NP_060297.1)	Domains	Ref.
	35A>G	1	Q12R(Q12fsX25)	MTS	[86]
	IVS2+5(A>G)	2	L13RfsX15	MTS	
	721T>A	9	W241R	CD	[91]
	35A>G	1	Q12R(Q12fsX25)	MTS	
	25A>G	1	19V	MTS	
	1586+3A>T	18	R504_L528del	ABD	
	734G>A	9	R245Q	CD	
	1406G>A	16	R469H	ABD	
	IVS2+5(A>G)	2	L13RfsX15	MTS	[92]
	IVS2+5(A>G)	2	L13RfsX15	MTS	
	1211T>A	14	M404K	CD	
	471-473del	7	K158del	CD	
YARS2 (XII)	Gene mutations (8,630 bp; gene ID: 51067)	Exons (5; NM_001040436.2)	Proteins (477 aa; NP_001035526.1)	Domains	Ref.
	1169A>G	13	D390G	CD	[108]
	1169A>G	13	D390G	CD	
	156C>G	1	F52L	CD	[97]
	156C>G	1	F52L	CD	
	137G>A	1	G46D	CD	[109]
	137G>A	1	G46D	CD	

Gene names of affected mt-aaRSs with chromosomal location under brackets are given on the left. Reported nucleotide mutations such as, e.g., replacement, insertion, deletion or depletion (Gene mutations), affected exons by the mutations (Exons), related amino acid modification (Proteins), affected domains of the protein (Domains), and referred literature (Ref.) are listed. Domains attributions were done on the basis of Pfam definition of the proteins. ABD stands for Anticodon Binding Domain, CD for Catalytic Domain, BED for Bacterial Extension Domain, BID for Bacterial Insertion Domain, HR for Hinge Region and MTS for predicted Mitochondrial Targeting Signal. Properties of each wild-type gene (length, gene accession number, total number of exons, mRNA accession number) and corresponding proteins (total length, protein accession number) are given on the first lane of each reported case

*Those three mutations are found on a same patient in a homozygous status

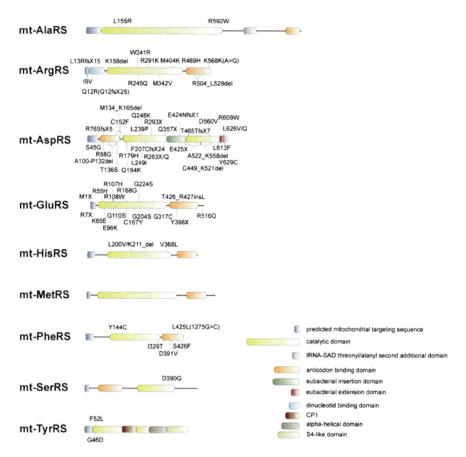

Fig. 6 Display of pathology-related mutations on modular organizations of mt-aaRSs. Color code of all domains is given

are distributed over nearly all exons and, thus, are found in all protein domains, including the predicted MTS (S45G) and the hinge region (C152F and M134_K165del). In addition to these mutations found on one allele of the gene, almost all LBSL patients have a mutation in a polypyrimidine tract at the 3′-end of intron 2, which is found on the second allele. This mutation affects correct splicing of the third exon, which leads to a frameshift and a premature stop (R76SfsX5). This frameshift mutation is "leaky," leading to a decrease but not zero expression of full-length mt-AspRS. Beside these typically compound heterozygous states of LBSL patients, homozygous *DARS2* mutations have been described more recently. Homozygous patients harbor either the R76SfsX5 mutation or the R609W mutation (in the bacterial insertion domain).

The *EARS2* gene codes for mt-GluRS and is located on chromosome XVI. It comprises 32,980 bp, nine exons and encodes a 523 aa long protein. Today, 16 different mutations are known: 3 nonsenses, 12 missenses, and 1 insertion.

The catalytic domain is dominantly affected. The two mutations (R7X and M1X) localized in the predicted MTS are nonsense mutation and certainly lead to truncated translation products. Among those, only one mutation (K65E) is in a homozygous state. The gene *RARS2* codes for mt-ArgRS, is located on chromosome VI, consists of 75,618 bp, 20 exons, and codes for a 562 aa long protein. Mutations in this gene affect all parts of the protein except the dinucleotide-binding domain. Ten different mutations are reported: a single nonsense (Q12fsX25), seven missenses, and two deletions (K158del, R504_L528del). In addition, three homozygous patients were found harboring a combination of two silent mutations (K568K and K291R) with one additional mutation (IVS2 + 5) causing exon 2 skipping. The patient's major transcript lacked exon 2, but a faint, normal-sized fragment was also seen. The gene *FARS2* is 403,026 bp long, codes for mt-PheRS, and is located on chromosome VI. It has six exons and codes for a 451 aa long protein. Today four different missense mutations have been reported. Just the Y144C one (in the anticodon binding domain) is in a homozygote state. The gene *HARS2* codes for mt-HisRS, is located on chromosome V, and consists of 6,900 bp and 13 exons. The corresponding protein is 506 aa long. There is presently only one reported case in which patients are compound heterozygotes. The mutation on one allele is a missense (V368L). The mutation on the second allele produces either a missense replacement (L200V) or creates an additional splice site, inducing the deletion of 11 aa (L200_K211del). The *YARS2* gene, coding for mt-TyrRS, is localized on chromosome XII, is 8,630 bp long, harbors five exons, and codes for a 477 aa long protein. For this aaRSs, only two homozygote mutations are known, both being localized in the catalytic domain. The *AARS2* gene codes for mt-AlaRS, is 13,672 bp long, localized on chromosome VI, has 22 exons, and codes for a 985 aa long protein. There are presently two-reported cases in which patients are harboring missense mutations (either homozygote or compound heterozygote), all localized in the catalytic domain. The *SARS2* gene codes for mt-SerRS, is located on chromosome XIX, is 15,127 bp long, and contains 16 exons. Only one homozygote mutation is reported, localized in the catalytic domain of the 518 aa long protein. It should be noted that the *MARS2* gene, located on chromosome II, is composed of just one exon of 1,779 bp length. It codes for a 593 aa long protein, named mt-MetRS. Interestingly, no "classical" mutation has been reported for this gene but complex rearrangements were shown to be the cause of ARSAL. The single-exon composition of the gene permits duplication events of either the full exon or part of it, leading to the homozygous state or compound heterozygous state of the patients. An additional situation with a large insertion in one of the alleles has also been reported.

In summary, none of the chromosomes is a "hot spot" for pathology-related mutations affecting mt-aaRSs, and neither exons nor protein domains have obvious favored mutation sites. Presently, *DARS2* is the most frequently hit gene. However this may not indicate a peculiar mutational exposure of this gene, but is more likely to be due to intensive investigations of this firstly reported example of an mt-aaRS gene correlated with a disease.

3.3 Compound Heterozygous vs. Homozygous States

Among the 64 reported combinations of mutations, 53 are compound heterozygous and 11 are homozygous (excluding the puzzling combinations found for *MARS2*; see below). In all cases, the parents are unaffected heterozygous carriers of one mutation, leading to autosomal recessive mutations that affect the two alleles in the children. Figure 7 schematically summarizes all observed situations that are combining splicing, missense, nonsense, deletion, and rearrangements defects.

Compound heterozygous status is dominantly observed. In most of the cases, mutations in the first allele produce a splicing defect, resulting in reduced expression of the protein. The second allele carries a missense mutation. Such combinations were reported for *DARS2* [82–85] and *RARS2* [86, 87]. In both situations the splicing defect is "leaky," so that a small amount of wild-type protein remains expressed, which is likely to be sufficient to support basal aminoacylation activity. The residual expression of wild-type protein is mandatory in the situation where the splicing defect is combined with nonsense mutations (which completely abolishes protein expression) as was reported for, e.g., *DARS2* (R76SfsX5/Q357X [88]) and *RARS2* (Q12fsX25/L13RfsX15 [86]). Other examples of nonsense mutations are found, but combined with missense mutations of likely moderate consequences. This is, for instance, reported for *EARS2* (R7X or Y398X combined with R108W [89]). Another possible impact of mutation is the deletion of one or more amino acids within the protein. Mt-HisRS harboring the L200_K211 deletion has been suggested to be unstable when mutants are transiently expressed in human cells [90]. Deletions have also been reported in *RARS2* and *DARS2* (combined in the latest with the abundant R76SfsX5 mutation). However, their possible impacts on protein expression level or stability remains unclear [82, 83, 91, 92]. As a last example, combinations of two different missense mutations have been reported for *EARS2*, *FARS2*, *RARS2*, *DARS2*, and *AARS2*. In these cases, it is assumed that proteins are expressed but folding, structure, stability, and/or activity could be affected (see below).

Figure 7 recalls the natural oligomeric status of mt-aaRSs: some are monomers (mt-PheRS, mt-GluRS, mt-ArgRS), some are dimers (mt-AspRS, mt-SerRS, mt-TyRS, mt-HisRS, mt-MetRS), and one (mt-AlaRS) is homotetramer. A consequence of heterozygosity is the production of two distinct polypeptide chains that can randomly associate to build a dimer (or a tetramer), which theoretically leads to an equal proportion between the two possible homodimers (each harboring the same mutation) and the heterodimer (where each constitutive monomer is harboring a different mutation). However, it has been reported that some of the mutations have an impact on the oligomerization rate, leaving out the random association of mutated polypeptide chains. For instance, mt-HisRS, having the V368L mutation, oligomerizes more efficiently than any combinations, including the wild-type polypeptide [90].

The discovery of *homozygous mutations* was quite unexpected. In fact, mutations in *DARS2* were found initially only in a compound heterozygous state, suggesting that the activity of mutant mt-AspRS homodimers may be incompatible

Pathogenic Implications of Human Mitochondrial Aminoacyl-tRNA Synthetases

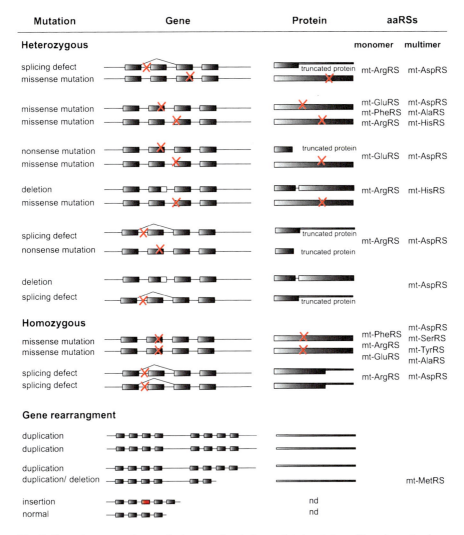

Fig. 7 From the gene to the protein: impact of pathology related mutations. The schematic views illustrate reported combinations of compound heterozygous and homozygous mutations (*red crosses*), and gene rearrangements on a genetic level, as well as possible impacts on corresponding proteins. Exonic organization is represented here in a simplified way (*gray boxes* schematize exons). Truncated proteins result from either nonsense mutations or splicing defects (reported to be "leaky" so that a reduced amount of full-length protein remains expressed, as shown by the *thinner bars*). The natural oligomeric status of mt-aaRSs is recalled on the right. *nd* stands for not determined

with life. It was then proposed that dimers carrying two different mutations would have more residual functional activity than those carrying the same mutation, and thus would not yield to a lethal phenotype [93]. The recent discovery of homozygous mutations of *DARS2* correlated with LBSL in a German patient and in a Japanese family [94, 95] ruled out this conviction. Therefore, the sole possibility for homozygous mutations to be compatible with the survival of the patient is that the mutation does not exert a too severe effect. As an example, the homozygous R76fsX5 mutation in *DARS2* from an LBSL patient induces a splicing defect that was demonstrated to be "leaky." This allows the expression of a small amount of wild-type protein likely to be sufficient to support some basal activity [96]. Today, homozygous missense mutations are found in *AARS2*, *DARS2*, *EARS2*, *FARS2*, *RARS2*, *YARS2*, and *SARS2*. Their "moderate" effect is not obvious since none of the observed-mutations (e.g., K65E in *EARS2*, Y144C in *FARS2*, D390G in *SARS2*, or F52L and G46D in *YARS2*) conserve any of the physico-chemical properties of the amino acids (isostericity, net charge, hydrophobicity, . . .). It has however been demonstrated that mt-TyrRS carrying the F52L mutation remains catalytically active, with only a twofold reduced aminoacylation rate [97], emphasizing that volume/polarity of the amino acid cannot be the only parameter to take into account. Possible neighborhood effect and structural impacts are discussed below.

In this discussion, *MARS2* is an exception. No "classical" missense or nonsense mutations are observed, but instead complex gene rearrangements have been reported. As already mentioned, this gene is composed of a single exon. A local genomic instability and/or recombination errors have been hypothesized to be the cause of template switching during DNA replication [98]. As a consequence, duplication events of either the full exon or part of it are observed in patients, leading to either homozygous or compound heterozygous states. An additional situation with a large insertion of 300 bp in one of the alleles has also been reported.

4 Diverse Molecular Impacts

A mutation in an mt-aaRS gene can have numerous molecular consequences, affecting either the biogenesis of the enzyme itself, and/or its import and maturation within the organelle, and/or its functional properties. Figure 1 (Part 1) schematized the link of these different steps with ATP synthesis. Biogenesis of mt-aaRSs involves expression and processing of the corresponding mRNA, protein expression, and stability in the cytosol and addressing to mitochondria. Importing into mitochondria requires several steps. It is followed by maturation steps of imported proteins upon entry into mitochondria. Housekeeping function of mature mt-aaRS corresponds to amino acid activation, tRNA recognition, and tRNA aminoacylation. Pathology-related mutations may thus have a direct effect on the mitochondrial translation machinery by impacting mt-aaRS biogenesis, localization (see Sect. 4.1) and/or function (see Sect. 4.2). As a consequence, the translation efficiency/rate of the full set of (or specific) mt-DNA-encoded subunits of respiratory chain complexes may be affected as well, impacting respiratory chain complexes activities and, finally, ATP synthesis (see Sect. 4.3).

4.1 Impact of Mutations on mt-aaRSs Biogenesis

4.1.1 Defects in mt-aaRS mRNAs Processing and Expression

The previously mentioned complex gene rearrangement of *MARS2* leads to an increased amount of transcript (due to duplication of full size gene and/or of regulatory elements). However, no increase in the amount of proteins is observed. It was suggested that corresponding mRNAs undergo transcriptional regulatory event(s) drastically lowering mRNA stability and thus leading to a reduction of 40–80% of the normal protein level [98]. In *DARS2* and *RARS2*, several intronic mutations were reported to affect pre-mRNAs processing. Most of the LBSL patients have mutations (of different types) within intron 2, which affects the correct splicing of exon 3. Similarly, some PCH6 patients have an IVS2 + 5(A > G) mutation, which leads to exon 2 skipping. Exclusion of exon 3 (in *DARS2*) or of exon 2 (in *RARS2*) causes a frameshift (since the two exons are asymmetric) and generates a premature stop codon. In the two situations, it is speculated that the leaky nature of the splicing defect allows for the synthesis of a small (sufficient) amount of wild-type mRNA in most tissues [94, 96, 99, 100]. However, the selective vulnerability within the nervous system is explained by tissue-specific differences in the concentration of the splicing factors (reduced in neural cell) and the presence of rather weak splice sites. In agreement, 5′-splice site of exon 3 in *DARS2* and 3′-splice site of exon 2 in *RARS2* were shown to deviate markedly from the consensus and to have low splicing scores (even in the absence of disease-causing mutations). As a consequence, the exclusion of related exons induced by the mutations is augmented in neural cells [96, 101]. Another example of mis-splicing concerns mutation Q12R(Q12fsX25) in *RARS2*. This mutation interferes with a splicing-enhancer element and causes the retention of 221 bp from intron 1, a consequent frameshift, and the truncation of protein after residue 25 [87].

4.1.2 Defects in mt-aaRSs Expression and Stability

A defect in protein expression and stability can straightforwardly be associated with surveillance pathways, such as, e.g., nonsense-mediated mRNA decay. The main function of these pathways is to reduce errors in gene expression by eliminating mRNA transcripts that contain deletions or premature stop codons (resulting from, for instance, mis-splicing events). However, the easy correlation between decreased mRNA stability and decreased protein stability is not obvious. Pierce and co-workers have identified in *HARS2* a mutation in a compound heterozygous state (L200V), which creates an alternative splice-site and leads to an in frame deletion of 12 codons in exon 6 (L200_K211). The level of the spliced-mRNA is significantly increased in the affected child compared to the unaffected (but carrier) father. Transient expression of the spliced-mRNA into HEK293T cells results in a poorly detectable mutant protein, suggesting its instability.

However, the mechanism involved in stability deficiency for the mutant protein remains unclear [90]. In three of the PCH6 patients, compound heterozygous mutations I9V/R504_L529, R245Q/R469H, and W241R/Q12R were found in *RARS2*. It was shown by western blot experiments on cultured fibroblasts that the expression level of the total proteins was reduced down to approximately 28% of the wild-type content. However, the level of mt-ArgRS-encoding mRNAs, as measured by quantitative PCR, remains low but normal in the patients, excluding transcripts instability induced by the mutations [91]. Three mutations (C152F, Q184K, and D560V) within *DARS2* were also shown to have an impact on the expression of mt-AspRS. Western blotting on transiently transfected HEK293T cells with mutated sequences indicates strongly reduced steady state levels of the mutant proteins. Further analyses on cycloheximide (to inhibit production of newly synthesized proteins) treated cells indicate a decreased stability of ~50% of the three mutant proteins as compared to the wild-type protein [102].

4.1.3 Defects in mt-aaRSs Import

Several pathology-related mutations are found within (or close to) the predicted MTS. Mutations M1X and R7X, in *EARS2*, are nonsense and likely lead to untranslated and truncated products, respectively. None will have the opportunity to be imported into mitochondria [89]. Two mutations, Q12R(Q12fsX25) and I9V, are located within the MTS-encoding sequence of *RARS2*. The mutation Q12R was predicted to enhance import efficiency of mt-ArgRS. As already discussed, this mutation also and mostly interferes with a splicing-enhancer element and causes the retention of 221 bp from intron 1, leading to a frameshift that truncates the protein after residue 25 (Q12fsX25) and likely prevents its import into the mitochondria [87]. The role of the I9V mutation on import of mt-ArgRS into mitochondria is difficult to anticipate since both amino acids are hydrophobic/aliphatic residues, with very similar physico-chemical properties. Predictions on the probability of the I9V mutant to be imported into mitochondria indeed suggest a modest effect of the mutation [91]. Confocal microscopy imaging revealed that mutation S45G located in the predicted MTS of mt-AspRS affects neither the targeting nor the binding of the protein to the mitochondria. However, by combining in vitro import and processing assays, the translocation step was found to be impaired by the mutation [103]. A more recent study investigating the impact of nine mutations found in *DARS2* on sub-cellular localization of mt-AspRS by immuno-cytochemistry (using antibodies against transiently expressed tagged mutant proteins in HEK293T cells) did not confirm a localization defect [102].

4.2 Impact of Mutations on mt-aaRSs Function

4.2.1 Impact on mt-aaRSs Enzyme Activities

The housekeeping activity of the aaRSs is to provide aminoacylated-tRNAs (aa-tRNAs) for translation. The effectiveness of aminoacylation can be detected either by in vitro or by in vivo methods (Table 2). ATP-PPi exchange assays and in vitro aminoacylation reactions can be used to establish kinetic parameters k_{cat}, K_M for ATP, amino acid, and/or tRNA. As examples, mutations L200V and V368L (*HARS2*) and I329T (*FARS2*) were shown to affect ATP-PPi exchange ability of mt-HisRS [90] and mt-PheRS [104], respectively. As an alternative procedure of the ATP-PPi exchange experiment, Cassandrini and co-workers applied a colorimetric-based measurement of Pi production [105]. The authors demonstrated that crude mitochondrial extracts, extracted from cultured skin fibroblasts of patients harboring either R245Q/R469H or W241R/Q12R mutations on mt-ArgRS, have residual activities in Pi formation of only 33% and 19%, respectively. However, they also demonstrate that the level of the protein itself is affected by the mutations [91]. Impacts of mutations on in vitro aminoacylation efficiency of recombinant mt-aaRSs have been investigated at several instances. It has been reported, for example, that mutations within mt-AspRS impact the enzymatic activity (measured as nmol of incorporated amino acid per milligram of enzyme per minute) in a maximum range of ~135-fold (for R263Q, [102]). Other studies have revealed impacts of the Y144C and the F52L mutations on respectively mt-PheRS and mt-TyrRS catalytic efficiencies (relative ratios of k_{cat}/K_M, expressed in $s^{-1} \mu M^{-1}$) of 2.3-fold [104] and 9-fold [97]. It should be noted that these effects remain in a moderate range as compared to what has been observed for pathology-related mutations on mt-tRNAs (e.g., >5,000-fold for variants of human mt-tRNA^{Lys}, [106]).

Measurement of in vivo steady state levels of aa-tRNAs is performed by separation of aminoacylated- and non-aminoacylated-tRNAs on acidic gels. In vivo steady state levels of aa-tRNAs were investigated using total RNA extracted from patients biopsies or cultured cells and northern blotting on acidic gels (reviewed in, e.g., [107]). The total amount of tRNA^{Arg} is reduced in patient cells by comparison to the amount found in cells from healthy individuals. They observed, however, that remaining tRNA^{Arg} were fully aminoacylated, probably by the few wild-type mt-ArgRS that escaped from the splicing defect engendered by the L13RfsX15 mutation. It is thus suggested that uncharged tRNA might undergo degradation [91, 101]. Belostostky and coworkers showed that the amount of tRNA^{Ser}(AGY) in cells harboring the D390G mutation in mt-SerRS was reduced to 10–20% as compared to unaffected control, and that the residual pool of tRNA^{Ser}(AGY) was not aminoacylation. Interestingly, the same mutation affected neither stability nor aminoacylation properties of tRNA^{Ser}(UCN) [108]. In contrast to these observations, no effect on the steady state level of Met-tRNA^{Met} was observed in patient cells, despite a clear reduced amount of mt-MetRS [98].

Additional methods were developed to investigate the possible impact of mutations on enzyme activity in vivo. For instance, human mutations were modeled on yeast strains, deleted from either the *MSR1* or the *HTS1* gene (homologues of human *RARS2* and *HARS2*, respectively). Homologue mutants of R469H and R245Q were able to complement *MSR1*-deleted strains under fermentable conditions (a situation where the respiratory chain is dispensable), but unable (R469H) or barely able (R245Q) to complement *MSR1*-deleted strain under respiratory conditions. In contrast, a homologue mutant of W241R fully complements the same yeast strain [91]. The homologue mutant L200_K211del was unable to complement a yeast *HTS1*-deleted strain, and the corresponding human sequence couldn't be expressed in bacteria or in human cells, suggesting that this mutant is likely to be unable to provide any activity in vivo and confirming the instability of the protein [90]. As an alternate experimental procedure, the retroviral expression of *YARS2* rescues the translation defect observed in patient muscle cells [109]. Finally, the correlation between mutations in *MARS2* and the ARSAL pathology was confirmed by using the fly as a model organism [98].

As an outcome, defects in the aminoacylation properties of the mt-aaRSs are not necessarily sufficient to explain the pathogenicity of a mutation. Therefore, the cellular environment and additional physiological conditions have to be considered to understand clearly their pathogenic impacts. As examples, tissue-related concentration of the different substrates (e.g., [104, 110]) and/or alternated yet unidentified functions of the mt-aaRSs (e.g., [82, 86, 109]) have been conjectured.

4.2.2 Structure–Function Connections

AaRSs are modular enzymes, composed of well-defined and organized domains, with conserved amino acid residues having either a structural or architectural role, or a function in the chemistry of substrate recognition or in the aminoacylation reaction. Evidence suggests that replacement of key conserved residues may alter essential physico-chemical properties (side chain length, net charge, polarity, hydrophobicity, hydrophilicity, . . .) and thus may have a key impact on the properties of the protein. Present-day knowledge of the 3D structures of mt-aaRSs, or of aaRSs from evolutionary related species, is of great help for connecting the structural impact of pathology-related mutations with its possible functional consequences. For instance, investigation of the crystallographic structures of prokaryotic HisRSs revealed that the mutation L200 and V368 are both implicated in packing interactions with highly conserved hydrophobic amino acids, involved respectively in ATP binding and in histidine recognition [90]. The two mutations (L200V and V368L) may destabilize the packing interactions, engendering movements that are perturbing some secondary structure elements and possibly reducing binding affinities for either ATP or histidine. In agreement, activities of both mutant proteins, measured by ATP-PPi exchange assay, are significantly reduced relative to the wild-type protein. In a second example, the structural and functional impacts of three pathology-related mutations affecting mt-PheRS could be connected [104]. The crystallographic structure of the enzyme [72, 77] reveals that I329 is located within the ATP-binding site. Replacement of this residue by the small and uncharged threonine should result in a widened ATP-binding site likely decreasing the affinity for the small substrate. In agreement, ATP-PPi exchange kinetic assay confirms a 2.5-fold decreased affinity for ATP for the mutant enzyme, while the binding for phenylalanine remains unaffected. In addition, D391 and Y144 are situated on both sides of the contact surface between the catalytic core and the anticodon-binding domain of mt-PheRS and are stabilizing (by forming hydrogen bounds with key conserved residues) the –closed– state of mt-PheRS. The enzyme was shown to undergo drastic conformational changes upon tRNA binding towards the functional –open– state [72] (Fig. 5). Mutations D391V and Y144C are likely to alter the rotation mechanisms upon tRNA binding and thus to affect the conformational stability of the protein. Along this line, a clear decrease for tRNA^{Phe} binding (increased K_M) was observed, but only for the Y144C mutant enzyme. Instead, a decreased affinity for phenylalanine was measured for the D391V mutant, despite D391 not being situated in the catalytic core. As an explanation, the authors are emphasizing that D391 is involved into a close network of interactions encompassing conserved residues of motif 2 (Y188) and near motif 3 (R330). The D391V replacement may cause R330 and other neighboring residues to adopt different conformations, leading to perturbations in a loop, which is critical for binding and coordination of phenylalanine [104]. The ninefold loss of catalytic efficiency (k_{cat}/K_M) measured for the F52L mutant of mt-TyrRS by in vitro aminoacylation assay [97] might also be explained by the localization of this residue near the catalytic center, as observed within the crystallographic structure of the enzyme [76].

In contrast, 3D representations of aaRSs are not always sufficient to explain or predict the molecular effect of pathology-related mutations. As previously underlined, the V368L mutation was shown to reduce the enzymatic activity of mt-HisRS, in agreement with its location inside the conserved HisB motif (which is specific to HisRSs and contributes to histidine binding pocket). However, this same mutation was also shown to influence the rate of protein dimerization, although it is not localized at the dimerization interface [90]. Similarly, R592 and L155 are respectively situated in 21.15 Å of a site that could have an editing activity, and in the surrounding of conserved catalytic residues of mt-AlaRS. However, their mutations neither affect editing activity nor aminoacylation properties of the enzyme, leaving the connection between structural predictions and functional mechanisms unclear [110]. Mt-SerRS has the functional peculiarity of being able to recognize two isoacceptor tRNAs (tRNA$^{Ser}_{AGY}$ and tRNA$^{Ser}_{UCN}$). Investigation of the crystallographic structure of bovine mt-SerRS revealed key residues responsible for recognition but also discrimination of the two isoacceptors. Those residues are situated within the helical arm of the synthetase and within or flanking the "distal helix" (Fig. 5), shown to be a structural peculiarity of the mitochondrial enzyme [75]. Analysis of aminoacylation properties revealed that the D390G mutation significantly impacts the acylation of tRNA$^{Ser}_{AGY}$ but does not alter that of tRNA$^{Ser}_{UCN}$. Unexpectedly, the D390 residue is not situated in

282   H. Schwenzer et al.

the isoacceptors discriminating area but in a beta-strand from the catalytic core, far away from the "distal helix" and the helical arm [108].

Despite these few examples where the connection between the structure and the function is not obvious, any future knowledge on crystallographic structures and/or on biophysical properties of mt-aaRSs will be of help to understand the mechanistic aspect and functional impacts of some of the pathology-related mutations. It will also be of help to predict and direct functional investigations. As an example, the crystallographic structure of the yeast cytosolic ArgRS was investigated to assess the possible consequences of the M404K and K158del mutations and to predict a tRNA binding deficiency for the first mutation, and an altered aminoacylation property for the second [92]. Similar predictions could then be drawn for other pathology-related mutations.

4.3 Impact of Mutations on Mitochondrial Translation and Activity of the Respiratory Chain Complexes

As stated above, ATP synthesis is dependent on the coordinated expression of nuclear and mitochondrial genes and requires precise recognition between tRNAs and mt-aaRSs to allow for accurate synthesis of aa-tRNAs. Accordingly, key links between the aminoacylation activity of mt-aaRSs in charge of the synthesis of the 13 mt-DNA-encoded proteins and the activity of respiratory chain (RC) complexes are foreseen. It can be anticipated that any dysfunction of a single macromolecule of the translation machinery may have severe impacts either on the translation and/or on the activity of the mt-DNA-encoded RC subunits (for all complexes except Complex II, of complete nuclear origin). However, this view appears too simplistic. Figure 8 summarizes observed molecular defects (at the levels of mRNAs expression and mt-aaRSs biosynthesis and functioning) and possible defects in the activity of the RC complexes for all reported mutations. It also schematizes that the routes linking the molecular impact of a mutation with its possible phenotypic effect are not yet always fully deciphered.

4.3.1 Translation and/or Activity of the Respiratory Chain Complexes Sub-units are Differentially Affected

Defects in the translation of mt-DNA-encoded RC subunits were reported to be correlated with mutations within *YARS2* [97, 111], *MARS2* [98], and *FARS2* [104]. In addition, defects in the activity of those complexes were reported to be correlated with mutations within *YARS2* [97, 111], *SARS2* [108], *RARS2* [92, 101], *MARS2* [98], *FARS2* [104], *EARS2* [89], *DARS2* [112], and *AARS2* [110]. However, affected subunits may vary from one case to another. Combinations of translation and/or activity defects are numerous. For example, the translation of all complexes is affected by mutation in *YARS2*, while solely the expression of Complex IV

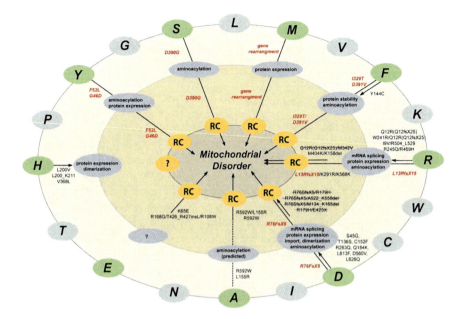

Fig. 8 Summary of the impacts of pathology-related mutations on the different steps of mt-aaRS life cycle and subsequent products activities. Mt-aaRSs (on the *outer circle*) are represented using the one-letter code. Mt-aaRSs affected by pathology-related mutations are in *green*; those yet unrelated to human pathologies are in *gray*. Observed molecular defects at the levels of mRNAs expression and mt-aaRSs biosynthesis and functioning are represented in *blue* in the *middle circle*. Possible defects on activity of the RC complexes are schematized in *orange* in the *inner circle*. The routes that link the molecular impact of a mutation with its possible phenotypic effect and lead to mitochondrial disorders are schematized (*plain* and *dashed arrows*). Mutations displayed in *red* were shown to engender obvious connections between mt-aaRS expression and/or aminoacylation defects with RC dysfunctioning. However, these routes are diverse, and interruptions indicate either the absence of full investigation, or the absence of clear molecular link (*discontinuous* or *dashed arrows*)

is impaired by mutations in *FARS2*. Similarly, the activity of all the complexes was shown to be impaired by a mutation in *SARS2*, while only Complex IV is moderately affected by the M404K/K158del mutations in *RARS2*, leaving unaffected the activities of Complexes I and III. To be noticed as well, gene rearrangement of *MARS2* engenders defects in the translation of all subunits, but noticeably causes activity defects solely in Complex I.

4.3.2 The Same Affected Gene, but Different Impacts on Translation and/or Activity of the Sub-units

A striking observation is that mutations affecting the same gene have different consequences. For instance, mutations M342V/Q12R and IVS2 + 5(A>G)/K291R/K568K, both located in *RARS2*, respectively do not affect and drastically impair respiratory chain complexes [92, 101]. More confusingly, the same combination of mutations (IVS2 + 5(A>G)/K291R/K568K) does not engender the same effects on three affected individuals. This combination of mutations reduced activities of Complexes I, III, and IV in muscles from patient II-2, but reduced activity of Complexes I in muscles from patient II-4, or activity of Complex IV from patient II-5 [87].

4.3.3 Tissues Specificity

Phenotypic manifestations of mutations have been investigated several times on mitochondria extracted from cell cultures established from biopsies specimens obtained from patients. Here again, discrepancies are observed depending on the investigated tissues. Fibroblasts are frequently used, although investigations performed on those cells showed effects on RC subunits solely for mutations within *EARS2* [89]. Conversely, measurements made on muscle, brain, or heart cells show defects on RC subunits at several instances (muscle cells affected by *DARS2*, *RARS2*, *YARS2*, *SARS2*, *AARS2*, *EARS2*, and *FARS2* mutations, brain cells affected by *AARS2* and *FARS2* mutations, and heart cells affected by *AARS2* mutations). Consequently, discrepancies can be observed when testing the effect of a given mutation on different types of cells. For instance, the F52L mutation found in *YARS2* does not show any impact on mitochondrial translation in fibroblasts, but engenders drastic impairment in skeletal muscle and myotubes [97]. Similarly, mutations in *AARS2* affect the RC activity in only muscle and brain cells, but not in liver cells [110]. Of note, it is not yet possible to make a direct link with the tissue prevalence of the disease (see Table 3) since cellular models haven't been systematically investigated.

In summary, the diverse effects of mutations on the respiratory chain complexes highlight the numerous routes linking the molecular impact of a mutation with its possible phenotypic effect. These routes are summarized and schematized in Fig. 8. Some examples show obvious connections between mt-aaRS expression and/or aminoacylation defects with RC dysfunctioning (e.g., mutations in *MARS2*, *YARS2*, and *SARS2*). Conversely, it can be seen that, for some of the reported cases, investigations are not yet complete so that it is not possible to draw the pathogenic pathway (e.g., mutations in *HARS2*, *EARS2*, *AARS2*, *DARS2*, *RARS2*, and *FARS2*). Another situation is visible in the scheme, where the molecular routes are diverse for the set of mutations affecting the same gene (mutations in, e.g., *DARS2*, *RARS2*, and *FARS2*). It can thus be hypothesized that a diversity of mechanisms, with tissue specificity, contributes to the heterogeneous manifestations of mitochondrial disorders.

5 Outlook

Human mitochondrial aminoacylation systems deserve specific attention as a consequence of their recently recognized connection with human pathologies. Understanding the fundamental properties/peculiarities of aminoacylation systems is critical to resolving the link between mutation and pathology. In the past 20 years a large number of human neuromuscular and neurodegenerative disorders have been reported to be correlated with point mutations in the mt-mRNAs, mt-rRNAs, and especially in mt-tRNAs (reviewed in, e.g., [5, 6, 10, 113–115]). Among the mutations leading to "mitochondrial disorders," more than 230 are distributed all over the 22 tRNAs. Numerous studies have attempted to unravel the molecular impacts of the mutations on the various properties of the tRNA that lead to a mosaïcity of phenotypic effects. Whilst there is no general rule, a trend towards a structural perturbation as initial molecular impact of mutations can be retained [116, 117]. However, the housekeeping function of tRNAs, namely their capacity to become esterified by an amino acid, is not systematically affected in mutated variants, so that alternative functions of tRNAs or at least alternative partnerships have to be considered [118]. The discovery of a new family of impacted genes, the mt-aaRSs, further extends the complexity of mitochondrial disorders.

The distance between the birth of an mt-aaRS and its final role in ATP synthesis is very long, so that the gap between the comprehension of the molecular impact of mutations in macromolecules and a dysfunction of the respiratory chain is very large. Also, considering the sole/common outcome from any defect in mt-aaRSs (or in mt-tRNAs) results from an effect on the translation rate of each of the 13 mt-DNA-encoded proteins is a view that rapidly appeared too simplistic. Indeed, the major outcome from the present review is that there is no common combination of affected steps that correlates the 64 reported cases (combining 65 mutations within nuclear-encoded mt-aaRS genes) to the various observed phenotypic expressions. There is obviously no "favored" mt-aaRS gene. Nine are reported today, but their recent, rapid and exponential correlations with human pathologies suggest as evidence that all mt-aaRS genes are likely to be affected by pathology-related mutations, but remain to be revealed. There are also no "favored" affected parts of the protein, which is in agreement with the fact that all steps of the mt-aaRS life cycle can be impacted. Finally, despite the fact that primary observations would suggest an exclusive connection of mt-aaRS disorders to the nervous system and to inherited neurological diseases, sporadic manifestations were lately observed in, e.g., skeletal muscle, kidney, lung, and/or heart. Thus, links between the activity of a given aaRS to mitochondrial translation on one hand and ATP production on the other, involve a number of issues that need to be further explored. Those issues may consider a possible combined effect of mutations affecting other gene (s), such as, for instance, affecting the tRNA modifying enzyme as observed in MLASA patients [109]. Those issues should also take into account the possibility that aminoacylation may turn out not to be the sole function of mt-aaRSs in a living cell, and that these enzymes may also participate in other processes and/or be implicated in various fine-tuning mechanisms, as already shown for various bacterial and eukaryal aaRSs; see below. Thus it becomes obvious that we have to integrate mt-aaRSs into a functional network at the cellular level. In other words, it is of outstanding interest to nail down all the potential interacting components of mt-aaRSs and study their dynamic location within the cell.

In support of this assumption, developments in genomics and post-genomics, associated with conventional biochemical studies, led to the finding of unexpected non-conventional auxiliary functions for human cytosolic aaRSs and connections to other cellular activities (reviewed in, e.g., [119]). Examples include enzymes secreted as procytokines that, after activation, operate in pathways linked to the immune system or angiogenesis (e.g., cyt-TyrRS, cyt-TrpRS [120, 121]), or involved in the vascular development (cyt-SerRS [122]). In addition, accumulating evidence indicates that disruption of non-canonical functions of cytosolic aaRSs connects to various types of diseases, including neural pathologic conditions and cancer [123]. For example, point mutations in human cytosolic TyrRS and cytosolic GlyRS are associated with Charcot-Marie-Tooth (CMT) diseases. Examination of the aminoacylation activities demonstrates that CMT disease can occur without loss of aminoacylation activity [124, 125]. Finally, nowadays evidence indicates that macromolecular assemblies might be sources for proteins with auxiliary functions. Multiprotein complexes containing aaRSs are widely found in all three domains of life, playing roles in apoptosis, viral assembly, and regulation of transcription and translation (reviewed in [126]. As an example, the cytosolic translational apparatus in human cells is highly organized. Nine of the cytosolic aaRS are assembled into the MARS complex (with three auxiliary proteins), which has been emphasized as an anchoring platform for multitasking proteins [127–129]. Those were shown to be recognizable not only by displaying atypical functional activities (possibly linked to structural inventions [69, 119, 130]), but also by their atypical cellular organization, atypical selection pressure, or by atypical "omics" behaviors. The organization of the aaRS within human mitochondria remains mostly unknown at present but will merit full attention in the near future.

Acknowledgements We thank Redmond Smyth for many stylistic improvements of the manuscript. Our work is supported by Centre National de la Recherche Scientifique (CNRS), Université de Strasbourg (UdS), and the French National Program "Investissement d'Avenir" (Labex MitCross), administered by the "Agence National de la Recherche," and referenced ANR-10-IDEX-002-02. The ADIRAL association is acknowledged. HS was supported by Région Alsace, Université de Strasbourg, Association Française contre les Mytopathies (AFM) and Fondation des Treilles.

References

1. Scheffler IE (2001) A century of mitochondrial research: achievements and perspectives. Mitochondrion 1:3–31
2. Anderson S, Bankier AT, Barrell BG et al (1981) Sequence and organization of the human mitochondrial genome. Nature 290:457–465
3. Ojala D, Montoya J, Attardi G (1981) tRNA punctuation model of RNA processing in human mitochondria. Nature 290:470–474
4. Christian BE, Spremulli LL (2012) Mechanism of protein biosynthesis in mammalian mitochondria. Biochim Biophys Acta 1819:1035–1054

5. Florentz C, Sohm B, Tryoen-Tóth P et al (2003) Human mitochondrial tRNAs in health and disease. Cell Mol Life Sci 60:1356–1375
6. Suzuki T, Nagao A, Suzuki T (2011) Human mitochondrial tRNAs: biogenesis, function, structural aspects, and diseases. Annu Rev Genet 45:299–329
7. Watanabe K (2010) Unique features of animal mitochondrial translation systems. The non-universal genetic code, unusual features of the translational apparatus and their relevance to human mitochondrial diseases. Proc Jpn Acad Ser B Phys Biol Sci 86:11–39
8. Bonnefond L, Fender A, Rudinger-Thirion J et al (2005) Toward the full set of human mitochondrial aminoacyl-tRNA synthetases: characterization of AspRS and TyrRS. Biochemistry 44:4805–4816
9. Sissler M, Pütz J, Fasiolo F, Florentz C (2005) Mitochondrial aminoacyl-tRNA synthetases. In: Ibba M, Francklyn C, Cusack S (eds) Aminoacyl-tRNA synthetases. Landes Biosciences, Georgetown, pp 271–284
10. Ylikallio E, Suomalainen A (2012) Mechanisms of mitochondrial diseases. Ann Med 44: 41–59
11. Shoffner JM, Lott MT, Lezza AM et al (1990) Myoclonic epilepsy and ragged-red fiber disease (MERRF) is associated with a mitochondrial DNA tRNA(Lys) mutation. Cell 61: 931–937
12. Goto Y, Nonaka I, Horai S (1990) A mutation in the tRNALeu(UUR) gene associated with the MELAS subgroup of mitochondrial encephalomyopathies. Nature 348:651–653
13. Dimauro S, Davidzon G (2005) Mitochondrial DNA and disease. Ann Med 37:222–232
14. Diaz F (2010) Cytochrome c oxidase deficiency: patients and animal models. Biochim Biophys Acta 1802:100–110
15. Stumpf JD, Copeland WC (2011) Mitochondrial DNA replication and disease: insights from DNA polymerase γ mutations. Cell Mol Life Sci 68:219–233
16. Scheper GD, Van der Knaap MS, Proud CG (2007) Translation matters: protein synthesis defects in inherited disease. Nat Rev Genet 8:711–723
17. Konovalova S, Tyynismaa H (2013) Mitochondrial aminoacyl-tRNA synthetases in human disease. Mol Genet Metab 108:206–211
18. Reeve AK, Krishnan KJ, Turnbull D (2009) Mitochondrial DNA mutations in disease, aging, and neurodegeneration. Ann N Y Acad Sci 1147:21–29
19. Wallace DC (2010) Mitochondrial DNA mutations in disease and aging. Environ Mol Mutagen 51:440–450
20. McFarland R, Elson JL, Taylor RW et al (2004) Assigning pathogenicity to mitochondrial tRNA mutations: when "definitely maybe" is not good enough. Trends Genet 20:591–596
21. Zeviani M, Di Donato S (2004) Mitochondrial disorders. Brain 127:2153–2172
22. Smits P, Smeitink J, Van den Heuvel L (2010) Mitochondrial translation and beyond: processes implicated in combined oxidative phosphorylation deficiencies. J Biomed Biotechnol 2010:737385
23. Fernández-Silva P, Acín-Pérez R, Fernández-Vizarra E et al (2007) In vivo and in organello analyses of mitochondrial translation. Methods Cell Biol 80:571–588
24. DiMauro S, Hirano M (2003) Mitochondrial DNA deletion syndromes. In: Pagon RA, Bird TD, Dolan CR, Stephens K, Adm MP (eds) GeneReviews, Seattle
25. Kunz WS, Kudin A, Vielhaber S et al (2000) Flux control of cytochrome c oxidase in human skeletal muscle. J Biol Chem 275:27741–27745
26. Munnich A, Rustin P (2001) Clinical spectrum and diagnosis of mitochondrial disorders. Am J Med Genet 106:4–17
27. Barrientos A (2002) In vivo and in organello assessment of OXPHOS activities. Methods 26:307–316
28. Brand MD, Nicholls DG (2011) Assessing mitochondrial dysfunction in cells. Biochem J 435:297–312
29. Chance B, Williams GR (1955) A simple and rapid assay of oxidative phosphorylation. Nature 175:1120–1121

30. N'Guessan B, Zoll J, Ribera F et al (2004) Evaluation of quantitative and qualitative aspects of mitochondrial function in human skeletal and cardiac muscles. Mol Cell Biochem 256–257:267–280
31. Veksler VI, Kuznetsov AV, Sharov VG et al (1987) Mitochondrial respiratory parameters in cardiac tissue: a novel method of assessment by using saponin-skinned fibers. Biochim Biophys Acta 892:191–196
32. Letellier T, Malgat M, Coquet M et al (1992) Mitochondrial myopathy studies on permeabilized muscle fibers. Pediatr Res 32:17–22
33. Bouitbir J, Charles A-L, Echaniz-Laguna A et al (2012) Opposite effects of statins on mitochondria of cardiac and skeletal muscles: a "mitohormesis" mechanism involving reactive oxygen species and PGC-1. Eur Heart J 33:1397–1407
34. Nijtmans LG, Henderson NS, Holt IJ (2002) Blue native electrophoresis to study mitochondrial and other protein complexes. Methods 26:327–334
35. Brown WM, George M Jr, Wilson AC (1979) Rapid evolution of animal mitochondrial DNA. Proc Natl Acad Sci USA 76:1967–1971
36. Castellana S, Vicario S, Saccone C (2011) Evolutionary patterns of the mitochondrial genome in Metazoa: exploring the role of mutation and selection in mitochondrial protein coding genes. Genome Biol Evol 3:1067–1079
37. Giegé R, Jühling F, Pütz J et al (2012) Structure of transfer RNAs: similarity and variability. Wiley Interdiscip Rev RNA 3:37–61
38. Helm M, Brulé H, Friede D et al (2000) Search for characteristic structural features of mammalian mitochondrial tRNAs. RNA 6:1356–1379
39. Mudge SJ, Williams JH, Eyre HJ et al (1998) Complex organisation of the 5'-end of the human glycine tRNA synthetase gene. Gene 209:45–50
40. Shiba K, Schimmel P, Motegi H, Noda T (1994) Human glycyl-tRNA synthetase. Wide divergence of primary structure from bacterial counterpart and species-specific aminoacylation. J Biol Chem 269:30049–30055
41. Tolkunova E, Park H, Xia J et al (2000) The human lysyl-tRNA synthetase gene encodes both the cytoplasmic and mitochondrial enzymes by means of an unusual alternative splicing of the primary transcript. J Biol Chem 275:35063–35069
42. Rinehart J, Krett B, Rubio MA et al (2005) Saccharomyces cerevisiae imports the cytosolic pathway for Gln-tRNA synthesis into the mitochondrion. Genes Dev 19:583–592
43. Ibba M, Soll D (2000) Aminoacyl-tRNA synthesis. Annu Rev Biochem 69:617–650
44. Pujol C, Bailly M, Kern D et al (2008) Dual-targeted tRNA-dependent amidotransferase ensures both mitochondrial and chloroplastic Gln-tRNAGln synthesis in plants. Proc Natl Acad Sci USA 105:6481–6485
45. Schön A, Kannangara CG, Gough S, Söll D (1988) Protein biosynthesis in organelles requires misaminoacylation of tRNA. Nature 331:187–190
46. Frechin M, Duchêne A-M, Becker HD (2009) Translating organellar glutamine codons: a case by case scenario? RNA Biol 6:31–34
47. Nagao A, Suzuki T, Katoh T et al (2009) Biogenesis of glutaminyl-mt tRNAGln in human mitochondria. Proc Natl Acad Sci USA 106:16209–16214
48. Frechin M, Senger B, Brayé M et al (2009) Yeast mitochondrial Gln-tRNA(Gln) is generated by a GatFAB-mediated transamidation pathway involving Arc1p-controlled subcellular sorting of cytosolic GluRS. Genes Dev 23:1119–1130
49. Alfonzo JD, Söll D (2009) Mitochondrial tRNA import–the challenge to understand has just begun. Biol Chem 390:717–722
50. Gray MW, Burger G, Lang BF (1999) Mitochondrial evolution. Science 283:1476–1481
51. Brindefalk B, Viklund J, Larsson D et al (2007) Origin and evolution of the mitochondrial aminoacyl-tRNA synthetases. Mol Biol Evol 24:743–756
52. Pfanner N (2000) Protein sorting: recognizing mitochondrial presequences. Curr Biol 10: R412–R415
53. Baker MJ, Frazier AE, Gulbis JM, Ryan MT (2007) Mitochondrial protein-import machinery: correlating structure with function. Trends Cell Biol 17:456–464

54. Becker T, Böttinger L, Pfanner N (2012) Mitochondrial protein import: from transport pathways to an integrated network. Trends Biochem Sci 37:85–91
55. Bolender N, Sickmann A, Wagner R et al (2008) Multiple pathways for sorting mitochondrial precursor proteins. EMBO Rep 9:42–49
56. Gakh O, Cavadini P, Isaya G (2002) Mitochondrial processing peptidases. Biochim Biophys Acta 1592:63–77
57. Van der Laan M, Hutu DP, Rehling P (2010) On the mechanism of preprotein import by the mitochondrial presequence translocase. Biochim Biophys Acta 1803:732–739
58. Neupert W, Herrmann JM (2007) Translocation of proteins into mitochondria. Annu Rev Biochem 76:723–749
59. Schmidt O, Pfanner N, Meisinger C (2010) Mitochondrial protein import: from proteomics to functional mechanisms. Nat Rev Mol Cell Biol 11:655–667
60. Vögtle F-N, Wortelkamp S, Zahedi RP et al (2009) Global analysis of the mitochondrial N-proteome identifies a processing peptidase critical for protein stability. Cell 139:428–439
61. Varshavsky A (2011) The N-end rule pathway and regulation by proteolysis. Protein Sci 20(8):1298–1345
62. Sissler M, Lorber B, Messmer M et al (2008) Handling mammalian mitochondrial tRNAs and aminoacyl-tRNA synthetases for functional and structural characterization. Methods 44: 176–189
63. Bullard JM, Cai YC, Spremulli LL (2000) Expression and characterization of the human mitochondrial leucyl-tRNA synthetase. Biochim Biophys Acta 1490:245–258
64. Yao Y-N, Wang L, Wu X-F, Wang E-D (2003) Human mitochondrial leucyl-tRNA synthetase with high activity produced from Escherichia coli. Protein Expr Purif 30:112–116
65. Gaudry A, Lorber B, Neuenfeldt A et al (2012) Re-designed N-terminus enhances expression, solubility and crystallizability of mitochondrial protein. Protein Eng Des Sel 25:473–481
66. Neuenfeldt A, Lorber B, Ennifar E et al (2012) Thermodynamic properties distinguish human mitochondrial aspartyl-tRNA synthetase from bacterial homolog with same 3D architecture. Nucleic Acids Res 41:2698–2708
67. Cusack S, Berthet-Colominas C, Härtlein M et al (1990) A second class of synthetase structure revealed by X-ray analysis of Escherichia coli seryl-tRNA synthetase at 2.5Å. Nature 347:249–255
68. Eriani G, Delarue M, Poch O et al (1990) Partition of tRNA synthetases into two classes based on mutually exclusive sets of sequence motifs. Nature 347:203–206
69. Guo M, Schimmel P, Yang X-L (2010) Functional expansion of human tRNA synthetases achieved by structural inventions. FEBS Lett 584:434–442
70. Bullard JM, Cai YC, Demeler B, Spremulli LL (1999) Expression and characterization of a human mitochondrial phenylalanyl-tRNA synthetase. J Mol Biol 288:567–577
71. Sanni A, Walter P, Boulanger Y et al (1991) Evolution of aminoacyl-tRNA synthetase quaternary structure and activity: Saccharomyces cerevisiae mitochondrial phenylalanyl-tRNA synthetase. Proc Natl Acad Sci USA 88:8387–8391
72. Klipcan L, Moor N, Finarov I et al (2012) Crystal structure of human mitochondrial PheRS complexed with tRNAPhe in the active "open" state. J Mol Biol 415:527–537
73. Kaiser E, Hu B, Becher S et al (1994) The human EPRS locus (formerly the QARS locus): a gene encoding a class I and a class II aminoacyl-tRNA synthetase. Genomics 19:280–290
74. Bhat TN, Blow DM, Brick P, Nyborg J (1982) Tyrosyl-tRNA synthetase forms a mononucleotide-binding fold. J Mol Biol 158:699–709
75. Chimnaronk S, Gravers Jeppesen M, Suzuki T et al (2005) Dual-mode recognition of non-canonical tRNAsSER by seryl-tRNA synthetase in mammalian mitochondria. EMBO J 24: 3369–3379
76. Bonnefond L, Frugier M, Touzé E et al (2007) Crystal structure of human mitochondrial tyrosyl-tRNA synthetase reveals common and idiosyncratic features. Structure 15: 1505–1516
77. Klipcan L, Levin I, Kessler N et al (2008) The tRNA-induced conformational activation of human mitochondrial phenylalanyl-tRNA synthetase. Structure 16:1095–1104

78. Fender A, Sauter C, Messmer M et al (2006) Loss of a primordial identity element for a mammalian mitochondrial aminoacylation system. J Biol Chem 281:15980–15986
79. Messmer M, Blais SP, Balg C et al (2009) Peculiar inhibition of human mitochondrial aspartyl-tRNA synthetase by adenylate analogs. Biochimie 91:596–603
80. Fender A, Gaudry A, Jühling F et al (2012) Adaptation of aminoacylation identity rules to mammalian mitochondria. Biochimie 94:1090–1097
81. Kumazawa Y, Himeno H, Miura K, Watanabe K (1991) Unilateral aminoacylation specificity between bovine mitochondria and eubacteria. J Biochem 109:421–427
82. Scheper GC, Van der Klok T, Van Andel RJ et al (2007) Mitochondrial aspartyl-tRNA synthetase deficiency causes leukoencephalopathy with brain stem and spinal cord involvement and lactate elevation. Nat Genet 39:534–539
83. Isohanni P, Linnankivi T, Buzkova J et al (2010) DARS2 mutations in mitochondrial leucoencephalopathy and multiple sclerosis. J Med Genet 47:66–70
84. Labauge P, Dorboz I, Eymard-Pierre E et al (2011) Clinically asymptomatic adult patient with extensive LBSL MRI pattern and DARS2 mutations. J Neurol 258:335–337
85. Lin J, Chiconelli Faria E, Da Rocha AJ et al (2010) Leukoencephalopathy with brainstem and spinal cord involvement and normal lactate: a new mutation in the DARS2 gene. J Child Neurol 25:1425–1428
86. Namavar Y, Barth PG, Kasher PR et al (2011) Clinical, neuroradiological and genetic findings in pontocerebellar hypoplasia. Brain 134:143–156
87. Rankin J, Brown R, Dobyns WB et al (2010) Pontocerebellar hypoplasia type 6: a British case with PEHO-like features. Am J Med Genet A 152A:2079–2084
88. Sharma S, Sankhyan N, Kumar A et al (2011) Leukoencephalopathy with brain stem and spinal cord involvement and high lactate: a genetically proven case without elevated white matter lactate. J Child Neurol 26:773–776
89. Steenweg ME, Ghezzi D, Haack T et al (2012) Leukoencephalopathy with thalamus and brainstem involvement and high lactate 'LTBL' caused by EARS2 mutations. Brain 135:1387–1394
90. Pierce SB, Chisholm KM, Lynch ED et al (2011) Mutations in mitochondrial histidyl tRNA synthetase HARS2 cause ovarian dysgenesis and sensorineural hearing loss of Perrault syndrome. Proc Natl Acad Sci USA 108:6543–6548
91. Cassandrini D, Cilio MR, Bianchi M et al (2012) Pontocerebellar hypoplasia type 6 caused by mutations in RARS2: definition of the clinical spectrum and molecular findings in five patients. J Inherit Metab Dis 36:43–53
92. Glamuzina E, Brown R, Hogarth K et al (2012) Further delineation of pontocerebellar hypoplasia type 6 due to mutations in the gene encoding mitochondrial arginyl-tRNA synthetase, RARS2. J Inherit Metab Dis 35:459–467
93. Antonellis A, Green ED (2008) The role of aminoacyl-tRNA synthetases in genetic diseases. Annu Rev Genomics Hum Genet 9:87–107
94. Miyake N, Yamashita S, Kurosawa K et al (2011) A novel homozygous mutation of DARS2 may cause a severe LBSL variant. Clin Genet 80:293–296
95. Synofzik M, Schicks J, Lindig T et al (2011) Acetazolamide-responsive exercise-induced episodic ataxia associated with a novel homozygous DARS2 mutation. J Med Genet 48:713–715
96. Van Berge L, Dooves S, Van Berkel CG et al (2012) Leukoencephalopathy with brain stem and spinal cord involvement and lactate elevation is associated with cell-type-dependent splicing of mtAspRS mRNA. Biochem J 441:955–962
97. Riley LG, Cooper S, Hickey P et al (2010) Mutation of the mitochondrial tyrosyl-tRNA synthetase gene, YARS2, causes myopathy, lactic acidosis, and sideroblastic anemia–MLASA syndrome. Am J Hum Genet 87:52–59
98. Bayat V, Thiffault I, Jaiswal M et al (2012) Mutations in the mitochondrial methionyl-tRNA synthetase cause a neurodegenerative phenotype in flies and a recessive ataxia (ARSAL) in humans. PLoS Biol 10:e1001288

99. Mierzewska H, Van der Knaap MS, Scheper GC et al (2011) Leukoencephalopathy with brain stem and spinal cord involvement and lactate elevation in the first Polish patient. Brain Dev 33:713–717

100. Yamashita S, Miyake N, Matsumoto N et al (2012) Neuropathology of leukoencephalopathy with brainstem and spinal cord involvement and high lactate caused by a homozygous mutation of DARS2. Brain Dev 35:312–316

101. Edvardson S, Shaag A, Kolesnikova O et al (2007) Deleterious mutation in the mitochondrial arginyl-transfer RNA synthetase gene is associated with pontocerebellar hypoplasia. Am J Hum Genet 81:857–862

102. Van Berge L, Kevenaar J, Polder E et al (2012) Pathogenic mutations causing LBSL affect mitochondrial aspartyl-tRNA synthetase in diverse ways. Biochem J 450:345–350

103. Messmer M, Florentz C, Schwenzer H et al (2011) A human pathology-related mutation prevents import of an aminoacyl-tRNA synthetase into mitochondria. Biochem J 433:441–446

104. Elo JM, Yadavalli SS, Euro L et al (2012) Mitochondrial phenylalanyl-tRNA synthetase mutations underlie fatal infantile Alpers encephalopathy. Hum Mol Genet 21:4521–4529

105. Chang GG, Pan F, Yeh C, Huang TM (1983) Colorimetric assay for aminoacyl-tRNA synthetases. Anal Biochem 130:171–176

106. Sissler M, Helm M, Frugier M et al (2004) Aminoacylation properties of pathology-related human mitochondrial tRNA(Lys) variants. RNA 10:841–853

107. Köhrer C, Rajbhandary UL (2008) The many applications of acid urea polyacrylamide gel electrophoresis to studies of tRNAs and aminoacyl-tRNA synthetases. Methods 44:129–138

108. Belostotsky R, Ben-Shalom E, Rinat C et al (2011) Mutations in the mitochondrial seryl-tRNA synthetase cause hyperuricemia, pulmonary hypertension, renal failure in infancy and alkalosis, HUPRA syndrome. Am J Hum Genet 88:193–200

109. Sasarman F, Nishimura T, Thiffault I, Shoubridge EA (2012) A novel mutation in YARS2 causes myopathy with lactic acidosis and sideroblastic anemia. Hum Mutat 33:1201–1206

110. Götz A, Tyynismaa H, Euro L et al (2011) Exome sequencing identifies mitochondrial alanyl-tRNA synthetase mutations in infantile mitochondrial cardiomyopathy. Am J Hum Genet 88:635–642

111. Sasarman F, Karpati G, Shoubridge EA (2002) Nuclear genetic control of mitochondrial translation in skeletal muscle revealed in patients with mitochondrial myopathy. Hum Mol Genet 11:1669–1681

112. Orcesi S, La Piana R, Uggetti C et al (2011) Spinal cord calcification in an early-onset progressive leukoencephalopathy. J Child Neurol 26:876–880

113. Rötig A (2011) Human diseases with impaired mitochondrial protein synthesis. Biochim Biophys Acta 1807:1198–1205

114. Wallace DC (1999) Mitochondrial diseases in man and mouse. Science 283:1482–1488

115. Yarham JW, Elson JL, Blakely EL et al (2010) Mitochondrial tRNA mutations and disease. Wiley Interdiscip Rev RNA 1:304–324

116. Levinger L, Mörl M, Florentz C (2004) Mitochondrial tRNA $3'$ end metabolism and human disease. Nucleic Acids Res 32:5430–5441

117. Wittenhagen LM, Kelley SO (2003) Impact of disease-related mitochondrial mutations on tRNA structure and function. Trends Biochem Sci 28:605–611

118. Jacobs HT, Holt IJ (2000) The np 3243 MELAS mutation: damned if you aminoacylate, damned if you don't. Hum Mol Genet 9:463–465

119. Guo M, Schimmel P (2013) Essential nontranslational functions of tRNA synthetases. Nat Chem Biol 9:145–153

120. Wakasugi K, Slike BM, Hood J et al (2002) Induction of angiogenesis by a fragment of human tyrosyl-tRNA synthetase. J Biol Chem 277:20124–20126

121. Wakasugi K, Slike BM, Hood J et al (2002) A human aminoacyl-tRNA synthetase as a regulator of angiogenesis. Proc Natl Acad Sci USA 99:173–177

122. Kawahara A, Stainier DYR (2009) Noncanonical activity of seryl-transfer RNA synthetase and vascular development. Trends Cardiovasc Med 19:179–182
123. Park SG, Schimmel P, Kim S (2008) Aminoacyl tRNA synthetases and their connections to disease. Proc Natl Acad Sci 105:11043–11049
124. Seburn KL, Nangle LA, Cox GA et al (2006) An active dominant mutation of glycyl-tRNA synthetase causes neuropathy in a Charcot-Marie-Tooth 2D mouse model. Neuron 51:715–726
125. Storkebaum E, Leitão-Gonçalves R, Godenschwege T et al (2009) Dominant mutations in the tyrosyl-tRNA synthetase gene recapitulate in Drosophila features of human Charcot-Marie-Tooth neuropathy. Proc Natl Acad Sci USA 106:11782–11787
126. Hausmann CD, Ibba M (2008) Aminoacyl-tRNA synthetase complexes: molecular multitasking revealed. FEMS Microbiol Rev 32:705–721
127. Han JM, Lee MJ, Park SG et al (2006) Hierarchical network between the components of the multi-tRNA synthetase complex: implications for complex formation. J Biol Chem 281:38663–38667
128. Kaminska M, Havrylenko S, Decottignies P et al (2009) Dissection of the structural organization of the aminoacyl-tRNA synthetase complex. J Biol Chem 284:6053–6060
129. Ray PS, Arif A, Fox PL (2007) Macromolecular complexes as depots for releasable regulatory proteins. Trends Biochem Sci 32:158–164
130. Guo M, Yang X-L, Schimmel P (2010) New functions of aminoacyl-tRNA synthetases beyond translation. Nat Rev Mol Cell Biol 11:668–674
131. Uluc K, Baskan O, Yildirim KA et al (2008) Leukoencephalopathy with brain stem and spinal cord involvement and high lactate: a genetically proven case with distinct MRI findings. J Neurol Sci 273:118–122
132. Tzoulis C, Tran GT, Gjerde IO et al (2012) Leukoencephalopathy with brainstem and spinal cord involvement caused by a novel mutation in the DARS2 gene. J Neurol 259:292–296
133. Talim B, Pyle A, Griffin H et al (2013) Multisystem fatal infantile disease caused by a novel homozygous EARS2 mutation. Brain 136:e228
134. Shamseldin HE, Alshammari M, Al-Sheddi T et al (2012) Genomic analysis of mitochondrial diseases in a consanguineous population reveals novel candidate disease genes. J Med Genet 49:234–241

Top Curr Chem (2014) 344: 293–330
DOI: 10.1007/128_2013_425
© Springer-Verlag Berlin Heidelberg 2013
Published online: 11 May 2013

Role of Aminoacyl-tRNA Synthetases in Infectious Diseases and Targets for Therapeutic Development

Varun Dewan, John Reader, and Karin-Musier Forsyth

Abstract Aminoacyl-tRNA synthetases (AARSs) play a pivotal role in protein synthesis and cell viability. These 22 "housekeeping" enzymes (1 for each standard amino acid plus pyrrolysine and o-phosphoserine) are specifically involved in recognizing and aminoacylating their cognate tRNAs in the cellular pool with the correct amino acid prior to delivery of the charged tRNA to the protein synthesis machinery. Besides serving this canonical function, higher eukaryotic AARSs, some of which are organized in the cytoplasm as a multisynthetase complex of nine enzymes plus additional cellular factors, have also been implicated in a variety of non-canonical roles. AARSs are involved in the regulation of transcription, translation, and various signaling pathways, thereby ensuring cell survival. Based in part on their versatility, AARSs have been recruited by viruses to perform essential functions. For example, host synthetases are packaged into some retroviruses and are required for their replication. Other viruses mimic tRNA-like structures in their genomes, and these motifs are aminoacylated by the host synthetase as part of the viral replication cycle. More recently, it has been shown that certain large DNA viruses infecting animals and other diverse unicellular eukaryotes encode tRNAs, AARSs, and additional components of the protein-synthesis machinery. This chapter will review our current understanding of the role of host AARSs and tRNA-like structures in viruses and discuss their potential as anti-viral drug targets. The identification and development of compounds that target bacterial AARSs, thereby serving as novel antibiotics, will also be discussed. Particular attention will be given to recent work on a number of tRNA-dependent

---

V. Dewan and K.-M. Forsyth (✉)
Department of Chemistry and Biochemistry, Ohio State Biochemistry Program, Center for RNA Biology, and Center for Retroviral Research, The Ohio State University, Columbus, OH 43210, USA
e-mail: musier@chemistry.ohio-state.edu

J. Reader
Department of Cell Biology and Physiology, The University of North Carolina at Chapel Hill School of Medicine, Chapel Hill, NC 27514, USA

AARS inhibitors and to advances in a new class of natural "pro-drug" antibiotics called Trojan Horse inhibitors. Finally, we will explore how bacteria that naturally produce AARS-targeting antibiotics must protect themselves against cell suicide using naturally antibiotic resistant AARSs, and how horizontal gene transfer of these AARS genes to pathogens may threaten the future use of this class of antibiotics.

Keywords Aminoacyl-tRNA synthetases · Horizontal gene transfer · Human immunodeficiency virus type-1 · Rous sarcoma virus · tRNA-like structure · Trojan Horse inhibitors

Contents

1 Introduction 295
2 tRNA-Like Structures in Plant Viruses 297
3 Aminoacyl-tRNA synthetases and tRNAs Encoded by Nucleocytoplasmic Large DNA Viruses 298
4 Host tRNAs and Aminoacyl-tRNA Synthetases in Retroviruses 299
    4.1 Packaging of Lysyl-tRNA Synthetase in HIV-1 299
    4.2 tRNA-Like Elements in the Genome of HIV-1 302
    4.3 Packaging of Tryptophanyl-tRNA Synthetase in Rous Sarcoma Virus 303
5 Aminoacyl-tRNA Synthetases as Targets for Therapeutic Development in Bacterial Infections 304
    5.1 Aminoacyl-tRNA Synthetase Inhibitor Mechanisms 305
    5.2 Trojan Horse tRNA Synthetase Inhibitors 313
    5.3 Antibiotic Resistant AARS Genes and the Threat of Horizontal Gene Transfer to Pathogenic Bacteria 316
6 Conclusions 319
References 320

Abbreviations

AA-AMP   Aminoacyl adenylate
AARS   Aminoacyl-tRNA synthetase
CA   Capsid
CA-CTD   Capsid C-terminal domain
cDNA   Complementary DNA
EPRS   Glutaminyl-prolyl-tRNA synthetase
HIV-1   Human immunodeficiency virus-1
IFN-γ Interferon-γ
IleRS   Isoleucyl-tRNA synthetase
LeuRS   Leucyl-tRNA synthetase
LysRS   Lysyl-tRNA synthetase
MetRS   Methionyl-tRNA synthetase
mRNA   Messenger RNA

MSC	Multi-synthetase complex
PBS	Primer binding site
PheRS	Phenylalanyl-tRNA synthetase
RSV	Rous sarcoma virus
SerRS	Seryl-tRNA synthetase
ThrRS	Threonyl-tRNA synthetase
TLE	tRNA-like element
TLS	tRNA-like structure
tRNA	Transfer RNA
TrpRS	Tryptophanyl-tRNA synthetase
TYMV	Turnip yellow mosaic virus
TyrRS	Tyrosyl-tRNA synthetase
ValRS	Valyl-tRNA synthetase

1 Introduction

Decoding the genetic material from DNA to RNA to protein is the major and most energy consuming process in the cell. In addition to ribosomes, mRNA, and tRNA, this process requires amino acids, nucleotides, and specialized proteins. The key link between transcription (DNA to mRNA) and translation (mRNA to proteins) is provided by aminoacyl-tRNA synthetases (AARSs), an essential family of enzymes responsible for the identification and aminoacylation of cognate tRNAs [1]. The two-step aminoacylation reaction involves activation of the amino acid with ATP to form an aminoacyl-adenylate intermediate followed by transfer of the activated amino acid to the $3'$end of the tRNA. Over the past decade, AARSs have emerged as multifaceted molecules participating in a variety of non-canonical functions in addition to their role in protein synthesis. Diverse functions such as signaling, stress response, and transcriptional regulation have been attributed to these "housekeeping" enzymes [2]. In addition to the emerging non-canonical and diverse roles of AARSs in the cell, their exploitation by viruses for replication is another well-known function of synthetases.

The link between tRNA biology and virology can be found at many levels. The occurrence of tRNA-like structures (TLSs) at the $3'$end of certain positive-strand RNA plant viruses is intriguing [3] and is discussed in Sect. 2. These TLSs are built on distinct design principles, yet they possess key properties of tRNAs with the incorporation of conserved features and aminoacylation identity elements. A possible role in supporting high rates of viral protein synthesis has been suggested for these TLSs [4].

Retroviruses also exploit tRNAs and tRNA-like elements for their replication (discussed in Sect. 4). In retroviruses and long terminal repeat retrotransposons, cellular tRNAs are used as primers for reverse transcription [5]. In some retroviruses a host cell tRNA is selectively packaged into the virion, where it is placed onto the complementary primer binding site (PBS) of the viral RNA genome. The tRNA is used to prime the reverse transcriptase-catalyzed synthesis of minus-strand strong-stop DNA, the first step in reverse transcription [6]. A select number of tRNAs have been identified as primer tRNAs in retroviruses. For example, $tRNA^{Trp}$ is the primer for all members of the avian sarcoma virus–avian leukosis virus group examined to date [7, 8]. $tRNA^{Lys1,2}$, representing two $tRNA^{Lys}$ isoacceptors differing by one base pair in the anticodon stem, is the primer tRNA for Mason-Pfizer monkey virus and human foamy virus [9]. On the other hand, $tRNA^{Lys,3}$ serves as the primer for mouse mammary tumor virus and lentiviruses, such as equine infectious anemia virus, feline immunodeficiency virus, simian immunodeficiency virus, and HIV-1 [10]. $tRNA^{Pro}$ is the common primer for Moloney murine leukemia virus [11]. In some cases the packaging of these primer tRNAs into retroviruses is quite selective in that there is an increase in the percentage of the primer tRNA in going from the cytoplasm to the virus. For example, in avian myeloblastosis virus, the relative concentration of $tRNA^{Trp}$ changes from 1.4% to 32% [12].

In HIV-1, both primer $tRNA^{Lys,3}$ and the other major $tRNA^{Lys}$ isoacceptors, $tRNA^{Lys1,2}$, are selectively packaged. The relative concentration of $tRNA^{Lys}$ changes from 5–6% in the cytoplasm to 50–60% in the virus [13]. In murine leukemia virus, selective packaging of primer $tRNA^{Pro}$ is also observed. A cytoplasmic concentration of 5–6% $tRNA^{Pro}$ increases to 12–24% of low molecular weight RNA [11]. Selective packaging of primer tRNAs in retroviruses is consistent with the importance of these molecules in catalyzing a crucial step in viral replication.

In addition to specific packaging of a primer tRNA from the host milieu, some retroviruses also package the corresponding cognate synthetase. In HIV-1, lysyl-tRNA synthetase (LysRS) is packaged with high specificity [14, 15]. There are approximately 20–25 molecules of LysRS packaged into HIV-1, which is in good agreement with the amount of $tRNA^{Lys}$ primer packaged per virion. The 1:1 stoichiometry of viral LysRS:$tRNA^{Lys}$ suggests that they enter the virus as a complex [9]. In this chapter we will explore how LysRS mediates $tRNA^{Lys}$ packaging, along with other recent discoveries that suggest additional roles for LysRS in HIV-1. The putative role of tryptophanyl-tRNA synthetase (TrpRS) as a "carrier" for the $tRNA^{Trp}$ primer in Rous sarcoma virus (RSV) will also be discussed (Sect. 4).

In addition to plant viruses and retroviruses, nucleocytoplasmic large DNA viruses have recently emerged as new players in exploiting the functional capabilities of AARSs and their associated translational machinery [16]. Some of the tRNA genes found in the genomes of *Chlorella*-infecting phycodnavirus contain intron-like insertions in the anticodon loop and have been observed to be aminoacylated during the infectious cycle, thereby aiding in protein synthesis [17]. The giant mimivirus of the *Mimiviridae* family is the largest known double-stranded DNA virus infecting *Acanthaamoeba polyphaga*. Its enormous genome of 1.2 million base-pairs encodes 6 tRNAs, 4 AARSs, along with a full set of translation initiation, elongation, and termination factors [18]. Although the significance of these factors for viral replication is not entirely clear, a brief description is included here (Sect. 3).

The role of AARSs in viruses suggests the possibility for the development of novel therapies aimed at targeting these host factors. Many challenges exist, however, due to their essential function in the host cell. On the other hand, due to species-specific differences in AARS domain structure and tRNA discrimination, this family of enzymes has long been recognized as a potential drug target in countering bacterial infections. With the tremendous progress in structural and biochemical characterization of synthetases in recent years, the repertoire of drugs from natural sources and synthetic libraries has increased. The emergence of multi-drug-resistant strains of pathogenic microorganisms has increased the need for new classes of antimicrobial agents. New strategies aimed at targeting the active sites are therefore being developed for bacterial protein synthesis systems, as discussed in Sect. 5 of this chapter.

2 tRNA-Like Structures in Plant Viruses

The genomes of positive strand RNA viruses simultaneously serve as mRNAs for expressing viral genes and as templates for RNA replication. The presence of TLSs within the genomes of positive strand plant viruses is considered a relic of a self-replicating RNA from a period prior to protein synthesis [3, 19, 179]. TLSs are divided into three types based on their aminoacylation by valine, histidine, or tyrosine. These TLSs are 3′-terminal elements that are built around a pseudoknotted aminoacyl-acceptor stem, which makes them ideal elements to be recognized by cognate AARSs in an infectious event [3, 20, 21]. Three tRNA-related properties of these TLSs include: aminoacylation, complex formation with eEF1A·GTP, and 3′adenylation. One of the best studied examples of a viral TLS is an 82-nucleotide sequence identified at the 3′ end of the turnip yellow mosaic virus (TYMV) genome (Fig. 1) [22]. It most closely resembles tRNA^{Val} and is therefore a strong substrate for aminoacylation by valyl-tRNA synthetase (ValRS). The structure consists of two arms, the 12-bp acceptor-T arm and the anticodon-D arm, joined by four nucleotides that mimic the variable loop. A valine anticodon and the C residue at the 3′-end of the anticodon loop constitutes this TLSs key valine identity element [23]. The most important identity nucleotide for aminoacylation is A56 in the middle of the anticodon, followed by C55 also in the anticodon. The three signature properties of valylation, eEF1A·GTP ternary complex formation, and 3′adenylation are displayed by this TLS, and valylation by ValRS imparts infectivity to TYMV [3, 19].

In contrast to the TYMV TLS, the histidine-specific TLSs of tobamoviral RNAs, which includes Tobacco mosaic virus (TMV) [24], possess an acceptor-T arm that is recognizable by histidyl-tRNA synthetase, while the other arm has no resemblance to the anticodon-D arm. The TMV TLS has an acceptor-T arm that is built in a similar way to that of TYMV, except that the arm is only 11-bp long. The rest of the TLS involves a considerably more complex structure than the TYMV TLS and has rather weak similarity to tRNA, except for a GUU histidine anticodon (GUG in

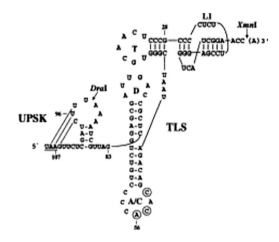

Fig. 1 The 3′ tRNA-like structure (TLS) and upstream pseudoknot (UPSK) present at the 3′ end of TYMV RNA. Figure taken from [178]. The circled nucleotides correspond to the critical identity elements for aminoacylation by ValRS. The loops corresponding to the D-, T-, and anticodon (A/C) loops are indicated

some strains). Even though some deviations exist, the presence of key tRNA^{His} identity elements at the end of the aminoacyl-acceptor stem makes the TMV TLS compatible for aminoacylation by the cognate synthetase [25].

Similarly, TLSs that mimic tRNA^{Tyr} and are recognized by tyrosyl-tRNA synthetase (TyrRS) have been reported for brome mosaic virus. This virus possesses a complex 134-nucleotide long TLS with a 6-bp aminoacyl-acceptor stem containing a pseudoknot that serves to bring together distant nucleotides [24]. Although a distinct anticodon domain is not present, aminoacylation by TyrRS is still possible due to the presence of tRNA^{Tyr} identity elements in the acceptor stem [3].

3 Aminoacyl-tRNA synthetases and tRNAs Encoded by Nucleocytoplasmic Large DNA Viruses

Nucleocytoplasmic large DNA viruses (NCLDVs) are eukaryotic viruses infecting animals and diverse unicellular eukaryotes. These viruses replicate either exclusively in the cytoplasm of the host, or possess both cytoplasmic and nuclear stages in their life cycle [26].

Chlorella viruses are NCLDVs that replicate in certain strains of the unicellular green alga *Chlorella*. Genome analysis of the *Chlorella* virus PBCV-1 revealed 702 open reading frames of which 377 were predicted to encode functional proteins and 121 of these resembled proteins in the databases [17]. Surprisingly, the predicted genes included components of the translational machinery including tRNA genes. For example, 14 tRNA genes have been detected in the genome of the CVK2 virus and some of these tRNAs have been reported to contain intron-like insertions in the anticodon loop. Seven of the 14 tRNAs have been observed to be aminoacylated during the infectious cycle and thus are presumably active in protein synthesis [17]. It has been proposed that the presence and functioning of these tRNAs contributes to the preferential translation of the viral proteins.

Mimivirus is one of the largest known DNA viruses with a genome length of 1.2 million bp [16]. The genome sequence of this virus, which infects *Acanthamoeba*, revealed the presence of numerous genes central to the steps of translation, thereby apparently violating the established dependence of viruses exclusively on the host translational machinery for protein synthesis [18]. Its genome encodes six tRNA-like genes (three Leu and one each of Trp, Cys, and His) in addition to homologs of ten proteins with functions central to protein synthesis. Encoded AARSs include ArgRS, CysRS, MetRS, and TyrRS. Mimivirus tRNAs have been shown to be functional in aminoacylation, with MetRS and TyrRS conforming to the tRNA identity rules for archaea/eukarya [18]. Phylogenic studies support an evolutionary scenario involving genome reduction as opposed to recent acquisition from cellular organisms [16, 18]. This conclusion is also supported by a recent genome analysis of a mimivirus relative with an even larger genome [27]. *Megavirus chilensis*, a giant virus isolated off the coast Chile, encodes the four AARSs found in the mimivirus genome plus three additional AARS genes (IleRS, TrpRS, and AsnRS). Although the presence of AARSs and TLSs encoded within the genomes of these viruses is well established, their function is not entirely clear. However, the lack of any unique structural features supports a role in the viral protein synthesis process in infected cells [18].

4 Host tRNAs and Aminoacyl-tRNA Synthetases in Retroviruses

Retroviruses are classified as infectious particles consisting of an RNA genome packaged in a protein capsid (CA) and surrounded by a lipid envelope. This lipid envelope contains polypeptide chains that bind to membrane receptors of the host cell to initiate the process of infection. Retrovirus particles also contain the enzyme reverse transcriptase, which synthesizes a DNA copy of the viral RNA template. The cDNA produced contains the genetic instructions that allow infection of the host cell to proceed. The initiation of reverse transcription requires a cellular tRNA primer, which in some cases is selectively incorporated during virion assembly [9]. In lentiviruses, including HIV-1, $tRNA^{Lys,3}$ serves as the primer tRNA. The mechanism of tRNA primer recruitment and the role of AARSs in this process have been most thoroughly studied in HIV-1, and will be comprehensively discussed here.

4.1 Packaging of Lysyl-tRNA Synthetase in HIV-1

HIV-1 is the causative agent of acquired immunodeficiency syndrome. Its intricate and complex replication cycle requires extensive involvement of host factors, which are specifically hijacked from their normal function in the cell.

Host cellular tRNA^{Lys,3} is required to prime the first step of reverse transcription in the viral life cycle [5]. The resultant proviral DNA is translocated into the nucleus of the infected cell where it integrates into the host cell's DNA and codes for viral RNA and proteins. The mature viral structure includes glycosylated envelope proteins and proteins resulting from processing of the large precursor proteins Gag and GagPol [28]. Both Gag and GagPol are translated from full-length viral RNA, which is also packaged into assembling virions where it serves as the genomic RNA. During maturation, Gag is cleaved by HIV-1 protease to yield matrix, CA, nucleocapsid, and p6 proteins. Protease, reverse transcriptase, and integrase are enzymes produced as a result of GagPol processing. During the assembly step of the viral life cycle, Gag, GagPol, vRNA, and specific cellular components are selectively packaged into the virion for initiating subsequent infectious cycles [29].

Host-encoded tRNA^{Lys,3}, which serves as the primer for reverse transcription, is selectively packaged into HIV-1, along with the other major human tRNA^{Lys} isoacceptors (tRNA^{Lys1,2}). Human LysRS, the enzyme that aminoacylates tRNA^{Lys}, is the only known cellular factor that specifically recognizes all tRNA^{Lys} species. The selective packaging of tRNA^{Lys} isoacceptors into HIV-1 raised the possibility that LysRS also participates in viral tRNA packaging. The packaging of tRNA^{Lys} requires Gag, GagPol [13], and LysRS, also selectively packaged into HIV-1 [14]. Indeed, overexpression of LysRS increases tRNA^{Lys} packaging into HIV-1 particles [30] and siRNA knockdown of LysRS decreases the tRNA^{Lys} amounts incorporated [31]. Thus, LysRS is the limiting factor for tRNA^{Lys} packaging [30]. Increasing the concentration of viral tRNA^{Lys,3} in HIV-1 by overexpressing exogenous tRNA^{Lys,3} also results in increased annealing of the tRNA onto the primer binding site and enhanced viral infectivity [30]. Furthermore, packaging of tRNA^{Lys} isoacceptors requires binding to LysRS [32], but aminoacylation of the tRNA is not required [33].

Human LysRS is a class II synthetase, forming a closely related sub-group (known as IIb) with aspartyl- and asparaginyl-tRNA synthetases [34, 35]. Class II synthetases are generally functional dimers or tetramers and their active site consists of an antiparallel β-sheet structure and three highly degenerate consensus sequences (motifs 1, 2 and 3). Motif 1, which is comprised of α-helices 5, 6, and 7 (H7) and part of a β-sheet (β6), constitutes the dimer interface, whereas motif 2 and 3 together constitute the aminoacylation active site [36–38]. Although an α_2 homodimer is the functional oligomerization state for aminoacylation, hLysRS crystallizes as an $\alpha_2\alpha_2$ tetramer [37]. Within the cytoplasm, LysRS is also part of the high molecular weight multisynthetase complex (MSC) present in higher eukaryotes [39]. Within the MSC, LysRS specifically interacts with the scaffold protein p38/AIMP2 and is present in a unique $\alpha_2\beta_1:\beta_1\alpha_2$ orientation, which is designed to control both retention and mobilization of LysRS from the MSC [40].

The source of LysRS packaged into HIV-1 is still unclear. One report suggests that the packaged viral LysRS is of host mitochondrial origin [41]. However, earlier studies suggested that packaged LysRS does not appear to originate from any of its identified steady-state cellular compartments, which include the cytoplasmic MSC, nuclei, mitochondria, or cell membrane. Instead, newly-synthesized cytoplasmic LysRS has been shown to interact with HIV-1 Gag before entering any of these compartments upon viral infection [15]. More recently it has been suggested that interactions also occur between LysRS and the Pol domain of the GagPol precursor [42].

The packaging of LysRS into HIV-1 is specific; out of nine AARSs and three additional components of the mammalian MSC tested, only LysRS has been shown to be packaged [15]. The domains critical for interaction between LysRS and Gag have been identified to include the motif 1 domain of LysRS and the C-terminal domain (CTD) of CA [43]. Deletions that extend into HIV-1 CA-CTD or LysRS motif 1 helix 7 (H7) abolish the interaction between Gag and hLysRS in vitro. Furthermore, deletion of motif 1 eliminates LysRS packaging into Gag viral-like particles [43]. Interestingly, both of these primarily helical regions (motif 1 of LysRS and HIV-1 CA-CTD) are critical for homodimerization of the individual proteins. The interaction between Gag and LysRS has previously been investigated by fluorescence anisotropy (FA) measurements and gel chromatography [44]. An apparent equilibrium binding constant (K_d) of 310 nM was measured for the Gag/LysRS interaction, and CA alone bound LysRS with a similar affinity (~400 nM) as full-length Gag. Gag and LysRS variants containing point mutations in their dimerization motifs that effectively eliminated homodimerization still interacted in vitro, suggesting that dimerization of each protein per se is not required for the interaction. A gel chromatography study conducted with the wild-type proteins is consistent with formation of a heterodimeric LysRS–Gag complex [44]. How the monomer–dimer equilibrium is perturbed in vivo to favor the packaging of the monomeric form of LysRS is unknown.

Nuclear magnetic resonance and mutagenesis studies mapped the CA residues critical for the interaction to the helix 4 (h4) region of CA-CTD [45]. More recently, an ab initio energy minimized "bridging monomer" model of the HIV-1 CA-CTD/LysRS/tRNA^{Lys} ternary complex has been proposed, which is also consistent with an interaction between h4 of CA-CTD and the H7 region of LysRS [46]. Circular dichroism experiments along with in silico studies also support this h4/H7 interaction [47]. Key residues within the helix 7 of LysRS that contribute to Gag interaction have also recently been more finely mapped. Residues along one face of this helix play a dual role in maintaining the dimer interface of LysRS, as well as mediating the binding to CA [48]. Serving as an attractive host-viral therapeutic target, cyclic peptide inhibitors have been discovered that inhibit the Gag–LysRS protein–protein interaction in vitro [49]. In this work, high-throughput screening of a combinatorial library initially identified novel cyclic peptide ligands against HIV-1 CA. The most promising peptides bound CA with ~500 nM affinity and inhibited the CA–LysRS interaction in vitro with IC_{50} values of ~1 μM. Furthermore, nuclear magnetic resonance studies of two peptides, along with mutational analyses and in silico docking studies, suggested that both bind to a site on the CA-CTD that has been previously identified as the interface for LysRS interaction.

AARSs function in a wide array of cellular processes that are distinct from aminoacylation and changes in oligomeric state, as well as post-translational modifications, have been shown to regulate these alternate functions [2, 50]. Eukaryotic LysRS, in particular, has been implicated in a wide variety of non-canonical roles. The dynamic nature of LysRS within the cell is illustrated by its mobilization to the nucleus. In response to an immunological challenge within mast cells, specific phosphorylation of S207 of LysRS has been shown to result in its release from the MSC and nuclear import. LysRS catalyzes diadenosine tetraphosphate synthesis in the nucleus and binding of this dinucleotide to the Hint-1 repressor protein results in transcriptional activation of genes via the microphthalmia-associated transcription factor [51]. It was recently shown that a single conformational change triggered by phosphorylation drives the switch of LysRS function from translation to transcription [52].

More recently, laminin-induced phosphorylation of LysRS at residue T52 was shown to result in release of LysRS from the MSC and trafficking to the membrane [53]. Membrane association of LysRS and association with the laminin receptor regulate the receptor and its effect on cell migration. Whether the post-translational modification state of LysRS is altered during HIV-1 infection, thereby modulating oligomerization and facilitating interaction with components of the tRNA^{Lys} packaging complex, remains to be tested.

4.2 tRNA-Like Elements in the Genome of HIV-1

Selective packaging of the primer tRNA is required for optimizing both tRNA^{Lys,3} annealing to the viral RNA PBS and infectivity of the HIV-1 population [9]. The nucleocapsid domain of HIV-1 Gag facilitates annealing of tRNA^{Lys,3} to highly conserved complementary sequences in the HIV-1 genomic RNA [54–57]. The longest of these sequences is the 18-nucleotide PBS, which is complementary to the 3′ end of tRNA^{Lys,3}. The PBS is located in the 5′-untranslated region of the ~9.4 kB HIV-1 RNA (Fig. 2). Although the entire length of the viral RNA contains secondary structural elements whose functions are in most cases still unknown [58], the 5′ UTR is especially rich in complex secondary structures with known functions in many steps of the virus life cycle [59]. In addition to the 18 nucleotides that hybridize with the tRNA acceptor-TΨC stem, additional interactions occur between viral RNA and complementary sequences in the variable arm and anticodon stem-loop of tRNA^{Lys,3} [60].

In addition to the PBS located in the 5′ UTR, a 9-nt long sequence complementary to nucleotides 38–46 of tRNA^{Lys,3} was identified near the 3′ end of the HIV-1 genomic RNA [61, 62]. It was suggested that this interaction stimulates the minus-strand transfer step of reverse transcription. Recently a much longer sequence element that resembles the entire tRNA^{Lys,3} gene embedded in the 3′ end of the HIV-1 genomic RNA and closely related lentiviral genomes has been discovered [63] (Fig. 2). This sequence includes the previously identified 9-nt sequence, is extensively complementary to tRNA^{Lys,3}, and even contains a vestigial intron inserted in approximately the position expected for a tRNA intron. The inclusion

Fig. 2 tRNA elements in HIV-1. Highlighted at the 5' end of the viral genome is the tRNA anticodon loop-like element (TLE, *red*) [64]. The TLE binds human LysRS and is proximal to the primer binding site (PBS, *green*). Shown at the 3' end is a sequence in the U3-R region that is complementary to the entire tRNA^{Lys,3} sequence (tRNA gene, *blue*) [63]

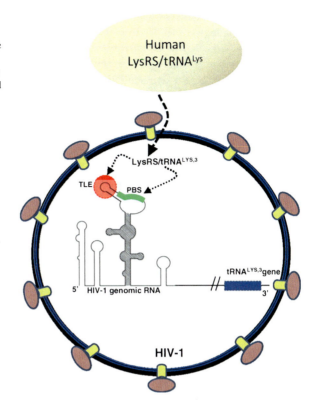

of this extended sequence in the acceptor template further enhances the efficiency of minus-strand transfer beyond that observed with the 9-nt segment.

Molecular mimicry of the human tRNA^{Lys} anticodon domain by HIV-1 genomic RNA has recently been reported [64]. This tRNA-like element (TLE) is located proximal to the PBS (Fig. 2) and high affinity binding of LysRS to the U-rich TLE occurs in the context of a larger LysRS binding domain that includes elements of the psi packaging signal. This LysRS–genome interaction increases the efficiency of tRNA^{Lys} annealing to the viral RNA and appears to be important for optimal viral infection. Additional studies are needed to determine whether molecular mimicry of the entire tertiary structure of the tRNA occurs in this region of the genome.

4.3 *Packaging of Tryptophanyl-tRNA Synthetase in Rous Sarcoma Virus*

RSV is an avian retrovirus and was the first oncovirus to be discovered. It causes sarcoma in chickens and, as with all retroviruses, it reverse transcribes its RNA genome into cDNA before integration into the host DNA [65]. RSV selectively packages host cell tRNA^{Trp}, raising the possibility that TrpRS might also be packaged [12]. Experiments performed with RSV isolated from infected turkey embryo fibroblasts revealed the presence of TrpRS in the viral lysate. Furthermore, the packaging of TrpRS was specific as LysRS and prolyl-tRNA synthetase were not detected. The number of TrpRS molecules as quantified by dot-blot analysis was found to be 12 molecules/virion [6]. The RT region of RSV Gag-Pol appears to be crucial for tRNA^{Trp} recruitment [6], but the interactions that dictate TrpRS packaging are currently unknown.

Although eukaryotic TrpRS is not part of the MSC, like LysRS it has been demonstrated to have a number of ex-translational functions. TrpRS is highly induced by interferon-γ (IFN-γ) [66–68] and it is secreted and inhibits vasculature development [69–71]. This angiostatic function is activated by removal of an appended N-terminal domain [70]. TrpRS has previously been shown to localize to the nucleus [72, 73]. Recently it has been shown that upon IFN-γ stimulation in mammalian cells, nuclear localization is increased. Nuclear TrpRS forms a complex with DNA-dependent protein kinase (DNA-PKcs) and with poly (ADP-ribose) polymerase 1 (PARP-1) via the appended eukaryote-specific WHEP domain of TrpRS. Nuclear TrpRS stimulates poly ADP-ribosylation of DNA-PKcs and this in turn leads to p53 phosphorylation and activation [69]. Interestingly, RSV Gag also traffics to the nucleus and is believed to interact with viral RNA in this cellular compartment [74, 75]. Whether the emerging ex-translational nuclear roles of LysRS and TrpRS, as well as their roles as signaling molecules, are linked to their retroviral function requires further investigation.

5   Aminoacyl-tRNA Synthetases as Targets for Therapeutic Development in Bacterial Infections

The use of antibiotics to fight infectious diseases has had a profound effect on human health and is thought to be a major contributing factor in doubling human lifespan over the last 100 years [76]. However, inappropriate use of anti-infectives in both medicine and animal husbandry has exacerbated the threat of microbial antibiotic resistance, ultimately leading to the evolution of bacterial pathogens with multiple-antibiotic resistance to a large number of first line of defense antibiotics [77, 78]. To combat the resistance threat, both chemical derivatives of existing antibiotics and novel anti-infectives have been actively sought. This task has been made difficult by the limited number of essential cellular processes that can be targeted. In this regard, inhibiting protein synthesis has been one of the most effective approaches with numerous antibiotics targeted towards the ribosome, such as chloramphenicol and tetracycline [79].

AARSs are essential cellular components of the protein translation machinery that ligate a specific amino acid to its canonical tRNA. The 20 standard AARS enzymes appear to represent an abundant group of proteins towards which small molecule anti-infectives could be developed. This view is supported by the large number of naturally occurring secondary metabolites that achieve antibiosis by targeting these enzymes [80]. Despite the fact that all AARSs share the same reaction mechanism, the high level of phylogenetic divergence between bacterial and eukaryote/archaea enzymes can allow AARS inhibitors to discriminate effectively between pathogen and host AARSs [76]. Yet despite the obvious attractive features, very few AARS inhibitors have been successfully utilized as anti-infectives [81]. In fact, only one AARS inhibitor to date is being used in the clinic, mupirocin [82, 83] (sold under the name Bactroban). Mupirocin is a very successful topical agent against *Staphylococcus aureus* infections, validating AARS enzymes as effective antibiotic targets [84].

In this section we will provide an overview of the enzymatic and biological properties of AARSs and examine the general inhibition mechanisms employed by both natural and synthetically derived AARS-targeting antibiotics. We pay particular attention to recent work on a number of tRNA-dependent AARS inhibitors. We also look at new advances in a new class of natural "pro-drug" or cleavable antibiotics, called the Trojan Horse AARS inhibitors, which employ "molecular subterfuge" to trick pathogens into actively taking up and processing these compounds into toxic moieties. Finally, we will explore how bacteria that naturally produce AARS-targeting antibiotics must protect themselves against cell suicide using naturally antibiotic resistant AARSs and how horizontal gene transfer of these AARS genes to pathogens may threaten the future use of this class of antibiotics.

5.1 Aminoacyl-tRNA Synthetase Inhibitor Mechanisms

AARSs catalyze the covalent attachment of each of the 20 amino acids to its cognate tRNA in the first and essential step of protein synthesis. Thus, AARSs create the rules of the genetic code, since each tRNA species contains the anticodon triplet of the code corresponding to the attached amino acid. All AARSs catalyze the aminoacylation reaction in two steps: in the first step, the amino acid (AA) is activated to form the tightly bound high energy aminoacyl-adenylate (AA-AMP); in the second step the AA-AMP reacts with the $2'$-OH or $3'$-OH of the terminal ribose at the $3'$-end of tRNA to form the aminoacylated-tRNA [85]. Each amino acid is activated by a particular AARS. The enzymes are divided into two classes of ten enzymes each. The classes, in turn, are defined by their active site architectures – a Rossmann nucleotide binding fold of alternating β-strands and α-helices characterize class I enzymes, and a seven-stranded β-structure flanked by α-helices characterize class II enzymes. Apart from a few exceptions [86], each amino acid-specific AARS belongs to a particular class throughout the tree of life.

AARSs have acquired numerous domains and insertions during the course of evolution that have enhanced their catalytic properties. For example, weakly conserved anticodon-binding domains increase the affinity and ability of AARSs to recognize and discriminate between different tRNA species [87]. AARSs have thermodynamic limitations on their specificity to distinguish between similar amino acids and therefore often employ conserved proofreading or editing domains to hydrolyze misacylated-tRNAs to ensure the fidelity of translation [88]. The occurrence of other domains in AARSs can be more idiosyncratic, sometimes even exhibiting species-specific differences. This variable domain composition is further complicated by significant phylogenetic divergence between AARS catalytic domains derived from a bacterial or an archael/eukaryotic lineage [89].

Importantly, a number of natural products that act as AARS inhibitors exist which validate AARSs as antibiotic targets. AARS inhibitors exploit an array of different mechanisms to inhibit their respective target enzymes that reflect the multi-domain nature of, and complex reactions catalyzed by, AARSs. Most inhibitors bind to the highly conserved synthetic active sites and act as competitive inhibitors of the respective substrate amino acid or mimics of the obligate reaction intermediate, the aminoacyl-adenylate. However, a number of inhibitors bind to other regions of the AARS outside the active site, either allosterically affecting the synthetic active site or binding to alternative active sites, such as the editing domain. We outline these basic inhibitor mechanisms below.

5.1.1 Amino Acid Analogs

A number of naturally derived AARS inhibitors bind to the catalytic site of AARSs and act as competitors for a particular amino acid. The most well studied of these compounds is indolymycin, a tryptophan analog produced by *Streptomyces griseus* (Fig. 3a), that is a ~400-fold more selective inhibitor towards bacterial TrpRSs than that from eukaryotes [90, 91]. Although indolmycin is not effective against *Enterobacteriaceae*, steptococci, and enterocci pathogens [84, 91], it is effective against *S. aureus* MRSA strains [92], opening the door for its development as a clinical anti-infective. Chaungxinmycin, another tryptophan analog with close similarity to indolmycin, has also been reported to inhibit Gram-negative and Gram-positive bacteria, including *Escherichia coli* and *Shigella dysenteriae* in mouse infection models [93]. However, chemical derivatives of this compound showed minimal improvement compared to wild-type chaungxinmycin, suggesting a limited number of binding interactions are possible between the inhibitor and the TrpRS active site [93].

AARS inhibitor amino acid analogs are not limited to tryptophan, but also include the isoleucine analog cispentacin [94] and synthetic analog icofungipen [95] that inhibit isoleucyl-tRNA synthetase (IleRS), proline analog halofuginone that inhibits eukaryotic proline aminoacylation activity in the fused EPRS enzyme [96], and phenylalanine analog ochratoxin A that is active against bacterial, yeast and higher eukaryote PheRSs [97]. Some of these compounds have high affinity for their respective AARS, such as SB-219383, a synthetic carbocyclic analog of tyrosine based on a microbial metabolite that has been reported to inhibit potently *S. aureus* TyrRS with low nanomolar potency [98, 99].

Role of Aminoacyl-tRNA Synthetases in Infectious Diseases and Targets for... 307

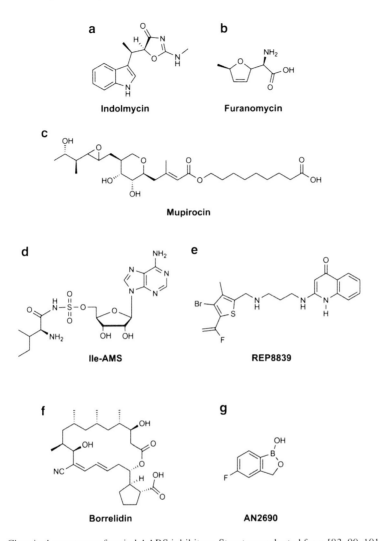

Fig. 3 Chemical structures of varied AARS inhibitors. Structures adapted from [83, 90, 101, 115, 118, 128]

In addition to the amino acid analogs that simply compete with an amino acid binding to its canonical AARS, a non-protein amino acid with antimicrobial activity can compete with an amino acid and also be misaminoacylated onto the tRNA substrate, even in the presence of a proofreading domain. The toxic effect of this type of compound is therefore both at the AARS level and also at the translated protein level. An example of this is furanomycin [100], which is an IleRS inhibitor from *Streptomyces* sp. [101] (Fig. 3b).

5.1.2 Aminoacyl-Adenylate Mimics

All AARSs catalyze the aminoacylation of a particular tRNA via an AA-AMP reaction intermediate tightly bound to the active site of the highly conserved catalytic domain. A substantial number of natural inhibitors mimic these reaction intermediates to form specific tight binding inhibitor–AARS complexes that compete with binding of both the ATP and amino acid substrates. Mupirocin or pseudomonic acid is the most prominent example of such inhibitors and is a potent inhibitor of IleRSs [82, 102, 103] (Fig. 3c). Sold under the name Bactroban, this compound is the only clinically approved antibiotic that targets AARSs. Mupirocin is used to treat topical infections caused by the bacterial pathogens *Haemophilus influenzae*, *Neisseria gonorrhoeae*, and *Neisseria meningitides*, and is the most effective topical anti-infective against methicillin resistant *S. aureus* [104]. Mupirocin inhibits IleRSs from bacteria, fungi, and archaea but not from higher eukaryotes [105], an essential trait for an effective AARSs-targeting antibiotic. Mupirocin binds to the IleRS active site by containing elements that mimic the Ile-AMP intermediate. Structural studies of the IleRs-mupirocin complex show that the pyran ring and C1–C3 of mupirocin binds to the enzyme's ATP binding site while the 14-methyl terminus of the inhibitor interacts with the isoleucine binding site [106]. A number of natural pseudomonic acid derivatives exist with different structures and affinities [105], while synthetic derivatives have been designed with femtomolar binding affinities [107]. Steady and pre-steady state kinetic studies on the inhibitory mechanism of mupirocin on *S. aureus* IleRS have shown that it behaves differently from stable Ile-AMP analogs, which inhibit the amino acid activation reaction by acting as simple competitor inhibitors of ATP and isoleucine (with a $K_i = \sim1$ nM for Ile-AMS – Fig 3d) [108]. Mupirocin binds to the IleRS in an initial encounter complex with a $K_i = 2$ nM before a slow isomerization of the EI complex occurs, leading to a tight binding EI* complex ($K_i = 50$ pM) [109]. Despite mupirocin's extreme potency and discrimination towards bacterial IleRS, stability and bioavailability issues associated with the compound have prevented its use as a systemic medication.

The Trojan Horse AARS toxins agrocin 84, albomycin, and microcin C are composed of two units: (1) a cleavable uptake moiety that allows the antibiotic to make use of bacterial transport systems and (2) nucleotide toxins that have been found to be stable mimics of an AA-AMP intermediate and inhibit a particular AARS. These fascinating inhibitors are described in more detail below.

The fact that all of the 20 AARS enzymes employ a common mechanism of tRNA aminoacylation via an enzyme-bound intermediate has obviously raised interest in using the basic AA-AMP structure as a platform to produce rationally designed antibiotics [76]. Direct AA-AMP analogs bind to the AARS synthetic active sites and act as simple competitors of both AA and ATP substrates as was found with Ile-AMS (Fig 3d) [109]. These analogs have stable linkages, such as sulfamoyls, between the nucleoside component and the amino acid group replacing the labile phosphoanhydride of the respective AA-AMP intermediate. These analogs have proven to be extremely potent inhibitors of AARSs in vitro, often binding to their respective enzymes with low nanomolar to high picomolar affinity. There are of course some drawbacks with these most basic mimics of AA-AMP: first, they can show a limited degree of phylogenetic discrimination between AARSs from bacterial and eukaryotic lineages and, second, their hydrophilic nature leads to a rather limited uptake by microbes [76]. The lack of phylogenetic discrimination is to be expected, as the ancient conserved active sites of these enzymes (both eukaryote and bacteria) catalyze the same reaction. One way of overcoming this feature has been to use the AA-AMP structure as a framework from which chemical libraries have been developed to create molecules that exploit subtle differences in the active sites of bacterial pathogen AARSs and those found in higher eukaryotes. One notable example is CB432, an Ile-AMP mimic in which the adenine ring has been replaced with a phenyltetrazole group. This molecule was a low nanomolar inhibitor showing a 60- to 1,100-fold discrimination between bacterial pathogen IleRSs compared to the human enzyme. This was in stark contrast with a simple Ile-AMP analog with a sulfamoyl linkage, CB138, which only showed a tenfold discrimination between bacterial and eukaryotic enzymes [76]. A large number of studies have been performed on the AARS inhibition properties of stable-AMP analogs derived from both rational design and chemical library screening approaches, as excellently reviewed elsewhere [84]. Unfortunately, these approaches have not yet been translated into any clinically relevant anti-infectives.

A second promising approach to overcome the discrimination issue for AA-AMP analogs is to focus on inhibitors that target AARSs not present in the cytoplasm of higher eukaryotes, but found in particular pathogens. One example of such an enzyme is the class I LysRSs found only in archae and, importantly, spirochetes [86]. The class I LysRSs still catalyze the aminoacylation of tRNA^{Lys} via a Lys-AMP intermediate as class II LysRSs do but there are clear differences between the active sites of the enzymes. The differences between the class I LysRS of the spirochete pathogens and class II LysRS of the eukaryotic host have been proposed as promising targets exploited for the development of anti-infectives against such diseases as syphilis (*Trepenoma pallidum*) or Lyme disease (*Borrelia burgdoferi*) [110, 111].

The potential of chemical libraries to contain AARS inhibitors has not been restricted to variants of molecules based on an AA-AMP framework. A number of compounds with AARS inhibitory properties have been isolated from libraries that do not resemble the standard AA-AMP analogs, yet have still been found to bind to the catalytic active site of AARS enzymes. AARS inhibitors with novel structural frameworks may not only be potent inhibitors but may also have improved uptake and bioavailability properties. A prominent example of such a molecule is REP8839 (Replidyne, CO) developed from high-throughput screening assays of chemical libraries targeting AARS enzymes from Gram-positive *S. aureus* strains. An initial MetRS inhibitor hit with a promising IC_{50} of 350 nM and little similarity to the Met-AMP reaction intermediate was significantly improved to produce REP8839 (Fig. 3e), a fluorovinylthiophene containing a diaryldiamine compound [112]. REP8839 was found to be a tight binding inhibitor of *S. aureus* MetRS with an IC_{50} value of <1.9 nM and also induced the formation of ppGpp, the signal for the cell's stringent response [113]. The inhibitor was found to prevent the first stage of the aminoacylation reaction – methionine activation – and was competitive with the methionine substrate, suggesting that the inhibitor binds to the MetRS active site [114]. Surprisingly, REP8839 is uncompetitive with respect to ATP, suggesting that, unlike AA-AMP analogs, the presence of ATP in the active site actively promotes the binding of the inhibitor. ATP was found to increase significantly the half-life of the REP8839-enzyme complex [114]. REP8839 shows very promising antibacterial properties against Gram-positive pathogens *Streptococcus pneumoniae* and *Streptococcus pyogenes* and is being evaluated as a topical treatment of methicillin resistant *S. aureus* skin infections [115, 116].

5.1.3 Allosteric AARS Inhibitors That Bind Outside the Active Site

Inhibition of AARS enzymes is not just limited to small molecules that bind to the catalytic active site. Borrelidin, an 18-membered macrolide-polyketide antimicrobial isolated from *Streptomyces* sp. [117] (Fig. 3f), has been shown to inhibit threonyl-tRNA synthetase (ThrRS) enzymes from *E. coli* and *Saccharomyces cerevisiae* as well as from Chinese hamster ovary cells. Work by Ruan et al. has revealed that borrelidin is a slow, tight binding inhibitor of the threonine activation reaction catalyzed by *E. coli* ThrRSs. Interestingly, it is non-competitive with respect to threonine and ATP [118]. This indicates that borrelidin must bind outside the active site. Screening of random *E. coli* ThrRS borrelidin-resistant mutants indicated that the antibiotic was bound to a unique hydrophobic patch near the active site close to a Zn^{2+} binding site on the protein [118]. Surprisingly, the compound has also been reported to inhibit a cyclin-dependent kinase in *S. cerevisiae* [119], causes apoptosis in select cancer cell lines [120], and has anti-angiogenic effects on human endothelial cells in vitro [121]. In addition, borrelidin has been shown to have anti-malarial activity [122]. These findings are being actively pursued with studies showing inhibition of *Plasmodium falciparum* when used in conjunction with artemisin [123] and mupirocin [124].

5.1.4 Novel tRNA-Dependent AARS Inhibitors

Another under explored approach to inhibit AARSs is to target the interaction of the enzyme with its canonical tRNA. Unlike amino acids and ATP, tRNAs are substantial macromolecules (~27 kDa) making numerous interactions with AARS enzymes, not only with the catalytic domain but also with the anticodon binding domain, editing domain, and a number of additional domains. These RNA–protein interactions can be critical for aminoacylation reactions but, importantly, they may well be domain- or even species-specific interactions. As a consequence, there is in all likelihood a number of RNA–protein domain interactions that could be targeted by novel antimicrobials [76]. Initial attempts to target AARSs in this manner were investigated with an aminoglycoside antibiotic tobramycin. This antibiotic normally acts by binding to the aminoacyl (A) site of the 16S ribosomal RNA subunit [125]. Walter et al. used in vivo aminoacylation assays to demonstrate that tobramycin inhibits the second stage of the aminoacylation reaction of yeast AspRS by binding to, and changing the conformation of, tRNA^{Asp} [126]. Future work might advance these studies to develop new antimicrobials that bind to AARSs or tRNA and prevent these key RNA–protein interactions.

The interaction of the tRNA substrate with its corresponding AARS is a dynamic one. There is substantial movement of the CCA 3′-end of a respective tRNA from its original binding position outside the catalytic active site to dipping into the active site during the second stage of the reaction, and allowing aminoacyl transfer of the amino acid to the tRNA to occur. In addition, for those enzymes with additional proofreading domains, the aminoacylated CCA 3′-end is required to translocate from the catalytic active site to the editing domain so the CCA 3′-end can be interrogated for misaminoacylated amino acids. A second exciting approach suggested by Schimmel et al. [76] towards inhibiting AARSs via their tRNA interaction is to design/discover small molecules that bind to the editing domains of AARSs with the intention of interfering with the binding of tRNA or trapping it in this domain and thus preventing its involvement in the aminoacylation reaction carried out in the catalytic domain. Recent studies have described two exciting new AARS inhibitors which target leucyl-tRNA synthetases and exploit dramatically different inhibition mechanisms, both of which employ tRNA and its translocation from the editing to the catalytic active site as an integral part of the inhibition mechanism [127, 128].

The first small compound is AN2690 [128] (5-fluoro-1,3-dihydro-1-hydroxy-2,1-benzoxaborole, Anacor Pharmaceuticals, CA), obtained from a chemical library screen of compounds for antifungal activity towards *S. cerevisiae*. Spontaneous AN2690-resistant mutants were found to have mutations in the *CDC60* gene, which will encode the cytoplasmic expressed LeuRS. The mutations found all mapped to the proofreading domain of this enzyme. In vitro enzyme kinetics and structural studies showed that AN2690 specifically inhibits yeast LeuRS activity by binding to the editing domain of the enzyme and trapping the 3′-end of the tRNA^{Leu} isoacceptor there, thus stopping subsequent rounds of catalysis [128]. AN2690 contains a boron atom in a benzoxaborole ring (Fig. 3g) which, after weak binding of the small molecule to the LeuRS editing active site, reacts with the 2′-OH and 3′-OH residues of a deacylated tRNA to form a stable adduct that is relatively long lived, if not irreversible, leading to a tight binding inhibitor of LeuRS [129]. This work validates AARS editing domains as bona fide targets and the contribution tRNA can make in AARS inhibition mechanisms. The small size of the compound and its surprising selectivity, are both extremely promising properties for an antibiotic. The AN2690 is now in clinical trials for the treatment of the fungal disease onychomycosis [130].

AN2690 derivatives with improved discrimination towards fungal pathogens, as well as the development of novel antibacterials based on this compound, are being actively sought. Structural studies of the inhibitor bound to the isolated editing domain of the cytoplasmic LeuRS from the fungal pathogen *Candida albicans* and

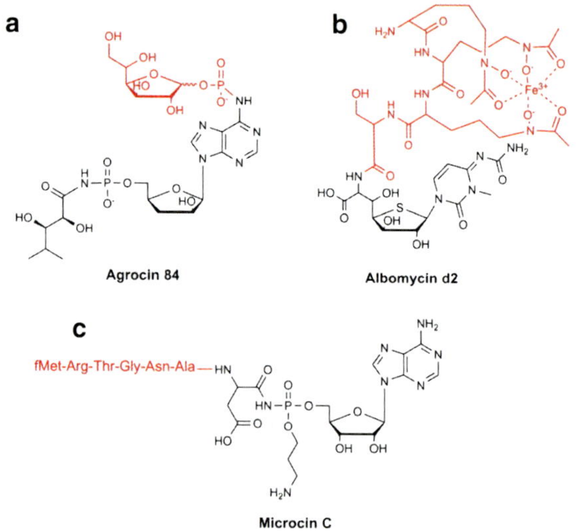

Fig. 4 Trojan Horse AARS inhibitors. Adapted structures of agrocin 84 [141], albomycin d2 [146], and microcin C [148] depicted with uptake moieties in *red*, and toxins in *black*

Agrocin 84

Albomycin d2

Microcin C

comparison with the corresponding domains from human and bacterial LeuRSs have revealed new insights into AN2690 specificity [131]. For these studies, tRNA^{Leu} could be dispensed with as an AN3018-AMP adduct (an AN2690-AMP derivative) could effectively mimic the covalently linked AN2690-tRNA^{Leu} 3'-end complex. The analysis revealed that bacterial editing domains lack a helix present in both fungal and human proteins, which in the eukaryotic domains forms a number of interactions with the adduct burying it in the active site. The lack of this helix may well explain the relatively weaker affinity of AN2690 towards some bacterial LeuRSs [132] and allow the development of more effective derivatives directed towards bacterial LeuRSs. Small differences between the *C. albicans* and *Homo sapiens* domains may also be exploited to improve discrimination of the compound [131].

The second recently identified LeuRS inhibitor agrocin 84 (TM84) (Fig. 4a) employs a quite different tRNA-dependent mechanism compared to AN2690. The actual inhibitor is TM84, which is the processed toxic moiety from the naturally occurring Trojan Horse inhibitor agrocin 84 (see below). TM84 shares a close structural similarity to Leu-AMP [127] but contains a stable 5'-N-acyl phosphoramidate bond rather than a labile phosphoanhydride. TM84 was found to be a potent inhibitor of its natural target, LeuRS from the plant pathogen *Agrobacterium tumefaciens* [127]. However, despite close similarity to a stable analog of Leu-AMP, TM84 has been found to be a very weak inhibitor of the first stage of the aminoacylation reaction, namely, the activation of leucine by ATP [133]. This was surprising since standard aminoacyl adenylate analogs have been found to compete actively with the amino acid and ATP substrates to inhibit amino acid activation [108]. Subsequent biochemical and biophysical analyses revealed that TM84 requires the active participation of the third substrate tRNA^{Leu} to form a tight binding LeuRS·tRNA^{Leu}·TM84 ternary inhibition complex. This is despite the fact that leucine activation is tRNA independent. Isothermal titration calorimetry shows that TM84 has a relatively weak affinity for the enzyme in the absence of tRNA ($K_{d \text{ TM84}} = 152 \pm 20$ nM) but this is enhanced by two orders of magnitude to ~ 1 nM in the presence of the nucleotide [133]. An X-ray crystal structure of this ternary complex provides an explanation for these results. TM84 does indeed bind to the synthetic active site but the tRNA^{Leu} adopts an "aminoacylation-like" conformation with the CCA 3'-end dipping into the catalytic domain. TM84 forms novel interactions with both the terminal adenosine 3'-OH and also with the second lysine of the class I defining KMSKS peptide. The placement of the tRNA acceptor stem and positioning of the interacting KMSKS loop over the active site effectively encapsulate the ligand [133]. The biological rationale for this unusual tRNA-dependent inhibition mechanism is currently being explored.

5.2 Trojan Horse tRNA Synthetase Inhibitors

Trojan Horse AARS inhibitors are members of a class of highly specific, hydrophilic antibiotics that gain access to a targeted bacterium through a specific transport system by mimicking a particular bacterial substrate. Having gained access to the interior of the cell by subterfuge, these antibiotics are then processed by existing enzymes inside the cell into a toxin which inhibits a particular AARS. Interestingly, despite the fact that members of this class of AARS inhibitors use a two-step "Trojan Horse" mechanism of action, the genes that biosynthesize these compounds are not related. The reason why nature has evolved these compounds to use these elaborate mechanisms to gain access inside the cell is an area of active research, but the fact is that these hydrophilic compounds can easily bypass the cell envelope barrier to enter bacterial cells. Natural Trojan Horse inhibitors are of particular interest because they have overcome the problem encountered by many synthetic AA-AMP analogs, namely high affinity for their enzymes in vitro but low uptake levels in vivo.

5.2.1 Agrocin 84

One example of a Trojan Horse AARS inhibitor is the highly specific anti-agrobacterial antibiotic agrocin 84 that is processed in pathogenic agrobacterial cells into a potent LeuRS inhibitor-TM84 [127]. Agrocin 84 is produced by the bacterial biocontrol agent *Agrobacterium radiobacter*, strain K84 [134, 135] that is used worldwide to prevent the economically important plant disease known as crown gall [136]. The role of agrocin 84 in the biocontrol of crown gall is to inhibit specifically the growth of the causative agent – pathogenic strains of *A. tumefaciens* [137]. *A. tumefaciens* induces the formation of plant tumors (galls) through the transfer of fragments of oncogenic DNA (T-DNA) from certain tumor-inducing (Ti) plasmids into infected plant cells [138]. The T-DNA encodes genes for host-cell proliferation and for enzymes that biosynthesize carbon compounds, called opines [139], which are exuded from infected cells and utilized as an energy source by the tumor-inducing agrobacteria that colonize the tumor surface. Agrocin 84 is selectively toxic to pathogenic *A. tumefaciens* strains because it is actively imported by the same Ti plasmid-encoded opine permease that transports a unique tumor-derived sugar phosphate opine-agrocinopine A [140]. The bacteriocin acts as a Trojan Horse: having gained access to the interior of the agrobacterium by subterfuge it is then processed into a toxic moiety that prevents growth of the pathogen by inhibiting the aminoacylation of tRNA^{Leu} [127].

Agrocin 84 contains a 9-(3'-deoxy-β-D-2,3-threopentafuranosyl) adenine nucleoside-like core linked to two substituents by unusual phosphoramidate bonds [141] (Fig. 4a). A 5'-phosphoramidate bond links the "nucleoside-like" core to a D-threo-2, 3-dihydroxy-4-methylpentanamide while a second phosphoramidate bond links a D-glucosyloxyphosphoryl group to the adenine base (Fig. 4a). This is the only known example of a 6-*N* phosphoramidate bond found in nature [142]. Although this moiety is required for selective uptake of agrocin 84 into susceptible cells, it is not required for toxicity [143]. After agrocin 84 is imported into the pathogenic *A. tumefaciens* cell, via the Ti plasmid encoded agrocinopine A permease, the phosphoramidate linkage of the N6-glucofuranosyl moiety of agrocin 84 is cleaved, leaving the toxic moiety of agrocin 84 (TM84). Interestingly, TM84 closely resembles in structure Leu-AMP, a critical enzyme-bound reaction intermediate for all LeuRSs, having a relatively stable 5'-phosphoramidate bond instead of the labile phosphoanhydride linkage. It was hypothesized that the stable TM84 imparts its antibiotic effect on the agrobacterium by binding to the catalytic domain of the *A. tumefaciens* genome encoded LeuRS$_{At}$ and acting as a simple Leu-AMP mimic. Recent work on TM84 (outlined above) now indicates that TM84 employs a unique tRNA-dependent inhibition mechanism [133].

What is particularly interesting about agrocin 84 for antibiotic research is that we know the exact bacterial pathogen that this AARS inhibitor targets in nature as well as the substrate the compound mimics and the biological rationale for the tripartite relationship between the plant host, the *A. tumefaciens* C58 pathogen and the *A. radiobacter* strain K84. In its natural environment *A. radiobacter* strain K84 lives alongside the pathogen on an established crown gall and benefits from the tumor's opine harvest, since it expresses an opine permease and metabolic enzymes to process agrocinopine A. A number of open questions still remain however about this relationship such as how production of agrocin 84 is regulated and why does it specifically target LeuRS rather than another AARS? Another interesting aspect of this system is that agrocin 84 specifically targets the pathogen, but does not inhibit the non-pathogenic commensal agrobacterium, which lacks the agrocinopine permease in its genome. Common broad range antibiotics do not discriminate between commensal and pathogenic bacterial strains, killing both. The inability of this antibiotic to target a commensal bacterium presumably limits the development of resistance in the agrobacterial population as a whole. This is a very desirable property for any antibiotic and is worthy of future investigation.

5.2.2 Albomycin

Another example of a Trojan Horse AARS inhibitor is albomycin, produced by *Streptomyces* sp. strain ATC700974 and effective against *E. coli*, *Salmonella typhimurium*, *S. pneumoniae*, *S. aureus*, and *Yersinia enterocolitica* [144, 145]. Albomycin contains a siderophore-like uptake moiety coupled to an aminoacyl-thioribosyl pyrimidine toxin (Fig. 4b) and is actively transported inside the targeted bacterial cell by a ferrichrome ABC-type transport system [146]. After transport, peptidase N cleaves the peptide bond of albomycin, releasing the aminoacyl-thioribosyl pyrimidine toxic moiety [144]. The albomycin active toxin was actually discovered in *Streptomyces* sp. fermentation media and named SB-217452, having presumably broken down after secretion. Importantly, SB-217452 was found to be a potent inhibitor of SerRS from *S. aureus* [147].

5.2.3 Microcin C

Microcin C is a post-translationally modified peptide produced by *Enterobacteriaceae* [148] and targets a range of Gram-negative bacteria including *E. coli*, *Klebsiella* sp., and *Salmonella* sp. [149]. Microcin C is composed of a heptapeptide with an adenosine nucleotide linked to the peptide's C-terminus by an *N*-acyl phosphoramidate bond (Fig. 4c). The phosphate of the nucleotide is further decorated by a 3-aminopropyl group. The six N-terminal amino acids in the peptide act as an uptake moiety that is recognized by a particular permease, the YejABEF ABC-type transporter [150], located on the inner membrane of sensitive cells. After transport into the cell, the peptide is deformylated and then processed by broad-specificity peptidases A, B, or N into a hexapeptide and a toxic moiety that was found to be a potent AspRS inhibitor [151]. The toxin resembles a stable Asp-AMP reaction intermediate and competes with Asp and ATP for binding to AspRS [151].

Six genes are responsible for biosynthesizing microcin C. The peptide part of the microcin C is produced from the smallest known gene (*mccA*), encoding a seven amino acid peptide [180]. After translation on the ribosome the terminal asparagine of the peptide is cyclized into *N*-succinimidyl ester by the MccB enzyme via an Asn-AMP intermediate that subsequently reacts with a second ATP molecule to form the *N*-acyl phosphoramidate bond linked adenosine unit and a peptide with a C-terminal aspartic acid [152]. This metabolite is then further decorated with the 3-aminopropyl group on the phosphate (MccD and MccE). Cell suicide is prevented by MccF, a serine protease homolog that cleaves the phosphoramidate bond of any microcin C and toxic moiety not secreted or MccE which acetylates the toxic moiety to make it inactive [153].

The ribosomal derived uptake moiety of microcin C is particularly amenable to mutational analysis and has allowed the molecular requirements of the transporter system and processing peptidases to be probed [154, 155]. Importantly, the C-terminal asparagine could not be altered as the R group of this amino acid is required for formation of the *N*-acyl phosphoramidate bond, ruling out the possibility of the biosynthesis of AA-AMP analogs directed against other AARSs. However, synthetic derivatives of microcin C with alternative terminal amino acids (Leu and Glu) linked by sulfamoyl groups to adenosine residues have been constructed. These modified microcins can be successfully transported and cleaved by sensitive *E. coli* cells and can inhibit their respective AARSs [156]. The exact nature of the peptide substrates that microcin C mimics is unclear at the present time, but as more is revealed about their biological function, the more promising redesign of this Trojan Horse inhibitor as an anti-infective becomes.

The Trojan Horse AARS inhibitors are genetically unrelated, yet nature has found a common theme by evolving potent hydrophilic antibiotics that can efficiently bypass the cell wall to target select bacterial strains. Understanding the biological working of these inhibitors may allow us to design more pathogen-selective antibiotics that might be more resistant to the ever-present threat of antibiotic resistance. Presumably more of these Trojan Horse AARS inhibitors remain to be discovered in the vast array of microbial secondary metabolites.

5.3 Antibiotic Resistant AARS Genes and the Threat of Horizontal Gene Transfer to Pathogenic Bacteria

5.3.1 Naturally Occurring Self-immunity AARS Genes and Antibiotic Biosynthesis

Many of the microbes biosynthesizing AARS inhibitors also require the presence of self-immunity AARS genes to prevent cell suicide. For example, the mupirocin gene cluster from *Pseudomonas fluorescens* encodes an IleRS gene *mupM* that, when transferred to *E. coli*, induces a resistance phenotype. The transcription of the mupirocin biosynthesis genes in *P. fluorescens* is regulated by a homoserine lactone signal quorum sensing system that ensures sufficient cells are present to produce enough toxin to kill competitor bacteria in the local soil environment. Importantly, the transcription of *mupM* is coordinated with the other biosynthesis genes, ensuring the MupM is present when the cells are making mupirocin, but not if the antibiotic is not produced [157]. A second IleRS in *P. fluorescens* NCIMB 10586 has also been found, although it only shares 29% sequence identity with MupM [103].

S. griseus NBRC 13350 and *S. griseus* ATCC 12648, producers of indolmycin, have been found to express high-level indolmycin resistant TrpRS enzymes, in addition to constitutively expressed, indolmycin-sensitive TrpRS enzymes (TrpRS-2) [158]. Interestingly, the resistance enzyme is also resistant to chunagximycin, another TrpRS analog produced from *Actinoplanes tsinanensis*. A number of other *Streptomyces* strains, such as *Steptomyces coelicolor,* that do not produce indolmycin, nonetheless encode dual resistance TrpRS genes (TrpRS-1) [90], consistent with the finding that actinomycetes, which produce two-thirds of all antibiotics share a large number of antibiotic resistant genes with each other. Interestingly, the TrpRS-1 gene is regulated via ribosome-mediated transcriptional attenuation [159, 160]. The authors of this study suggest that the common regulatory elements found here may be found in the regulation of other antibiotic resistant AARSs.

Streptomyces sp. strain ATC700974 encodes the biosynthetic gene cluster to make the Trojan Horse SerRS inhibitor albomycin [161]. The organism's genome contains one *serRS* housekeeping gene encoding a protein called SerRS1, while a second gene *alb10* in the biosynthetic gene cluster encodes a second SerRS protein, SerRS2. Both genes are capable of complementing a temperature-sensitive SerRS *E. coli* strain, but only SerRS2 is capable of imparting the cell with albomycin resistance [161]. While the SerRS1 protein is phylogenetically similar to other SerRS proteins belonging to closely related *Streptomyces* species, SerRS2 is more distantly related, probably due to horizontal gene transfer. Analysis of the enzymatic properties of both enzymes indicates that, while both have similar overall catalytic efficiencies, SerRS2 has a 20-fold weaker K_m for serine, whereas the binding of tRNA^{Ser} substantially increases the K_m for SerRS1 in the aminoacylation reaction. It has been suggested that the difference in serine affinities of sensitive and resistant enzymes may be important in regulating the levels of the albomycin produced by the *Streptomyces* sp. cell. Interestingly, both SerRS1 and SerRS2 were inhibited by serine hydroxamate, a serine analog, indicating that SerRS2 resistance was specific towards SB-217452 [161].

A. radiobacter strain K84, the organism that produces agrocin 84, also encodes a second, plasmid borne self-immunity gene, *agnB2*, that encodes an active LeuRS enzyme [127, 162, 163], in addition to the *A. radiobacter* genomically-encoded LeuRS. The *agnB2* gene is not essential for growth of *A. radiobacter* [164], and transfer of this gene to pathogenic strains of *A. tumefaciens* C58 imparts resistance to agrocin 84 [163]. In vitro aminoacylation assays of the AgnB2-LeuRS show the enzyme is approximately 1,000-fold less sensitive to the toxic moiety of agrocin 84 (TM84) than the antibiotic sensitive LeuRS$_{At}$ enzyme encoded in the pathogen *A. tumefaciens*. This work supports the hypothesis that the bioconrol organism carries a second, self-protective copy of a LeuRS to avoid cell suicide [127].

Many of the self-immunity AARS enzymes appear to have been obtained from phylogenetically more distant organisms and then further refined by evolution. This would indicate that, despite the fact that AARS synthetic active sites are highly conserved, these enzymes are sufficiently phylogenetically diverse to exhibit significant resistance to AARS inhibitors. This is of course a double-edged sword, because although it implies that AARS targeting antibiotics may not inhibit AARS enzymes from more distant organisms such as higher eukaryotic hosts, it may also imply that divergent bacteria in the local vicinity can contribute AARS resistant genes to the pan-genome of a targeted pathogen. Horizontal gene transfer of AARSs seems to be particularly prevalent in the phylogenetic history of AARS, which may be due to the selective pressure of naturally occurring AARS inhibitors [165–167]. This might be explained by the observation that AARS enzymes are more separate and modular members of the protein translation machinery and so may be more prone to horizontal gene transfer than, say, the segmental transfer of ribosomal proteins, which make more protein–protein and protein–RNA interactions with more conserved elements in the protein translation machinery [165–167].

The large number of genomes from soil-borne bacteria that contain multiple copies of AARSs has raised the possibility that these AARS isozymes may, to some degree, represent AARSs that are resistant to antibiotics present in the environment [161]. Finding organisms that live in the same environment with duplicated AARS genes might therefore be a promising strategy for discovering new AARS antibiotics [161].

5.3.2 Development of Bacterial Resistance to AARS Inhibitors in Pathogenic Strains

An inevitable consequence of the use of antibiotics for the treatment of infectious disease is the evolution of antibiotic resistance in pathogenic strains. Antibiotics based on AARS inhibition would not be expected to be an exception to this rule, although some inhibitors may be more prone than others. The majority of work in this area has been performed on mupirocin, the only clinically used AARS inhibitor. Cultures from patients infected with *S. aureus* MRSA appear to have evolved two types of resistance (mupirocin MIC for sensitive *S. aureus* MRSA $\leq$ 4 µg/mL). Low-level antibiotic resistance (MIC values from 8 to 64 µg/mL) and high-level resistance (MIC values from $\geq$512 µg/mL)) [168]. The low level resistance is characterized by point mutations in the target IleRS located around the ATP binding regions of the enzyme, in close proximity to one of the class defining catalytic motifs – the KMSKS loop [103]. High-level mupirocin resistant strains of *S. aureus* are characterized by the presence of a second resistant IleRS obtained by horizontal gene transfer from either eukaryotes or a bacterial gene mupA (not related to the gene of the same name in the mupirocin biosynthesis cluster) [157, 169].

The prevalence of antibiotic resistant AARS genes present in both bacteria that produce AARS targeting antibiotics, and those that simply co-habit the same environment as these organisms, underscores the extent to which horizontal gene transfer occurs in nature. Screening of small molecule libraries for inhibitors of methionyl-tRNA synthetase (MetRS) has revealed that subpopulations of the pathogen *S. pneumoniae* can even encode secondary MetRS enzymes (MetRS2) that are resistant to synthetically-derived inhibitors including REP8839 [112, 114, 170]. The *S. pneumoniae* species appear to have obtained the *metRS2* gene by horizontal gene transfer from the Gram-positive pathogen *Bacillus anthracis* or close relatives. *B. anthracis* also seems to have transferred the mupirocin-resistant gene encoding the IleRS2 protein to *S. aureus* strains [171].

A. radiobacter strain K84 is the most successful bacterial biocontrol agent discovered to date, having successfully protected an array of plant varieties against crown gall. Production of the LeuRS inhibitor agrocin 84 is thought to play a major role in plant protection against *A. tumefaciens* C58 infection. However, the biocontrol organism is not immune to the development of agrocin 84 resistance in the agrobacterium pathogen. Agrocin 84 resistant strains of *A. tumefaciens* C58 have already arisen in the field after pre-inoculating the biocontrol bacteria on the roots of young plants [172]. Closer analysis of these strains revealed that the pAgK84 plasmid from *A. radiobacter* strain K84, which encodes the genes required for agrocin 84 biosynthesis and the *agrocin 84* self-immunity LeuRS gene, *agnB2*, had been transferred to the pathogen [136, 173]. This plasmid transfer threatened the future viability of the *A. radiobacter* strain K84 as a successful biocontrol agent. Importantly, the pAgK84 plasmid was engineered to prevent conjugation of the biocontrol agent with the pathogen [174] and so stopped the transfer of the agrocin 84 self-immunity LeuRS gene, *agnB2*. This modified bacteria (*A. radiobacter* strain K1026) was the first bioengineered bacteria released into the environment, preventing the facile horizontal gene transfer of an antibiotic resistant AARS gene to a pathogen. With the help of this engineered strain, the biocontrol of crown gall has been successfully maintained for almost 40 years [175]. It is noteworthy that the *agnB2* self-immunity gene and the pAgK84 plasmid are only maintained in the *A. radiobacter* strain K84, and not the pathogen, when the bacteria live side by side in their natural setting. Only when the bacterial colonization of the plant roots is altered, when *A. radiobacter* strain K84 is used as a biocontrol agent, is there a selective advantage for the displaced pathogen to maintain a copy of the *agnB2* gene and/or pAgK84 plasmid obtained from the local environment. The use of antibiotics outside their natural environment can have unforeseen implications such as the evolution of multiple antibiotic resistant pathogens. The limited knowledge about the true biological targets, concentrations, and context in which antimicrobials act [176, 177] can acerbate these issues, especially when a large number of co-habiting bacteria in complex environments contain antibiotic resistant genes. This appears to be the case for AARSs [167]. Horizontal gene transfer is the main mechanism by which bacteria exchange/obtain modified genetic material when under selective pressure. When selective pressure is applied by an antibiotic, the targeted pathogen will readily take up a resistance gene encoded in the metagenome of any organisms in the nearby environment, assuming the fitness cost imposed by that gene is not too great. Identifying novel AARS inhibitors that do not have associated self-immunity genes or natural resistance genes found in the nearby environment would be a promising approach to forestall the evolution of AARS antibiotic-resistant pathogens.

6 Conclusions

AARSs are promising targets for the development of new anti-infectives. Although there are no known anti-viral agents targeting AARSs or their tRNA substrates used in the clinic, the discovery that specific host cell AARS/tRNA pairs are packaged into retroviruses such as HIV and RSV, and exploited for viral replication, offers an exciting opportunity for the future development of new agents targeting these factors. Gaining a better understanding of host factor/virus interactions and dissecting the pathway of virus recruitment will be key to the success of this approach, which is challenging due to the essential function of AARSs in the host cell. On the other hand, the essential nature of AARSs makes them attractive targets for development of new therapeutics against bacterial infection. In principle, the high level of phylogenetic diversity of AARS domain structure and species-specific tRNA recognition should facilitate effective inhibitor discrimination between pathogen and host. In practice, mupirocin, a very successful topical agent against *S. aureus* infections, is the only AARS inhibitor used in the clinic. Recent discoveries of tRNA-dependent inhibitors and studies of natural cleavable antibiotics known as Trojan Horse inhibitors are two of the promising new approaches being pursued to tackle this challenging problem.

Acknowledgements The authors acknowledge funding from National Institutes of Health Grants GM049928 and AI077387 (to K.M.-F.) and National Science Foundation Grant MCB-1158488 (to J.S.R.).

References

1. Delarue M (1995) Aminoacyl-tRNA synthetases. Curr Opin Struct Biol 5:48–55
2. Park SG, Schimmel P, Kim S (2008) Aminoacyl tRNA synthetases and their connections to disease. Proc Natl Acad Sci U S A 105:11043–11049
3. Dreher TW (2010) Viral tRNAs and tRNA-like structures. Wiley Interdiscip Rev RNA 1:402–414
4. Miller ES, Kutter E, Mosig G, Arisaka F, Kunisawa T, Ruger W (2003) Bacteriophage T4 genome. Microbiol Mol Biol Rev 67:86–156 (table of contents)
5. Mak J, Kleiman L (1997) Primer tRNAs for reverse transcription. J Virol 71:8087–8095
6. Cen S, Javanbakht H, Kim S, Shiba K, Craven R, Rein A, Ewalt K, Schimmel P, Musier-Forsyth K, Kleiman L (2002) Retrovirus-specific packaging of aminoacyl-tRNA synthetases with cognate primer tRNAs. J Virol 76:13111–13115
7. Dahlberg JE, Harada F, Sawyer RC (1975) Structure and properties of an RNA primer for initiation of Rous sarcoma virus DNA synthesis in vitro. Cold Spring Harb Symp Quant Biol 39(Pt 2):925–932
8. Peters GG, Hu J (1980) Reverse transcriptase as the major determinant for selective packaging of tRNA's into Avian sarcoma virus particles. J Virol 36:692–700
9. Kleiman L, Jones CP, Musier-Forsyth K (2010) Formation of the tRNALys packaging complex in HIV-1. FEBS Lett 584:359–365
10. Jiang M, Mak J, Ladha A, Cohen E, Klein M, Rovinski B, Kleiman L (1993) Identification of tRNAs incorporated into wild-type and mutant human immunodeficiency virus type 1. J Virol 67:3246–3253
11. Harada F, Peters GG, Dahlberg JE (1979) The primer tRNA for Moloney murine leukemia virus DNA synthesis. Nucleotide sequence and aminoacylation of tRNAPro. J Biol Chem 254:10979–10985
12. Waters LC, Mullin BC (1977) Transfer RNA into RNA tumor viruses. Prog Nucleic Acid Res Mol Biol 20:131–160

Role of Aminoacyl-tRNA Synthetases in Infectious Diseases and Targets for...

13. Mak J, Jiang M, Wainberg MA, Hammarskjold ML, Rekosh D, Kleiman L (1994) Role of Pr160gag-pol in mediating the selective incorporation of tRNA(Lys) into human immunodeficiency virus type 1 particles. J Virol 68:2065–2072

14. Cen S, Khorchid A, Javanbakht H, Gabor J, Stello T, Shiba K, Musier-Forsyth K, Kleiman L (2001) Incorporation of lysyl-tRNA synthetase into human immunodeficiency virus type 1. J Virol 75:5043–5048

15. Halwani R, Cen S, Javanbakht H, Saadatmand J, Kim S, Shiba K, Kleiman L (2004) Cellular distribution of Lysyl-tRNA synthetase and its interaction with Gag during human immunodeficiency virus type 1 assembly. J Virol 78:7553–7564

16. Raoult D, Audic S, Robert C, Abergel C, Renesto P, Ogata H, La Scola B, Suzan M, Claverie JM (2004) The 1.2-megabase genome sequence of Mimivirus. Science 306:1344–1350

17. Nishida K, Kawasaki T, Fujie M, Usami S, Yamada T (1999) Aminoacylation of tRNAs encoded by Chlorella virus CVK2. Virology 263:220–229

18. Abergel C, Rudinger-Thirion J, Giege R, Claverie JM (2007) Virus-encoded aminoacyl-tRNA synthetases: structural and functional characterization of mimivirus TyrRS and MetRS. J Virol 81:12406–12417

19. Dreher TW (2009) Role of tRNA-like structures in controlling plant virus replication. Virus Res 139:217–229

20. Haenni AL, Joshi S, Chapeville F (1982) tRNA-like structures in the genomes of RNA viruses. Prog Nucleic Acid Res Mol Biol 27:85–104

21. Mans RM, Guerrier-Takada C, Altman S, Pleij CW (1990) Interaction of RNase P from Escherichia coli with pseudoknotted structures in viral RNAs. Nucleic Acids Res 18:3479–3487

22. Dreher TW, Goodwin JB (1998) Transfer RNA mimicry among tymoviral genomic RNAs ranges from highly efficient to vestigial. Nucleic Acids Res 26:4356–4364

23. Dreher TW, Tsai CH, Florentz C, Giege R (1992) Specific valylation of turnip yellow mosaic virus RNA by wheat germ valyl-tRNA synthetase determined by three anticodon loop nucleotides. Biochemistry 31:9183–9189

24. Rietveld K, Linschooten K, Pleij CW, Bosch L (1984) The three-dimensional folding of the tRNA-like structure of tobacco mosaic virus RNA. A new building principle applied twice. EMBO J 3:2613–2619

25. Rudinger J, Felden B, Florentz C, Giege R (1997) Strategy for RNA recognition by yeast histidyl-tRNA synthetase. Bioorg Med Chem 5:1001–1009

26. Iyer LM, Aravind L, Koonin EV (2001) Common origin of four diverse families of large eukaryotic DNA viruses. J Virol 75:11720–11734

27. Arslan D, Legendre M, Seltzer V, Abergel C, Claverie JM (2011) Distant Mimivirus relative with a larger genome highlights the fundamental features of Megaviridae. Proc Natl Acad Sci U S A 108:17486–17491

28. Swanstrom R, Wills JW (1997) Synthesis, assembly, and processing of viral proteins. Cold Spring Harbor Lab Press, Cold Spring Harbor

29. Kleiman L, Halwani R, Javanbakht H (2004) The selective packaging and annealing of primer tRNALys3 in HIV-1. Curr HIV Res 2:163–175

30. Gabor J, Cen S, Javanbakht H, Niu M, Kleiman L (2002) Effect of altering the tRNA(Lys)(3) concentration in human immunodeficiency virus type 1 upon its annealing to viral RNA, GagPol incorporation, and viral infectivity. J Virol 76:9096–9102

31. Guo F, Cen S, Niu M, Javanbakht H, Kleiman L (2003) Specific inhibition of the synthesis of human lysyl-tRNA synthetase results in decreases in tRNA(Lys) incorporation, tRNA(3) (Lys) annealing to viral RNA, and viral infectivity in human immunodeficiency virus type 1. J Virol 77:9817–9822

32. Javanbakht H, Cen S, Musier-Forsyth K, Kleiman L (2002) Correlation between tRNALys3 aminoacylation and its incorporation into HIV-1. J Biol Chem 277:17389–17396

33. Cen S, Javanbakht H, Niu M, Kleiman L (2004) Ability of wild-type and mutant lysyl-tRNA synthetase to facilitate tRNA(Lys) incorporation into human immunodeficiency virus type 1. J Virol 78:1595–1601

34. Eriani G, Delarue M, Poch O, Gangloff J, Moras D (1990) Partition of tRNA synthetases into two classes based on mutually exclusive sets of sequence motifs. Nature 347:203–206
35. Eriani G, Dirheimer G, Gangloff J (1990) Aspartyl-tRNA synthetase from Escherichia coli: cloning and characterisation of the gene, homologies of its translated amino acid sequence with asparaginyl- and lysyl-tRNA synthetases. Nucleic Acids Res 18:7109–7118
36. Cusack S, Yaremchuk A, Tukalo M (1996) The crystal structures of T. thermophilus lysyl-tRNA synthetase complexed with E. coli tRNA(Lys) and a T. thermophilus tRNA(Lys) transcript: anticodon recognition and conformational changes upon binding of a lysyl-adenylate analogue. EMBO J 15:6321–6334
37. Guo M, Ignatov M, Musier-Forsyth K, Schimmel P, Yang XL (2008) Crystal structure of tetrameric form of human lysyl-tRNA synthetase: implications for multisynthetase complex formation. Proc Natl Acad Sci U S A 105:2331–2336
38. Onesti S, Miller AD, Brick P (1995) The crystal structure of the lysyl-tRNA synthetase (LysU) from Escherichia coli. Structure 3:163–176
39. Park SG, Ewalt KL, Kim S (2005) Functional expansion of aminoacyl-tRNA synthetases and their interacting factors: new perspectives on housekeepers. Trends Biochem Sci 30:569–574
40. Fang P, Zhang HM, Shapiro R, Marshall AG, Schimmel P, Yang XL, Guo M (2011) Structural context for mobilization of a human tRNA synthetase from its cytoplasmic complex. Proc Natl Acad Sci U S A 108:8239–8244
41. Kaminska M, Shalak V, Francin M, Mirande M (2007) Viral hijacking of mitochondrial lysyl-tRNA synthetase. J Virol 81:68–73
42. Kobbi L, Octobre G, Dias J, Comisso M, Mirande M (2011) Association of mitochondrial Lysyl-tRNA synthetase with HIV-1 GagPol involves catalytic domain of the synthetase and transframe and integrase domains of Pol. J Mol Biol 410:875–886
43. Javanbakht H, Halwani R, Cen S, Saadatmand J, Musier-Forsyth K, Gottlinger H, Kleiman L (2003) The interaction between HIV-1 Gag and human lysyl-tRNA synthetase during viral assembly. J Biol Chem 278:27644–27651
44. Kovaleski BJ, Kennedy R, Hong MK, Datta SA, Kleiman L, Rein A, Musier-Forsyth K (2006) In vitro characterization of the interaction between HIV-1 Gag and human lysyl-tRNA synthetase. J Biol Chem 281:19449–19456
45. Kovaleski BJ, Kennedy R, Khorchid A, Kleiman L, Matsuo H, Musier-Forsyth K (2007) Critical role of helix 4 of HIV-1 capsid C-terminal domain in interactions with human lysyl-tRNA synthetase. J Biol Chem 282:32274–32279
46. Guo M, Shapiro R, Morris GM, Yang XL, Schimmel P (2010) Packaging HIV virion components through dynamic equilibria of a human tRNA synthetase. J Phys Chem B 114:16273–16279
47. Na Nakorn P, Treesuwan W, Choowongkomon K, Hannongbua S, Boonyalai N (2011) In vitro and in silico binding study of the peptide derived from HIV-1 CA-CTD and LysRS as a potential HIV-1 blocking site. J Theor Biol 270:88–97
48. Dewan V, Wei M, Kleiman L, Musier-Forsyth K (2012) Dual role for motif 1 residues of human lysyl-tRNA synthetase in dimerization and packaging into HIV-1. J Biol Chem 287:41955–41962
49. Dewan V, Liu T, Chen KM, Qian Z, Xiao Y, Kleiman L, Mahasenan KV, Li C, Matsuo H, Pei D, Musier-Forsyth K (2012) Cyclic peptide inhibitors of HIV-1 capsid-human lysyl-tRNA synthetase interaction. ACS Chem Biol 7:761–769
50. Park SG, Kim HJ, Min YH, Choi EC, Shin YK, Park BJ, Lee SW, Kim S (2005) Human lysyl-tRNA synthetase is secreted to trigger proinflammatory response. Proc Natl Acad Sci U S A 102:6356–6361
51. Yannay-Cohen N, Carmi-Levy I, Kay G, Yang CM, Han JM, Kemeny DM, Kim S, Nechushtan H, Razin E (2009) LysRS serves as a key signaling molecule in the immune response by regulating gene expression. Mol Cell 34:603–611

52. Ofir-Birin Y, Fang P, Bennett SP, Zhang HM, Wang J, Rachmin I, Shapiro R, Song J, Dagan A, Pozo J, Kim S, Marshall AG, Schimmel P, Yang XL, Nechushtan H, Razin E, Guo M (2013) Structural switch of lysyl-tRNA synthetase between translation and transcription. Mol Cell 49:30–42

53. Kim DG, Choi JW, Lee JY, Kim H, Oh YS, Lee JW, Tak YK, Song JM, Razin E, Yun SH, Kim S (2012) Interaction of two translational components, lysyl-tRNA synthetase and p40/37LRP, in plasma membrane promotes laminin-dependent cell migration. FASEB J 26:4142–4159

54. Feng YX, Campbell S, Harvin D, Ehresmann B, Ehresmann C, Rein A (1999) The human immunodeficiency virus type 1 Gag polyprotein has nucleic acid chaperone activity: possible role in dimerization of genomic RNA and placement of tRNA on the primer binding site. J Virol 73:4251–4256

55. Jones CP, Datta SA, Rein A, Rouzina I, Musier-Forsyth K (2011) Matrix domain modulates HIV-1 Gag's nucleic acid chaperone activity via inositol phosphate binding. J Virol 85:1594–1603

56. Levin JG, Guo J, Rouzina I, Musier-Forsyth K (2005) Nucleic acid chaperone activity of HIV-1 nucleocapsid protein: critical role in reverse transcription and molecular mechanism. Prog Nucleic Acid Res Mol Biol 80:217–286

57. Levin JG, Mitra M, Mascarenhas A, Musier-Forsyth K (2010) Role of HIV-1 nucleocapsid protein in HIV-1 reverse transcription. RNA Biol 7:754–774

58. Watts JM, Dang KK, Gorelick RJ, Leonard CW, Bess JW Jr, Swanstrom R, Burch CL, Weeks KM (2009) Architecture and secondary structure of an entire HIV-1 RNA genome. Nature 460:711–716

59. Bolinger C, Boris-Lawrie K (2009) Mechanisms employed by retroviruses to exploit host factors for translational control of a complicated proteome. Retrovirology 6:8

60. Wilkinson KA, Gorelick RJ, Vasa SM, Guex N, Rein A, Mathews DH, Giddings MC, Weeks KM (2008) High-throughput SHAPE analysis reveals structures in HIV-1 genomic RNA strongly conserved across distinct biological states. PLoS Biol 6:e96

61. Brule F, Bec G, Keith G, le Grice SF, Roques BP, Ehresmann B, Ehresmann C, Marquet R (2000) In vitro evidence for the interaction of tRNA(3)(Lys) with U3 during the first strand transfer of HIV-1 reverse transcription. Nucleic Acids Res 28:634–640

62. Song M, Balakrishnan M, Gorelick RJ, Bambara RA (2009) A succession of mechanisms stimulate efficient reconstituted HIV-1 minus strand strong stop DNA transfer. Biochemistry 48:1810–1819

63. Piekna-Przybylska D, Dichiacchio L, Mathews DH, Bambara RA (2010) A sequence similar to tRNA 3 Lys gene is embedded in HIV-1 U3-R and promotes minus-strand transfer. Nat Struct Mol Biol 17:83–89

64. Jones CP, Saadatmand J, Kleiman L, Musier-Forsyth K (2013) Molecular mimicry of human tRNA^{Lys} anti-codon domain by HIV-1 RNA genome facilitates tRNA primer annealing. RNA 19(2):219–229

65. Temin HM, Mizutani S (1970) RNA-dependent DNA polymerase in virions of Rous sarcoma virus. Nature 226:1211–1213

66. Bange FC, Flohr T, Buwitt U, Bottger EC (1992) An interferon-induced protein with release factor activity is a tryptophanyl-tRNA synthetase. FEBS Lett 300:162–166

67. Fleckner J, Rasmussen HH, Justesen J (1991) Human interferon gamma potently induces the synthesis of a 55-kDa protein (gamma 2) highly homologous to rabbit peptide chain release factor and bovine tryptophanyl-tRNA synthetase. Proc Natl Acad Sci U S A 88:11520–11524

68. Rubin BY, Anderson SL, Xing L, Powell RJ, Tate WP (1991) Interferon induces tryptophanyl-tRNA synthetase expression in human fibroblasts. J Biol Chem 266:24245–24248

69. Sajish M, Zhou Q, Kishi S, Valdez DM Jr, Kapoor M, Guo M, Lee S, Kim S, Yang XL, Schimmel P (2012) Trp-tRNA synthetase bridges DNA-PKcs to PARP-1 to link IFN-gamma and p53 signaling. Nat Chem Biol 8:547–554

70. Wakasugi K, Slike BM, Hood J, Otani A, Ewalt KL, Friedlander M, Cheresh DA, Schimmel P (2002) A human aminoacyl-tRNA synthetase as a regulator of angiogenesis. Proc Natl Acad Sci U S A 99:173–177

71. Zhou Q, Kapoor M, Guo M, Belani R, Xu X, Kiosses WB, Hanan M, Park C, Armour E, Do MH, Nangle LA, Schimmel P, Yang XL (2010) Orthogonal use of a human tRNA synthetase active site to achieve multifunctionality. Nat Struct Mol Biol 17:57–61

72. Popenko VI, Cherni NE, Beresten SF, Zargarova TA, Favorova OO (1989) Immune electron microscope determination of the localization of tryptophanyl-tRNA-synthetase in bacteria and higher eukaryotes. Mol Biol (Mosk) 23:1669–1681

73. Popenko VI, Cherny NE, Beresten SF, Ivanova JL, Filonenko VV, Kisselev LL (1993) Immunoelectron microscopic location of tryptophanyl-tRNA synthetase in mammalian, prokaryotic and archaebacterial cells. Eur J Cell Biol 62:248–258

74. Garbitt-Hirst R, Kenney SP, Parent LJ (2009) Genetic evidence for a connection between Rous sarcoma virus gag nuclear trafficking and genomic RNA packaging. J Virol 83:6790–6797

75. Kenney SP, Lochmann TL, Schmid CL, Parent LJ (2008) Intermolecular interactions between retroviral Gag proteins in the nucleus. J Virol 82:683–691

76. Schimmel P, Tao J, Hill J (1998) Aminoacyl tRNA synthetases as targets for new anti-infectives. FASEB J 12:1599–1609

77. Alekshun MN, Levy SB (2007) Molecular mechanisms of antibacterial multidrug resistance. Cell 128:1037–1050

78. Levy SB (2002) The antibiotic paradox: how the misuse of antibiotic destroys their curative powers. Perseus, Cambridge

79. Walsh CT (2003) Antibiotics: actions, origins, resistance. Amer Society for Microbiology, Washington, DC, p 345

80. Ataide SF, Ibba M (2006) Small molecules: big players in the evolution of protein synthesis. ACS Chem Biol 1:285–297

81. Vondenhoff GH, van Aerschot A (2011) Aminoacyl-tRNA synthetase inhibitors as potential antibiotics. Eur J Med Chem 46:5227–5236

82. Fuller AT, Mellows G, Woolford M, Banks GT, Barrow KD, Chain EB (1971) Pseudomonic acid: an antibiotic produced by Pseudomonas fluorescens. Nature 234:416–417

83. Sutherland R, Boon RJ, Griffin KE, Masters PJ, Slocombe B, White AR (1985) Antibacterial activity of mupirocin (pseudomonic acid), a new antibiotic for topical use. Antimicrob Agents Chemother 27:495–498

84. Ochsner UA, Sun X, Jarvis T, Critchley I, Janjic N (2007) Aminoacyl-tRNA synthetases: essential and still promising targets for new anti-infective agents. Expert Opin Investig Drugs 16:573–593

85. Ibba M, Soll D (2000) Aminoacyl-tRNA synthesis. Annu Rev Biochem 69:617–650

86. Ibba M, Morgan S, Curnow AW, Pridmore DR, Vothknecht UC, Gardner W, Lin W, Woese CR, Soll D (1997) A euryarchaeal lysyl-tRNA synthetase: resemblance to class I synthetases. Science 278:1119–1122

87. Beuning PJ, Musier-Forsyth K (1999) Transfer RNA recognition by aminoacyl-tRNA synthetases. Biopolymers 52:1–28

88. Fersht AR (1981) Enzymic editing mechanisms and the genetic code. Proc R Soc Lond B Biol Sci 212:351–319

89. Woese CR, Olsen GJ, Ibba M, Soll D (2000) Aminoacyl-tRNA synthetases, the genetic code, and the evolutionary process. Microbiol Mol Biol Rev 64:202–236

90. Kitabatake M, Ali K, Demain A, Sakamoto K, Yokoyama S, Soll D (2002) Indolmycin resistance of Streptomyces coelicolor A3(2) by induced expression of one of its two tryptophanyl-tRNA synthetases. J Biol Chem 277:23882–23887

91. Werner RG, Thorpe LF, Reuter W, Nierhaus KH (1976) Indolmycin inhibits prokaryotic tryptophanyl-tRNA ligase. Eur J Biochem 68:1–3

92. Hurdle JG, O'Neill AJ, Chopra I (2004) Anti-staphylococcal activity of indolmycin, a potential topical agent for control of staphylococcal infections. J Antimicrob Chemother 54:549–552
93. Brown MJ, Carter PS, Fenwick AS, Fosberry AP, Hamprecht DW, Hibbs MJ, Jarvest RL, Mensah L, Milner PH, O'Hanlon PJ, Pope AJ, Richardson CM, West A, Witty DR (2002) The antimicrobial natural product chuangxinmycin and some synthetic analogues are potent and selective inhibitors of bacterial tryptophanyl tRNA synthetase. Bioorg Med Chem Lett 12:3171–3174
94. Konishi M, Nishio M, Saitoh K, Miyaki T, Oki T, Kawaguchi H (1989) Cispentacin, a new antifungal antibiotic. I. Production, isolation, physico-chemical properties and structure. J Antibiot (Tokyo) 42:1749–1755
95. Hasenoehrl A, Galic T, Ergovic G, Marsic N, Skerlev M, Mittendorf J, Geschke U, Schmidt A, Schoenfeld W (2006) In vitro activity and in vivo efficacy of icofungipen (PLD-118), a novel oral antifungal agent, against the pathogenic yeast Candida albicans. Antimicrob Agents Chemother 50:3011–3018
96. Keller TL, Zocco D, Sundrud MS, Hendrick M, Edenius M, Yum J, Kim YJ, Lee HK, Cortese JF, Wirth DF, Dignam JD, Rao A, Yeo CY, Mazitschek R, Whitman M (2012) Halofuginone and other febrifugine derivatives inhibit prolyl-tRNA synthetase. Nat Chem Biol 8:311–317
97. Konrad I, Roschenthaler R (1977) Inhibition of phenylalanine tRNA synthetase from Bacillus subtilis by ochratoxin A. FEBS Lett 83:341–347
98. Jarvest RL, Berge JM, Brown P, Hamprecht DW, McNair DJ, Mensah L, O'Hanlon PJ, Pope AJ (2001) Potent synthetic inhibitors of tyrosyl tRNA synthetase derived from C-pyranosyl analogues of SB-219383. Bioorg Med Chem Lett 11:715–718
99. Jarvest RL, Berge JM, Houge-Frydrych CS, Mensah LM, O'Hanlon PJ, Pope AJ (2001) Inhibitors of bacterial tyrosyl tRNA synthetase: synthesis of carbocyclic analogues of the natural product SB-219383. Bioorg Med Chem Lett 11:2499–2502
100. Tanaka K, Tamaki M, Watanabe S (1969) Effect of furanomycin on the synthesis of isoleucyl-tRNA. Biochim Biophys Acta 195:244–245
101. Kohno T, Kohda D, Haruki M, Yokoyama S, Miyazawa T (1990) Nonprotein amino acid furanomycin, unlike isoleucine in chemical structure, is charged to isoleucine tRNA by isoleucyl-tRNA synthetase and incorporated into protein. J Biol Chem 265:6931–6935
102. Hughes J, Mellows G (1980) Interaction of pseudomonic acid A with Escherichia coli B isoleucyl-tRNA synthetase. Biochem J 191:209–219
103. Yanagisawa T, Lee JT, Wu HC, Kawakami M (1994) Relationship of protein structure of isoleucyl-tRNA synthetase with pseudomonic acid resistance of Escherichia coli. A proposed mode of action of pseudomonic acid as an inhibitor of isoleucyl-tRNA synthetase. J Biol Chem 269:24304–24309
104. Baines, PJ, Jackson D, Mellows G, Swaisland AJ, Tasker TCG (1984) Mupirocin: its chemistry and metabolism. The Royal Society of Medicine International Congress and Symposium Series No. 80: mupirocin—a novel topical antibiotic, pp 13–22
105. Nakama T, Nureki O, Yokoyama S (2001) Structural basis for the recognition of isoleucyl-adenylate and an antibiotic, mupirocin, by isoleucyl-tRNA synthetase. J Biol Chem 276:47387–47393
106. Silvian LF, Wang J, Steitz TA (1999) Insights into editing from an ile-tRNA synthetase structure with tRNAile and mupirocin. Science 285:1074–1077
107. Brown MJ, Mensah LM, Doyle ML, Broom NJ, Osbourne N, Forrest AK, Richardson CM, O'Hanlon PJ, Pope AJ (2000) Rational design of femtomolar inhibitors of isoleucyl tRNA synthetase from a binding model for pseudomonic acid-A. Biochemistry 39:6003–6011
108. Pope AJ, Lapointe J, Mensah L, Benson N, Brown MJ, Moore KJ (1998) Characterization of isoleucyl-tRNA synthetase from Staphylococcus aureus. I: Kinetic mechanism of the substrate activation reaction studied by transient and steady-state techniques. J Biol Chem 273:31680–31690

109. Pope AJ, Moore KJ, McVey M, Mensah L, Benson N, Osbourne N, Broom N, Brown MJ, O'Hanlon P (1998) Characterization of isoleucyl-tRNA synthetase from Staphylococcus aureus. II. Mechanism of inhibition by reaction intermediate and pseudomonic acid analogues studied using transient and steady-state kinetics. J Biol Chem 273:31691–31701

110. Levengood J, Ataide SF, Roy H, Ibba M (2004) Divergence in noncognate amino acid recognition between class I and class II lysyl-tRNA synthetases. J Biol Chem 279:17707–17714

111. Raczniak G, Ibba M, Soll D (2001) Genomics-based identification of targets in pathogenic bacteria for potential therapeutic and diagnostic use. Toxicology 160:181–189

112. Jarvest RL, Berge JM, Berry V, Boyd HF, Brown MJ, Elder JS, Forrest AK, Fosberry AP, Gentry DR, Hibbs MJ, Jaworski DD, O'Hanlon PJ, Pope AJ, Rittenhouse S, Sheppard RJ, Slater-Radosti C, Worby A (2002) Nanomolar inhibitors of Staphylococcus aureus methionyl tRNA synthetase with potent antibacterial activity against Gram-positive pathogens. J Med Chem 45:1959–1962

113. Ochsner UA, Young CL, Stone KC, Dean FB, Janjic N, Critchley IA (2005) Mode of action and biochemical characterization of REP8839, a novel inhibitor of methionyl-tRNA synthetase. Antimicrob Agents Chemother 49:4253–4262

114. Green LS, Bullard JM, Ribble W, Dean F, Ayers DF, Ochsner UA, Janjic N, Jarvis TC (2009) Inhibition of methionyl-tRNA synthetase by REP8839 and effects of resistance mutations on enzyme activity. Antimicrob Agents Chemother 53:86–94

115. Critchley IA, Ochsner UA (2008) Recent advances in the preclinical evaluation of the topical antibacterial agent REP8839. Curr Opin Chem Biol 12:409–417

116. Critchley IA, Young CL, Stone KC, Ochsner UA, Guiles J, Tarasow T, Janjic N (2005) Antibacterial activity of REP8839, a new antibiotic for topical use. Antimicrob Agents Chemother 49:4247–4252

117. Hutter R, Poralla K, Zachau HG, Zahner H (1966) Metabolic products of microorganisms. 51. On the mechanism of action of borrelidin-inhibition of the threonine incorporation in sRNA. Biochem Z 344:190–196

118. Ruan B, Bovee ML, Sacher M, Stathopoulos C, Poralla K, Francklyn CS, Soll D (2005) A unique hydrophobic cluster near the active site contributes to differences in borrelidin inhibition among threonyl-tRNA synthetases. J Biol Chem 280:571–577

119. Tsuchiya E, Yukawa M, Miyakawa T, Kimura KI, Takahashi H (2001) Borrelidin inhibits a cyclin-dependent kinase (CDK), Cdc28/Cln2, of Saccharomyces cerevisiae. J Antibiot (Tokyo) 54:84–90

120. Habibi D, Ogloff N, Jalili RB, Yost A, Weng AP, Ghahary A, Ong CJ (2012) Borrelidin, a small molecule nitrile-containing macrolide inhibitor of threonyl-tRNA synthetase, is a potent inducer of apoptosis in acute lymphoblastic leukemia. Invest New Drugs 30:1361–1370

121. Kawamura T, Liu D, Towle MJ, Kageyama R, Tsukahara N, Wakabayashi T, Littlefield BA (2003) Anti-angiogenesis effects of borrelidin are mediated through distinct pathways: threonyl-tRNA synthetase and caspases are independently involved in suppression of proliferation and induction of apoptosis in endothelial cells. J Antibiot (Tokyo) 56:709–715

122. Otoguro K, Ui H, Ishiyama A, Kobayashi M, Togashi H, Takahashi Y, Masuma R, Tanaka H, Tomoda H, Yamada H, Omura S (2003) In vitro and in vivo antimalarial activities of a non-glycosidic 18-membered macrolide antibiotic, borrelidin, against drug-resistant strains of Plasmodia. J Antibiot (Tokyo) 56:727–729

123. Ishiyama A, Iwatsuki M, Namatame M, Nishihara-Tsukashima A, Sunazuka T, Takahashi Y, Omura S, Otoguro K (2011) Borrelidin, a potent antimalarial: stage-specific inhibition profile of synchronized cultures of Plasmodium falciparum. J Antibiot (Tokyo) 64:381–384

124. Jackson KE, Pham JS, Kwek M, de Silva NS, Allen SM, Goodman CD, McFadden GI, de Pouplana LR, Ralph SA (2012) Dual targeting of aminoacyl-tRNA synthetases to the apicoplast and cytosol in Plasmodium falciparum. Int J Parasitol 42:177–186

Role of Aminoacyl-tRNA Synthetases in Infectious Diseases and Targets for... 327

125. Vicens Q, Westhof E (2002) Crystal structure of a complex between the aminoglycoside tobramycin and an oligonucleotide containing the ribosomal decoding a site. Chem Biol 9:747–755
126. Walter F, Putz J, Giege R, Westhof E (2002) Binding of tobramycin leads to conformational changes in yeast tRNA(Asp) and inhibition of aminoacylation. EMBO J 21:760–768
127. Reader JS, Ordoukhanian PT, Kim JG, De Crecy-Lagard V, Hwang I, Farrand S, Schimmel P (2005) Major biocontrol of plant tumors targets tRNA synthetase. Science 309:1533
128. Rock FL, Mao W, Yaremchuk A, Tukalo M, Crepin T, Zhou H, Zhang YK, Hernandez V, Akama T, Baker SJ, Plattner JJ, Shapiro L, Martinis SA, Benkovic SJ, Cusack S, Alley MR (2007) An antifungal agent inhibits an aminoacyl-tRNA synthetase by trapping tRNA in the editing site. Science 316:1759–1761
129. Baker SJ, Tomsho JW, Benkovic SJ (2011) Boron-containing inhibitors of synthetases. Chem Soc Rev 40:4279–4285
130. Barak O, Loo DS (2007) AN-2690, a novel antifungal for the topical treatment of onychomycosis. Curr Opin Investig Drugs 8:662–668
131. Seiradake E, Mao W, Hernandez V, Baker SJ, Plattner JJ, Alley MR, Cusack S (2009) Crystal structures of the human and fungal cytosolic leucyl-tRNA synthetase editing domains: a structural basis for the rational design of antifungal benzoxaboroles. J Mol Biol 390:196–207
132. Tan M, Zhu B, Zhou XL, He R, Chen X, Eriani G, Wang ED (2010) tRNA-dependent pre-transfer editing by prokaryotic leucyl-tRNA synthetase. J Biol Chem 285:3235–3244
133. Chopra S, Palencia A, Virus C, Tripathy A, Temple BR, Velazquez-Campoy A, Cusack S, Reader JS (2013) Plant tumour biocontrol agent employs a tRNA-dependent mechanism to inhibit leucyl-tRNA synthetase. Nat Commun 4:1417
134. Ellis JG, Kerr A, van Montagu M, Schell J (1979) Agrobacterium: genetic studies on agrocin 84 production and the biological control of crown gall. Physiol Plant Pathol 15:311–319
135. Kerr A, Htay K (1974) Biological control of crown gall through bacteriocin production. Physiol Plant Pathol 4:37–44
136. Kerr A (1980) Biological control of crown gall through production of agrocin 84. Plant Dis 64:25–30
137. Zhu J, Oger PM, Schrammeijer B, Hooykaas PJ, Farrand SK, Winans SC (2000) The bases of crown gall tumorigenesis. J Bacteriol 182:3885–3895
138. Chilton MD, Drummond MH, Merio DJ, Sciaky D, Montoya AL, Gordon MP, Nester EW (1977) Stable incorporation of plasmid DNA into higher plant cells: the molecular basis of crown gall tumorigenesis. Cell 11:263–271
139. Bevan MW, Chilton MD (1982) T-DNA of the Agrobacterium Ti and Ri plasmids. Annu Rev Genet 16:357–384
140. Ryder MH, Tate ME, Jones GP (1984) Agrocinopine A, a tumor-inducing plasmid-coded enzyme product, is a phosphodiester of sucrose and L-arabinose. J Biol Chem 259:9704–9710
141. Tate ME, Murphy PJ, Roberts WP, Kerr A (1979) Adenine N6-substituent of agrocin 84 determines its bacteriocin-like specificity. Nature 280:697–699
142. Roberts WP, Tate ME, Kerr A (1977) Agrocin 84 is a 6-N-phosphoramidate of an adenine nucleotide analogue. Nature 265:379–381
143. Murphy PJ, Tate ME, Kerr A (1981) Substituents at N6 and C-5′ control selective uptake and toxicity of the adenine-nucleotide bacteriocin, agrocin 84, in Agrobacteria. Eur J Biochem 115:539–543
144. Braun V, Gunthner K, Hantke K, Zimmermann L (1983) Intracellular activation of albomycin in Escherichia coli and Salmonella typhimurium. J Bacteriol 156:308–315
145. Pramanik A, Stroeher UH, Krejci J, Standish AJ, Bohn E, Paton JC, Autenrieth IB, Braun V (2007) Albomycin is an effective antibiotic, as exemplified with Yersinia enterocolitica and Streptococcus pneumoniae. Int J Med Microbiol 297:459–469
146. Pramanik A, Braun V (2006) Albomycin uptake via a ferric hydroxamate transport system of Streptococcus pneumoniae R6. J Bacteriol 188:3878–3886

147. Stefanska AL, Fulston M, Houge-Frydrych CS, Jones JJ, Warr SR (2000) A potent seryl tRNA synthetase inhibitor SB-217452 isolated from a Streptomyces species. J Antibiot (Tokyo) 53:1346–1353
148. Guijarro JI, Gonzalez-Pastor JE, Baleux F, San Millan JL, Castilla MA, Rico M, Moreno F, Delepierre M (1995) Chemical structure and translation inhibition studies of the antibiotic microcin C7. J Biol Chem 270:23520–23532
149. Garcia-Bustos JF, Pezzi N, Mendez E (1985) Structure and mode of action of microcin 7, an antibacterial peptide produced by Escherichia coli. Antimicrob Agents Chemother 27:791–797
150. Novikova M, Metlitskaya A, Datsenko K, Kazakov T, Kazakov A, Wanner B, Severinov K (2007) The Escherichia coli Yej transporter is required for the uptake of translation inhibitor microcin C. J Bacteriol 189:8361–8365
151. Metlitskaya A, Kazakov T, Kommer A, Pavlova O, Praetorius-Ibba M, Ibba M, Krasheninnikov I, Kolb V, Khmel I, Severinov K (2006) Aspartyl-tRNA synthetase is the target of peptide nucleotide antibiotic Microcin C. J Biol Chem 281:18033–18042
152. Roush RF, Nolan EM, Lohr F, Walsh CT (2008) Maturation of an Escherichia coli ribosomal peptide antibiotic by ATP-consuming N-P bond formation in microcin C7. J Am Chem Soc 130:3603–3609
153. Novikova M, Kazakov T, Vondenhoff GH, Semenova E, Rozenski J, Metlytskaya A, Zukher I, Tikhonov A, van Aerschot A, Severinov K (2010) MccE provides resistance to protein synthesis inhibitor microcin C by acetylating the processed form of the antibiotic. J Biol Chem 285:12662–12669
154. Kazakov T, Metlitskaya A, Severinov K (2007) Amino acid residues required for maturation, cell uptake, and processing of translation inhibitor microcin C. J Bacteriol 189:2114–2118
155. Kazakov T, Vondenhoff GH, Datsenko KA, Novikova M, Metlitskaya A, Wanner BL, Severinov K (2008) Escherichia coli peptidase A, B, or N can process translation inhibitor microcin C. J Bacteriol 190:2607–2610
156. van de Vijver P, Vondenhoff GH, Kazakov TS, Semenova E, Kuznedelov K, Metlitskaya A, van Aerschot A, Severinov K (2009) Synthetic microcin C analogs targeting different aminoacyl-tRNA synthetases. J Bacteriol 191:6273–6280
157. Thomas CM, Hothersall J, Willis CL, Simpson TJ (2010) Resistance to and synthesis of the antibiotic mupirocin. Nat Rev Microbiol 8:281–289
158. Vecchione JJ, Sello JK (2009) A novel tryptophanyl-tRNA synthetase gene confers high-level resistance to indolmycin. Antimicrob Agents Chemother 53:3972–3980
159. Vecchione JJ, Sello JK (2008) Characterization of an inducible, antibiotic-resistant aminoacyl-tRNA synthetase gene in Streptomyces coelicolor. J Bacteriol 190:6253–6257
160. Vecchione JJ, Sello JK (2010) Regulation of an auxiliary, antibiotic-resistant tryptophanyl-tRNA synthetase gene via ribosome-mediated transcriptional attenuation. J Bacteriol 192:3565–3573
161. Zeng Y, Roy H, Patil PB, Ibba M, Chen S (2009) Characterization of two seryl-tRNA synthetases in albomycin-producing Streptomyces sp. strain ATCC 700974. Antimicrob Agents Chemother 53:4619–4627
162. Kim JG, Park BK, Kim SU, Choi D, Nahm BH, Moon JS, Reader JS, Farrand SK, Hwang I (2006) Bases of biocontrol: sequence predicts synthesis and mode of action of agrocin 84, the Trojan horse antibiotic that controls crown gall. Proc Natl Acad Sci U S A 103:8846–8851
163. Ryder MH, Slota JE, Scarim A, Farrand SK (1987) Genetic analysis of agrocin 84 production and immunity in Agrobacterium spp. J Bacteriol 169:4184–4189
164. Shim J-S, Farrand SK, Kerr A (1987) Biological control of crown gall: construction and testing of new biocontrol agents. Phytopathology 77:463–466
165. Andam CP, Fournier GP, Gogarten JP (2011) Multilevel populations and the evolution of antibiotic resistance through horizontal gene transfer. FEMS Microbiol Rev 35:756–767
166. Andam CP, Gogarten JP (2011) Biased gene transfer and its implications for the concept of lineage. Biol Direct 6:47

167. Andam CP, Gogarten JP (2011) Biased gene transfer in microbial evolution. Nat Rev Microbiol 9:543–555
168. Patel JB, Gowitz RJ, Jernigan JA (2009) Mupirocin resistance. Clin Infect Dis 49:935–941
169. Brown JR, Zhang J, Hodgson JE (1998) A bacterial antibiotic resistance gene with eukaryotic origins. Curr Biol 8:R365–R367
170. Gentry DR, Ingraham KA, Stanhope MJ, Rittenhouse S, Jarvest RL, O'Hanlon PJ, Brown JR, Holmes DJ (2003) Variable sensitivity to bacterial methionyl-tRNA synthetase inhibitors reveals subpopulations of Streptococcus pneumoniae with two distinct methionyl-tRNA synthetase genes. Antimicrob Agents Chemother 47:1784–1789
171. Brown JR, Gentry D, Becker JA, Ingraham K, Holmes DJ, Stanhope MJ (2003) Horizontal transfer of drug-resistant aminoacyl-transfer-RNA synthetases of anthrax and Gram-positive pathogens. EMBO Rep 4:692–698
172. Panagopoulos CG, Psallidas PG, Alivizatos AS (1979) Evidence of a breakdown in the effectiveness of biological control of crown gall. In: Shippers B, Gams W (eds) Soil-borne plant pathogens. Academic, London
173. Ellis JG, Kerr A (1979) Transfer of agrocin 84 production from strain 84 to pathogenic recipients: a comment on a previous paper. In: Shippers B, Gams W (eds) Soil-borne plant pathogens. Academic, London
174. Jones DA, Ryder MH, Clare BG, Farrand SK, Kerr A (1988) Construction of the Tra- deletion mutant of pAgK84 to safeguard the biological control of crown gall. Mol Gen Genet 212:207–214
175. Kerr A, Tate M (2004) Biological control of crown gall. In: Nester E, Gordon MP, Kerr A (eds) *Agrobacterium tumefaciens:* from plant pathology to biotechnology. APS Press, St. Paul
176. Davies J (2006) Are antibiotics naturally antibiotics? J Ind Microbiol Biotechnol 33:496–499
177. O'Brien J, Wright GD (2011) An ecological perspective of microbial secondary metabolism. Curr Opin Biotechnol 22:552–558
178. Matsuda D, Dreher TW (2004) The tRNA-like structure of turnip yellow mosaic virus RNA is a $3'$-translational enhancer. Virology 321:36–46
179. Weiner AM, Maizels N (1987) tRNA-like structures tag the $3'$ ends of genomic RNA molecules for replication: implications for the origin of protein synthesis. Proc Natl Acad Sci USA 84:7383–7387
180. González-Pastor JE, San Millán JL, Moreno F (1994) The smallest known gene. Nature 369 (6478):281

Top Curr Chem (2014) 344: 331–346
DOI: 10.1007/128_2013_421
© Springer-Verlag Berlin Heidelberg 2013
Published online: 12 March 2013

Flexizymes, Their Evolutionary History and Diverse Utilities

Toby Passioura and Hiroaki Suga

Abstract In contemporary organisms the aminoacylation of tRNAs is performed exclusively by protein aminoacyl-tRNA synthetases. However, in vitro selection experiments have identified RNA enzymes that exhibit the necessary characteristics to charge tRNA molecules with acyl groups in a way that is compatible with ribosomal translation, suggesting that such ribozymes may have fulfilled this function prior to the evolution of proteinaceous life. The current chapter provides a review of the history, structure, and function of these RNA aminoacyl synthetases, and discusses their practical application to "genetic reprogramming" and other biotechnologies.

Keywords Flexizyme · Genetic code reprogramming · Ribozyme

Contents

1 Introduction 332
2 Evolutionary History of Aminoacylation by Ribozymes 333
3 Flexizymes 335
   3.1 Flexizyme Structure 337
4 Utilities of Flexizymes 338
   4.1 Flexible In Vitro Translation System 338
   4.2 Random Non-standard Peptide Integrated Display 340

T. Passioura
Department of Chemistry, Graduate School of Science, The University of Tokyo,
7-3-1 Bunkyo-ku, Tokyo 113-0033, Japan

H. Suga (✉)
Department of Chemistry, Graduate School of Science, The University of Tokyo,
7-3-1 Bunkyo-ku, Tokyo 113-0033, Japan

Japan Science and Technology Agency (JST), Core Research for Evolutional Science and
Technology (CREST), Saitama 332-0012, Japan
e-mail: hsuga@chem.s.u-tokyo.ac.jp

4.3	Site-Specific Protein Modification	340
4.4	Investigation of Translation Mechanisms	340
5	Future Directions and Conclusions	342
References		343

Abbreviations

ABT	2-Aminoethyl)amidocarboxybenzyl thioester
AMP	Adenosine monophosphate
ARS	Aminoacyl-RNA synthetase
cab	2-Amino-4-(2-chloroacetamido)butanoic acid
CBT	Chlorobenzyl thioester
CME	Cyanomethyl ester
DBE	3,5-Dinitro-benzyl ester
FIT	Flexible in vitro translation
Gln	Glutamine
Phe	Phenylalanine
PheEE	Phenylalanine ethyl ester
RNase	Ribonuclease
tRNA	Transfer RNA
Tyr	Tyrosine

1 Introduction

Aminoacylation of the 3'-terminus of tRNA is a fundamental biochemical reaction common to all forms of life. In contemporary organisms the catalysis of these reactions appears to be performed exclusively by protein aminoacyl-tRNA synthetases (ARSs). However, a wide range of chemical reactions is potentially catalyzed by RNA enzymes (ribozymes), including protein synthesis itself, which is catalyzed by a ribozyme component of the ribosome. This notion implies that organisms in which RNA was both the genetic and catalytic material may have pre-dated proteinaceous life [1]. The evolution of programmed protein synthesis from this so-called "RNA world" would have necessitated the existence of ribozymes capable of catalyzing both the peptidyl-transferase activity evident in the ribosome and the tRNA aminoacylation reaction that is performed by protein ARSs in modern organisms. To date, no naturally occurring ribozymes which catalyze the second reaction have been identified. However, in vitro evolution and selection techniques have yielded a number of different ribozymes capable of catalyzing RNA aminoacylation [2–5]. The identification of such ribozymes is of significant biochemical interest and provides proof-of-principal evidence for the "RNA world" hypothesis. Moreover, improved variants of these enzymes are now exhibiting substantial utility in cell biology as well as biotechnological and chemical synthesis applications. The present chapter details the history, structure and utility of these ribozyme aminoacyl-tRNA synthetases.

2 Evolutionary History of Aminoacylation by Ribozymes

The earliest report of an RNA capable of catalyzing an amino-acylation reaction was from Illangasekare and coworkers, who performed in vitro selection experiments to isolate several ribozymes which self-aminoacylated their own CCG 3'-terminii at the 2'/3'-hydroxyl using phenylalanyl-adenosinemonophosphate (Phe-AMP) or tyrosinyl-AMP (Tyr-AMP) as a donor substrate [2]. Subsequently, a ribozyme that catalyzed self-aminoacylation at an internal 2'-OH using N-biotinylated-Phe-AMP as a donor substrate was also identified [3]. These ribozymes demonstrated the feasibility of RNA catalyzed aminoacylation of RNA; however, they did not yield products that could act as substrates for ribosomal translation. For this to occur, aminoacylation would be required to occur on the 3'-terminal (at either the 2' or 3' hydroxyl) of a tRNA molecule.

A ribozyme capable of acylating tRNA in a manner compatible with translation was ultimately identified as a consequence of the discovery of a ribozyme that catalyzed the reverse reaction, that is, acyl transfer from the 3'-OH of a tRNA. Through the use of in vitro selection, Lhose and Szostak had identified the acyl-transferase ribozyme (ATRib), which catalyzed the transfer of N-biotinylated-Phe from the 3'-OH of a 3' tRNA fragment (5'-CAACCA-3') to its own 5'-OH [6]. Since this reaction is energetically neutral, it was possible to exploit its reversibility to evolve a ribozyme, referred to as AD02, which catalyzed a two-step reaction which involved self-aminoacylation of its own 5'-OH using an activated N-biotinylated Gln, followed by transfer of the acyl group to the 3'-OH of a tRNA substrate [4]. AD02 exhibited a high degree of specificity for Gln as a substrate, but aminoacylated tRNAs in a relatively non-selective fashion. Another in vitro evolution based on ATRib led to the development of the BC28 ribozyme, which is able to transfer various amino acids charged on its own 5'-hydroxyl group to specific tRNAs through interactions with the tRNA 3'-terminus and anti-codon loop [7]. This demonstrated proof-of-principle for the specific aminoacylation of a specific tRNA by a ribozyme; however, the complex, two-step mechanism of AD02 and BC28 (and consequent modest activity) precluded further development of these enzymes.

In order to derive a ribozyme with robust activity which (1) catalyzes the aminoacylation of specific tRNA using a specific amino acid substrate and (2) selectively aminoacylates the tRNA 3' terminus, a modular RNA library was employed based on tRNA processing in modern prokaryotes. Prokaryotic precursor tRNAs are transcribed with a variable 5'-leader sequence which is cleaved from the mature tRNA through the activity of RNase P [8]. To date, no biological function has been ascribed to these leader sequences, and it seems plausible that they are vestigial ribozymes from the "RNA world" which previously functioned to self-aminoacylate each tRNA at the

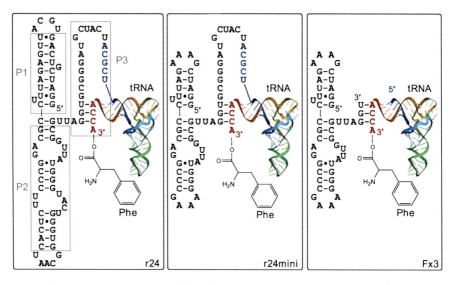

Fig. 1 Structural comparison of r24, r24mini, and Fx3. The r24 catalytic core (*left panel*) was isolated by selection for self-aminoacylation of a contiguous 3' tRNA sequence using CME activated Phe as a substrate. Structural analyses demonstrated that elements of the P1 and P2 domains of r24 were not required for activity, leading to isolation of the truncated r24mini (*middle panel*). Selection experiments based on a semi-randomized r24mini showed that the majority of the P3 domain, with the exception of the GGU guide sequence, was also not required for activity, giving rise to the first flexizyme, Fx3 (*right panel*)

3'-terminus prior to processing by RNase P. On the basis of this hypothesis, an RNA library was constructed in which a random sequence was incorporated into the 5'-leader region of a tRNA, and selection was performed based on self-aminoacylation using *N*-biotinylated-Phe activated with cyanomethyl ester (CME – chosen for its small size and absence of hydrogen bonding potential in order to avoid selection for recognition of the leaving group rather than the amino acid moiety) as an aminoacyl donor [5]. This selection yielded a single clone termed pre-24 which included the 90 nucleotide catalytic domain r24 (Fig. 1). Importantly, the r24 domain could be cleaved from the tRNA by the activity of RNase P, yielding an aminoacylated mature tRNA and thus demonstrating the feasibility of a precursor tRNA with self-aminoacylating capacity as an ARS.

Structural and biochemical characterization of the r24 domain showed that it could be truncated to 57-nucleotides (r24mini) with no loss of activity by the removal of unnecessary sequences in stems P1 and P2 [9] (Fig. 1). r24 and r24mini both contain an internal GGU guide sequence in loop 3 which hybridizes to the ACC trinucleotide motif at positions 73–75 of the tRNA (i.e., the three nucleotides immediately preceding the terminal adenosine which is the point of attachment for aminoacylation). Moreover, in addition to its activity in cis, the r24 domain catalyzed the aminoacylation of tRNA in trans when each component RNA (i.e., the r24mini RNA and tRNA) was transcribed separately [5]. Unsurprisingly (given the initial in vitro selection conditions), catalysis by r24 was found to be highly Mg^{2+} dependent [10]. The site of tRNA aminoacylation was confirmed to be specific to the 3'-OH and not the 2'-OH at the 3' terminus of the tRNA and characterization of specificity for the activated amino acid substrate showed that r24 recognized the aromatic ring of the Phe moiety [9, 11]. Consequently, r24 showed selectivity for L-Phe as a substrate although aminoacylation using D-Phe and L-Tyr was also observed with lower activity. Moreover, its activity was not affected by substitution of the leaving group, i.e., L-Phe activated by a thioester or AMP instead of CME also acted as a substrate [5].

3 Flexizymes

The r24 family of ribozymes described above had shown that RNA enzymes located in the 5'-leader sequences of precursor tRNAs could selectively aminoacylate the tRNA 3'-terminus with a specific activated amino acid prior to processing into a mature tRNA, thus demonstrating the capacity of ribozymes to act as true ARSs in pre-proteinaceous life. However, the high efficiency and in trans activity of the r24 ribozymes raised the possibility of utilizing these, or similar, RNAs to misacylate tRNAs as a tool for genetic code reprogramming. For such an application, ARS ribozymes would ideally have low specificity for both the tRNA and the activated amino acid substrate to allow for greater versatility of reprogramming.

In view of this, r24mini was used as the starting point for selection of ARS ribozymes that retained high activity but exhibited broader substrate specificity. Initially, the selection scheme was geared towards expanding the tRNA substrate specificity. Randomizing the majority of bases in P3 and L3 and selecting for aminoacylation (using activated N-biotinylated-Phe as a substrate) demonstrated that no nucleotides in the P3 or L3 regions except for the GGU guide motif were required for activity [12]. The resulting 45 nucleotide ribozyme was termed flexizyme3 (Fx3) and had the ability to aminoacylate tRNAs with A, G, or U nucleotides at position 73 using activated aromatic amino acid CMEs (Fig. 1).

While Fx3 was relatively promiscuous with respect to recognition of its tRNA substrate, it exhibited strong selectivity for amino acids bearing aromatic side chains. In order to identify an Fx3 derivative which accepted a wider range of amino acid side chains, selection experiments were performed using a partially randomized Fx3 RNA library and an amino acid substrate in which the aromatic moiety was embedded in the leaving group (in this case the leaving group was 3,5-dinitro-benzyl ester, DBE) rather than the side chain [13]. The product of this selection was the so-called dinitro-flexizyme (dFx), a 46 nucleotide ribozyme that is capable of acylating a range of tRNA molecules using an extremely wide range of acyl donor substrates, provided that they are activated with DBE (Fig. 2). To date, amino acids with non-standard side chains, D-amino acids, β-amino acids, N-alkyl-amino acids, N-acyl-amino acids, and α-hydroxy acids have all been shown to be efficient substrates for dFx [14–18].

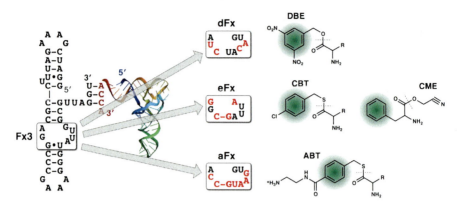

Fig. 2 Structural comparison of Fx3, dFx, eFx, and aFx. The three second-generation flexizymes, dFx, eFx, and aFx, were each derived from Fx3 through selection experiments that altered the aromatic moiety binding pocket of the ribozyme (boxed sequence in the Fx3 structure). The binding pockets for dFx (which recognizes DBE in the leaving group), eFx (which recognizes CBT in the leaving group or an aromatic moiety in a CME activated amino acid), and aFx (which recognizes ABT in the leaving group) are all shown, along with the structures of their respective activated amino acid substrates. Nucleotides that differ from the parental Fx3 sequence are *highlighted*. The scissile bond between the leaving group and the acyl donor in each chemical structure is indicated by a *dashed line*, and the aromatic moiety in each structure is *highlighted*

In parallel with the development of dFx, selection experiments designed to increase the activity of the original Fx3 flexizyme led to the isolation of the enhanced flexizyme (eFx), which exhibits greater tRNA aminoacylation activity than Fx3 using amino acid substrates bearing aromatic side chains and activated with CME [13] (Fig. 2). eFx is also capable of accepting substrates with non-aromatic side chains activated with chlorobenzyl thioester (CBT) through recognition of the aromatic moiety in this leaving group, and can therefore catalyze tRNA aminoacylation using a wide range of CBT activated amino acid-like substrates.

Between them, dFx and eFx are capable of catalyzing the acylation of tRNAs with a very wide range of amino acids. However, because all of the substrates must necessarily contain an aromatic moiety, there are some instances where the desired activated substrate is poorly soluble in the aqueous buffer system required for flexizyme-catalyzed acylation. To circumvent this limitation a third flexizyme (aFx) was evolved to recognize (2-aminoethyl)amidocarboxybenzyl thioester (ABT) as a leaving group (the amino group in ABT is protonated at neutral pH and therefore increases the aqueous solubility of the activated substrate) [19] (Fig. 2). With the improved utility provided by aFx, and by careful consideration of which combination of flexizyme and leaving group to employ, the three flexizymes discussed above (aFx, dFx, and eFx) are capable of catalyzing the acylation of tRNAs with a near infinite range of amino acids or similar activated substrates (specific examples are discussed below). In fact, the range of potential acyl substrates for the existing set of flexizymes appears to be substantially larger than the range of acyl substrates that are accepted by the ribosome. Since the

Flexizymes, Their Evolutionary History and Diverse Utilities

majority of applications of flexizymes involve subsequent ribosomal translation, the current set of flexizymes (aFx, dFx, and eFx) is sufficient for genetic reprogramming with an extremely diverse set of substrates.

3.1 Flexizyme Structure

In collaboration with the Ferre-D'Amare group, we were able to determine the crystal structure of the original Fx3 flexizyme [20]. To achieve this, Fx3 was conjugated to a micro-helix RNA (analogous to the flexizyme binding site of tRNA) and a functionally dispensable section of the P2 loop region was replaced with a U1A protein-binding motif as a means of facilitating crystallization by co-crystallizing with U1A. Additionally, in order to co-crystallize the flexizyme with an activated amino acid substrate, the hydrolytically stable L-Phe-CME analogue L-Phe-ethyl ester (PheEE – which acts as an inhibitor of flexizyme) was used. Through this methodology, a crystal structure for the active form of Fx3 bound to both its amino acid and tRNA substrates was obtained.

The flexizyme core is composed of one irregular and three A-form helices. A hairpin bend near the 3'-terminus of the flexizyme allows the terminal 3 nucleotides to protrude away from the helical stack and makes them available for base pairing to the 3'-CCA-terminus of the tRNA substrate (forming helix P3 in the process). This positions the acylation site of the tRNA adjacent to the minor groove of the irregular helix. Protein ARSs can be classified on the basis of structure into class I enzymes (approach the tRNA acceptor stem from the minor groove and aminoacylate the 2'-OH) and class II enzymes (approach the tRNA acceptor stem from the major groove and aminoacylate the 3'-OH) [21]. The crystal structure of Fx3 showed that it was structurally analogous to class II enzymes, consistent with its preference for aminoacylation at the 3'-OH.

The active site of Fx3 is located in the irregular helix, which is composed of three non-canonical (one A:G and two G:U) base pairs, unpaired A and U nucleotides and a hydrated Mg^{2+} ion. This structure contains the amino acid binding pocket of Fx3, which interacts with the aromatic moiety of the activated Phe but not the amino moiety or leaving group. Although the crystal structures of the later generation flexizymes (dFx, eFx, and aFx) have not been resolved, it seems likely that the amino acid binding pocket in each of these enzymes retains the interaction with the aromatic moiety present in the relevant substrate. As such, it is probable that in eFx, substrates with aromatic side chains activated with CME are bound by the flexizyme in the same orientation as was observed for Fx3 complexed with Phe-EE. By contrast, dFx, and aFx (as well as eFx when used with substrates activated by CBT) probably bind the amino acid substrate in an inverted position relative to that observed for Fx3 complexed with Phe-EE, projecting the side chain of the substrate away from the flexizyme core. This is consistent with the diversity of bulky side chains that can be accommodated in these reactions.

4 Utilities of Flexizymes

4.1 Flexible In Vitro Translation System

As described above, flexizyme technology allows for the acylation of tRNAs with a diverse range of exotic amino acids and their derivative acyl donors. By introducing these tRNAs into a fully reconstituted *Escherichia coli* in vitro translation system it is possible to synthesize a wide range of structurally diverse peptides and proteins [13–18, 22–33]. The combination of these two technologies (flexizyme and in vitro translation) is referred to as the flexible in vitro translation (FIT) system (Fig. 3). In this instance, the reconstituted translation system is comprised of a complex cocktail of recombinant initiation factors, elongation factors, release factors, ribosome recycling factor, ARSs for all 20 canonical amino acids, methionyl-tRNA formyltransferase, creatine kinase, myokinase, inorganic pyrophosphatase, nucleotide diphosphate kinase, T7 RNA polymerase, total tRNA, purified ribosomes, the 4 canonical ribonucleotides, and the 20 canonical amino acids [34]. Introduction of an appropriate DNA template containing a T7 promoter site (or an mRNA) into this reaction mix thus leads to the translation of the introduced open reading frame. The modularity and defined composition of this system (as compared to translation in vivo or in cell lysate systems) means that, by simply omitting an amino acid from the reaction, "vacant" codons are generated in the open reading frame. If an appropriate tRNA that has been charged with an exotic amino acid by a flexizyme is included in the reaction, it is thus possible to substitute the omitted canonical amino acid with an exotic one. In this way, peptides or proteins containing multiple non-canonical amino acids can be synthesized.

The FIT system has some distinctive advantages compared to other "genetic reprogramming" technologies. Such technologies can generally be classified into those which (1) employ non-enzymatic organic synthesis to create a tRNA charged with a non-canonical acyl group and (2) employ enzymatic synthesis for the same reaction. Whilst the first approach is technically feasible, such synthesis is laborious and complex, and consequently has not been broadly applied. By contrast, (non-flexizyme) enzymatic synthesis of acylated tRNAs has been used by a number of different research groups to charge tRNAs with non-canonical amino acids both in vitro and in vivo. Whilst a detailed discussion of this field is beyond the scope of this review, these studies generally rely on the capacity of naturally occurring protein ARSs to recognize close analogues of their endogenous amino acid substrates [35–37]. This allows for the synthesis of tRNAs conjugated to homologous, though not identical, amino acids to the twenty canonical species. In some cases, molecular engineering of the ARS has been used to broaden further the possible range of amino acid substrates [38]. This has proven to be a useful tool for a variety of applications. However, the range of potential non-canonical amino acids that can be used in these systems remains limited by the requirement for homology to

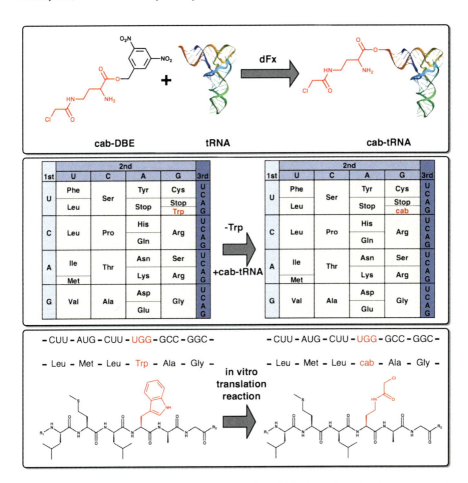

Fig. 3 The FIT system. Aminoacylation of the desired tRNA is performed using an appropriate combination of flexizyme and activated amino acid, in this example dFx and 2-amino-4-(2-chloroacetamido)butanoic acid-DBE (cab-DBE, *top panel*). Omitting a canonical amino acid from an in vitro translation reaction opens up a "vacant" codon, which can be replaced with the pre-acylated non-canonical amino acid-tRNA (in this example tryptophan is omitted, *middle panel*). Translation of a template sequence in such a reaction results in the replacement of the omitted amino acid with the non-canonical one in the resulting polypeptide (*bottom panel*)

canonical amino acids. Moreover, because of the selectivity of protein ARSs for specific tRNAs, reassignment of codons (e.g., replacing the initiating Met with a Phe) is problematic. The promiscuous nature of flexizymes with respect to both amino acid and tRNA substrates largely obviates these limitations, and, to date, more than 300 non-canonical amino acids have been incorporated into peptide chains using this technology, making this an exceptionally powerful tool for synthetic peptide and protein chemistry.

4.2 Random Non-standard Peptide Integrated Display

One of the most powerful applications for the FIT system is random non-standard peptide integrated display (RaPID). This technique has been reviewed in detail recently [39, 40] and so will not be covered extensively here. However, in brief, this technique combines the FIT system with mRNA display (the conjugation of in vitro transcribed peptides and proteins to their cognate mRNAs through a puromycin linkage) to produce and screen extremely diverse (greater than 10^{12}) libraries of non-canonical and cyclic peptides for binding to targets of interest. In this way, potent peptide inhibitors of a range of target proteins have been identified [24, 26, 33].

4.3 Site-Specific Protein Modification

Techniques such as site-specific mutagenesis and alanine scanning mutagenesis (in which molecular engineering is used to substitute specific amino acid residues in a protein of interest) have proven to be invaluable tools in the analysis of protein structure and function. However, such techniques have been fundamentally limited, to date, by the available choice of amino acids, with only the 20 naturally occurring amino acid side chains available for chemical probing. Consequently, in many cases relatively gross alterations of the protein of interest have been necessitated for functional and structural studies (e.g., replacing a charged side chain with a methyl group in the case of alanine scanning mutagenesis). Since the FIT system allows for the site-specific integration of virtually any conceivable amino acid side chain, far more nuanced probing of the roles of specific amino acid residues is possible. For example, Öjemalm and coworkers performed scanning mutagenesis on a transmembrane alpha-helix using a range of linear aliphatic (ethyl through to octyl), cyclic aliphatic (cyclopropyl, cyclopentyl, and cyclohexyl), and aromatic (variously substituted benzene, naphthalene, phenol, and heterocyclic groups) side chains [41] (Fig. 4). By measuring the apparent free energy of membrane insertion for each of the resulting proteins, they were able to relate subtle changes in the physico-chemical characteristics of the helix to its membrane partitioning behavior in a way that would have been impossible using only the canonical amino acids, demonstrating the power of this approach to provide "high-resolution" information about structure–function relationships.

4.4 Investigation of Translation Mechanisms

In the preceding section the utility of the FIT system for the subtle chemical probing of protein function was discussed. Similarly, the capacity of flexizymes to acylate tRNA with a diverse range of compounds has proven useful for determining the

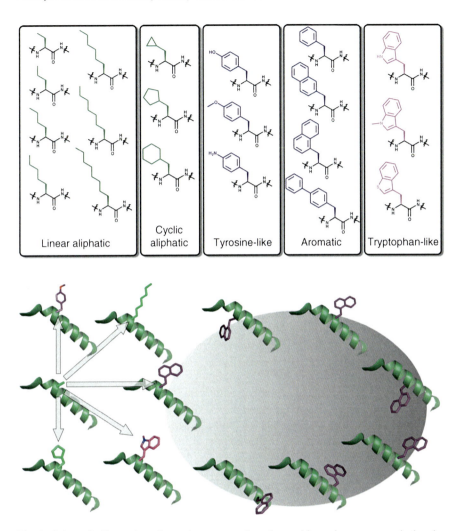

Fig. 4 Schematic illustration of protein structure–function probing using non-canonical amino acids. Through the use of the FIT system, it is possible to vastly increase the diversity of amino acid side chains in scanning mutagenesis. Examples of linear aliphatic, cyclic, aromatic, and heterocyclic side chains that can be used in such studies are shown in the *upper panels* (tyrosine, phenylalanine, and tryptophan are included for comparison). By introducing such non-canonical side chains at a specific site in a polypeptide (the *lower left* of the figure demonstrates the substitution of an alanine in an alpha-helix by five of the non-canonical amino acids from the *upper panels*) it is possible to determine the functional effects of subtle changes in protein structure. By combining this approach with scanning mutagenesis, the effect of introducing non-canonical side chains at different positions in the polypeptide can also be determined, as exemplified by seven different naphthalene-moiety containing residues in the *lower right*

precise chemical and structural requirements for ribosome mediated acyl transfer. This has proven to be especially true for the study of translation initiation. Because flexizymes can be used to acylate any tRNA species, a diverse set of acyl donors can be charged onto the translation initiating tRNA$^{\text{fMet}}_{\text{CAU}}$ [14, 23, 42]. In this way, the structural limitations of the acyl group initiating translation have been extensively investigated. Surprisingly, translation initiation appears to accept a wider range of acyl substrates than elongation, with D-amino acids, N-methyl-amino acids, N-acyl-amino acids, β-amino acids, and even short peptides (up to pentapeptides) serving as efficient initiators of translation.

5 Future Directions and Conclusions

In addition to their academic value as proof-of-principle, aminoacylating ribozymes that provide support for the "RNA world" hypothesis, flexizyme ARSs have proven to be exceptionally useful tools for a variety of studies. However, the broad range of potential applications for these ribozymes (a consequence of their substrate promiscuity) leads us to believe that their usefulness is yet to be exploited fully.

The use of flexizymes to introduce non-canonical amino acids site-specifically into proteins for structure–function studies (as described above) is still in its infancy, and is likely to be particularly useful for the study of protein–protein and protein–drug interactions. By comparison, the use of flexizymes to investigate the structural requirements of acyl donors during translation (also described above) has been more thoroughly explored, although we foresee further studies in this area. A similar, although to the best of our knowledge as yet unexplored, approach is use of flexizymes for the investigation of aminoacyl-tRNA proof-reading (whereby some protein ARSs hydrolyze the acyl linkage of mischarged tRNAs). In this context, we expect that the ability of flexizymes to acylate diverse tRNAs with diverse amino acids will allow for the elucidation of the specific structural features that define aminoacyl-tRNA proof-reading. This in turn may facilitate the related field of protein ARS rational engineering.

The use of the FIT system for organic synthesis has already proved its practical value through application to the RaPID system (discussed briefly above and reviewed elsewhere). Looking further ahead, we envisage that improvements in existing applications of flexizymes will lead to revolutionary synthetic biochemical technologies. Further engineering of the FIT system through a combination of rational design and directed evolution is likely to produce ribosomes capable of catalyzing polymerization reactions other than acyl transfer. Similar approaches will allow for the "splitting" of codon boxes during in vitro translation, so that non-canonical amino acids can be translated without the need to omit canonical ones. This diversity of amino acids in a single translation reaction could potentially be further expanded through the use of ribosomes that recognize 4-base codons, as have already been described [43]. Combined with improved post-translational modification techniques, these advances will lead to the development of systems capable of the genetically programmed organic synthesis of a diverse range of organic molecules not restricted to amide polymers.

To summarize, the studies of aminoacylating ribozymes over the last two decades have provided convincing evidence that ribozymes capable of the specific aminoacylation of tRNAs are possible, thus supporting the hypothesis that proteinaceous organisms evolved from earlier RNA-based life. Moreover, a detailed understanding of the structures and mechanisms of these ribozymes has led to the development of molecular technologies of significant utility. The maturation of these technologies is likely to facilitate substantially the study of protein and RNA biochemistry, and the development of bioorganic synthetic chemistry and synthetic biology.

Acknowledgements This work was supported by the JSPS Grant-in-Aid for the Specially Promoted Research (21000005) and the NRF (R31-2008-000-10103-0) to H.S. and partly supported by JST-CREST, Molecular Technologies.

References

1. Cech TR (2009) Evolution of biological catalysis: ribozyme to RNP enzyme. Cold Spring Harb Symp Quant Biol 74:11–16
2. Illangasekare M, Sanchez G, Nickles T, Yarus M (1995) Aminoacyl-RNA synthesis catalyzed by an RNA. Science 267:643–647
3. Jenne A, Famulok M (1998) A novel ribozyme with ester transferase activity. Chem Biol 5:23–34
4. Lee N, Bessho Y, Wei K, Szostak JW, Suga H (2000) Ribozyme-catalyzed tRNA aminoacylation. Nat Struct Biol 7:28–33
5. Saito H, Kourouklis D, Suga H (2001) An in vitro evolved precursor tRNA with aminoacylation activity. EMBO J 20:1797–1806
6. Lohse PA, Szostak JW (1996) Ribozyme-catalysed amino-acid transfer reactions. Nature 381:442–444
7. Bessho Y, Hodgson DR, Suga H (2002) A tRNA aminoacylation system for non-natural amino acids based on a programmable ribozyme. Nat Biotechnol 20:723–728
8. Guerrier-Takada C, Gardiner K, Marsh T, Pace N, Altman S (1983) The RNA moiety of ribonuclease P is the catalytic subunit of the enzyme. Cell 35:849–857
9. Saito H, Watanabe K, Suga H (2001) Concurrent molecular recognition of the amino acid and tRNA by a ribozyme. RNA 7:1867–1878
10. Saito H, Suga H (2002) Outersphere and innersphere coordinated metal ions in an aminoacyl-tRNA synthetase ribozyme. Nucleic Acids Res 30:5151–5159
11. Saito H, Suga H (2001) A ribozyme exclusively aminoacylates the 3′-hydroxyl group of the tRNA terminal adenosine. J Am Chem Soc 123:7178–7179
12. Murakami H, Saito H, Suga H (2003) A versatile tRNA aminoacylation catalyst based on RNA. Chem Biol 10:655–662
13. Murakami H, Ohta A, Ashigai H, Suga H (2006) A highly flexible tRNA acylation method for non-natural polypeptide synthesis. Nat Methods 3:357–359
14. Goto Y, Murakami H, Suga H (2008) Initiating translation with D-amino acids. RNA 14:1390–1398

15. Goto Y, Ohta A, Sako Y, Yamagishi Y, Murakami H, Suga H (2008) Reprogramming the translation initiation for the synthesis of physiologically stable cyclic peptides. ACS Chem Biol 3:120–129
16. Kawakami T, Murakami H, Suga H (2008) Messenger RNA-programmed incorporation of multiple N-methyl-amino acids into linear and cyclic peptides. Chem Biol 15:32–42
17. Kawakami T, Murakami H, Suga H (2008) Ribosomal synthesis of polypeptoids and peptoid-peptide hybrids. J Am Chem Soc 130:16861–16863
18. Ohta A, Murakami H, Higashimura E, Suga H (2007) Synthesis of polyester by means of genetic code reprogramming. Chem Biol 14:1315–1322
19. Niwa N, Yamagishi Y, Murakami H, Suga H (2009) A flexizyme that selectively charges amino acids activated by a water-friendly leaving group. Bioorg Med Chem Lett 19:3892–3894
20. Xiao H, Murakami H, Suga H, Ferre-D'Amare AR (2008) Structural basis of specific tRNA aminoacylation by a small in vitro selected ribozyme. Nature 454:358–361
21. Eriani G, Delarue M, Poch O, Gangloff J, Moras D (1990) Partition of tRNA synthetases into two classes based on mutually exclusive sets of sequence motifs. Nature 347:203–206
22. Goto Y, Iwasaki K, Torikai K, Murakami H, Suga H (2009) Ribosomal synthesis of dehydrobutyrine- and methyllanthionine-containing peptides. Chem Commun (Camb) 3419–3421
23. Goto Y, Suga H (2009) Translation initiation with initiator tRNA charged with exotic peptides. J Am Chem Soc 131:5040–5041
24. Hayashi Y, Morimoto J, Suga H (2012) In vitro selection of anti-Akt2 thioether-macrocyclic peptides leading to isoform-selective inhibitors. ACS Chem Biol 7:607–613
25. Kawakami T, Ohta A, Ohuchi M, Ashigai H, Murakami H, Suga H (2009) Diverse backbone-cyclized peptides via codon reprogramming. Nat Chem Biol 5:888–890
26. Morimoto J, Hayashi Y, Suga H (2012) Discovery of macrocyclic peptides armed with a mechanism-based warhead: isoform-selective inhibition of human deacetylase SIRT2. Angew Chem Int Ed Engl 51:3423–3427
27. Nakajima E, Goto Y, Sako Y, Murakami H, Suga H (2009) Ribosomal synthesis of peptides with C-terminal lactams, thiolactones, and alkylamides. Chembiochem 10:1186–1192
28. Ohshiro Y, Nakajima E, Goto Y, Fuse S, Takahashi T, Doi T, Suga H (2011) Ribosomal synthesis of backbone-macrocyclic peptides containing gamma-amino acids. Chembiochem 12:1183–1187
29. Ohta A, Murakami H, Suga H (2008) Polymerization of alpha-hydroxy acids by ribosomes. Chembiochem 9:2773–2778
30. Sako Y, Goto Y, Murakami H, Suga H (2008) Ribosomal synthesis of peptidase-resistant peptides closed by a nonreducible inter-side-chain bond. ACS Chem Biol 3:241–249
31. Sako Y, Morimoto J, Murakami H, Suga H (2008) Ribosomal synthesis of bicyclic peptides via two orthogonal inter-side-chain reactions. J Am Chem Soc 130:7232–7234
32. Yamagishi Y, Ashigai H, Goto Y, Murakami H, Suga H (2009) Ribosomal synthesis of cyclic peptides with a fluorogenic oxidative coupling reaction. Chembiochem 10:1469–1472
33. Yamagishi Y, Shoji I, Miyagawa S, Kawakami T, Katoh T, Goto Y, Suga H (2011) Natural product-like macrocyclic N-methyl-peptide inhibitors against a ubiquitin ligase uncovered from a ribosome-expressed de novo library. Chem Biol 18:1562–1570
34. Goto Y, Katoh T, Suga H (2011) Flexizymes for genetic code reprogramming. Nat Protoc 6:779–790
35. Forster AC, Tan Z, Nalam MN, Lin H, Qu H, Cornish VW, Blacklow SC (2003) Programming peptidomimetic syntheses by translating genetic codes designed de novo. Proc Natl Acad Sci U S A 100:6353–6357
36. Hartman MC, Josephson K, Lin CW, Szostak JW (2007) An expanded set of amino acid analogs for the ribosomal translation of unnatural peptides. PLoS One 2:e972
37. Johnson JA, Lu YY, van Deventer JA, Tirrell DA (2010) Residue-specific incorporation of non-canonical amino acids into proteins: recent developments and applications. Curr Opin Chem Biol 14:774–780

38. Datta D, Wang P, Carrico IS, Mayo SL, Tirrell DA (2002) A designed phenylalanyl-tRNA synthetase variant allows efficient in vivo incorporation of aryl ketone functionality into proteins. J Am Chem Soc 124:5652–5653
39. Hipolito CJ, Suga H (2012) Ribosomal production and in vitro selection of natural product-like peptidomimetics: the FIT and RaPID systems. Curr Opin Chem Biol 16:196–203
40. Reid PC, Goto Y, Katoh T, Suga H (2012) Charging of tRNAs using ribozymes and selection of cyclic peptides containing thioethers. Methods Mol Biol 805:335–348
41. Ojemalm K, Higuchi T, Jiang Y, Langel U, Nilsson I, White SH, Suga H, von Heijne G (2011) Apolar surface area determines the efficiency of translocon-mediated membrane-protein integration into the endoplasmic reticulum. Proc Natl Acad Sci U S A 108:E359–E364
42. Goto Y, Ashigai H, Sako Y, Murakami H, Suga H (2006) Translation initiation by using various N-acylaminoacyl tRNAs. Nucleic Acids Symp Ser (Oxf) 293–294
43. Neumann H, Wang K, Davis L, Garcia-Alai M, Chin JW (2010) Encoding multiple unnatural amino acids via evolution of a quadruplet-decoding ribosome. Nature 464:441–444

Index

A

Acyltransferase ribozyme (ATRib), 333
Adenocarcinoma, 213, 222
Adenylyl cyclase (AC), 191
Adhesion protein-mediated mechanism, 155
Agrobacterium radiobacter, 313
Agrocin 84, 313
AIMPs. *See* ARS-interacting multifunctional protein (AIMPs)
Alanyl-tRNA synthetase (AlaRS), 3, 5, 27, 167
Alarmones, 189
AlaRS. *See* Alanyl-tRNA synthetase (AlaRS)
AlaX proteins, 29
Albomycin, 315
Amino acids, 43
    analogs, 306
    recognition, 62
    specificity, 62
Aminoacyl adenylate, 1
    mimics, 308
Aminoacylation, 1, 6, 45, 93, 124, 168, 189, 212, 223, 253, 279, 295, 339
Aminoacyl-tRNA synthesis, 4
Aminoacyl-tRNA synthetase (AARS/ARS), 43, 89, 145, 167, 189, 247, 293
    class I, 1, 9, 49
    class II, 1, 14, 50
    classification, 7, 46
    metamorphosis, 106
Amyotrophic lateral sclerosis (ALS), 134
AN2690, 311
Anaplastic lymphoma kinase (ALK), 216
Angiogenesis, 211, 221
Antibiotics, 293
Ap$_3$A/Ap$_4$A, 189
Arginine, 11, 15, 260
Arginyl-tRNA synthetase (ArgRS), 8

ARS-interacting multifunctional protein (AIMPs) 119, 153, 207, 213, 230
Artemisin, 310
Asparaginase, 53, 214
Asparagine, 70, 214, 315
Asparagine synthetase A (AsnA), 70
Asparaginyl-tRNA synthetase (AsnRS/NRS), 52, 152, 208, 300
Ataxia telangiectasia mutated (ATM/ATR), 130
ATP synthesis, mitochondrial, 249

B

Bacterial resistance, 318
Blue native polyacrylamide (or agarose) gel electrophoresis (BN-PAGE), 257
Borrelidin, 233, 236, 307, 310
Breast cancer, 212

C

Cadherin, 97, 107, 138, 140, 156, 221, 226
Cancer, 207
Cancer associated genes (CAGs), 209
Carcinogenesis, 207
Caspase-7, 221
CD23, 126
Cell–cell communication, 145
Cell surface receptor, 67-kDa laminin receptor (67LR), 132
Ceruloplasmin (Cp), 139
Chromosomal rearrangement, 216
Coenzyme Q, 249
Colorectal cancer, 211, 214, 227
Connective peptide I (CPI), 4, 50
Crown gall, 313

347

Index

Cu, Zn-superoxide dismutase, 134
CXC receptor (CXCR), 158
Cyclic adenosine monophosphate (cAMP), 191
Cyclic guanosine monophosphate (cGMP), 191
Cyclic nucleotide phosphodiesterase (PDE), 191
Cysteinyl-tRNAsynthetase (CysRS/CRS), 54, 216
Cytochrome c oxidase (COX), 256
Cytokines, 93, 105, 107, 126, 138, 145, 221, 227, 230

D

Dendritic cells (DCs), 154
Diacylglycerol (DAG), 191
Diadenosine polyphosphates (Ap$_n$A), 190, 191
Diadenosine tetraphosphate (Ap$_4$A), 109, 133, 171, 189, 199, 302
  hydrolase, 196
Dinitro-flexizyme (dFx), 335
Dinucleotide polyphosphates, 189
DNA polymerase K associated protein, 196
DNA-dependent protein kinase (DNA-PKcs), 137, 304
Domain expansion, 90

E

Editing enzymes, 19, 64
Elongation factor-1 (EF-1), 135
Endothelial-monocyte activating polypeptide 2, 67, 93, 149, 153, 169, 221.
  See Endothelial-monocyte activating polypeptide 2
Enzyme kinetics, 1
Enzyme specificity, 43
EPRS. *See* Glutamyl-prolyl-tRNA synthetase (EPRS)
Eukaryotic initiation factor 2γ (eIF2γ), 131
Evolution, 43, 89, 258, 331
Expression levels, 146, 209, 214, 223, 274

F

Far upstream element binding protein 1 (FBP1), 127
Febrifugine, 233
Flexizymes, 331, 335
Flexizyme3 (Fx3), 334
Formylmethionyl-tRNA, 53
Fragile histidine triad (Fhit) protein, 189, 196, 201
Furanomycin, 307

G

Gag, 134
Gamma-IFN-activated inhibitor of translation (GAIT), 96, 109, 123, 139, 174, 217, 227
Gastric cancer, 214
Gene expression, 68, 70, 123, 167, 170, 199, 210, 277
Genetic code reprogramming, 331
Glioblastomas, 213
Gliomas, 213
GluProRS. *See* Glutamyl-proryl-tRNA synthetase
Glutaminyl-tRNA synthetase (GlnRS/QRS), 3, 8, 46, 52, 136
Glutamyl-prolyl-tRNA synthetase GluProRS/EPRS, 139, 167, 174, 218
Glutamyl-tRNA synthetase (GluRS), 7, 8, 14, 46, 50, 69, 103, 109, 174, 258
Glutathione *S*-transferase (GST), 91, 94
Glyceraldehyde-3-phosphate dehydrogenase, 196
Glycogen phosphorylase, 196
Glycyl-tRNA synthetase (GRS), 151
Gp96, 125
GST. *See* Glutathione *S*-transferase (GST)
GTPase, 68, 99, 131, 137, 193, 201, 223, 234
GTPase-activating protein (GAP), 223

H

Halofuginone, 233
Heat shock protein 90 (HSP90), 140
Hemoglobin, 196
Hepatocarcinoma, 213
Histidine triad (HIT) superfamily, 195
  nucleotide-binding protein-1 (Hint-1), 171, 196, 198
Histidyl-tRNA synthetase (HRS), 152
Histiocytomas, 213
HIV-1. *See* Human immunodeficiency virus type-1 (HVI-1)
Horizontal gene transfer, 294, 305, 316
HSP90. *See* Heat shock protein 90 (HSP90)
Human immunodeficiency virus type-1 (HVI-1), 63, 134, 293, 296, 299, 302
Human mitochondrial disorders, 248

I

IleRS. *See* Isoleucyl-tRNA synthetase (IleRS)
Immune response, 171
Indolmycin, 307
Inhibitor mechanisms, 305

Index 349

Inositol 1,4,5-trisphosphate (IP3), 191
Integrin alpha 5 beta 1, 156
Interferon-γ (IFN-γ), 211
Intercellular communication, 145
Isoleucyl-tRNA synthetase (IleRS), 3

K
KRS (lysyl-tRNA synthetase), 45, 55, 122, 132, 148, 152, 208, 222, 234

L
Lamin A, 131
Last universal common ancestor (LUCA), 44, 53, 59
Leucine zippers, 97, 122, 127, 171
Leucyl-tRNAsynthetase (LeuRS/LRS), 136, 167
  mitochondrial, 179
Leukemias, 214
Leukoencephalopathy, 264
Lymphoma, 213
Lyme disease, 309
Lysyl-tRNA synthetase (LysRS/KRS), 45, 55, 122, 132, 148, 152, 167, 171, 189, 208, 222, 234
  MITF signaling pathway, 200

M
Mammalian target of rapamycin (mTOR), 99, 136, 211, 223
MAPK. *See* Mitogen-activated protein kinases (MAPK)
Mediators, 145
Methionyl-tRNA synthetase (MetRS), 3, 131, 167
Microcin C, 315
Microphthalmia-associated transcription factor (MITF), 133, 171, 189, 198
MicroRNA, 212
Mitochondria, disorders, 253
  translation defects, 254
  translation machinery, 252
Mitochondrial targeting sequence (MTS), 258
Mitogen-activated protein kinases (MAPK), 136, 152, 154, 198, 222
MRS. *See* Methionyl-tRNA synthetase (MetRS)
MSCp43, 152
Multisynthetase complex (MSC), 67, 89, 152
Multi-tRNA synthetase complex, 119

Mupirocin, 233, 307, 310
Mutations, 214
  pathology-related, 247
Mutation-induced structural change, 110

N
N-terminal helix, 92
Neovasculization, 234
Noncanonical function, 167
Norvaline, 23
Novel functions, 90
NS1-associated protein (NSAP1), 139
Nucleocytoplasmic large DNA viruses (NCLDVs), 298
Nucleoside diphosphate linked moiety X (Nudix) superfamily, 195

O
Onychomycosis, 233, 311
Opine permease, 314
Osteosarcoma, 213, 234
Ovarian cancer, 213
Oxidative phosphorylation pathway (OXPHOS), 247, 249, 253

P
P2 purinergic receptors, 196
p53, 128
Phenylalanyl-tRNA synthetase (PheRS), 3
Phosphatidylinositol 4,5-bisphosphate, 191
Phosphodiesterase (PDE), 191
Phosphoinositide 3-kinase (PI3K), 136
Phospholipase C (PLC), 191
Phosphorylation, 222
O-Phosphoserine, 293
Phosphoseryl-tRNA synthetase (SepRS), 58
Plant viruses, 297
Point mutations, 214
Polarography, 256
Poly(ADP-ribose) polymerase 1 (PARP-1), 137, 304
Posttransfer editing, 24
Posttranslational modification, 109, 221
Pretransfer editing, 21
Proofreading, 43
Prolyl-tRNA synthetase (ProRS), 4
Prostate cancer, 211
Protein kinases, 191
Protein modification, 340
Protein synthesis, 43

350 Index

Protein–protein interaction, 119
Pyrolysyl-tRNA synthetase (PylRS), 56
Pyrrolysine, 293

Q

QRS (glutaminyl-tRNA synthetase), 3, 8, 46, 52, 136

R

r24, 334
  mini, 334
RagD GTPase, 223
Random non-standard peptide integrated display (RaPID), 340
Rapamycin, 136, 211
Receptors, 145
REP8839, 309
Respiratory chain defects, 248
Retroviruses, 299
Ribonucleoprotein, 1
Ribosomal proteins, 89
Ribozyme, 178, 234, 331, 335
RNA, microRNA, 212
  splicing, 178
  tRNA, 1, 43, 167
  world, 342
Rossmann fold, 4
Rous sarcoma virus, 294, 303

S

S6 kinase 1 (S6K1), 136
Second messengers, 189, 191
SecRS, 53
Selenocysteine synthase (SelA), 54
Selenocysteine-tRNA, 53
Serine hydroxamate, 21
Seryl-tRNA synthetase (SRS), 152
Signaling, 145
Site-specific protein modification, 340
Small-angle X-ray scattering (SAXS), 120
Smurf2, 127
SOD1. See Superoxide dismutase (SOD1)
Splicing, 65, 167, 217, 258, 272
  alternative, 107
Structural metamorphosis, 90
Substrate-assisted catalysis, 1
Succinate dehydrogenase (SDH), 256
Superoxide dismutase (SOD1), 134
Synthetases, functional evolution, 59
  paralogs, 68
  specificity, 60
Syphilis, 309

T

Tavaborole (AN2690), 233
Therapeutics, 207
Threonyl-tRNA synthetase (ThrRS), 4, 167, 172
TNF-receptor associated factor 2 (TRAF2), 128
Tobacco mosaic virus (TMV), 297
Tobramycin, 310
TRAF2. See TNF-receptor associated factor 2 (TRAF2)
Transamidation, 43, 52
Transamidosome, 52
Transcriptional control, 167
Transcript-selective translational silencing, 174
Transediting factors, 29
Transition state, 1
Translation, 43, 340
  control, 167
Translocase complexes, 260
Transmembrane receptors, 155
tRNA, 1
  recognition, 60
tRNA synthetases, 90, 167
Trojan Horse inhibitors, 294, 312
Tryptophanyl-tRNA synthetase (TrpRS/WRS), 7, 137, 149, 195, 198, 296, 303, 316
Tumor necrosis factor (TNF)-α, 129, 146, 213, 223
Tyrosyl-tRNA synthetase (TyrRS/YRS), 7, 149, 225
  mitochondrial, 178

U

UNE-X, 98
Upstream stimulatory factor 2 (USF2), 198

V

Valyl-tRNA synthetase (ValRS/VRS), 3, 27, 135
Vascular development, 171
Vascular endothelial growth factor A (VEGFA), 171
VE-cadherin, 138

W

WHEP, 67, 96, 221, 227, 304
WRS (tryptophanyl-tRNA synthetase), 7, 137, 149, 195, 198, 226, 296, 303, 316

Y

YRS (tyrosyl-tRNA synthetase), 7, 149, 225